Biology and Breeding
of Crucifers

Biology and Breeding of Crucifers

Edited by

Surinder Kumar Gupta

CRC Press
Taylor & Francis Group
Boca Raton London New York

CRC Press is an imprint of the
Taylor & Francis Group, an **informa** business

CRC Press
Taylor & Francis Group
6000 Broken Sound Parkway NW, Suite 300
Boca Raton, FL 33487-2742

First issued in paperback 2017

ISBN 13: 978-1-138-11345-9 (pbk)
ISBN 13: 978-1-4200-8608-9 (hbk)

Library of Congress Cataloging-in-Publication Data

Biology and breeding of crucifers / editor: Surinder Kumar Gupta.
    p. cm.
    Includes bibliographical references and index.
    ISBN 978-1-4200-8608-9 (hardcover : alk. paper)
    1. Cruciferae. 2. Cruciferae--Breeding. I. Gupta, Surinder Kumar. II. Title.

QK495.C9B56 2009
583'.64--dc22

                                 2009005503

Visit the Taylor & Francis Web site at
http://www.taylorandfrancis.com

and the CRC Press Web site at
http://www.crcpress.com

Contents

Preface

Despite recent advances made in rapeseed-mustard breeding, the need and opportunities to increase its production, productivity, oil content and quality, and protein yield are as great today as they have ever been. The alien variations available in wild crucifers have been utilized enormously by the technique of chromosome and genetic engineering to develop noble varieties. Realizing the importance of crucifer crops in Europe and the rest of the countries of the world, there is an urgent need to search for new gene pools with special reference to wild species and to update the knowledge of the recent technologies developed thus far in enhancing rapeseed-mustard production at the global level. At present, no single publication available deals exclusively with crucifers, with the primary emphasis on wild species.

This book includes 18 chapters that have been well prepared by leading *Brassica* scientists around the world with extensive experience, and their contributions are well recognized worldwide. Chapters 1 and 2 deal with the systematics and phylogenies of wild crucifers, while Chapters 3 and 4 describe the major wild relatives of crucifers accompanied by beautiful photographs to assist in recognizing diagnostic characteristics and to aid in the identification of the species. This is followed by chapters on breeding methods, self-incompatibility, cytoplasmic male sterility, germination and viability, and plant-insect interactions in crucifers. Chapters 10 and 11 provide a detailed account of comparative cytogenetics and distant hybridizations involving wild crucifers. The phytoalexins and their role are discussed in Chapter 12, followed by chapters on introgression of genes from wild crucifers (Chapter 13), biotechnology (Chapter 14), and microspore culture and haploidy breeding (Chapter 15). Another chapter (Chapter 16) on genetic improvement of vegetable crucifers using wild species has been added to enhance the beauty of this book. Finally, Chapter 17 provides a brief account of industrial products from wild crucifers and Chapter 18 describes the preservation and maintenance of crucifer plant genetic resources at the global level. Although a few chapters may overlap with each other to some extent with regard to subject matter, this has been dealt with in depth by each of the contributors.

I am highly indebted to Professor Nagendra Sharma, Honorable Vice-Chancellor, Sher-e-Kashmir University of Agricultural Sciences and Technology of Jammu, India, for encouraging me to carry out research on crucifer crops with all the required modern facilities. I am also thankful to all the contributors of the various chapters for their ready response.

Help rendered by Professor Marcus Koch, Heidelberg University, Germany; Ihsan A. Al-Shehbaz, Head, Missouri Botanical Garden, Missouri, United States; Professor Suzanne I. Warwick, Eastern Cereal and Oilseeds Research Centre, Ontario, Canada; Professor Martin A. Lysak, Institute of Experimental Biology, Masaryk University, Czech Republic; Professor M. Soledade C. Pedras, University of Saskatchewan, Canada; Professor Y. Takahata, Iwate University, Japan; Professors Shyam Prakash and P.R. Kalia, Indian Agricultural Research Institute, New Delhi, India, is also gratefully acknowledged.

I am indeed grateful to Professor W.J. Zhou, Crop Science Institute, Hangzhou, China, and to Dr. César Gómez Campo, Unversidad Politécnica de Madrid, Spain, for providing technical input and critically reviewing some of the chapters. Ms. Randy Brehm, assistant editor, CRC Press, deserves special thanks for bringing this book to life. Dr. Aditya Pratap, assistant professor, Division of Plant Breeding & Genetics, SKUAST-Jammu, also deserves sincere thanks for an outstanding and formidable volume of correspondence with regard to this book. Shri Madan Mohan Gupta, my fatherin-law, has been the instrumental force behind this book. Unfortunately, he has left for heavenly abode. I owe so very much to this departed soul and also to my better half, Dr. Neena Gupta, for their unstinting help and patience during the preparation of this manuscript.

S.K. Gupta
Editor

About the Editor

Dr. S.K. Gupta, born in 1959, is currently working as professor and head/chief scientist (Oilseeds) in the Division of Plant Breeding and Genetics, SK University of Agricultural Sciences and Technology, FOA, Chatha, Jammu, India, and holds a brilliant academic and service record. For almost two decades he has devoted his research interests to the area of oilseed brassicas.

Dr. Gupta obtained his post-graduate degrees (M.Sc., Ph.D.) from Punjab Agricultural University, Ludhiana, India, in 1984 and 1987, respectively. He is the recipient of a post-doctoral fellowship in plant biotechnology, and has published more than 100 research papers in esteemed national and international journals, mostly on brassicas. He has already developed five varieties of rapeseed-mustard. In addition, he has written two books on plant breeding and edited two volumes: one on *Recent Advances in Oilseed Brassicas*, Kalyani Publishers, New Delhi, India, and the second on *Rapeseed Breeding-Advances in Botanical Research*, Vol. 45, Academic Press, Elsevier Publishers. For his excellent scientific endeavors, he has been conferred with the Young Scientist Award: 1993–1994 by the State Department of Science and Technology.

Contributors

D.P. Abrol
Division of Entomology
Sher-e-Kashmir University of Agricultural
  Sciences and Technology
Chatha, Jammu, India

Ihsan A. Al-Shehbaz
Head, Department of Asian Botany
Missouri Botanical Garden
St. Louis, Missouri, USA

San Woo Bang
Laboratory of Plant Breeding
Faculty of Agriculture
Utsunomiya University
Utsunomiya, Japan

S.R. Bhat
National Research Centre on Plant
  Biotechnology
Indian Agricultural Research Institute
New Delhi, India

Vinitha Cardoza
BASF Corporation
26 Davis Drive
Durham, North Carolina, USA

Mei Desheng
Institute of Oil Crops Research
Chinese Academy of Agricultural Science
Wuhan, P.R. China

Ting-Dong Fu
The National Key Laboratory of Crop Genetic
  Improvement
Huazhong Agricultural University
Wuhan, P.R. China

César Gómez-Campo
Universidad Politécnica de Madrid
Departamento de Biología Vegetal
E.T.S. Ingenieros Agrónomos
Ciudad Universitaria
Madrid, Spain

S.K. Gupta
Division of Plant Breeding and Genetics
Sher-e-Kashmir University of Agricultural
  Sciences and Technology
Chatha, Jammu, India

Jocelyn C. Hall
Department of Biological Sciences
University of Alberta
Edmonton, Alberta, Canada

Pritam Kalia
Division of Vegetable Science
Indian Agricultural Research Institute (IARI)
New Delhi, India

H. Kitashiba
Graduate School of Agricultural Science
Tohoku University
Sendai, Japan

Marcus Koch
Heidelberg Institute of Plant Sciences
Heidelberg University
Heidelberg, Germany

D. Liu
Institute of Crop Science
Zhejiang University
Hangzhou, P.R. China

Martin A. Lysak
Department of Functional Genomics and
  Proteomics
Masaryk University
Brno, Czech Republic

Yasuo Matsuzawa
Laboratory of Plant Breeding
Faculty of Agriculture
Utsunomiya University
Utsunomiya, Japan

Peter McVetty
Department of Plant Science
University of Manitoba
Winnipeg, Manitoba, Canada

M.S. Naeem
Institute of Crop Science
Zhejiang University
Hangzhou, P.R. China

Takeshi Nishio
Graduate School of Agricultural Science
Tohoku University
Sendai, Japan

M. Soledade C. Pedras
Department of Chemistry
University of Saskatchewan
Saskatoon, Saskatchewan, Canada

Shyam Prakash
National Research Centre on Plant
    Biotechnology
Indian Agricultural Research Institute
New Delhi, India

Aditya Pratap
Division of Crop Improvement
Indian Institute of Pulses Research (ICAR)
Kanpur, India

Hu Qiong
Institute of Oil Crops Research
Chinese Academy of Agricultural Science
Wuhan, P.R. China

R. Raziuddin
Institute of Crop Science
Zhejiang University
Hangzhou, P.R. China

C. Neal Stewart, Jr.
Department of Plant Sciences
University of Tennessee
Knoxville, Tennessee, USA

Y. Takahata
Faculty of Agriculture
Iwate University
Morioka, Japan

G.X. Tang
Institute of Crop Science
Zhejiang University
Hangzhou, P.R. China

G.L. Wan
Institute of Crop Science
Zhejiang University
Hangzhou, P.R. China

Suzanne I. Warwick
Eastern Cereal and Oilseeds Research Centre
Agriculture and Agri-Food Canada
Ottawa, Ontario, Canada

Kaneko Yukio
Laboratory of Plant Breeding
Faculty of Agriculture
Utsunomiya University
Utsunomiya, Japan

Li Yunchang
Institute of Oil Crops Research
Chinese Academy of Agricultural Science
Wuhan, P.R. China

Carla Zelmer
Department of Plant Science
University of Manitoba
Winnipeg, Manitoba, Canada

Qingan Zheng
Department of Chemistry
University of Saskatchewan
Saskatoon, Saskatchewan, Canada

W.J. Zhou
Institute of Crop Science
Zhejiang University
Hangzhou, P.R. China

1 Molecular Systematics and Evolution

Marcus A. Koch and Ihsan A. Al-Shehbaz

CONTENTS

INTRODUCTION

The past two decades can be characterized by the tremendously increasing number of studies focusing on the systematics, development, phylogenetics, and phylogeography of cruciferous plants (mustards). *Brassicaceae* (Cruciferae) is a large plant family (338 genera and 3709 species; see Warwick et al., 2006b) of major scientific and economic importance. Almost a century after Hayek's (1911) major taxonomic account, which was followed by the more thorough monograph of Schulz (1936), we are now closer to the first comprehensive phylogenetic system of the mustard family. The increasing importance of *Arabidopsis* and *Brassica* species as model organisms in the plant sciences has greatly advanced research into the systematics, taxonomy, evolution, and development of the entire family, including the cultivated taxa and their wild relatives.

The first attempt to summarize knowledge of the family was provided more than 30 years ago (Vaughan et al., 1976). It was followed by Tsunoda et al. (1980), who dealt with the biology and breeding of *Brassica* crops and their wild allies. During the past 20 years, molecular biology and DNA techniques have revolutionized plant systematics and evolution; and because of the selection of *Arabidopsis thaliana* as the model flowering plant, the *Brassicaceae* have been at the forefront of scientific research. Except for the highly specific monograph on *Brassica* (Gómez-Campo, 1999), no family-wide symposium or textbook was devoted to its systematics and evolution. That gap was bridged in a special symposium organized by Koch and Mummenhoff (2006) during the XVII International Botanical Congress in Vienna. The symposium, entitled "Evolution and Phylogeny of the *Brassicaceae*," and dedicated to Herbert Hurka's 65th birthday and his contributions to evolutionary studies in the family, addressed diversified fields such as phylogeny, systematics, phylogeography, polyploidy, hybridization, comparative genomics, and developmental genetics. The contributed papers appeared in a special issue of *Plant Systematics and Evolution* (Volume 259(2–4), 2006) that included a comprehensive checklist of all species of the family (Warwick et al., 2006b) and a compilation of chromosome numbers to that date (Warwick and Al-Shehbaz, 2006).

Many recent contributions (e.g., Koch et al., 2000, 2001, 2003a; Koch, 2003; Appel and Al-Shehbaz, 2003; Al-Shehbaz et al., 2006; Bailey et al., 2006; Beilstein et al., 2006; Koch and Mummenhoff, 2006; Warwick and Al-Shehbaz, 2006; Warwick et al., 2006; Koch et al., 2007; Warwick et al., 2007) have paved the way toward a better understanding of the phylogenetic relationships within the *Brassicaceae* and to the delimitations of the major lineages based on comprehensive morphological and taxonomical treatments in light of molecular data. As a result, a phylogenetically based tribal classification of the family emerged and has been refined (e.g., Al-Shehbaz et al., 2006; Al-Shehbaz and Warwick, 2007; German and Al-Shehbaz, 2008).

What were the most important milestone accomplishments during the past two decades? In principle, and aside from the wealth of knowledge on the model organisms in *Arabidopsis, Brassica,* and *Capsella,* there are four: (1) achieving a new infrafamiliar classification based on phylogenetically circumscribed new tribes; (2) recognition and assignment of monophyletic genera; (3) unraveling the principles in crucifer evolution and exploring detailed examples for species- or genus-specific evolutionary histories; and (4) phylogenetic circumscription of the order Capparales and the determination of Cleomaceae as the closest and sister family to the *Brassicaceae*. This introductory chapter deals mainly with the first issue. Some of these issues are discussed in more detail in subsequent chapters; others are outlined in various contributions presented in this book.

RECOGNITION OF INFRAFAMILIAR TAXA: THE TRIBAL SYSTEM

The history of tribal classification systems is long, and is well summarized in various reviews (e.g., Appel and Al-Shehbaz, 2003; Koch, 2003; Koch et al., 2003a; Mitchell-Olds et al., 2005; Al-Shehbaz et al., 2006) and need not be repeated here. Prior to 2005, the most important conclusion reached in phylogenetic studies was that except for the Brassiceae, the other tribes are artificially delimited and do not reflect the phylogenetic relationships of their component genera.

The other exception was thought to be the tribe Lepidieae (e.g., Zunk et al., 1999), but that too was shown to be artificially circumscribed (Al-Shehbaz et al., 2006). Of the 49 infrafamiliar taxa (19 tribes and 30 subtribes) recognized by Schulz (1936), 9 tribes (Alysseae, Arabideae, Brassiceae, Euclidieae, Heliophileae, Hesperideae, Lepidieae, Schizopetaleae, and Sisymbrieae) were maintained by Al-Shehbaz et al. (2006), although the limits of all except the Brassiceae and Heliophileae were substantially altered. These authors also recognized 16 additional tribes that were either described as new or reestablished. The first comprehensive phylogeny of the *Brassicaceae*, in which 101 genera were sampled was based on the plastidic gene *ndh*F (Beilstein et al., 2006). It identified three significantly supported major clades (Figure 1.1). The study provided the main foundation on which a new tribal classification was introduced (Al-Shehbaz et al., 2006). A subsequent internal transcribed spacer (ITS)-based study (Bailey et al., 2006) provided substantial support for the new system. In a more recent analysis focusing primarily on the evolution of plastid *trn*F pseudogene in the mustard family, a supernetwork was reconstructed based on nuclear alcohol dehydrogenase (*adh*), chalcone synthase (*chs*), and an ITS of nuclear ribosomal DNA and plastidic maturase (*mat*K) sequence data (Koch et al., 2007). In that article, the corresponding *trn*L-F derived phylogeny was largely in congruence with this supertree, and all three major lineages identified by Beilstein et al. (2006) were confirmed. The supertree approach clearly demonstrated that there is a substantial conflicting "phylogenetic signal" at the deeper nodes of the family tree, resulting in virtually unresolved phylogenetic trees at the genus level. However, other similarities are in congruence when comparing the supertree with the *ndh*F phylogeny of Beilstein et al. (2006). For example, the tribes Arabideae, Thlaspideae, Eutremeae, and Isatideae are closely related to lineage II comprising the tribes Schizopetaleae, Sisymbrieae, and Brassiceae. Furthermore, the tribe Alysseae is more closely related to lineage I (all have the *trn*F pseudogenes) (Figure 1.1). On the other hand, some results are contradictory, such as the ancestral position of Cochlearieae (Koch et al., 2007), which is not confirmed by the results from *ndh*F (Beilstein et al., 2006), or ITS data (Bailey et al., 2006). Remarkably, a phylogenetic study focusing on the mitochondrial *nad*4 (Franzke et al., 2008) is highly congruent with ITS and *ndh*F studies. In a recent Bayesian analysis (Franzke et al., 2008), the Heliophileae fell in lineage III, which disagrees with the *ndh*F data. In ongoing research on genome-size evolution in the family (Lysak and Koch, unpublished), the Heliophileae was also placed in lineage III, using a supertree approach (*adh*, *chs*, ITS, *mat*K, *trn*L-F). In summary, most of the tribes recognized by Al-Shehbaz et al. (2006) are clearly delimited, but strong support for the intertribal relationships is still lacking.

Despite the use of multigene phylogenies, the lack of resolution in the skeletal backbone of the family is not yet understood, and two hypotheses explain that. First, early radiation events were quite rapid and were characterized by low levels of genetic variation separating the different lineages. Second, reticulate evolution (e.g., as found in the tribe Brassiceae) resulted in conflicting gene trees that did not reflect species phylogenies. The mitochondrial *nad*4 intron data presented by Franzke et al. (2008) perhaps favor the first hypothesis. This scenario was also favored by Koch et al. (2007), who found that the microstructural evolutionary changes may be useful for inferring early events of divergence. In fact, the two structural rearrangements described by Koch et al. (2007) for the *trn*L-F region identify ancient patterns of divergence supported by phylogenetic analysis of that region excluding the microstructural mutations. Further support is also found from analyses of the nuclear ITS sequence data (Bailey et al., 2006) and is discussed below.

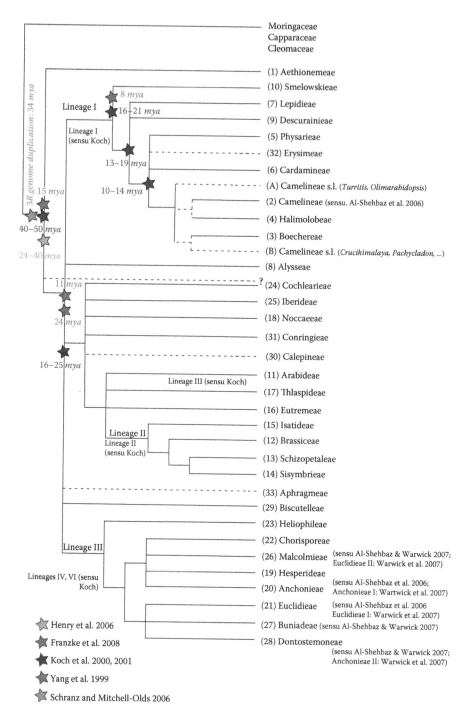

FIGURE 1.1 (See color insert following page 128.) Synopsis of phylogenetic hypothesis from various sources of tribal relationships in the *Brassicaceae* family (for details, refer to the text). Lineages I–III are described in Beilstein et al. (2006). Koch et al. (2007) used different numbers given also as "sensu Koch," and we suggest the use of Beilstein's version only to avoid future confusion. Dashed lines indicate uncertain phylogenetic position. However, it should be kept in mind that this synopsis is not derived from one single phylogenetic analysis.

The recently proposed tribal classification of Al-Shehbaz et al. (2006) recognized 25 tribes (1–25, see below). More recently, Franzke et al. (2008) presented a family phylogeny based on the mitochondrial *nad*4 intron. Although the sampling in the latter study was smaller, both cpDNA (Beilstein et al., 2006) and mtDNA (Franzke et al., 2008) phylogenies were totally congruent with each other. However, it is still unclear why there are major inconsistencies between these two phylogenies and those generated from the nuclear genome, such as the ITS by Bailey et al. (2006) or the *adh* and *chs* by Koch et al. (2000, 2001).

Additional studies have shown that some the tribes proposed by Al-Shehbaz et al. (2006) were broadly delimited or paraphyletic and needed further splitting. For example, the tribes Euclidieae and Anchonieae were shown by Warwick et al. (2007) and Al-Shehbaz and Warwick (2007) to consist of more than one lineage, and they recognized the new tribes Malcolmieae and Dontostemoneae and reestablished the tribe Buniadeae (tribes 26–28). The studies by German et al. (2008) of primarily Asian taxa have also resulted in the description of the new tribes Aphragmeae and Conringieae, as well as the reestablishment of the tribes Calepineae, Biscutelleae, and Erysimeae (tribes 29–33). The ITS studies of Bailey et al. (2006) and Koch (unpublished) justify the recognition of the last tribe. They also demonstrated that the tribe Camelineae sensu Al-Shehbaz et al. (2006) is paraphyletic and requires further division, herein recognized as tribes 34 (A) and 35 (B) (Table 1.1). An overview of these various tribes and a synopsis of the relationships among them are presented in Figure 1.1. However, this figure does not represent the outcome of an overall analysis, and a family-wide phylogenetic study is needed to achieve that. Furthermore, it is important to emphasize that phylogenetic hypotheses based individually on one marker (e.g., plastid, mitochondrial, or nuclear) would be of limited value (Koch et al., 2001; Koch et al., 2007). To have a comprehensive phylogeny of the entire family, several problematical genera must be sampled and adequately assigned to tribes (Al-Shehbaz et al., 2006).

The present study does not deal with generic-level delimitations, and the interested reader should consult Appel and Al-Shehbaz (2003) and the database of Warwick et al. (2006b). As for the prior tribal assignments of various genera and tribal limits, the reader is advised to consult Al-Shehbaz et al. (2006).

1. TRIBE AETHIONEMEAE

This unigeneric tribe of about 45 spp. consists of *Aethionema,* including *Moriera.* The vast majority of species are endemic to Turkey, and only a few grow as far east as Turkmenistan and west into Spain and Morocco.

Aethionema was previously placed in the tribe Lepidieae (e.g., Hayek, 1911; Schulz, 1936). However, Al-Shehbaz et al. (2006) placed it in its own tribe because molecular data consistently show its sister position to the rest of the *Brassicaceae.* The genus is highly variable in habit, fruit and floral morphology, and chromosome number (Appel and Al-Shehbaz, 2003; Al-Shehbaz et al., 2006). Knowledge of genome size and duplication, base chromosome number, evolutionary trends, most basal taxa, and monophyly of *Aethionema* is undoubtedly valuable in understanding the evolution and early radiation of the entire family.

2. TRIBE CAMELINEAE

As delimited by Al-Shehbaz et al. (2006), the Camelineae have recently been shown to be paraphyletic and consist of a heterogeneous assemblage of genera (Bailey et al., 2006; Warwick et al., 2007; Koch et al., 2007; German and Al-Shehbaz 2008; Koch, unpublished). Indeed, the tribe should be subdivided into at least four monophyletic tribes, of which the unigeneric Erysimeae (ca. 180 spp.) is now recognized (German and Al-Shehbaz, 2008). Therefore, the species total in the Camelineae s.str. (following the removal of *Erysimum, Turritis, Olimarabidopsis, Crucihimalaya, Transberingia,* and *Pachycladon*) would be about 35 species. The tribe includes the genera *Arabidopsis* (10 spp.),

TABLE 1.1
Overview on the Tribes and the Number of Genera and Species of the
***Brassicaceae* as Scored Herein**

Tribe	Genera	Species	Ref.
1. Aethionemeae	1	45	This chapter
2. Camelineae	7	35	This chapter
3. Boechereae	7	118	Al-Shehbaz et al. (2006)
4. Halimolobeae	5	39	Bailey et al. (2007)
5. Physarieae	7	133	This chapter
6. Cardamineae	9	333	This chapter
7. Lepidieae	4	235	This chapter
8. Alysseae	15	283	This chapter; Warwick et al. (2008)
9. Descurainieae	6	57	Al-Shehbaz et al. (2006)
10. Smelowskieae	1	25	Al-Shehbaz et al (2006)
11. Arabideae	8	470	This chapter
12. Brassiceae	46	230	Al-Shehbaz et al. (2006)
13. Schizopetaleae s.l.	28	230	Al-Shehbaz et al. (2006)
14. Sisymbrieae	1	40	Al-Shehbaz et al. (2006)
15. Isatideae	2	65	This chapter
16. Eutremeae	1	26	Al-Shehbaz and Warwick (2006)
17. Thlaspideae	7	27	Al-Shehbaz et al. (2006)
18. Noccaeeae	3	90	This chapter
19. Hesperideae	1	45	Al-Shehbaz et al. (2006)
20. Anchonieae	8	68	Al-Shehbaz and Warwick (2007)
21. Euclidieae	13	115	Al-Shehbaz and Warwick (2007)
22. Chorisporeae	3	47	Al-Shehbaz and Warwick (2007)
23. Heliophileae	1	80	Al-Shehbaz et al. (2006)
24. Cochlearieae	1	21	Al-Shehbaz et al. (2006)
25. Iberideae	1	27	Al-Shehbaz et al. (2006)
26. Malcolmieae	8	37	Al-Shehbaz and Warwick (2007)
27. Buniadeae	1	3	Al-Shehbaz and Warwick (2007)
28. Dontostemoneae	3	28	Al-Shehbaz and Warwick (2007)
29. Biscutelleae	1	53	German and Al-Shehbaz (2008)
30. Calepineae	3	8	German and Al-Shehbaz (2008)
31. Conringieae	2	9	German and Al-Shehbaz (2008)
32. Erysimeae	1	180	German and Al-Shehbaz (2008)
33. Aphragmeae	1	11	German and Al-Shehbaz (2008)
34. Unnamed-I (A)	2	5	This chapter
35. Unnamed-II (B)	3	20	This chapter
Total	**212**	**3249**	

Capsella (3 spp.), *Catolobus* (1 sp.), *Camelina* (8 spp.), *Neslia* (2 spp.), *Pseudoarabidopsis* (1 sp.), and perhaps the Australian-endemic *Stenopetalum* (10 spp.). The tribe is primarily Eurasian, and only two species of *Arabidopsis* are native to North America.

Due to the extensive use of *Arabidopsis thaliana* in basically every field of experimental biology, the genus and its relatives above received considerable study (e.g., Mummenhoff and Hurka, 1994, 1995; Price et al., 1994, 2001; O'Kane and Al-Shehbaz, 1997, 2003; O'Kane et al., 1997; Al-Shehbaz et al., 1999; Koch et al., 1999a, 2000, 2001, 2007, unpublished; Mitchell and Heenan, 2000; Al-Shehbaz and O'Kane, 2002a; Heenan and Mitchell, 2003; Heenan et al., 2002).

3. TRIBE BOECHEREAE

This tribe of 7 genera and 118 species is almost exclusively North American, and only *Boechera furcata* grows in the Russian Far East (Al-Shehbaz, 2005). Except for *Boechera* (110 species), the remaining genera are either monospecific (*Anelsonia*, *Nevada*, *Phoenicaulis*, *Polyctenium*) or bispecific (*Cusickiella*, *Sandbergia*).

All members of the tribe typically have a base chromosome number of $x = 7$, mostly entire leaves (except *Polyctenium* and one *Sandbergia*), and branched trichomes (absent or in few *Boechera* and simple in *Nevada*). The majority are perennials with well-defined basal rosette.

Rollins (1993) treated all species of *Boechera* as members of *Arabis*, but extensive molecular studies (summarized in Al-Shehbaz, 2003, and Al-Shehbaz et al., 2006) suggest that the two genera belong to different tribes.

4. TRIBE HALIMOLOBEAE

The Halimolobeae is a New World tribe of 5 genera and 39 species mostly distributed in northern and central Mexico (Bailey et al., 2007), although genera such as *Exhalimolobos* (9 spp.), *Mancoa* (8 spp.), and *Pennellia* (10 spp.) are also disjunctly distributed in northern Argentina, Bolivia, and Peru (Bailey et al., 2002; Fuentes-Soriano, 2004). Three species of *Halimolobos* (8 spp.) grow in the southern United States, whereas *Sphaerocardamum* (4 spp.) is endemic to Mexico.

Members of the Halimolobeae have branched trichomes, white (rarely purplish) flowers, seeds mucilaginous when wetted, ebracteate racemes (except two *Mancoa*), often spreading sepals, and a base number of $x = 8$.

5. TRIBE PHYSARIEAE

The tribe consists of 7 genera and 133 species distributed primarily in North America. *Physaria* (105 spp.) is disjunct into South America (5 spp., northern Argentina and southern Bolivia) and has one species, *P. arctica,* distributed from northern Canada and Alaska into arctic Russia. The tribe also includes *Dimorphocarpa* (4 spp.), *Dithyrea* (2 spp.), *Lyrocarpa* (3 spp.), *Nerisyrenia* (9 spp.), *Paysonia* (8 spp.), and *Synthlipsis* (2 spp.). *Lesquerella* is paraphyletic and within which is nested the previously published *Physaria,* which necessitated their union into one genus (Al-Shehbaz and O'Kane, 2002b).

The Physarieae are readily separated from the rest of the *Brassicaceae* by having pollen with four or more colpi (the rest of *Brassicaceae* are tricolpate). The only exception is *Lyrocarpa coulteri,* in which a reversal to the tricolpate state apparently occurred. Other features of the tribe, none unique, are discussed by Al-Shehbaz et al. (2006).

6. TRIBE CARDAMINEAE

The tribe includes 333 species, most of which belong to the genera *Cardamine,* including *Dentaria* (ca. 200 spp.), *Rorippa* (86 spp.), and *Barbarea* (25 spp.). Except for *Barbarea,* which does not occur in South America, the genera are represented by native species on all other continents. The other genera are *Nasturtium* (5 spp.; 2 native to Mexico and the United States), and the North American *Iodanthus* (1 sp.), *Leavenworthia* (8 spp.), *Ornithocarpa* (2 spp.), *Planodes* (1 sp.), and *Selenia* (5 spp.).

Species of the Cardamineae grow predominantly in mesic or aquatic habitats, and *Subularia* (2 spp., one in Africa and the other in North America, northern Europe, and northern Russia), which occupies such habitats, should be checked molecularly to determine whether or not it belongs here. The majority of species are glabrous or with simple trichomes only, and have divided leaves, accumbent cotyledons, and a base chromosome number of $x = 8$.

7. TRIBE LEPIDIEAE

The Lepidieae (235 species) consist of *Lepidium,* a genus recently expanded by Al-Shehbaz et al. (2002), to include *Cardaria, Coronopus*, and *Stroganowia*. It is represented by native species on all continents except Antarctica. The monospecific *Acanthocardamum* (Afghanistan) and the Middle Eastern and Central Asian *Winklera* (3 spp.) and *Stubendorffia* (8 spp.) most likely also belong here.

The tribe is distinguished by the angustiseptate fruits (secondarily inflated in two species formerly assigned to *Cardaria*), one ovule per locule, often mucilaginous seeds, and simple or no trichomes.

Schulz (1936) artificially delimited the Lepidieae based solely on the presence of angustiseptate fruits and included genera assigned by Al-Shehbaz et al. (2006) to some 12 tribes. Evidently, the independent evolution of angustiseptate fruits in the *Brassicaceae* took place in the majority of tribes.

8. TRIBE ALYSSEAE

Dudley and Cullen (1965) expanded the limits of Alysseae to include genera now assigned to different tribes. For example, *Ptilotrichum* is removed to the Arabideae (Al-Shehbaz et al., 2006; Warwick et al., 2008). The Alysseae are distributed in Eurasia and North Africa, and only one species (*Alyssum obovatum*) extends its distribution to Canada and Alaska. The tribe includes some 253 species in the genera *Alyssum* (ca. 180 spp.), *Alyssoides* (6 spp.), *Aurinia* (13 spp.), *Berteroa* (5 spp.), *Bornmuellera* (7 spp.), *Clastopus* (2 spp.), *Clypeola* (10 spp.), *Degenia* (1 sp.), *Fibigia* (16 spp.), *Galitzkya* (3 spp.), *Hormathophylla* (7 spp.), *Physoptychis* (2 spp.), and *Strausiella* (1 sp.).

The majority of species in the tribe have stellate trichomes, latiseptate or terete (rarely angustiseptate); mostly few-seeded silicles; often winged seeds; and usually winged, toothed, or appendaged filaments. *Farsetia* (26 spp.) and *Lobularia* (4 spp.) are somehow distinct but were retained in this tribe (Warwick et al., 2008). *Farsetia* is distributed from northern and eastern Africa through Southwest Asia into Pakistan and western India, whereas *Lobularia* is restricted to northwestern Africa and Macaronesia (Appel and Al-Shehbaz, 2003). All members of *Farsetia* and *Lobularia* are pubescent with exclusively malpighiaceous trichomes. However, this type of trichome occurs sporadically in species of other tribes, although often in combination with other trichome types. From these, they are distinguished by have latiseptate silicles or sometimes siliques, often winged seeds, petiolate, often entire cauline leaves, and accumbent cotyledons.

9. TRIBE DESCURAINIEAE

The tribe consists of 6 genera and some 57 species. *Descurainia* (47 spp.), including *Hugueninia*, is distributed in three centers — North American (17 spp.), South American (ca. 20 spp.), and Canarian (7 spp.) — plus three species in Eurasia. The tribe also includes the European *Hornungia* (3 spp.), central Asian *Ianhedgea* (1 sp.), North-South American *Tropidocarpum* (4 spp.), and (if distinct from *Descurainia*) the monospecific Middle Eastern *Robeschia* and Patagonian *Trichotolinum.*

The tribe is characterized by the petiolate, one- to three-pinnatisect stem leaves, dendritic or rarely only forked trichomes, incumbent cotyledons, and mostly yellow flowers. *Descurainia* is unique in the *Brassicaceae* for the presence in some species of unicellular, glandular papillae.

10. TRIBE SMELOWSKIEAE

This unigeneric tribe consists of *Smelowskia* (25 spp.), a genus with 7 species in North America and 18 species in central and eastern Asia. Based on molecular studies by Warwick et al. (2004b), the genus was expanded by Al-Shehbaz and Warwick (2006) to include *Gordokovia*, *Hedinia*, *Redowskia*, *Sinosophiopsis*, and *Sophiopsis*.

Members of the Smelowskieae have branched trichomes, petiolate, pinnatisect cauline leaves, white to purple (rarely cream) flowers, nonmucilaginous seeds, and incumbent cotyledons.

11. TRIBE ARABIDEAE

The tribe consists of at least 8 genera and some 470 species. *Draba* (370 spp.), which includes *Drabopsis*, *Erophila*, and *Schivereckia*, is the largest genus in the family. It is represented by 119 spp. in North America, 70 in South America, and over 100 in the Himalayas and neighboring central Asia, but it is absent in Australia and all except northwestern Africa. *Arabis* (70 spp.) is primarily Eurasian, with 15 spp. in North America, and only a few in northwestern and alpine tropical Africa. Other genera of the tribe are the Eurasian *Aubrieta* (15 spp.), Eurasian *Ptilotrichum* (ca. 10 spp.), Chinese *Baimashania* (2 spp.), western North American *Athysanus* (2 spp.), European *Pseudoturritis* (1 sp.), and central Asian *Berteroella* (1 sp.).

Species of the Arabideae primarily have branched trichomes, accumbent cotyledons, latiseptate or terete fruits, nonmucilaginous seeds, and mostly a base number of $x = 8$.

Prior to the molecular studies of Koch et al. (1999, 2000) and O'Kane and Al-Shehbaz (2003), *Arabis* was so broadly delimited that it was estimated to include about 180 species (Al-Shehbaz, 1988). Subsequent studies (e.g., Al-Shehbaz, 2003, 2005) led to the removal of many of its species to the genera *Arabidopsis*, *Boechera*, *Catolobus*, *Fourraea*, *Pennellia*, *Pseudoturritis*, *Rhammatophyllum*, *Streptanthus*, and *Turritis*, which are presently assigned to at least five tribes. Obviously, the characters on which *Arabis* was delimited (latiseptate fruits, accumbent cotyledons, branched trichomes) evolved independently numerous times in the *Brassicaceae*. *Arabis* is much in need of comprehensive molecular studies; and despite the removal of nearly 65% of its species to other genera, it remains paraphyletic because its type species (*A. alpina*) is sister to *Draba* and *Aubrieta*, rather than to most species still assigned to it (Koch et al., 2003a; Koch et al., unpublished).

12. TRIBE BRASSICEAE

This tribe of 46 poorly defined "genera" and some 230 species includes the most economically important plants in the family (e.g., species of *Brassica*, *Eruca*, *Raphanus*, *Sinapis*). It has been subjected to extensive molecular (Warwick and Black, 1997a, 1997b; Warwick and Sauder, 2005, and references therein), taxonomic, and other studies (Tsunoda et al., 1980; Gómez-Campo, 1999).

The vast majority of species in the Brassiceae have conduplicate cotyledons and/or segmented (heteroarthrocarpic) fruits. The tribe is distributed primarily in the Mediterranean region, adjacent southwestern Asia, and South Africa, and only four species of *Cakile* are native to North America.

Molecular studies (see the review chapter by Warwick and Hall) on the tribe amply show that the traditional generic boundaries recognized by Schulz (1936) and Gómez-Campo (1999) do not hold. Only a few genera (e.g., *Cakile*, *Vella*, *Crambe*) are monophyletic (Warwick and Black, 1994, 1997b; Francisco-Ortega, 1999, 2002), but the majority of them form two groups (the rapa and nigra clades) that are well supported by chloroplast but not nuclear data and are basically indistinguishable morphologically. The component "genera" of both clades exhibit tremendous fruit diversity, which are the main characters used in their delimitation. To have a taxonomy that reflects phylogenetic relationships, the generic boundaries in the Brassiceae need radical revision. As a result, some genera (e.g., *Diplotaxis*, *Eruca*, *Erucastrum*, *Hemicrambe*, *Hirschfeldia*, *Raphanus*, *Rapistrum*, *Sinapidendron*) may have to be abandoned.

13.   TRIBE SCHIZOPETALEAE

The tribe was broadly delimited by Al-Shehbaz et al. (2006) to consist of some 230 species in over 28 genera, including those previously assigned to the tribe Thelypodieae. However, molecular studies in progress (Warwick et al.) show that both tribes should be maintained. As a result, the Schizopetaleae will have many fewer species and genera all restricted to South America, whereas the Thelypodieae include genera in both North and South America. Therefore, the Thelypodieae sensu Al-Shehbaz (1973), minus *Macropodium*, should be expanded to include all the North American genera placed in the Schizopetaleae by Al-Shehbaz et al. (2006). The South American genera to be restored in the Thelypodieae will be added following the completion of research by Warwick and colleagues. Little else can be gained herein by speculating any further about the limits of both tribes.

The combined Schizopetaleae and Thelypodieae exhibit enormous floral diversity not observed elsewhere in the *Brassicaceae*. This aspect is further discussed by Al-Shehbaz et al. (2006); and in these tribes, the floral features are far more useful than fruit characters in the delimitation of genera.

14.   TRIBE SISYMBRIEAE

Based on extensive molecular data (Warwick et al. 2002, 2005), this tribe was delimited by Al-Shehbaz et al. (2006) to consist of about 40 species of *Sisymbrium* (including *Lycocarpus* and *Schoenocrambe*). Except for the North American *S. linifolium*, the remaining tribe species are distributed in Eurasia and Africa. This is in contrast to Schulz's (1924, 1936) delimitation of the Sisymbrieae, which included 70 genera and about 400 species.

Species of the Sisymbrieae have yellow flowers, pinnately divided basal and lowermost stem leaves, two-lobed stigmas, terete siliques, a base chromosome number of $x = 7$, and simple or no trichomes (only the South African *Sisymbrium bruchellii* has branched trichomes).

15.   TRIBE ISATIDEAE

This tribe of about 65 species and 2 genera consists of the monospecific *Myagraum* and *Isatis* (ca. 64 spp.), including *Boreava, Pachypterygium, Sameraria,* and *Tauscheria*. The union of the last four genera with *Isatis* is based on extensive morphological and molecular studies (Moazzeni et al., 2007, unpublished). Further studies are needed to determine if *Chartoloma, Tauscheria, Glastaria,* and *Schimpera* belong to this tribe. Members of the Isatideae have indehiscent, often pendulous, one- or two-seeded fruits; yellow or rarely white flowers; auriculate stem leaves; and simple or no trichomes.

16.   TRIBE EUTREMEAE

This unigeneric tribe comprises *Eutrema* (26 spp.), a genus distributed primarily in Asia, especially the Himalayas and neighboring central Asia, with two species extending their ranges into North America (Al-Shehbaz and Warwick, 2005). Molecular studies by Warwick et al. (2004a, 2006a) strongly suggested that the limits of *Eutrema* be expanded to include the genera *Neomartinella, Platycraspedum, Taphrospermum,* and *Thellungiella*. Members of the Eutremeae are glabrous or with simple trichomes and have white flowers, incumbent cotyledons, and often palmately veined basal leaves.

17.   TRIBE THLASPIDEAE

This European and Southwest Asian tribe includes 27 species in the genera *Alliaria* (2 spp.), *Graellsia* (8 spp.), *Pachyphragma* (1 sp.), *Parlatoria* (2 spp.), *Peltaria* (4 spp.), *Pseudocamelina* (4 spp.), and *Thlaspi* (6 spp.). Further studies are needed to establish if the Southwest Asian *Sobolewskia* (4 spp.) belongs here. Species of the tribe have striate or coarsely reticulate seeds, undivided cauline leaves, often palmately veined basal leaves, and simple or no trichomes.

Thlaspi used to include about 90 species but seed anatomy (Meyer, 1973, 1979, 2001a) and extensive molecular studies (Koch and Mummenhoff, 2001; Mummenhoff et al., 1997a, 1997b, 2001; Beilstein et al., 2006) have shown that it consists of only six species, and the bulk of its previous members should be assigned to *Noccaea* (see below).

18. TRIBE NOCCAEEAE

The tribe includes some 90 species, of which four belong to *Microthlaspi* (Meyer, 2003), three to *Neurotropis* (Meyer, 2001b), the rest to *Noccaea*. The last genus includes 67 species in Europe, Africa, and Southwest Asia (Meyer, 2006), but it also includes 4 species in the New World (Koch and Al-Shehbaz, 2004), 5 species in the Himalayas (Al-Shehbaz, 2002), and others to be transferred from *Aethionema* and other genera, including all of the other segregates (Meyer, 1973).

Members of the Noccaeeae were subjected (as *Thlaspi* or *Microthlaspi*) to extensive molecular studies (see Koch, 2003; Koch et al., 1998; Koch and Hurka, 1999; Koch and Bernhardt, 2004). They are glabrous plants with angustiseptate fruits, smooth seeds, and often auriculate cauline leaves.

19. TRIBE HESPERIDEAE

This unigeneric tribe includes about 45 spp. in *Hesperis*, a genus much in need of systematic and molecular studies. It is distributed primarily in the Middle East and Europe, with fewer species in central Asia and northwestern Africa. The Hesperideae are unique in the *Brassicaceae* for their unicellular glands on uniseriate, few-celled stalks.

20. TRIBE ANCHONIEAE

As delimited by Al-Shehbaz et al. (2006), the Anchonieae included 12 genera and approximately 130 species. However, Warwick et al. (2007) have shown the tribe to be polyphyletic, and Al-Shehbaz and Warwick (2007) redefined its limits to include 8 genera and 68 species. The genera are *Anchonium* (2 spp.), *Iskandera* (2 spp.), *Matthiola* (48 spp.), *Microstigma* (3 spp.), *Oreoloma* (3 spp.), *Sterigmostemum* (7 spp.), *Synstemon* (2 spp.), and *Zerdana* (1 sp.). *Matthiola* and *Sterigmostemum* are in need of thorough study because Warwick et al. (2007) demonstrated that they are polyphyletic.

The Anchonieae is distributed primarily in Eurasia and eastern and northern Africa. It is distinguished by the presence of multicellular glands on multicellular-multiseriate stalks, two-lobed stigmas, erect sepals, and often branched trichomes.

21. TRIBE EUCLIDIEAE

This tribe was also broadly delimited by Al-Shehbaz et al. (2006) to include some 25 genera and more than 150 species. It was also found to be polyphyletic (Warwick et al., 2007). As a result, Al-Shehbaz and Warwick (2007) and Yue et al. (2008) adjusted its boundaries to include only 13 genera and 115 species distributed primarily in Eurasia and northern and eastern Africa. The tribe includes *Braya* (17 spp., 7 in North America), *Cryptospora* (3 spp.), *Leiospora* (6 spp.), *Neotorularia* (11 spp.), *Rhammatophyllum* (10 spp.), *Sisymbriopsis* (5 spp.), *Solms-laubachia* (26 spp.), *Strigosella* (23 spp.), *Tetracme* Bunge (10 spp.), and the monospecific *Dichasianthus*, *Euclidium*, *Leptaleum*, and *Shangrilaia*. *Desideria* is nested within *Solms-laubachia* and is united herein with the latter (Yue et al., 2006, 2008). Both *Neotorularia* and *Sisymbriopsis* are polyphyletic (Warwick et al., 2004a), and their boundaries need to be redefined.

With the removal of several genera from the Euclidieae to the Malcolmieae (see below), the former become monophyletic and can easily be distinguished from the latter by the presence of simple and two- to several-rayed (vs. sessile stellate) trichomes.

22. TRIBE CHORISPOREAE

This tribe of 3 genera and 47 species is primarily Asian and only 4 of the 35 species of *Parrya* are North American. The other genera are *Chorispora* (11 spp.) and *Diptychocarpus* (1 sp.). Molecular data (Warwick et al., 2007) strongly support the assignment of *Parrya* to this tribe.

The Chorisporeae are distinguished by the presence of multicellular glands on multicellular-multiseriate stalks, connivent stigmas, and erect sepals, and by the lack of branched trichomes.

23. TRIBE HELIOPHILEAE

The tribe was defined by Appel and Al-Shehbaz (1997) to include six genera but based on molecular studies (Mummenhoff et al., 2005), Al-Shehbaz and Mummenhoff (2005) united all genera into *Heliophila* (80 spp.). The Heliophileae are exclusively South African and are easily distinguished by the diplecolobal cotyledons, often appendaged petals and/or staminal filaments, and simple or no trichomes.

24. TRIBE COCHLEARIEAE

This unigeneric tribe consists of *Cochlearia* (21 spp., including five of *Ionopsidium*). *Cochlearia* is distributed primarily in Europe, with the ranges of three species extending into northern North America and Asia and one into Northwest Africa. The genus received detailed molecular studies (Koch, 2002; Koch et al., 1996, 1999b, 2003b), and further work is needed on *Bivonaea* and *C. aragonensis* to determine if they belong in this tribe.

Members of the Cochlearieae have rosulate, undivided basal leaves; white petals; often sessile cauline leaves; terete or angustiseptate silicles; entire stigmas; biseriate seeds; ebracteate racemes; and no trichomes.

25. TRIBE IBERIDEAE

This tribe consists only of *Iberis* (27 spp.), a genus centered mainly in Europe, with a few species in Northwest Africa, and Southwest and Central Asia.

Species of the Iberideae are glabrous or with simple trichomes and have angustiseptate, two-seeded fruits; zygomorphic flowers; and corymbose infructescences.

26. TRIBE MALCOLMIEAE

This newly established tribe (Al-Shehbaz and Warwick, 2007) was segregated from the Euclidieae sensu Al-Shehbaz et al. (2006). It includes 37 species in 8 primarily Mediterranean genera, although some are distributed into Southwest Asia, the Canary Islands, and Africa. The genera are *Cithareloma* (3 spp.), *Diceratella* (11 spp.), *Eremobium* (1 sp.), *Malcolmia* (10 spp.), *Maresia* (3 spp.), *Morettia* (3 spp.), *Notoceras* (1 sp.), and *Parolinia* (5 spp.).

The Malcolmieae are characterized by having often sessile stellate trichomes, decurrent stigmas, and mostly accumbent cotyledons.

27. TRIBE BUNIADEAE

This unigeneric tribe includes only *Bunias* (3 spp.), a genus distributed exclusively in Eurasia, although two species are weeds naturalized in North America. Molecular studies (Beilstein et al., 2006; Koch, unpublished) show that *Bunias* groups close to the tribes Euclidieae and Anchonieae but should be excluded from the latter, as was done by Al-Shehbaz and Warwick (2007).

The Buniadeae have multicellular glands on multicellular-multiseriate stalks, indehiscent silicles, and spiral cotyledons.

28. Tribe Dontostemoneae

Members of the tribe are distributed exclusively in central and eastern Asia. It comprises 28 species in the genera *Clausia* (6 spp.), *Dontostemon* (12 spp.), and *Pseudoclausia* (10 spp.).

The Dontostemoneae differ from other tribes with multicellular glands on multiseriate-multicellular stalks by the lack of branched trichomes and the presence of often united or winged filaments, entire stigmas, and rounded repla.

29. Tribe Biscutelleae

This unigeneric tribe comprises the genus *Biscutella* L. (53 spp.), a primarily North African-European genus but with only a few species reaching the Middle East. Although established by Dumortier more than 180 years ago, the Biscutelleae was not recognized by subsequent authors and has only recently been reinstated by German and Al-Shehbaz (2008). It is distinguished from the other tribes by its didymous, angustiseptate, two-seeded fruits; long styles; entire stigmas; simple trichomes; and auriculate cauline leaves.

30. Tribe Calepineae

The tribe was first established by Horaninow some 160 years ago and was not recognized since then. As delimited by German and Al-Shehbaz (2008), the tribe includes eight Asian species in *Goldbachia* (6 spp.) and the monospecific *Spirorrhynchus* and *Calepina*. The last genus was previously assigned to the Brassiceae (Schulz, 1936; Gomez-Campo, 1999), but recent molecular studies (Anderson and Warwick, 1999; Francisco-Ortega, 1999; Lysak et al., 2005; Beilstein et al., 2006; German et al., unpublished) clearly support its exclusion from this tribe.

The tribe includes annuals with indehiscent, woody, one- to three-seeded fruits; entire stigmas; simple or no trichomes; and undivided, often auriculate cauline leaves.

31. Tribe Conringieae

Based on molecular studies (German et al., unpublished), German and Al-Shehbaz (2008) established this new tribe. It consists of nine, primarily Southwest Asian species in the genera *Conringia* (6 spp.) and *Zuvanda* (3 spp.), although the range of *C. planisiliqua* extends into the Himalayas and *C. orientalis* is a naturalized Eurasian weed.

As in *Calepina*, *Conringia* was previously included in the Brassiceae (Schulz, 1936; Gomez-Campo, 1999), but molecular data (see references under *Calepina*) clearly support its removal from that tribe. Species of the Conringieae are glabrous or with simple trichomes, and have sessile auriculate cauline leaves, linear fruits, capitate or conical and decurrent stigmas, and often incumbent cotyledons.

32. Tribe Erysimeae

This unigeneric tribe consists of *Erysimum* (ca. 180 spp.), a genus centered primarily in Eurasia, with 8 species in northern Africa and Macaronesia and 15 in North America. The genus was placed in the broadly circumscribed Camelineae sensu Al-Shehbaz et al. (2006), but molecular studies (Bailey et al., 2006; German et al., unpublished) clearly support it placement in a distinct tribe. Another genus, *Chrsyocamela* (3 spp.) should perhaps be added to the Erysimeae (Koch et al., unpublished).

The tribe is distinguished by the exclusively sessile, stellate and/or malpighiaceous trichomes, often yellow or orange flowers, and many-seeded siliques.

33. TRIBE APHRAGMEAE

This tribe includes only *Aphragmus* (11 spp.), a genus distributed primarily in the Himalayas and central Asia, with only *A. eschscholtzianus* growing in the Russian Far East and arctic Alaska and adjacent Canada. The Aphragmeae tribe has recently been described as new by German and Al-Shehbaz (2008) based on molecular studies by German et al. (unpublished).

The tribe includes herbaceous annuals or perennials with minute, forked or simple trichomes; bracteate racemes; non-auriculate cauline leaves; entire stigmas; incumbent cotyledons; and white to deep purple petals.

OTHER TRIBES

As discussed above, the tribe Camelineae sensu Al-Shehbaz et al. (2006) is polyphyletic. After the removal of *Erysimum* into the Erysimeae, the Camelineae remains polyphyletic (Koch et al., unpublished; German et al., unpublished). We suggest that the genera *Turritis* (2 spp.) and *Olimarabidopsis* (3 spp.) be placed in one tribe, and that *Crucihimalaya* (9 spp.), *Pachycladon* (10 spp.), and *Transberingia* (1 sp.) be placed in another. Studies by the present authors are underway to recognize these two tribes.

RECOGNITION AND ASSIGNMENT OF GENERA

Although a complete tribal classification system of the *Brassicaceae* is not yet available, we are gradually approaching that goal. Following the first phylogenetic tribal classification of the family (Al-Shehbaz et al., 2006), subsequent molecular studies (e.g., Bailey et al., 2006; Warwick et al., 2006a, 2007, 2008; Koch et al., 2007, unpublished) led to the tribal adjustments recently proposed by Al-Shehbaz and Warwick (2007) and German and Al-Shehbaz (2008). Table 1.1 summarizes and updates our present knowledge of the tribal placement of nearly two-thirds (62.7%) of the 338 genera, and 87.6% of the 3709 species compiled by Warwick et al. (2006b).

An ongoing comprehensive phylogenetic study of the family (involving Warwick, Al-Shehbaz, Mummenhoff, and Koch) aims to cover more than 95% of all accepted genera. The major difficulty lies in obtaining adequate material for molecular studies on species of numerous monospecific or oligospecific genera (see Figure 2 in Koch and Kiefer, 2006, and the estimates by Al-Shehbaz et al., 2006). Many of these are known only from the type collections of their species. Although most of the larger genera of the family (e.g., *Draba, Lepidium, Cardamine, Erysimum, Heliophila, Rorippa*) are reasonably well surveyed molecularly and are shown to be monophyletic, it is the smaller and medium-sized genera (especially of the tribes Brassiceae and Schizopetaleae s.l.) that need further studies. We suspect that many of these genera will be merged with others, and the total number of genera in the family will be substantially reduced.

FAMILY LIMITS AND AGE ESTIMATES

Based on strictly morphological studies, Judd et al. (1994) indicated that the *Brassicaceae* are nested within the paraphyletic Capparaceae (including Cleomaceae) and suggested their union as one family, *Brassicaceae* s.l. However, molecular studies (Hall et al., 2002, 2004; Schranz and Mitchell-Olds, 2006) clearly demonstrated that the *Brassicaceae* are sister to Cleomaceae and both are sister to Capparaceae. As a result, three families are currently recognized.

Divergence time estimates (Figure 1.1) are still controversial. The usage of Ks values, as presented by Schranz and Mitchell-Olds (2006) and Maere et al. (2005), are more reliable because they do not make any assumptions about molecular clocks. Schranz and Mitchell-Olds (2006) estimated a divergence time and very early radiation of the *Brassicaceae* at 34 mya (million years ago). This was based on a genome-wide estimated Ks average (Ks = 0.67) reflecting the last and third major genome duplication event (3R or α duplication) and using *Arabidopsis thaliana* as a reference

(Bowers et al., 2003; Simillion et al., 2002; De Bodt et al., 2005). Genome-wide comparison of Ks values from Cleomaceae and *Brassicaceae* suggest that the corresponding mean Ks value is 0.82, which refers to 41 mya as the divergence time estimate between these two families, provided that the same evolutionary mutational rate is applied.

REFERENCES

Al-Shehbaz, I.A. 1973. The biosystematics of the genus *Thelypodium*, Cruciferae. *Contr. Gray Herb.*, 204:3–148.

Al-Shehbaz, I.A. 1988. The genera of Arabideae (Cruciferae; *Brassicaceae*), the southeastern United States. *Arnold Arbor.*, 69:85–166.

Al-Shehbaz, I.A. 2002. *Noccaea nepalensis*, a new species from Nepal, and four new combinations in *Noccaea* (*Brassicaceae*). *Adansonia*, 24:89–91.

Al-Shehbaz, I.A. 2003. Transfer of most North American species of *Arabis* to *Boechera* (*Brassicaceae*). *Novon*, 13:381–391.

Al-Shehbaz, I.A. 2005. Nomenclatural notes on Eurasian *Arabis* (*Brassicaceae*). *Novon*, 15:519–524.

Al-Shehbaz, I.A. and K. Mummenhoff. 2005. Transfer of the South African genera *Brachycarpaea, Cycloptychis, Schlechteria, Silicularia,* and *Thlaspeocarpa* to *Heliophila* (*Brassicaceae*). *Novon*, 15:385–389.

Al-Shehbaz, I.A. and S.L. O'Kane, Jr. 2002a. Taxonomy and phylogeny of *Arabidopsis* (Brassicaceae). In: *The Arabidopsis Book*, Eds. C.R. Somerville and E.M. Meyerowitz, American Society of Plant Biologists, Rockville, MD, pp. doi/10.1199/tab.0001, http://www.aspb.org/publications/arabidopsis/.

Al-Shehbaz, I.A. and S.L. O'Kane, Jr. 2002b. *Lesquerella* is united with *Physaria* (*Brassicaceae*). *Novon*, 12:319–329.

Al-Shehbaz, I.A. and S.I. Warwick. 2005. A synopsis of *Eutrema* (*Brassicaceae*). *Harvard Pap. Bot.*, 10:129–135.

Al-Shehbaz, I.A. and S.I. Warwick. 2006. A synopsis of *Smelowskia* (*Brassicaceae*). *Harvard Pap. Bot.*, 11:91–99.

Al-Shehbaz, I.A. and S.I. Warwick. 2007. Two new tribes (Donstostemoneae and Malcolmieae) in the *Brassicaceae* (Cruciferae). *Harvard Pap. Bot.*, 12(2):429–433.

Al-Shehbaz, I.A., M.A. Beilstein, and E.A. Kellogg. 2006. Systematics and phylogeny of the *Brassicaceae* (Cruciferae): an overview. *Pl. Syst. Evol.*, 259:89–120.

Al-Shehbaz, I.A., S.L. O'Kane, Jr., and R.A. Price. 1999. Generic placement of species excluded from *Arabidopsis* (*Brassicaceae*). *Novon*, 9:296–307.

Al-Shehbaz, I.A., K. Mummenhoff, and O. Appel. 2002. The genera *Cardaria, Coronopus* and *Stroganowia* are united with *Lepidium* (*Brassicaceae*). *Novon*, 12:5–11.

Anderson, J.K. and S.I. Warwick. 1999. Chromosome number evolution in the tribe Brassiceae (*Brassicaceae*): evidence from isozyme umber. *Pl. Syst. Evol.*, 215:255–285.

Appel, O. and I.A. Al-Shehbaz. 1997. Re-evaluation of tribe Heliophileae (*Brassicaceae*). *Mitt. Inst. Allg. Bot. Hamburg*, 27:85–92.

Appel, O. and I.A. Al-Shehbaz. 2003. Cruciferae. In: *The Families and Genera of Vascular Plants*, Ed. K. Kubitzki and C. Bayer, p. 75–174. Springer-Verlag, Berlin.

Bailey, C.D., R.A. Price, and J.J. Doyle. 2002. Systematics of the halimolobine *Brassicaceae*: evidence from three loci and morphology. *Syst. Bot.*, 27:318–332.

Bailey, C.D., M.A. Koch, M. Mayer, K. Mummenhoff, S.L. O'Kane, S.I. Warwick, M.D. Windham, and I.A. Al-Shehbaz. 2006. A Global nrDNA ITS phylogeny of the *Brassicaceae*. *Mol. Biol. Evol.*, 23:2142–2160.

Bailey, C.D., I.A. Al-Shehbaz, and G. Rajanikanth. 2007. Generic limits in tribe Halimolobeae and description of the new genus *Exhalimolobos* (*Brassicaceae*). *Syst. Bot.*, 32:140–156.

Beilstein, M.A., I.A. Al-Shehbaz, and E.A. Kellogg. 2006. *Brassicaceae* phylogeny and trichome evolution. *Am. J. Bot.*, 93:607–619.

Bowers, J.E., B.A. Chapman, J.D. Rong, and A.H. Paterson. 2003. Unravelling angiosperm genome evolution by phylogenetic analysis of chromosomal duplication events. *Nature*, 422:433–438.

De Bodt, S., S. Maere, and Y. Van de Peer. 2005. Genome duplication and the origin of angiosperms. *Trends Ecol. Evol.*, 20:591–597.

Dudley, T.R. and J. Cullen. 1965. Studies in the Old World Alysseae Hayek. *Feddes Repert.*, 71:218–228.

Francisco-Ortega, J., J. Fuertes-Aguilar, C. Gómez-Campo, A. Santos-Guerra, and R.K. Jansen. 1999. Internal transcribed spacer sequence phylogeny of *Crambe* (*Brassicaceae*): molecular data reveal two Old World disjunctions. *Mol. Phylog. Evol.*, 11:361–380.

Francisco-Ortega, J., J. Fuertes-Aguilar, S.C. Kim, A. Santos-Guerra, D.J. Crawford, and R.K. Jansen. 2002. Phylogeny of the Macaronesian endemic Crambe section Dendrocrambe (*Brassicaceae*) based on internal transcribed spacer sequences of nuclear ribosomal DNA. *Am. J. Bot.*, 89:1984–1990.

Franzke, A., D. German, I.A. Al-Shehbaz, and K. Mummenhoff. 2008. *Arabidopsis* family ties: molecular phylogeny and age estimates in the *Brassicaceae* (in press).

Fuentes-Soriano, S. 2004. A taxonomic revision of *Pennellia* (*Brassicaceae*). *Harvard Pap. Bot.*, 8:173–202.

German, D.A. and I.A. Al-Shehbaz. 2008. Five additional tribes (Aphragmeae, Biscutelleae, Calepineae, Conringieae, and Erysimeae) in the *Brassicaceae*. *Harvard Pap. Bot.*, 13(1):165–170.

Gómez-Campo, C. (Ed.). 1999. *Biology of Brassica Coenospecies*. 489 pp. Elsevier, Amsterdam.

Hall, J.C., K.J. Sytsma, and H.H. Iltis. 2002. Phylogeny of Capparaceae and *Brassicaceae* based on chloroplast sequence data. *Am. J. Bot.*, 89:1826–1842.

Hall, J.C., H.H. Iltis, and K.J. Sytsma. 2004. Molecular phylogenetics of core brassicales, placement of orphan genera *Emblingia*, *Forchhammeria*, *Tirania*, and character evolution. *Syst. Bot.*, 29:654–669.

Hayek, A. 1911. Entwurf eines Cruciferensystems auf phylogenetischer Grundlage. *Beih. Bot. Centralbl.*, 27:127–335.

Heenan, P.B. and A.D. Mitchell. 2003. Phylogeny, biogeography and adaptive radiation of *Pachycladon* (*Brassicaceae*) in the mountains of South Island, New Zealand. *J. Biogeogr.*, 30:1737–1749.

Heenan, P.B., A.D. Mitchell, and M. Koch. 2002. Molecular systematics of the New Zealand *Pachycladon* (*Brassicaceae*) complex: generic circumscription and relationship to *Arabidopsis* sens. lat. and *Arabis* sens. lat. *New Zealand J. Bot.*, 40:543–562.

Henry, Y., M. Bedhomme, and G. Blanc. 2006. History, protohistory and prehistory of the *Arabidopsis thaliana* chromosome complement. *Trends in Pl. Sci.*, 11:267–273.

Judd, W.S., R.W. Sanders, and M.J. Donoghue. 1994. Angiosperm family pairs: preliminary phylogenetic analyses. *Harvard Pap. Bot.*, 5:1–51.

Koch, M. 2002. Genetic differentiation and speciation in prealpine *Cochlearia*: Allohexaploid *Cochlearia bavarica* Vogt (*Brassicaceae*) compared to its diploid ancestor *Cochlearia pyrenaica* DC. in Germany and Austria. *Pl. Syst. Evol.*, 232:35–49.

Koch, M. 2003, Molecular phylogenetics, evolution and population biology in *Brassicaceae*. In: Plant Genome: Biodiversity and Evolution, Vol. 1a (phanerogams), Eds. A. K.Sharma and A. Sharma, p. 1–35. Science Publishers, Enfield, NH.

Koch, M. and I.A. Al-Shehbaz. 2004. Taxonomic and phylogenetic evaluation of the American "*Thlaspi*" Species: identity and relationship to the Eurasian genus *Noccaea* (*Brassicaceae*). *Syst. Bot.*, 29:375–384.

Koch, M. and K.-G. Bernhardt. 2004. Comparative biogeography of the cytotypes of annual *Microthlaspi perfoliatum* (*Brassicaceae*) in Europe using isozymes and cpDNA data: refugia, diversity centers, and postglacial colonization. *Am. J. Bot.*, 91:114–124.

Koch, M. and H. Hurka. 1999. Isozyme analysis in the polyploid complex *Microthlaspi perfoliatum* (L.) F.K. Meyer: morphology, biogeography and evolutionary history. *Flora*, 194:33–48.

Koch, M. and C. Kiefer. 2006. Molecules and migration: biogeographical studies in cruciferous plants. *Pl. Syst. Evol.*, 259(2–4):121–142.

Koch, M. and K. Mummenhoff. 2001. *Thlaspi* s.str. (*Brassicaceae*) versus *Thlaspi* s.l.: morphological and anatomical characters in the light of ITS nrDNA sequence data. *Pl. Syst. Evol.*, 227:209–225.

Koch, M. and K. Mummenhoff. 2006. Evolution and phylogeny of the *Brassicaceae*. *Pl. Syst. Evol.*, 259:81–258.

Koch, M.A., C. Dobes. C. Kiefer, R. Schmickl, L. Klimes, and M.A. Lysak. 2007. SuperNetwork identifies multiple events of plastid *trn*F (GAA) pseudogene evolution in the *Brassicaceae*. *Mol. Biol. Evol.*, 24:63—73.

Koch, M., H. Hurka, and K. Mummenhoff. 1996. Chloroplast DNA restriction site variation and RAPD-analyses in *Cochlearia* (*Brassicaceae*): biosystematics and speciation. *Nord. J. Bot.*, 16:585–603.

Koch, M., M. Huthmann, and H. Hurka. 1998. Molecular biogeography and evolution of the *Microthlaspi perfoliatum* s.l. polyploid complex (*Brassicaceae*): chloroplast DNA and nuclear ribosomal DNA restriction site variation. *Can. J. Bot.*, 76:382–396.

Koch, M., J. Bishop, and T. Mitchell-Olds. 1999a. Molecular systematics of *Arabidopsis* and *Arabis*. *Pl. Biol.*, 1:529–537.

Koch, M., K. Mummenhoff, and H. Hurka. 1999b. Molecular phylogenetics of *Cochlearia* (*Brassicaceae*) and allied genera based on nuclear ribosomal ITS DNA sequence analysis contradict traditional concepts of their evolutionary relationship. *Pl. Syst. Evol.*, 216:207–230.

Koch, M., B. Haubold, and T. Mitchell-Olds. 2000. Comparative evolutionary analysis of chalcone synthase and alcohol dehydrogenase loci in *Arabidopsis*, *Arabis*, and related genera (*Brassicaceae*). *Mol. Biol. Evol.*, 17:1483–1498.

Koch, M., B. Haubold, and T. Mitchell-Olds. 2001. Molecular systematics of the *Brassicaceae*: evidence from coding plastidic *matK* and nuclear *Chs* sequences. *Am. J. Bot.*, 88:534–544.

Koch, M., I.A. Al-Shehbaz, and K. Mummenhoff. 2003a. Molecular systematics, evolution, and population biology in the mustard family (*Brassicaceae*). *Ann. Missouri Bot. Gard.*, 90:151–171.

Koch, M., K.G. Bernhardt, and J. Kochjarová. 2003b. *Cochlearia macrorrhiza* (*Brassicaceae*): a bridging species between *Cochlearia* taxa from the Eastern Alps and the Carpathians. *Pl. Syst. Evol.*, 242:137–147.

Lysak, M.L., M.A. Koch, A. Pecinka, and I. Schubert. 2005. Chromosome triplication found across the tribe Brassiceae. *Genome Res.*, 15:516–525.

Maere, S., S. De Bodt, J. Raes, T. Casneuf, M.V. Montagu, M. Kuiper, and Y. Van de Peer. 2005. Modeling gene and genome duplications in eukaryotes. *Proc. Natl. Acad. Sci. U.S.A.*, 102:5454–5459.

Meyer, F.K. 1973. Conspectus der *"Thlaspi"*-Arten Europas, Afrikas und Vorderasiens. *Feddes Repert.*, 84:449–470.

Meyer, F.K. 1979. Kritische Revision der *"Thlaspi"*-Arten Europas, Afrikas und Vorderasiens. I. Geschichte, Morphologie und Chorologie. *Feddes Repert.*, 90:129–154.

Meyer, F.K. 2001a. Kritische Revision der *"Thlaspi"*-Arten Europas, Afrikas und Vorderasiens, Spezieller Tiel, I. *Thlaspi* L. *Haussknechtia*, 8:3–42.

Meyer, F.K. 2001b. Kritische Revision der *"Thlaspi"*-Arten Europas, Afrikas und Vorderasiens, Spezieller Tiel, II. *Neurotropis* F.K. Mey. *Haussknechtia*, 8:43–58.

Meyer, F.K. 2003. Kritische Revision der *"Thlaspi"*-Arten Europas, Afrikas und Vorderasiens, Spezieller Tiel, III. *Microthlaspi* F.K. Mey. *Haussknechtia*, 9:3–59.

Meyer, F.K. 2006. Kritische Revision der *"Thlaspi"*-Arten Europas, Afrikas und Vorderasiens, Spezieller Tiel, IX. *Noccaea* Moench. *Haussknechtia*, 12:1–341.

Mitchell, A.D. and P.B. Heenan. 2000. Systematic relationships of New Zealand endemic *Brassicaceae* inferred from nrDNA ITS sequence data. *Syst. Bot.*, 25:98–105.

Mitchell-Olds, T., I.A. Al-Shehbaz, M.A. Koch, and T.F. Sharbel. 2005. Crucifer evolution in the post-genomic era. In: *Plant Diversity and Evolution*, Ed. R.J. Henry, p. 119–137. CAB International, Oxfordshire, U.K.

Moazzeni, H., S. Zarre, I.A. Al-Shehbaz, and K. Mummenhoff. 2007. Seed-coat microsculpturing and its systematic application in *Isatis* (*Brassicaceae*) and allied genera in Iran. *Flora*, 202:447–454.

Mummenhoff, K. and H. Hurka. 1994. Subunit polypeptide composition of rubisco and the origin of allopolyploid *Arabidopsis suecica* (*Brassicaceae*). *Biochem. Syst. Ecol.*, 22:807–812.

Mummenhoff, K. and H. Hurka. 1995. Allopolyploid origin of *Arabidopsis suecica* (Fries) Norrlin — evidence from chloroplast and nuclear genome markers. *Bot. Acta*, 108:449–456.

Mummenhoff, K., A. Franzke, and M. Koch. 1997a. Molecular data reveal convergence in fruit characters used in the classification of *Thlaspi* s.l. (*Brassicaceae*). *Bot. J. Linn. Soc.*, 125:183–199.

Mummenhoff, K., A. Franzke, and M. Koch. 1997b. Molecular phylogenetics of *Thlaspi* sl (*Brassicaceae*) based on chloroplast DNA restriction site variation and sequences of the internal transcribed spacers of nuclear ribosomal DNA. *Can. J. Bot.*, 75:469–482.

Mummenhoff, K., U. Coja, and H. Brüggemann. 2001. *Pachyphragma* and *Gagria* (*Brassicaceae*) revisited: molecular data indicate close relationship to *Thlaspi* s.str. *Folia Geobot.*, 36:293–302.

Mummenhoff, K., I.A. Al-Shehbaz, F.T. Bakker, H.P. Linder, H.P., and A. Mühlhaussen. 2005. Phylogeny, morphological evolution, and speciation of endemic *Brassicaceae* genera in the Cape flora of southern Africa. *Ann. Missouri Bot. Gard.*, 92:400–424.

O'Kane, S.L. Jr. and I.A. Al-Shehbaz. 1997. A synopsis of *Arabidopsis* (*Brassicaceae*). *Novon*, 7:323–327.

O'Kane S.L. Jr. and I.A. Al-Shehbaz. 2003. Phylogenetic position and generic limits of *Arabidopsis* (*Brassicaceae*) based on sequences of nuclear ribosomal DNA. *Ann. Missouri Bot. Gard.*, 90:603–612.

O'Kane, Jr., S.L. B.A. Schaal, and I.A. Al-Shehbaz. 1997. The origins of *Arabidopsis suecica* (*Brassicaceae*) as indicated by nuclear rDNA sequences. *Syst. Bot.*, 21:559–566.

Price, R.A., Palmer, J.D., and I.A. Al-Shehbaz. 1994. Systematic relationships of *Arabidopsis*: a molecular and morphological approach. In: *Arabidopsis*, Ed. E. Meyerowitz and C. Somerville, p. 7–19. Cold Spring Harbor Press, Cold Spring Harbor, NY.

Price, R.A., I.A. Al-Shehbaz, and S.L. O'Kane, Jr. 2001. *Beringia* (*Brassicaceae*), a new genus of Arabidopsoid affinities from Russia and North America. *Novon*, 11:332–336.

Rollins, R.C. 1993. *The Cruciferae of Continental North America*, Stanford University Press, Stanford, CA.

Schranz M.E. and T. Mitchell-Olds. 2006. Independent ancient polyploidy events in the sister families *Brassicaceae* and Cleomaceae. *Plant Cell*, 18:1152–1165.

Schulz, O.E. 1924. Cruciferae-Sisymbrieae. In: *Pflanzenreich IV. 105 (Heft 86)*, Ed. A. Engler, p. 1–388. Verlag von Wilhelm Engelmann, Leipzig.

Schulz, O.E. 1936. Cruciferae. In: *Die natürlichen Pflanzenfamilien*, Vol. 17B, Eds. A. Engler and H. Harms, p. 227–658. Verlag von Wilhelm Engelmann, Leipzig.

Simillion, C., K. Vandepoele, M.C.E. Van Montagu, M. Zabeau, and Y. Van de Peer. 2002. The hidden duplication past of *Arabidopsis thaliana*. *Proc. Natl. Acad. Sci. U.S.A.*, 99:7719–7723.

Tsunoda, S, K. Hinata, and C. Gómez-Campo. 1980. *Brassica Crops and Wild Allies*. p. 1–354. Japan Scientific Societies Press, Tokyo.

Vaughan, J.G., A.J. Macleod, and B.M.G. Jones. 1976. *The Biology and Chemistry of the Cruciferae*, p. 1–355. Academic Press, London.

Warwick, S.I. and I.A. Al-Shehbaz. 2005. *Brassicaceae*: chromosome number index and database on CD-ROM. *Pl. Syst. Evol.*, 259:237–248.

Warwick, S.I. and L.D. Black. 1994. Evaluation of the subtribes Moricandiinae, Savignyinae, Vellinae, and Zillinae (*Brassicaceae*, Tribe Brassiceae) using chloroplast DNA restriction site variation. *Can. J. Bot.*, 72:1692–1701.

Warwick, S.I. and L.D. Black. 1997a. Molecular phylogenies from theory to application in *Brassica* and allies (Tribe Brassiceae, *Brassicaceae*). *Opera Bot.*, 132:159–168.

Warwick, S.I. and L.D Black. 1997b. Phylogenetic implications of chloroplast DNA restriction site variation in subtribes Raphaninae and Cakilinae (*Brassicaceae*, tribe Brassiceae). *Can. J. Bot.*, 75:960–973.

Warwick, S.I. and C. Sauder. 2005. Phylogeny of tribe Brassiceae (*Brassicaceae*) based on chloroplast restriction site polymorphisms and nuclear ribosomal internal transcribed spacer and chloroplast *trn*L intron sequences. *Can. J. Bot.*, 83:467–483.

Warwick, S.I., I.A. Al-Shehbaz, R.A. Price, and C. Sauder. 2002. Phylogeny of *Sisymbrium* (*Brassicaceae*) based on ITS sequences of nuclear ribosomal DNA. *Can. J. Bot.*, 80:1002–1017.

Warwick, S.I., I.A. Al-Shehbaz, C. Sauder, J.G. Harris, and M. Koch. 2004a. Phylogeny of *Braya* and *Neotorularia* (*Brassicaceae*) based on nuclear ribosomal internal transcribed spacer and chloroplast trnL intron sequences. *Can. J. Bot.*, 82:376–392.

Warwick, S.I., I.A. Al-Shehbaz, C. Sauder, D.F. Murray, and K. Mummenhoff. 2004b. Phylogeny of *Smelowskia* and related genera (*Brassicaceae*) based on ITS sequences of nuclear ribosomal DNA and *trn*L intron of Chloroplast DNA. *Ann. Missouri Bot. Garden*, 91:99–123.

Warwick, S.I., I.A. Al-Shehbaz, and C.A. Sauder. 2005. Phylogeny and cytological diversity of *Sisymbrium* (*Brassicaceae*). In: *Plant Genome: Biodiversity and Evolution. 1C: Phanerogams (Angiosperm-Dicotyledons)*, Eds. A.K. Sharma and A. Sharma, p. 219–250. Oxford & IBH Publishing Co. Pvt. Ltd., New Delhi, with Science Publishers, USA.

Warwick, S.I., I.A. Al-Shehbaz, and C.A. Sauder. 2006a. Phylogenetic position of *Arabis arenicola* and generic limits of *Eutrema* and *Aphragmus* (*Brassicaceae*) based on sequences of nuclear ribosomal DNA. *Can. J. Bot.*, 84:269–281.

Warwick, S.I., A. Francis, and I.A. Al-Shehbaz. 2006b. *Brassicaceae*: species checklist and database on CD-ROM. *Pl. Syst. Evol.*, 259:249–258.

Warwick, S.I., C.A. Sauder, and I.A. Al-Shehbaz. 2008. Phylogenetic relationships in the tribe Alysseae (*Brassicaceae*) based on nuclear ribosomal ITS DNA sequences. *Can. J. Bot.*, 86:315–336.

Warwick, SI., C.A. Sauder, I.A. Al-Shehbaz, and F. Jacquemoud. 2007. Phylogenetic relationships in the tribes Anchonieae, Chorisporeae, Euclidieae, and Hesperideae (*Brassicaceae*) based on nuclear ribosomal ITS DNA sequences. *Ann. Miss. Bot. Gard.*, 94:56–78.

Yang, Y.W., K.N. Lai, P.Y. Tai, and W.H. Li. 1999. Rates of nucleotide substitution in angiosperm mitochondrial DNA sequences and dates of divergence between *Brassica* and the other angiosperm lineages. *J. Mol. Evol.*, 48:597–604.

Yue, J.P., H. Sun, I.A. Al-Shehbaz, and J.H. Li. 2006. Support for an expanded *Solms-laubachia* (*Brassicaceae*): evidence from sequences of chloroplast and nuclear genes. *Ann. Missouri Bot. Gard.*, 93:402–411.

Yue, J.P., H. Sun, J.H. Li, and I.A. Al-Shehbaz. 2008. A synopsis of an expanded *Solms-laubachia* (*Brassicaceae*), and the description of four new species from western China. *Ann. Missouri Bot. Gard.*, Vol. 95 (in press).

Zunk, K., K. Mummenhoff, and H. Hurka. 1999. Phylogenetic relationships in tribe Lepidieae (*Brassicaceae*) based on chloroplast DNA restriction site variation. *Can. J. Bot.*, 77:1504–1512.

2 Phylogeny of *Brassica* and Wild Relatives

Suzanne I. Warwick and Jocelyn C. Hall

CONTENTS

INTRODUCTION

The genus *Brassica* and its wild relatives are included in the tribe Brassiceae, one of approximately 25 to 30 tribes in the *Brassicaceae* or Cruciferae family (Al-Shehbaz et al., 2006; Al-Shehbaz and Warwick, 2007). The tribe Brassiceae has long been considered a monophyletic group (Hedge, 1976; Al-Shehbaz, 1985; Koch et al., 2001, 2003; Appel and Al-Shehbaz, 2003). The Brassiceae comprise 48 genera and approximately 240 species (Table 2.1, revised from Warwick and Sauder, 2005; Warwick et al., 2006). Except for the four species of *Cakile* that are native to North America, the tribe is primarily distributed in the Mediterranean and southwestern Asia, with a range extension southward into South Africa. It is geographically centered in the southwestern Mediterranean region (Algeria, Morocco, and Spain), where approximately 40 genera are either endemic or exhibit maximum diversity (Hedge, 1976; Gómez-Campo, 1980, 1999; Al-Shehbaz, 1985; Al-Shehbaz et al., 2006).

Tribal members are morphologically characterized by having conduplicate cotyledons (i.e., the cotyledons longitudinally folded around the radicle in the seed), and/or transversely segmented

TABLE 2.1
List of 48 Genera in the Tribe Brassiceae (*Calepina* Adanson and *Conringia* Heist. ex Fabr. excluded)

Genus	No. of Species	Base Chromosome No. (*n*)
Ammosperma Hook. f.	2	—
Brassica L.*	39	7, 8, 9, 10, 11
Cakile Mill.*	6	9
Carrichtera DC.	1	8
Ceratocnemum Coss. & Balansa*	1	8
Chalcanthus Boiss.	1	7
Coincya Porta & Rigo ex Rouy*	6	12
Cordylocarpus Desf.*	1	8
Crambe L.*	34	15
Crambella Maire*	1	11
Didesmus Desv.*	2	8
Diplotaxis DC.*	32	7, 8, 9, 10, 11, 13
Douepia Cambess. ex Jacquem.	2	8
Enarthrocarpus Labill.*	5	10
Eremophyton Bég.*	1	-
Eruca Mill.	4	11
Erucaria Gaertn.*	10	6, 7, 8
Erucastrum C. Presl*	25	7, 8, 9
Fezia Pit. ex Batt.*	1	11
Foleyola Maire	1	16
Fortuynia Shuttlw. ex Boiss.*	2	16
Guiraoa Coss.*	1	9
Hemicrambe Webb*	3	9
Henophyton Coss. & Durieu	2	42
Hirschfeldia Moench*	1	7
Kremeriella Maire*	1	12
Moricandia DC.	8	11, 14
Morisia J. Gay*	1	7
Muricaria Desv.*	1	12
Orychophragmus Bunge	2	12
Otocarpus Durieu*	1	8
Physorhynchus Hook.*	2	14, 16
Pseuderucaria (Boiss.) O.E. Schulz	2	14
Pseudofortuynia Hedge	1	7
Psychine Desf.	1	15
Quezeliantha H. Scholz ex Rauschert	1	-
Raffenaldia Godr.*	2	7
Raphanus L.*	3	9

—continued

TABLE 2.1 (continued)
List of 48 Genera in the Tribe Brassiceae (*Calepina*
Adanson and *Conringia* Heist. ex Fabr. excluded)

Genus	No. of Species	Base Chromosome No. (*n*)
Rapistrum Crantz*	2	8
Rytidocarpus Coss.	1	14
Savignya DC.	1	15
Schouwia DC.	1	18
Sinapidendron Lowe	4	9, 10
Sinapis L.*	4	7, 8, 9, 12
Succowia Medik.	1	18
Trachystoma O.E. Schulz*	3	7, 8
Vella L.	7	17
Zilla Forssk.	2	16

Note: Recent generic changes include placement of *Boleum* Desv. and *Euzomodendron* Coss. in *Vella* (Warwick and Al-Shehbaz, 1998); *Dolichorhynchus* Hedge & Kit Tan in *Douepea* (Appel and Al-Shehbaz, 2001); *Quidproquo* Greuter & Burdet in *Raphanus* (Al-Shehbaz and Warwick, 1997); *Nesocrambe* A.G. Mill. (Miller et al., 2002) in *Hemicrambe* (Al-Shehbaz, 2004); *Brassica* includes subgenus *Brassicaria* Gómez-Campo [= *Guenthera* Andr. (Gómez-Campo, 2003)].

* Indicates the presence of heteroarthrocarpic fruit.

Source: Adapted from Warwick and Sauder, 2005, taxonomic literature included therein; and Warwick et al., 2006, with base chromosome numbers from Warwick and Al-Shehbaz, 2006.

fruits that have seeds or rudimentary ovules in both segments (heteroarthrocarpic; Appel, 1999) and, if present, only simple trichomes or hairs (Gómez-Campo 1980, 1999; Al-Shehbaz, 1985). The first two features are unknown elsewhere in the family. The few exceptions to this character combination are the genera *Ammosperma* and *Pseuderucaria*, neither of which has the conduplicate cotyledons or the segmented fruits. Classical taxonomic delimitation in the tribe Brassiceae has depended mainly upon fruit characters, with considerable debate centered on the circumscription and relationships among subtribes and genera. In the most comprehensive taxonomic treatment of the tribe, Schulz (1919, 1923, 1936) recognized seven subtribes: Brassicinae, Cakilinae, Moricandiinae, Raphaninae, Savignyinae, Vellinae, and Zillinae. Gómez-Campo (1980) proposed a reduction to six subtribes, by including the Savignyinae in the Vellinae. The Brassicinae and Moricandiinae are characterized by elongated, siliquose fruit, whereas the other subtribes generally have reduced, shortened fruit; the morphological distinctness of subtribes Brassicinae, Moricandiinae, and Raphaninae is not well substantiated (Al-Shehbaz, 1985; Warwick and Black, 1994). The Moricandiinae, for example, were separated from the Brassicinae on the basis of two characters: (1) beak (a sterile upper segment, distinct from the upper segment of heteroarthrocarpic fruits), and (2) absence of median nectaries; the latter are present in the Brassicinae and the seeds are usually present in the distal segment. As we see below, recent molecular-based phylogenetic data have provided support for alternative tribal, subtribal, and generic circumscriptions.

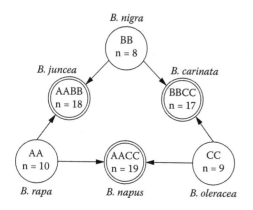

FIGURE 2.1 Relationships among the six *Brassica* crop species; Triangle of U. (*Source:* From U, 1935)

Because of its economic importance, extensive molecular-based phylogenetic studies have been conducted on members of the tribe. Earlier studies focused on *Brassica* crops and relatives. Relationships between the three diploid *Brassica* crop species [*B. nigra* (n = 8, BB), *B. rapa* (n = 10, AA), and *B. oleracea* (n = 9, CC)] and related amphidiploid species [*B. napus* (n = 19, AACC), *B. carinata* (n = 17, BBCC), and *B. juncea* (n = 18, AABB)] were first proposed by U in 1935 (Figure 2.1). Palmer et al. (1983) and Erickson et al. (1983) were the first to use restriction site data from the chloroplast DNA (cpDNA) to document the origins of the amphidiploid taxa, the results of which were later confirmed by Song et al. (1988) using nuclear RFLP markers. Yanagino et al. (1987) compared the cpDNA of 11 species in the tribe and found that the genus *Brassica* was not monophyletic, as *Brassica* taxa were intermixed with five allied genera rather than with each other. Song et al. (1990) studied the nuclear RFLPs of some 15 *Brassica* species and three additional genera, and found similar incongruities with traditional taxonomy.

Since these initial studies, sampling of the tribe has been extensive; all but seven genera (Table 2.2) have now been included in phylogenetic studies. In a series of studies based on the presence/absence of cpDNA restriction sites (Warwick and Black, 1991, 1993, 1994, 1997a; Warwick et al., 1992), analyses of relationships were extended to the whole tribe. Pradhan et al. (1992) also evaluated relationships in more than 60 species from ten genera of the tribe using cpDNA and mitochondrial DNA (mtDNA) RFLP data. The remaining studies analyzed DNA sequence variation, although they vary with regard to taxa and region sampled. ITS (internal transcribed spacers ITS-1 and ITS-2 of nuclear DNA, and the 5.8 rRNA gene) sequence-based phylogenetic studies were conducted on subtribe Vellinae (Crespo et al., 2000) and the genus *Crambe* (Francisco-Ortega et al., 1999, 2002). Focusing on 21 species, Lysak et al. (2005) analyzed sequences from the *trn*L (UAA)-*trn*F (GAA) region. Recent phylogenetic studies include sequences from the chloroplast gene maturaseK (*mat*K) and nuclear gene phytochrome A (*phy*A; Hall et al., unpublished) with a particular focus on subtribe Cakilinae. The study by Warwick and Sauder (2005) represents the most extensive taxonomic sampling of the tribe to date. These analyses were based on the ITS region and *trn*L sequence data. ITS sequences were obtained from 86 species of the tribe Brassiceae, representing subtribes Brassicinae (includes Moricandiinae and Raphaninae), Cakilinae, Vellinae, and Zillinae, and controversial tribe members *Calepina*, *Conringia*, and *Orychophragmus*. *trn*L sequences were obtained for 95 tribal species.

PHYLOGENETIC RELATIONSHIPS

TRIBAL LIMITS

In general, molecular studies support a monophyletic origin for the tribe (e.g., Anderson and Warwick, 1999; Francisco-Ortega, 1999; Lysak et al., 2005; Warwick and Sauder, 2005; Bailey et al., 2006; Beilstein et al., 2006). In fact, the delimitation of the tribe has not changed drastically since the detailed work of Schulz (1919, 1923; Al-Shehbaz et al., 2006). The Chinese genus *Spryginia* was excluded early from the tribe (Gómez-Campo, 1980). However, the placement

TABLE 2.2

Assignment of Genera to Phylogenetic Lineages

Rapa/Oleracea Lineage	Nigra Lineage	Cakile Lineage	Vella Lineage	Zilla Lineage	Savignya Lineage	NEW: Crambe Lineage
Brassica	*Brassica*	*Cakile*	*Carrichtera*	*Foleyola*	*Psychine*	*Crambe*
Diplotaxis	*Ceratocnemum*	*Crambella*	*Vella*	*Fortuynia*	*Savignya*	
Enarthrocarpus	*Coincya*	*Didesmus*		*Physorhynchus*	*Succowia*	**NEW**
Eruca	*Cordylocarpus*	*Erucaria*		*Schouwia*		*Henophyton*
Erucastrum	*Diplotaxis*			*Zilla*		*Pseuderucaria*
Moricandia	*Erucastrum*					
Morisia	*Guiraoa*					**NEW**
Raphanus	*Hemicrambe*					*Orychophragmus*
Rapistrum	*Hirschfeldia*					
Rytidocarpus	*Kremeriella*		**Not Studied:**	*Ammosperma*		
	Muricaria			*Chalcanthus*		
	Otocarpus			*Douepia*		
	Raffenaldia			*Eremophyton*		
	Sinapidendron			*Fezia*		
	Sinapis			*Pseudofortuynia*		
	Trachystoma			*Quezeliantha*		

of *Calepina* (1 sp.), *Conringia* (6 spp.), and *Orychophragmus* (2 spp.) (Gómez-Campo, 1980; Al-Shehbaz, 1985) within the tribe remains controversial and, as a result, has been the subject of several recent studies. *Calepina* and *Conringia* were once included in the Brassiceae (Schulz, 1936; Al-Shehbaz, 1985; Gómez-Campo, 1999). Gómez-Campo (1980) excluded these two genera from the tribe, but then tentatively re-included *Calepina* and *Orychophragmus* (Gómez-Campo, 1999). Parsimony analyses of the cpDNA, ITS, and combined ITS/*trn*L sequence data support a monophyletic origin for the tribe, including the controversial members *Calepina*, *Conringia*, and *Orychophragmus* (Warwick and Sauder, 2005). In all four data sets (cpDNA, ITS, *trn*L, and ITS/*trn*L), *Calepina* and *Conringia* formed a separate and well-supported clade (with bootstrap values of 91%, 88%, 73%, and 95%, respectively) that was sister to the rest of the tribe. Based on bootstrap support for the broader tribal clade (85% in combined ITS/*trn*L) and low (<50%) bootstrap support for the remaining Brassiceae (Figure 2.4), Warwick and Sauder (2005) retained *Calepina* and *Conringia* in the tribe. In contrast, other recent molecular studies (Anderson and Warwick, 1999; Francisco-Ortega, 1999; Lysak et al., 2005; Beilstein et al., 2006) clearly support their exclusion from the Brassiceae. For example, *Calepina* formed a clade with the four outgroup taxa rather than with the rest of the tribe in the ITS-based phylogenetic analysis of Francisco-Oretga et al. (1999). Phylogenetic relationships based on the chloroplast *trn*L-*trn*F region and estimated divergence times based on sequence data of the chalcone synthase gene are congruent with comparative chromosome painting data in placing *Calepina* and *Conringia* outside the clade of Brassiceae species with triplicated genomes (Lysak et al., 2005; see Chapter 10 in this book). Earlier evidence of isozyme duplication for *Pgm*-2 and *Tpi*-1 (Anderson and Warwick, 1999) in all tribal members, except *Calepina* and *Conringia*, also supported their exclusion from the tribe. Al-Shehbaz et al. (2006) suggested that both genera should be removed from the Brassiceae, and further that the alleged conduplicate cotyledons present in *Calepina* and one of the six species of *Conringia* are likely not homologous to those of typical members of the tribe.

The last problematic genus, *Orychophragmus* (2 spp., China), has been retained in the tribe (Warwick and Sauder, 2005; Lysak et al., 2005). Earlier isozyme duplication studies (Anderson and Warwick, 1999) supported its inclusion in the tribe as it had both *Pgm*-2 and *Tpi*-1 duplications, like all other genera in the tribe. Hybridization data (reviewed in Warwick et al., 2000; Warwick and Sauder, 2005) between *Orychophragmus violaceus* and the six cultivated *Brassica* species also support its inclusion in the tribe. The position of *Orychophragmus* within the tribe, however, has not been resolved; it was sister to the *Calepina*/*Conringia* clade in one of two most parsimonious cpDNA trees, but most closely associated with the Vellinae clade in the ITS/*trn*L analyses (Warwick and Sauder, 2005).

MAJOR MOLECULAR LINEAGES IN THE TRIBE

As indicated above, six or seven lineages are currently recognized in the tribe, some of which are consistent with traditional subtribal delimitations. Recent morphological, hybridization, and molecular data sets have provided support for alternative subtribal and generic circumscriptions. Restriction site analyses of cpDNA (Figures 2.2 and 2.3; Warwick and Black, 1993, 1994, 1997a) and *mat*K sequence data (Hall et al., unpublished) provided support for the recognition of subtribes Cakilinae, Vellinae, and Zillinae, but little support for the Brassicinae, Moricandiinae, and Raphaninae. Results from the cpDNA analyses divided the latter three subtribes into two clades, designated the Rapa/Oleracea and Nigra lineages (Warwick and Black, 1991, 1993, 1994, 1997a; Warwick et al., 1992), also referred to as the *Brassica* and *Sinapis* lineages, respectively, by other authors. Lysak et al. (2005) observed the two lineages in a *trn*L-*trn*F -based phylogeny and dated the split of the two lineages at 7.9 mya. Results from the ITS- and ITS/*trn*L-based clades (Figure 2.4; Warwick and Sauder, 2005) are similar to those obtained with cpDNA restriction site data (Figure 2.2; Warwick and Black, 1994, 1997a), which provided support for the recognition of

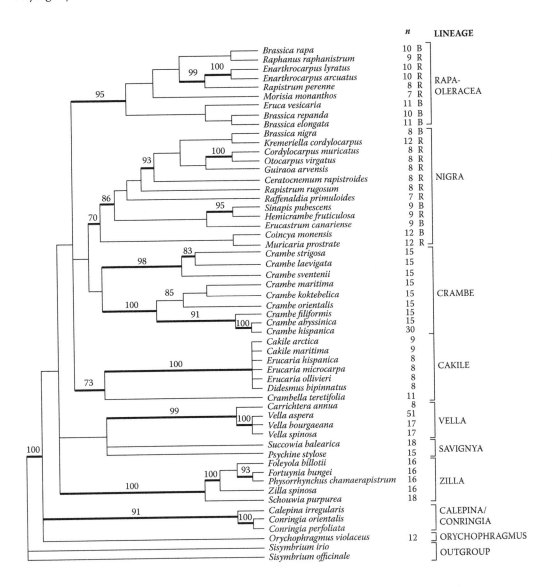

FIGURE 2.2 Strict consensus tree of the tribe Brassiceae based on maximum parsimony analysis of chloroplast DNA restriction site polymorphisms. Bolded branches have bootstrap support 70% or higher. Major lineages indicated to the right include *Rapa-Oleracea* lineage, *Nigra* lineage, *Crambe* lineage, *Cakile* lineage, *Vella* lineage, *Savignya* lineage, *Zilla* lineage, *Calepina/Conringia*, *Orychophragmus*, and the outgroup. Assignment to subtribes: Brassicinae (B) and Raphaninae (R). Chromosome numbers (*n*) are indicated at right. (*Source:* Adapted from Warwick and Black, 1997a, and Warwick and Sauder, 2005.)

taxonomic subtribes Cakilinae, Vellinae, and Zillinae; but as with previous cpDNA studies, there was little support for subtribes Brassicinae, Moricandiinae, and Raphaninae. The close genetic relatedness of the latter three subtribes is consistent with hybridization data where genetic exchange is possible among members of these subtribes (reviewed in Warwick et al., 2000).

Despite significant progress in identifying lineages within Brassiceae, our knowledge of how these lineages are related to one another is limited. Phylogenetic analyses of the Brassiceae based on nucleotide sequences of the S-locus related gene *SLR1* showed a close relationship between members of the Brassiceae and Raphaninae (Inaba and Nishio, 2002). Analyses of *matK* sequence

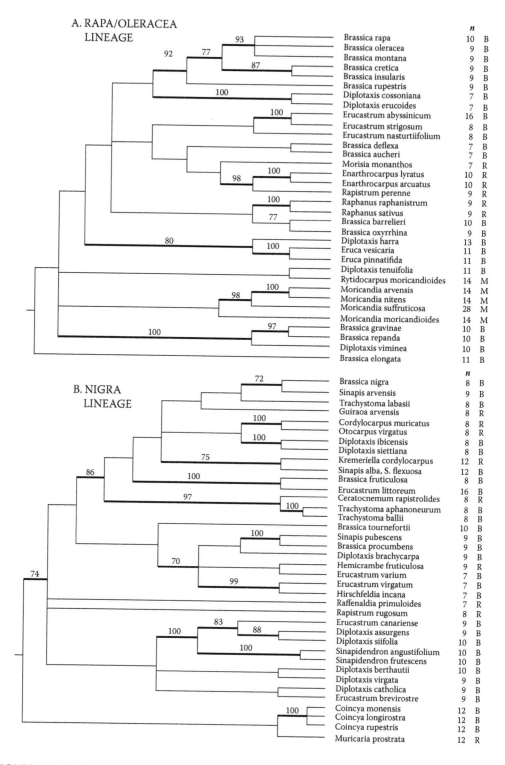

FIGURE 2.3 Strict consensus tree from parsimony analysis of chloroplast DNA restriction site polymorphisms for (A) *Rapa-Oleracea* lineage and (B) *Nigra* lineage in the tribe Brassiceae. Bolded branches have bootstrap values 70% or higher. Assignment to subtribes: Brassicinae (B), Raphaninae (R), and Moricandiinae (M). Chromosome numbers (*n*) are listed to the right. (*Source:* Adapted from Warwick and Black, 1997a.)

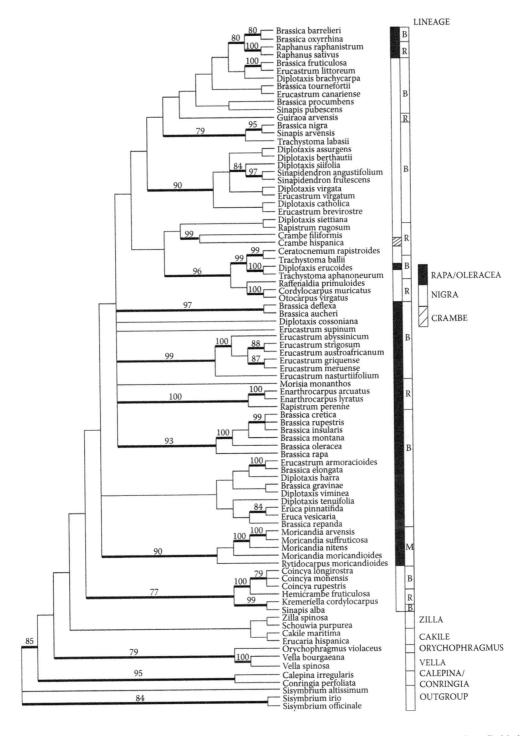

FIGURE 2.4 Strict consensus tree from parsimony analysis of combined ITS/*trnL* sequence data. Bolded branches have bootstrap values 70% or higher. Major clades are indicated to the far right and include combined Brassicinae (B) /Raphaninae (R) /Moricandiinae (M) lineage, *Zilla* lineage, *Cakile* lineage, *Orychophragmus*, *Vella* lineage, *Calepina/Conringia*, and the outgroup. *Rapa-Oleracea, Nigra,* and *Crambe* cpDNA clades are shown in the legend and placement indicated in inner column at right. (*Source:* Adapted from Warwick and Sauder, 2005.)

data (Hall et al., unpublished) support the *Zilla* lineage (*Schouwia* and *Zilla*) as sister to all other Brassiceae. There is little or no support for other relationships among lineages, which is surprising given the number of taxa sampled and the range of molecular markers utilized to examine relationships in the tribe. The unresolved backbone may be due to a lack of appropriate variation in molecular markers studied, or due to a lack of variation as the result of rapid radiation of the tribe. Although more molecular and morphological data are required, there is some evidence that this striking pattern may be the result of rapid radiation near the base of the tribe. The lineage is young, based on molecular dating information (ca. 7.9–14.6 mya; Lysak et al., 2005; Koch et al., 2001). An intriguing hypothesis is that the proposed radiation is perhaps due to either the evolution of the heteroarthrocarpic fruit and/or to the genome duplication that occurred at the base of the Brassiceae.

Each lineage shown in Table 2.2 is discussed in turn.

CAKILE LINEAGE

Phylogenetic analyses (Figure 2.2; Warwick and Black, 1997a) based on cpDNA provided evidence for the inclusion of the Cakilinae genera *Cakile* (*n* = 9), *Erucaria* (includes *Reboudia*; *n* = 6,7,8), and former Raphaninae genera *Crambella* (*n* = 11) and *Didesmus* (*n* = 8), in a *Cakile* lineage. ITS /*trn*L sequence data for *Cakile* and *Erucaria* (Figure 2.4; Warwick and Sauder, 2005), and more recently *mat*K- and *phy*A-based phylogenies based on extensive sampling for all four genera (Hall et al., unpublished), confirmed support for the lineage. Interestingly, the former study indicated that *Crambella* is sister to all other genera, whereas the latter study indicated that the monotypic genus is more derived in the clade. *Cakile, Didesmus,* and *Erucaria* were very closely related and, indeed, formed a single intermixed clade rather than three distinct generic clades.

The *Cakile* lineage uniformly has heteroarthrocarpic fruits with one to a few seeds in the upper segment, but the cotyledonary position is variable. *Crambella, Didesmus,* and some *Erucaria* (species formerly in *Reboudia*) have conduplicate cotyledons, a defining trait for the tribe, but *Cakile* has incumbent or accumbent cotyledons and the remaining *Erucaria* have incumbent or spiral cotyledons (Schulz, 1919, 1923, 1936).

ZILLA LINEAGE

Phylogenetic analyses (Warwick and Black, 1994) based on cpDNA provided evidence for the inclusion of the *n* =16 Zillinae genera *Foleyola, Fortuynia, Physorhynchus,* and *Zilla,* and the former Vellinae genus *Schouwia* (*n* = 18) in the *Zilla* lineage. The latter genus was the sister group to other Zillinae genera in these analyses. A number of studies support the *Zilla* lineage: ITS sequences for *Fortuynia, Schouwia,* and *Zilla* (Crespo et al., 2000); ITS and *trn*L sequences for *Fortuynia* and *Zilla* (Warwick and Sauder, 2005); combined ITS/*trn*L sequences for *Zilla* plus *Schouwia* (Figure 2.4; Warwick and Sauder, 2005); and more recently *mat*K and *phy*A sequences (Hall et al., unpublished data) for *Zilla* plus *Schouwia*.

Morphologically, members of the *Zilla* lineage are all adapted to xeromorphic conditions; share a similar cotyledon shape, which is large with a shallow obtuse notch; and have heteroarthrocarpic fruit, the upper segment of which is always fertile. Chromosome numbers are high — *n* = 16, 18.

VELLA AND SAVIGNYA LINEAGES

Phylogenetic analyses (Warwick and Black, 1994) based on cpDNA provided evidence for the inclusion of the Vellinae genera *Boleum* (*n* = 51), *Carrichtera* (*n* = 8), *Euzomodendron* (*n* = 17), and *Vella* (*n* = 17) in the *Vella* lineage. This result led to the transfer of monotypic genera *Boleum* and *Euzomodendron* to the genus *Vella* (Warwick and Al-Shehbaz, 1998). The ITS-sequence-based phylogeny of subtribe Vellinae (Crespo et al., 2000) generally confirmed the cpDNA-based Vellinae lineage, as did the ITS and ITS/*trn*L sequences for *Carrichtera* and *Vella* in studies by Warwick and

Sauder (2005; Figure 2.4). The ITS and *trn*L sequence data also provided strong support for the placement of *Vella aspera* (formerly *Boleum*) and *Vella bourgaeana* (formerly *Euzomodendron*) in *Vella* (Warwick and Sauder, 2005), as did *mat*K and *phy*A for *Vella aspera* (Hall et al., unpublished).

Morphologically, members of the *Vella* lineage have fruits with flattened beaks and a gynophore, and are either annual herbs (*Carrichtera*) or small, often spiny, shrubs (*Vella*). *Vella* (including *Boleum* and *Euzomodendron*) is also distinguished by large, showy petals and filaments of the inner stamens united in pairs (*Carrichtera* has small flowers with free filaments), winged seeds or with vestigial wings on seeds, a similar Vellinae cotyledon shape (acutely notched), and $n = 17$, which is unique within the tribe.

Analyses based on ITS sequence data (Crespo et al., 2000; Warwick and Sauder, 2005) support, along with other molecular data cited in Gómez-Campo (1999), the separate recognition of the *Vella* and *Savignya* lineages. The latter lineage includes *Savignya* ($n = 15$) and former Vellinae genera *Psychine* ($n = 15$) and *Succowia* ($n = 18$). Although *Savignya* was not included in *mat*K and *phy*A phylogenetic analyses (Hall et al., unpublished) and *trn*L-*trn*F (Lysak et al., 2005), *Psychine* was distinct from the *Vella* lineage. The presence of winged seeds was the only morphological trait used to distinguish subtribe Savignyinae from the Vellinae. *Psychine* and *Succowia* are morphologically similar to members of the Vellinae. For example, both are annual herbs, like *Carrichtera*. The valves of the fruit of *Psychine* are winged, similar to members of the Zillinae. *Psychine* is endemic to North Africa while *Succowia* is more widely distributed in the western Mediterranean, including North Africa. *Psychine*, like several members of the Vellinae (*Carrichtera*, and some species of *Vella*), has hairs on the cotyledon, whereas *Succowia* does not (reviewed in Warwick and Black, 1994). Finally, both *Succowia* and *Carrichtera* have very deeply notched — almost bilobed — cotyledons (i.e., most extreme in the tribe).

The tribal position of the highly polyploid taxa *Henophyton deserti* ($n = 21$), a tentative Savignyinae member, was not resolved in cpDNA analyses (Warwick and Black, 1994) or in a chloroplast gene *mat*K-based phylogeny (Hall et al., unpublished); but was sister to *Pseuderucaria* ($n = 14$) in nuclear gene *phy*A-based phylogeny (Hall et al., unpublished).

Rapa/Oleracea Lineage

The *Rapa/Oleracea* lineage, illustrated in Figure 2.3A (Warwick and Black, 1997a), includes all or part of ten genera belonging to subtribes Brassicinae, Raphaninae, and Moricandiinae: *Brassica* [in part], *Diplotaxis* [in part], *Enarthrocarpus*, *Eruca*, *Erucastrum* [in part], *Moricandia*, *Morisia*, *Raphanus*, *Rapistrum*, and *Rytidocarpus*. *Brassica* species include *B. aucheri*, *B. balearica*, *B. barrelieri*, *B. bourgeaui*, *B. cretica*, *B. deflexa*, *B. desnottesii*, *B. drepanensis*, *B. elongata*, *B. gravinae*, *B. hilarionis*, *B. incana*, *B. insularis*, *B. macrocarpa*, *B. montana*, *B. oleracea*, *B. oxyrrhina*, *B. rapa*, *B. repanda*, *B. rupestris*, and *B. villosa*. *Diplotaxis* species include *D. cossoniana*, *D. cretacea*, *D. erucoides*, *D. harra*, *D. muralis*, *D. simplex*, *D. tenuifolia*, and *D. viminea*. *Erucastrum* species include *E. abyssinicum*, *E. austroafricanum*, *E. gallicum*, *E. griquense*, *E. leucanthum*, *E. meruense*, *E. nasturtiifolium*, *E. palustre*, *E. strigosum*, and *E. supinum*. Chromosome numbers for diploids in the lineage range from $n = 7$ to 11, 13, and 14, as indicated in Figure 2.3A.

There was no evidence from the ITS, *trn*L, and the ITS/*trn*L data to support separate recognition of the subtribes Brassicinae, Raphaninae, or Moricandiinae (Figure 2.4; Warwick and Sauder, 2005) — nor was there support from the ITS and ITS/*trn*L data (and only slight indication from *trn*L data where two clades were evident in the strict tree, but with <50% bootstrap support) for the division of Brassicinae/Raphaninae/Moricandiinae taxa into the two cpDNA-based *Rapa/Oleracea* and *Nigra* lineages (Warwick and Black, 1991, 1993, 1994, 1997a; Warwick et al., 1992). However, many of the cpDNA subclades recognized in each of the two lineages were also recovered in the ITS- and ITS/*trn*L-based phylogeny. These included *Brassica barrelieri* and *B. oxyrrhina*; *B. deflexa* and *B. aucheri*; *B. oleracea* and other C genome species (*B. cretica*, *B. insularis*, *B. montana*, and *B. rupes-*

tris), and the A genome species *B. rapa*; *Enarthrocarpus* spp., and *Rapistrum perenne*; *Erucastrum* spp. (*E. abyssinicum*, *E. strigosum*, and *E. nasturtiifolium*); and *Moricandia* spp. and *Rytidocarpus*.

The two *Brassica* crops — *B. rapa* (*n* =10) and *B. oleracea* (along with other interfertile members of a group of *n* = 9 CC genome taxa) — were most closely related, as indicated by several molecular data sets including nuclear RFLPs (Song et al., 1988), cpDNA restriction sites (Warwick and Black, 1991, 1993; Pradhan et al., 1992), simple repetitive sequences (Poulsen et al., 1994), and ITS/*trn*L sequences (Warwick and Sauder, 2005). The closest wild relatives to these were the *n* = 7 *Diplotaxis* species, *D. cossoniana* and *D. erucoides* (Warwick and Black, 1991, 1993).

NIGRA LINEAGE

The *Nigra* lineage, illustrated in Figure 2.3B (Warwick and Black, 1997a), includes all or part of 16 genera belonging to subtribes Brassicinae and Raphaninae: *Brassica* [in part], *Ceratocnemum*, *Coincya*, *Cordylocarpus*, *Diplotaxis* [in part], *Erucastrum* [in part], *Guiraoa*, *Hemicrambe*, *Hirschfeldia*, *Kremeriella*, *Muricaria*, *Otocarpus*, *Raffenaldia*, *Sinapidendron*, *Sinapis*, and *Trachystoma*. *Brassica* species include *B. fruticulosa*, *B. maurorum*, *B. nigra*, *B. procumbens*, *B. spinescens*, and *B. tournefortii*. *Diplotaxis* species include *D. assurgens*, *D. berthautii*, *D. brachycarpa*, *D. brevisiliqua*, *D. catholica*, *D. ilorcitana*, *D. ibicensis*, *D. siettiana*, *D. siifolia*, *D. tenuisiliqua*, and *D. virgata*. *Erucastrum* species include *E. brevirostre*, *E. canariense*, *E. cardaminiodes*, *E. elatum*, *E. littoreum*, *E. rifanum*, *E. varium*, and *E. virgatum*. Chromosome numbers for diploids in the lineage range from *n* = 7 to 10, and 12, as indicated in Figure 2.3B.

Several of the subclades evident in the cpDNA *Nigra* lineage (Figure 2.3B) were also present in the ITS/*trn*L-based phylogeny (Figure 2.4; Warwick and Sauder, 2005). These included *Brassica fruticulosa* and *Erucastrum littoreum*; *Brassica procumbens* and *Sinapis pubescens*; *Brassica nigra*, *Sinapis arvensis*, and *Trachystoma labasii*; *Ceratocnemum rapistroides* and *Trachystoma ballii*; *Cordylocarpus muricatus* and *Otocarpus virgatus*; and *Sinapis alba* and *Kremeriella cordylocarpus*.

The *Brassica* crop species *B. nigra* (*n* = 8) was most closely related to *Sinapis arvensis* (*n* = 9) and other *Sinapis* spp.: *S. alba* (*n* = 12), *S. flexuosa* (*n* = 12), and *S. pubescens* (*n* = 9), or *Sinapis*-like, *n* = 7 species *Hirschfeldia incana* (= *S. incana*). The close relationship of this important *Brassica* crop species and *S. arvensis* has been confirmed by several molecular data sets, including nuclear RFLPs (Song et al., 1988), cpDNA restriction sites (Warwick and Black, 1991, 1993; Pradhan et al., 1992), simple repetitive sequences (Poulsen et al., 1994; Kapila et al., 1996) or shared presence of three pairs of satellite chromosomes (Cheng and Heneen, 1995), and ITS and *trn*L sequences (Warwick and Sauder, 2005).

CRAMBE LINEAGE

Several molecular data sets — including cpDNA restriction site data (Figure 2.2; Warwick and Black, 1997a), ITS sequence data (Figure 2.4; Francisco-Ortega et al., 1999, 2002; Warwick and Sauder, 2005), and *phy*A and *mat*K sequences (Hall et al., unpublished) — provided strong evidence of the evolutionary distinctness of a former Raphaninae genus, *Crambe,* from other subtribes or lineages in the tribe. A number of features support the establishment of subtribe Crambinae, as suggested by Gómez-Campo (1980, 1999): its isolated phylogenetic position; heteroarthrocarpic fruits that are globose with a one- or two-seeded upper segment and seedless, stalk-like lower segment; filaments that are often toothed; and a unique (*n* = 15) chromosome number. The *Crambe* lineage appears to have clear-cut limits that correspond to the generic limits of *Crambe*.

GENERIC CIRCUMSCRIPTION

Phylogenetic analyses based on restriction site cpDNA data (Figures 2.2 and 2.3; Warwick and Black, 1991, 1993; Warwick et al., 1992), ITS sequences (Warwick and Sauder, 2005), and *mat*K and *phy*A sequences (Hall et al., unpublished) all indicated polyphyletic origins for the four largest and/or most important genera in the Brassiceae (*Brassica, Diplotaxis, Erucastrum,* and *Sinapis*). The polyphyletic nature of *Sinapis* was corrected, in part, with the transfer of *S. aucheri*, a member of the *Rapa/Oleracea* lineage, to *Brassica* (Al-Shehbaz and Warwick, 1997). The remaining *Sinapis* species are all members of the *Nigra* lineage but do not form a single monophyletic group within this lineage. In fact, only a few genera formed monophyletic groups based on the ITS and ITS/*trn*L data: *Coincya, Crambe, Enarthrocarpus, Eruca, Raphanus,* and *Vella* (Warwick and Sauder, 2005). Polyphyletic origins for *Diplotaxis* have also been suggested in recent molecular studies using inter-simple sequence repeat markers (Martin and Sánchez-Yélamo, 2000) and for *Cakile, Didesmus,* and *Erucaria* (Hall et al., unpublished).

BRASSICA CROPS AND OTHER AMPHIDIPLOID TAXA

Relationships of the three diploid *Brassica* crop species were the same in both the ITS/*trn*L and cpDNA-based phylogenies (Figures 2.2 through 2.4; Warwick and Black, 1991, 1993; Warwick and Sauder, 2005). The B genome species *Brassica nigra* and *Sinapis arvensis* formed a well-supported clade separate from the well-supported clade of *Brassica rapa* (A genome) plus *Brassica oleracea* and other *Brassica* C genome species (*B. bourgeaui, B. cretica, B. drepanensis, B. hilarionis, B. incana, B. insularis, B. macrocarpa, B. montana, B. rupestris,* and *B. villosa*). As discussed previously, the parental origins of the three amphidiploid *Brassica* crop species, *B. carinata, B. juncea* and *B. napus,* have been confirmed by various molecular data sets including, most recently, molecular cytogenetic fluorescence *in situ* hybridization (FISH) data, which reliably identified the A, B, and C genome chromosomes in the amphidiploids (Snowdon et al., 2002; Schelfhout et al., 2004; Snowdon, 2007). Molecular studies can detect polyphyletic or multiple origins of a species, as shown, for example, for *Brassica napus* (Song and Osborne, 1992).

The origins of other well-known amphidiploid taxa in the tribe — *Erucastrum gallicum, E. elatum,* and *Diplotaxis muralis* — have also been confirmed by molecular studies. *E. gallicum* ($n = 15, 7 + 8$) is a natural amphidiploid between the $n = 7$ *Diplotaxis erucoides* and $n = 8$ *E. nasturtiifolium*, with the former serving as the maternal parent (Warwick and Black, 1993). Similarly, cpDNA data (Warwick and Black, 1993) confirmed *E. virgatum* as the maternal parent of the $n = 15$ *E. elatum* amphidiploid (*E. virgatum* $n = 7 \times E. littoreum$ $n = 8$), and *D. viminea* as the maternal parent of the $n = 21$ *D. muralis* amphidiploid (*D. viminea* $n = 10 \times D. tenuifolia$ $n = 11$). Isozyme data confirmed the proposed parentage of *E. gallicum* and *D. muralis* (Warwick and Anderson, 1997a, b).

CHLOROPLAST VS. NUCLEAR PHYLOGENIES

It is, of course, possible that gene trees, which hypothesize relationships among genes or genomes, may not be fundamentally congruent with the species phylogeny because of various biological phenomena such as introgression or lineage sorting due to stochastic events. Phylogenies based on the chloroplast genome face the additional challenge of tracing only the maternal lineages of the organisms. The advantage of the chloroplast genome, however, is its highly conserved nature, which allows for a broader comparison, such as across genera from different subtribes. Very close parallels were found in levels of genome similarity in the subtribe Brassicinae (Warwick and Black, 1993), as indicated by the comparison of the maternally inherited chloroplast genome and other genomic data sets more reflective of the nuclear genome, including cytological data and hybridization and meiotic pairing used to define cytodemes in the tribe.

Obvious discrepancies in phylogenies derived from a maternal cpDNA lineage, such as the *trn*L intron or *mat*K vs. the biparentally inherited nuclear ITS or *phy*A-based lineages, may be due to past hybridization and reticulation events. We know that with more than 50% of the family reported to be polyploid, hybridization and polyploidization are important mechanisms in the evolution of the

Brassicaceae (Koch, 2003; Koch et al., 2003). The Brassiceae are known to be ancient polyploids, with extensive aneuploid evolution (Table 2.1; Lysak et al.). Widespread hybridization within the tribe has been described (Warwick et al., 2000) and recurrent formation of polyploid species, either allo- or autoploids, may be the rule rather than the exception. Evidence for recent hybridization events include the presence of multiple ITS sequences within a single individual, as demonstrated in other *Brassicaceae* (e.g. *Braya* (Warwick et al., 2004)). Base additivity at variable sites may indicate a more recent hybrid origin in which insufficient time has elapsed for ITS sequences to have homogenized, a process known as concerted evolution. Alternatively, the presence of divergent ITS sequence types may be due to the fixation of different parental sequence types. Reticulation and hybridization in *Brassica* might thus be divided into young or old hybridization events. *Nigra* and *Rapa/Oleracea* lineages may represent ancient reticulation and chloroplast capture given the major incongruencies between the nuclear and chloroplast DNA-based phylogenies, and the lack of multiple ITS copies is consistent with the concept of concerted evolution. Accessions with extensive base additivity in more recent amphidiploid taxa probably reflect more recent, or even ongoing, hybridization and reticulation.

APPLICATION OF MOLECULAR PHYLOGENIES

A molecular-based cladogram serves as a testable hypothesis of the organismal phylogeny of a group, is critical for its taxonomic revision, and also provides an evolutionary framework for related research. The molecular-based phylogeny of the *Brassicaceae* has and continues to play a key role in investigating other evolutionary questions in the family (reviewed in Warwick and Black, 1997b; Koch et al., 2003), including the (1) evolution of chromosome number and genome triplication in the tribe and family (Lysak et al., 2005, 2007; Schranz and Mitchell-Olds, 2006); (2) evolution of other gene families (e.g., SINE retroposons (Lenoir et al., 1997; Tatout et al., 1999)); (3) origins of physiological and reproductive traits (e.g., glucosinolate evolution (Vioque et al., 1994; Windsor et al., 2005; Agerbirk, Warwick, Hansen, and Olsen, manuscript in preparation)); and (4) implications for wild germplasm preservation in cruciferous crops.

EVOLUTION OF CHROMOSOME NUMBERS AND GENOME EVOLUTION

Although the basic genomic structure of the six *Brassica* crop species was unraveled early, only recently has the origin and evolution of karyotypes in the tribe been revealed as a result of genomic tools, including genetic mapping and chromosome painting techniques. Variation in the base chromosome number of genera in the tribe ($n = 6$–18) is summarized in Table 2.1. The genomes of members of the tribe, like all of the family *Brassicaceae*, are made up of conserved genomic building blocks known as ancestral chromosomal blocks (see Chapter 10 by Lysak in this book). These blocks, however, have been triplicated in the tribe (Lysak et al., 2005, 2007) as the result of an ancient hexaploidization event estimated to have taken place approximately 7.9 to 14.6 million years ago. Rearrangement of these triplicated blocks accompanied by chromosome fusions/fissions is believed to have led to the present-day chromosome number variation within the Brassiceae (Lysak et al., 2005).

TAXONOMIC REVISION

No other group in the entire family shows as much fruit diversity as that of the Brassiceae (Al-Shehbaz et al., 2006). The majority of genera are readily recognizable by their fruits, and only poorly distinguished on the basis of vegetative and floral morphology. Al-Shehbaz et al. (2006, p. 103) stated that "rapid evolutionary bursts of fruit morphology, which are likely controlled by a relatively few genes, have most likely occurred independently of other aspects of morphology, and therefore obstructed the true relationships within the tribe and led to inadequate taxonomy."

The extensive molecular studies indicated above on the tribe have demonstrated that generic boundaries, as traditionally recognized (Schulz, 1936; Gómez-Campo, 1999), are problematic and need revision. Except for a few genera such as *Cakile* (perhaps including *Erucaria* and *Didesmus*), *Vella* (including *Euzomodendron* and *Boleum*), and *Crambe* (see Warwick and Black, 1994, 1997b; Francisco-Ortega, 1999, 2002), the remainder of the Brassiceae fall into two groups somewhat weakly defined molecularly but not morphologically: the *Nigra* and *Rapa/Oleracea* lineages. Al-Shehbaz et al. (2006) suggested that generic limits, if indeed possible to establish within these two lineages, should reflect the extensive molecular data available. This would require the abandonment of some of the most commonly known genera of the Brassiceae (e.g., *Brassica*, *Diplotaxis*, *Eruca*, *Erucastrum*, *Hemicrambe*, *Hirschfeldia*, *Raphanus*, *Rapistrum*, *Sinapidendron*, *Sinapis*) — not a trivial matter as many are "traditional" and include economically important or weedy taxa. Naturally, traditionalists would resist such major alterations but the vast majority of botanists believe that taxonomy must reflect phylogenetic data, and nomenclatural changes will have to be made sooner or later.

WILD GERMPLASM

The genetic relatedness of wild species has important implications for both germplasm preservation and utilization strategies for cruciferous crops. The molecular results indicate the artificial separation of *Brassica* from several related genera, suggesting that the genetic resource base is, therefore, much broader than originally predicted from traditional taxonomic circumscriptions, and revised taxonomic information systems on these closely related genera is equally relevant. Such information has been summarized in a guide to the wild germplasm (Warwick et al., 2000). This guide lists all species in the tribe and provides information on their cytodeme status, chromosome number, hybridization data, life cycle, growth form, ecology, and geographical distribution. In addition, a robust phylogeny also increases our ability to predict which wild/weedy relatives might hybridize with the *Brassica* crops, a major concern with regard to the release of transgenic varieties, and which may not be taxonomically evident if taxa are placed in separate genera.

CONCLUSION

The knowledge base associated with a natural taxonomic classification of the Brassiceae will continue to play a pivotal role in multidisciplinary studies of this economically important plant group, addressing both academic research questions and the needs of *Brassicaceae* crop breeding programs.

REFERENCES

Al-Shehbaz, I.A. 1985. The genera of Brassiceae (Cruciferae; *Brassicaceae*) in the southeastern United States. *J. Arnold Arbor. Harv. Univ.,* 66:279–351.

Al-Shehbaz, I.A. and S.I. Warwick. 1997. The generic disposition of *Quidproquo confusum* and *Sinapis aucheri* (*Brassicaceae*). *Novon,* 7:219–220.

Al-Shehbaz, I.A. and S.I. Warwick. 2007. Two new tribes (Dontostemoneae and Malcolmieae) in the *Brassicaceae* (Cruciferae). *Harv. Papers Bot.,* 121:429–433.

Al-Shehbaz, I.A., M.A. Beilstein, and E.A. Kellogg. 2006. Systematics and phylogeny of the *Brassicaceae* (Cruciferae): an overview. *Pl. Syst. Evol.,* 259:89–120.

Anderson, J.K. and S.I. Warwick. 1999. Chromosome number evolution in the tribe Brassiceae (*Brassicaceae*): evidence from isozyme number. *Plant Syst. Evol.,* 215:255–285.

Appel, O. 1999. The so-called 'beak', a character in the systematics of *Brassicaceae*? *Bot. Jahr. Syst . Pflanzengesch. Pflanzengeogr.,* 121:85–98.

Appel, O. and I.A. Al-Shehbaz. 2001. *Dolichorhynchus* is united with *Douepea* (*Brassicaceae*). *Novon,* 11:296–297.

Appel, O. and I.A. Al-Shehbaz. 2003. Cruciferae. In *Families and Genera of Vascular Plants*. Vol. 5., Ed. K. Kubitzki, p. 75–174. Berlin: Springer-Verlag.

Bailey, C.D., M.A. Koch, M. Mayer, K. Mummenhoff, S.L. O'Kane Jr., S.I. Warwick, M.D. Windham, and I.A. Al-Shehbaz. 2006. Toward a global phylogeny of the *Brassicaceae. Mol. Biol. Evol.*, 23:2142–2160.

Beilstein, M.A., I.A. Al-Shehbaz, and E.A. Kellogg. 2006. *Brassicaceae* phylogeny and trichome evolution. *Am. J. Bot.*, 93:607–619.

Cheng, B.F. and W.K. Heneen. 1995. Satellited chromosome nucleolus organizer regions and nucleoli of *Brassica campestris* L., *B. nigra* (L.) Koch and *Sinapis arvensis* L. *Hereditas*, 122:113–118.

Crespo, M.B., M.D. Lledo, M.F. Fay, and M.W. Chase. 2000. Subtribe Vellinae (Brassiceae, *Brassicaceae*): a combined analysis of ITS nrDNA sequences and morphological data. *Ann. Bot. (London)*, 86:53–62.

Erickson, L.R., N.A. Straus, and W.D. Beversdorf. 1983. Restriction patterns reveal origins of chloroplast genomes in *Brassica* amphidiploids. *Theor. Appl. Genet.*, 65:201–206.

Franciso-Ortega, J., J. Fuertes-Aguilar, C. Gómez-Campo, A. Santos-Guerra, and R.K. Jansen. 1999. Internal transcribed spacer sequence phylogeny of *Crambe* L. (*Brassicaceae*): molecular data revealed two old world disjunctions. *Mol. Phylogenet. Evol.*, 11:361–380.

Francisco-Ortega J., J. Fuertes-Aguilar, S.C. Kim, A. Santos-Guerra, D.J. Crawford, and R.K. Jansen. 2002. Phylogeny of the Macaronesian endemic *Crambe* section *Dendrocrambe* (*Brassicaceae*) based on internal transcribed spacer sequences of nuclear ribosomal DNA. *Am. J. Bot.*, 89:1984–1990.

Gómez-Campo, C. 1980. Morphology and morphotaxonomy of the tribe Brassiceae. In *Brassica Crops and Wild Allies*, Eds. S. Tsunoda, K. Hinata, and C. Gómez-Campo, p. 3-31. Tokyo: Japan Science Societies Press.

Gómez-Campo, C. 1999. Taxonomy. In *The Biology of Brassica coenospecies*, Ed. C. Gómez-Campo, p. 3-32. Amsterdam: Elsevier Science B.V.

Gómez-Campo, C. 2003. The genus *Guenthera* Andr. in Bess. (*Brassicaceae*, Brassiceae). *An. Jard. Bot. Madr.*, 60:301–307.

Hedge, I.C. 1976. A systematic and geographical survey of the Old World Cruciferae. In *The Biology and Chemistry of the Cruciferae*, Eds. J.G. Vaughn, A.J. MacLeod, and B.M.G. Jones, p. 1–45. London: Academic Press.

Inaba, R. and T. Nishio. 2002. Phylogenetic analysis of Brassiceae based on the nucleotide sequences of the S-locus related gene, SLR1. *Theor. Appl. Genet.*, 105:1159–1165.

Kapila, R., M.S. Negi, P. This, M. Delseny, P.S. Srivastra, and M. Lakshmikumaran. 1996. A new family of dispersed repeats from *Brassica nigra*: characterization and localization. *Theor. Appl. Genet.*, 93:1123–1129.

Koch, M. 2003. Molecular phylogenetics, evolution and population biology in the *Brassicaceae*. In *Plant Genome: Biodiversity and Evolution. Vol. 1: Phanerogams*, Eds. A.K. Sharma, and A Sharma, p. 1–35. Enfield, NH: Science Publishers Inc.

Koch, M., B. Haubold, and T. Mitchell-Olds. 2001. Molecular systematics of the *Brassicaceae*: evidence from coding plastidic *matK* and nuclear *Chs* sequences. *Am. J. Bot.*, 88:534–544.

Koch, M., I.A. Al-Shehbaz, and K. Mummenhoff. 2003. Molecular systematics, evolution, and population biology in the mustard family *Brassicaceae*: a review of a decade of studies. *Ann. Mo. Bot. Gard.*, 90:151–171.

Lenoir, A., B. Cournoyer, S.I. Warwick, G. Picard, and J.M. Deragon. 1997. Evolution of SINE S1 retroposons in Cruciferae plant species. *Mol. Biol. Evol.*, 14:934–941.

Lysak, M.L., M.A. Koch, A. Pecinka, and I. Schubert. 2005. Chromosome triplication found across the tribe Brassiceae. *Genome Res.*, 15:516–525.

Lysak, M.A., K. Cheung, M. Kitschke, and P. Bureš. 2007. Ancestral chromosomal blocks are triplicated in Brassiceae species with varying chromosome number and genome size. *Pl. Physiol.*, 145:402–410.

Martin, J.P. and M.D. Sánchez-Yélamo. 2000. Genetic relationships among species of the genus *Diplotaxis* (*Brassicaceae*) using inter-simple sequence repeat markers. *Theor. Appl. Genet,.* 101:1234–1241.

Miller, A.G., R. Atkinson, A. Wali Al Khulaidi, and N. Taleeb. 2002. *Nesocrambe*, a new genus of Cruciferae (Brassiceae) from Soqotra, Yemen. *Willdenowia*, 32:61–67.

Palmer, J.D., C.R. Shields, D.B. Cohen, and T.J. Orton. 1983. Chloroplast DNA evolution and the origin of amphidiploid *Brassica* species. *Theor. Appl. Genet.*, 65:181–189.

Poulsen, G.B., G. Kahl, and K. Weising. 1994. Differential abundance of simple repetitive sequences in species of *Brassica* and related Brassiceae. *Pl. Syst. Evol.*, 190:21–30.

Pradhan, A.K., S. Prakash, A. Mukhopadyay, and D. Pental. 1992. Phylogeny of *Brassica* and allied genera based on variation in chloroplast and mitochrondrial DNA patterns. Molecular and taxonomic classifications are incongruous. *Theor. Appl. Genet.*, 85:331–340.

Schelfhout, C.J., R. Snowdon, W.A. Cowling, and J.M. Wroth. 2004. PCR based B-genome-specific marker in *Brassica* species. *Theor. Appl. Genet.*, 109:917–921.

Schranz, M.E. and T. Mitchell-Olds. 2006. Independent ancient polyploidy events in the sister families Brassicaceae and Cleomaceae. *Plant Cell*, 18:1152–1165.

Schulz, O.E. 1919. Cruciferae - Brassiceae. Part 1. Subtribes I. Brassicinae and II. Raphaninae. In *Pflanzenr. 68–70 (IV. 105)*, Ed. A. Engler, p. 1–290. Leipzig: Verlag von Wilhelm Engelmann.

Schulz, O.E. 1923. Cruciferae – Brassiceae. Part II. Subtribes Cakilinae, Zillinae, Vellinae, Savignyinae, and Moricandiinae. In *Pflanzenr. 82–85 (IV. 105)*, Ed. A. Engler, p. 1–100. Leipzig: Verlag von Wilhelm Engelmann.

Schulz, O.E. 1936. Cruciferae. In *Die Natürlichen Pflanzenfamilien, 2nd ed.* Vol. 17B., Eds. A. Engler and H. Harms, p. 227–658. Leipzig: Verlag von Wilhelm Engelmann.

Snowdon, R.J. 2007. Cytogenetics and genome analysis in *Brassica* crops. *Chromosome Res.*, 15:85–95.

Snowdon, R.J., T. Friedrich, W. Friedt, and W. Köhler. 2002. Identifying the chromosomes of the A- and C-genome diploid *Brassica* species *B. rapa* (syn. *campestris*) and *B. oleracea* in their amphidiploid *B. napus. Theor. Appl. Genet.*, 104:533–538.

Song, K. and T.C. Osborne. 1992. Polyphyletic origins of *Brassica napus*: new evidence based on organelle and nuclear RFLP analyses. *Genome*, 35:992–1001.

Song, K.M., T.C. Osborne, and P.H. Williams. 1988. *Brassica* taxonomy based on nuclear restriction fragment length polymorphisms (RFLPs). 1. Genome evolution of diploid and amphidiploid species. *Theor. Appl. Genet.*, 75:784–794.

Song, K.M., T.C. Osborne, and P.H. Williams. 1990. *Brassica* taxonomy based on nuclear restriction fragment length polymorphisms (RFLPs). 3. Genome relationships in *Brassica* and related genera and the origin of *B. oleracea* and *B. rapa* (syn. *campestris*). *Theor. Appl. Genet.*, 79:497–506.

Tatout, C., S.I. Warwick, A. Lenoir, and J.-M. Deragon. 1999. Sine insertions as clade markers for wild crucifer species. *Mol. Biol. Evol.*, 16:1614–1621.

Vioque, J., J.E. Pastor, M. Alaiz, and E. Vioque. 1994. Chemotaxonomic study of seed glucosinolate composition in *Coincya. Bot. J. Linn. Soc.*, 116:343–350.

U, N. 1935. Genome analysis in *Brassica* with special reference to the experimental formation of *B. napus* and peculiar mode of fertilization. *Jpn. J. Bot.*, 7:389–452.

Warwick, S.I. and I.A. Al-Shehbaz. 1998. Generic evaluation of *Boleum, Euzomodendron* and *Vella* (*Brassicaceae*). *Novon*, 8:321–325.

Warwick, S.I. and I.A. Al-Shehbaz. 2006. *Brassicaceae*: Chromosome Number Index and database on CD-ROM. *Pl. Syst. Evol.*, 259:237–248.

Warwick, S.I. and J.K. Anderson, 1997a. Isozyme analysis of parentage in allopolyploid *Erucastrum gallicum* (L.) DC. (*Brassicaceae*). *Eucarpia Cruciferae Newslett.*, 19:37–38.

Warwick, S.I. and J.K. Anderson. 1997b. Isozyme analysis of parentage in allopolyploid *Diplotaxis muralis* (L.) DC. (*Brassicaceae*). *Eucarpia Cruciferae Newslett.*, 19:35–36.

Warwick, S.I. and L.D. Black. 1991. Molecular systematics of *Brassica* and allied genera (subtribe Brassicinae, Brassiceae) — chloroplast genome and cytodeme congruence. *Theor. Appl. Genet.*, 82:81–92.

Warwick, S.I. and L.D. Black. 1993. Molecular relationships in subtribe Brassicinae (Cruciferae, tribe Brassiceae). *Can. J. Bot.*, 71:906–918.

Warwick, S.I. and L.D. Black. 1994. Evaluation of the subtribes Moricandiinae, Savignyinae, Vellinae and Zillinae (*Brassicaceae*, tribe Brassiceae) using chloroplast DNA restriction site variation. *Can. J. Bot.*, 72:1692–1701.

Warwick, S.I. and L.D. Black. 1997a. Phylogenetic implications of chloroplast DNA restriction site variation in subtribes Raphaninae and Cakilinae (*Brassicaceae*, tribe Brassiceae). *Can. J. Bot.*, 75: 960–973.

Warwick, S.I. and L.D. Black. 1997b. Molecular phylogenies from theory to application in *Brassica* and allies (tribe Brassiceae, *Brassicaceae*). *Opera Bot.*, 132:159–168.

Warwick, S.I. and C. Sauder. 2005. Phylogeny of tribe Brassiceae based on chloroplast restriction site polymorphisms and nuclear ribosomal internal transcribed spacer (ITS) and chloroplast *trn*L intron sequences. *Can. J. Bot.*, 83:467–483.

Warwick, S.I., I.A. Al-Shehbaz, C. Sauder, J.G. Harris, and M. Koch. 2004. Phylogeny of *Braya* and *Neotorularia* (*Brassicaceae*) based on nuclear ribosomal DNA and chloroplast *trn*L intron sequences. *Can. J. Bot.*, 82:376–392.

Warwick, S.I., L.D. Black, and I. Aguinagalde. 1992. Molecular systematics of *Brassica* and allied genera (subtribe Brassicinae, Brassiceae) — chloroplast DNA variation in the genus *Diplotaxis. Theor. Appl. Genet.*, 83:839–850.

Warwick, S.I., A. Francis, and J. La Flèche. 2000. *Guide to the Wild Germplasm of Brassica and Allied Crops (Tribe Brassiceae, Brassicaceae)* [online]. 2nd ed. Agriculture and Agri-Food Canada Research Branch Publication, ECORC, Ottawa, Ontario, Canada. Contribution No. 991475. Available from http://www.brassica.info/information.htm [accessed April 2008].

Warwick, S.I., A. Francis, and I.A. Al-Shehbaz. 2006. *Brassicaceae*: Species Checklist and Database on CD-ROM. *Pl. Syst. Evol.,* 259:249–258.

Windsor, A.J., M. Reichelt, A. Figuth, A. Svatoš, J. Kroymann, D.J. Kliebenstein, J. Gershenzon, and T. Mitchell-Olds. 2005. Geographic and evolutionary diversification of glucosinolates among near relatives of *Arabidopsis thaliana* (*Brassicaceae*). *Phytochemistry,* 66:1321–1333.

Yanagino, T., Y. Takahata, and K. Hinata. 1987. Chloroplast DNA variations among diploid species in *Brassica* and allied genera. *Jpn. J. Genet.,* 62:119–125.

3 Biology and Ecology of Wild Crucifers

Aditya Pratap and S.K. Gupta

CONTENTS

INTRODUCTION

Species representing the tribe Brassiceae of family *Brassicaceae* (Cruciferae) are of great economic importance because they have provided sources of edible and industrial oils, condiments, vegetables, ornamentals, and salads in addition to having tremendous potential for other industrial and non-industrial products (Gómez-Campo, 1980, 1999; Gómez-Campo and Prakash, 1999). Some of the species have great importance in breeding programs of crop brassicas, either as donors of useful nuclear genes and/or to supply cytoplasmic androsterility (Gómez-Campo; 1980; Prakash and Hinata, 1980), while others are commonly found as invasive weeds in crop fields and therefore provide an interesting subject of study.

The tribe Brassiceae comprises 48 genera and about 240 species (Warwick et al., 2006; see Table 2.1 in Chapter 2). The wild species belonging to these genera, as well as some other crucifers, are a repository of useful agronomic traits such as resistance to various biotic and abiotic stresses, yield and quality traits, intermediate C_3-C_4 activity, and genes for inducing nuclear and cytoplasmic male sterility. Furthermore, some of these species could also be used as new crops. Using these diverse sources, a good number of alloplasmics of crop brassicas having desirable genes could be developed. This could be achieved through combination and introgression breeding efforts, for which hybridization between the cultivated and wild crucifers must be accomplished. For successful and purposeful gene transfer to occur between the wild and cultivated crucifers, desirable characters, genes controlling them and their expression behavior must be known in the species of interest that have to be utilized in a breeding program (Pratap et al., 2008). In addition, a number of pre- and post-fertilization requirements must be met. The pre-fertilization requirements include the physical proximity of two species, pollen movement, pollen longevity, synchrony in flowering, compatible breeding systems, floral characteristics, and competitiveness of foreign pollen. The post-fertilization factors include sexual compatibility, hybrid fertility, viability and fertility of progeny through generations of back crossings, and successful introgression of the genes into the chromosome recipient species (Salisbury, 2006). However, the above-mentioned pre- and post-fertilization factors can be taken care of only when the researcher has a thorough knowledge of the biology and breeding behavior of the wild crucifers. Although a lot of work has been done in cultivated *Brassica* species and sufficient literature is available, such work on wild crucifers is very limited. The present chapter primarily focuses on the biology and breeding aspects of wild crucifers, based upon the authors' personal observations as well as on those available in the literature, and different wild crucifers are discussed under 12 prominent genera.

DIPLOTAXIS

The genus *Diplotaxis* is a well-known and important member of the tribe Brassiceae because of its well-established constituent species that are widely distributed across several countries around the world. *Diplotaxis* includes 32 species (Warwick et al., 2006), some of which are cosmopolitan weeds while others are important sources of cytoplasmic male sterility in *Brassica* crops such as *D. muralis* in *B. rapa* (Hinata and Konno, 1979) and *B. napus* (Pellan-Delourne and Renard, 1987); *D. siifolia* in *B. juncea* (Rao et al., 1994) and *B. napus* (Rao and Shivanna, 1996); *D. erucoides* in *B. juncea* (Mallik et al., 1999; Bhat et al., 2008); *D. berthautii* in *B. juncea* (Mallik et al., 1999; Bhat et al., 2008); and *D. catholica* in *B. juncea* (Pathania et al., 2003, 2007). Chromosome numbers and ploidy status are well known for all the *Diplotaxis* species (Al-Shehbaz, 1978; Gómez-Campo and Hinata, 1980; Martínez-Laborde, 1991; Romano et al., 1986).

Considerable morphological variability is found among different *Diplotaxis* species for various characters (Table 3.2a, Figure 3.1). The plants can be annuals, biennials, or perennials with leafy or subscapose stems (Martínez-Laborde, 1997). Stems in this genus, in general, are erect or ascending with leaves green in color, ranging from almost entire to pinnatifid or pinnatisect. Trichomes on leaves and stems are generally absent or are simple. The flowers are borne in racemes with oblong or linear sepals, which may be erect or spreading in different species. Petals are usually yellow in color although white to purple variants are also found, as in the case of *Diplotaxis acris*. Stamens are six in number and tetradynamously arranged. Fruits are dehiscent siliques, linear in shape, glabrous, unsegmented or segmented, and have numerous seeds. Although most *Diplotaxis* species seeds produce mucilage when wet, some exceptions are found in the species that have seedless beaks (Gómez-Campo, 1980). Seeds are generally arranged in two rows. However, in *D. harra* and *D. siettiana*, these can be found in three or four rows; and in *D. siifolia*, these may be arranged in a single row. The terminal segment of silique, known as the beak, also shows variation in shape and size and can be either more or less compact and seedless as in case of *D. tenuifolia* and *D. muralis,* or hollow, which can have one or two seeds as in the case of *D. virgata* and *D. erucoides* (Martínez-Laborde, 1997). Seeds of this genus are ovoid to ellipsoid in shape, slightly flattened, and wingless with conduplicate cotyledons. They may also be subspherical in a few species, such as in the case of *D. siifolia*. Seed germination is highly sensitive to stress conditions, particularly salt stress (Khatri et al., 1991).

Tahahata and Hinata (1986), on the basis of 53 morphometric traits on 12 representative species, obtained a grouping of four clusters, one corresponding to the *Diplotaxis tenuifolia* group, another to the *D. harra* group, and the remaining two to the third group. Later, on the basis of numerical analysis using 47 morphological characters and 30 operational taxonomic units, Martínez-Laborde's work (1988, 1991, 1997) led to the classification of the *Diplotaxis* genus in three groups with one of them having three subgroups.

FIGURE 3.1 (See color insert following page 128.) Morpho-physiological variation in *Diplotaxis*: (A) pre-flowering plants of *D. cretacea;* (B) silique arrangement in *D. cretacea;* (C) flowers of *D. cretacea;* (D) silique structure of *D. cretacea;* (E) a young plant of *D. berthautii;* (F) a young plant of *D. harra;* (G) a young plant of *D. harra* ssp. *Crassifolia;* (H) silique structure of *D. harra* ssp. *Crassifolia;* (I) flowering plants of *D. muralis;* (J) silique arrangement in *D. muralis;* (K) leaf structure of *D. muralis;* (L) silique structure of *D. muralis;* (M) a flowering plant of *D. siettiana;* (N) silique arrangement in *D. siettiana;* (O) silique structure in *D. siettiana;* (P) flowers of *D. siifolia;* (Q) flowering plants of *D. tenuifolia;* (R) leaf structure of *D. tenuifolia;* (S) a flower of *D. tenuifolia;* and (T) silique structure of *D. tenuifolia*.

The *Diplotaxis tenuifolia* group includes taxa that have seedless beaks and the petals are brochododromous, with most of them being annuals having subscapose stems. However, *D. tenuifolia* itself in this group is an exception; it is a perennial and has leafy stems (Martínez-Laborde, 1997). The petal extracts of the *D. tenuifolia* group lack the phytochemical aglycone kaempferol, which is commonly found in their leaves (Sánchez-Yelamo, 1994). A strong rocky smell emerges from their leaves, which might be attributed to volatile isothiocyanates (Martínez-Laborde, 1997).

Commonly referred to as sand rocket (Lazarides et al., 1997) and also sand mustard, and Lincoln weed, *Diplotaxis tenuifolia* has a great weedy significance. It is believed to have originated in southern and central Europe and Asia Minor, and further spread to many warm-temperate climatic areas with porous and calcareous soils. It is found endemic in most of the Mediterranean countries and northeastern Europe (Bianco and Boari, 1997). Found in unused places, ballasts, railway beds, and along roadsides (Mulligan, 1959), it is a weed of economic importance in all of Europe, the western United States, Australia, New Zealand, and Argentina (Parsons and Cuthberstob, 1992).

Diplotaxis tenuifolia is a perennial, herbaceous plant with a deep tap root system and elongated leaves. It generally reproduces from seeds. The seedlings germinate in the presence of adequate moisture and form a slowly growing rosette in the initial growth stage, simultaneously developing a tap root system. The mature plant is erect and its height ranges from 30 cm up to 1 m. However, under natural field conditions at Jammu, India, its height has been observed to be 64 to 75 cm, with the main shoot length varying between 30 and 41 cm. Leaves are slender and deeply lobed with a pointed apex, glabrous, and medium green in color. Flowers are bright yellow in color and medium in size. Siliques are cylindrical, 2.5 to 4.0 cm long, and semi-appressed along the stem. These have an intermediately constricted surface texture. Seeds are many (40 to 86) in number in each silique and are borne in two rows. These are ovoid in shape, slightly pitted, and beige to deep brown in color. *D. tenuifolia* has been reported as a good multifarious plant (Caso, 1972), and its limited medicinal uses have also been reported in Italy (De Feo et al., 1993). Its use as a vegetable has been reported for at least 100 years in France (Ibarra and La Porte, 1947) and also in Italy where it is used as salad and also in other dishes (Bianco, 1995).

Diplotaxis muralis, commonly known as Wallrocket, is also considered a part of the *D. tenuifolia* group. However, it is smaller in size and has an annual, sometimes biennial spreading rosette. It may be found coexisting with *D. tenuifolia*. *D. muralis* can be distinguished from *D. tenuifolia* by its pale yellow flowers, constricted and slightly flattened pods, and small bristle-like hairs on the lower part of the stem.

The second group consists of two subspecies of *Diplotaxis harra* (Martínez-Laborde, 1997), which are perennials with leafy stems and their siliques are pendent with a seedless beak. Other species — viz. *D. gracilis, D. hirta, D. glauca,* and *D. vogeli* — are morphologically similar to *D. harra* but their siliques are not pendent. The same occurs for other relatives of *D. harra* such as *D. villosa, D. kohlaanensis, D. nepalensis,* and *D. pitardiana* (Martínez-Laborde, 1997). Two other species, *D. acris* and *D. griffithii*, although having different chromosome numbers, also belong to this group.

The third group of *Diplotaxis* consists of three subgroups. Taxa included in this are annuals with leafy stems and a seeded beak (Martínez-Laborde, 1997). In the second subgroup are species with $n = 8$ chromosomes having seminiferous or occasionally non-seminiferous beak (*D. ibicensis, D. brevisiliqua, D. ilorcitana,* and *D. siettiana*) and a stem with more or less appressed hairs. Among these, *D. siettiana* is endemic to a small island south of Spain. It is an annual herb 20- to 40-cm tall with sparse hairs. Initially forming a rosette, its deeply lobed leaves are little fleshy, 5- to 15-cm long. Flowers are yellow in color with petal lengths between 0.9 and 1.0 cm. The siliques are comparatively smaller in size, deeply flattened, smooth, and have prominent beaks. In this group, *D. siifolia* has spherical seeds and pinnate leaves. *D. catholica* has much divided foliage and *D. virgata* is quite hairy. The other species in this group are *D. tenuisiliqua* and *D. catholica*.

Diplotaxis tenuifolia, D. muralis, and *D. erucoides* are C_3-C_4 intermediate species character-ized by a high concentration of mitochondria and chloroplasts in the bundle sheath cells and a high potential for reassimilation of photorespired CO_2 (Apel et al., 1996). The C_3-C_4 intermediate traits, especially low photorespiration activity, may be valuable for the breeding of cultivated bras-sicas (Bang et al., 2003). The intergeneric hybrids with various genome constitutions and mono-somic addition lines (MALs) between the C_3-C_4 intermediate species and cultivated crops could also provide valuable information in understanding the evolution and genetic system of the C_3-C_4 intermediate traits. Such intergeneric hybrids between *D. tenuifolia* and cultivated *Brassica rapa, B. oleracea,* and *Eruca sativa* have been produced (Harberd and McArthur, 1980; Takahata and Hinata, 1983; Salisbury, 1989).

In addition to their use in intergeneric hybridization, *Diplotaxis* species have also found various other uses. Looking at ancient history, the Romans used these as vegetable and condiment plants (Blangiforti and Venora, 1997). Baldrati (1950) reported that in central Italy, seedlings of rocket with cotyledon leaves intact and two to four leaves were sold in the markets. Having aphrodisiac properties (Fernald, 1993), the popularity of *Diplotaxis* species is now increasing remarkably for edible purposes (Bianco and Boari, 1997). The leaves of *D. tenuifolia* are eaten raw in salads or cooked in many dishes (Bianco and Boari, 1997). Flowers are used as a garnish (Bianco, 1995), and the older leaves are pureed and added to sauces and soups (Facciola, 1990). In addition, several other uses of *Diplotaxis* have also been described in traditional medicare for many different properties such as astringent, depurative, digestive, diuretic, stimulant and tonic, laxative, anti-inflammatory, and digestive (Aritti, 1965; Ellison et al., 1980; Anonymous, 1991; De Feo et al., 1993). The peculiar pungent taste of its leaves increases its acceptability as a green salad vegetable. Such a taste might be related to the presence of volatile isothiocynates (Martínez-Laborde, 1997). Al-Shehbaz and Al-Shammary (1987) found two glucosinolate components in the seeds of *D. harra* and one (allyl) in those of *D. erucoides*.

Some species of *Diplotaxis* are also good sources of forage for animals, and camels and sheep are reported to graze on *D. harra* in Iraq (Hedge et al., 1980) and *D. villosa* in Jordan (Boulos, 1977). There are also reports of *D. assurgens, D. catholica,* and *D. virgata* being grazed by animals in Morocco (Negre, 1961; Martínez-Laborde, 1997). Simultaneously, medicinal uses of *D. harra* in the Near East were also reported (Yaniv, 1995).

BRASSICA

The genus *Brassica* consists of the most important and commercially valuable species, the major being *B. rapa* (syn. *B. campestris*), *B. juncea, B. napus, B. oleracea,* and *B. nigra,* which yield industrial and edible oils, vegetables, condiments, and many other products. Due to contributions from all these species, oilseed rapes are the world's third most important source of vegetable oils after palm and soybean (Beckman, 2005; Gupta and Pratap, 2007). Cultivated oleiferous *Brassica* taxa have been grown since antiquity in various parts of the world, and these have been an impor-tant part of the agrarian systems of various countries (Gupta and Pratap, 2007).

Genus *Brassica* has 39 constituent species with chromosome numbers (*n*) varying from 7 to 11 (Table 3.1). These species have an amalgam of breeding systems ranging from complete cross-pol-lination to a high level of self-pollination, and therefore they are quite interesting from a breeding point-of-view (Rai et al., 2007).

In addition to the cultivated members, the genus also consists of several other untapped wild species that have tremendous potential for desirable gene transfer (Table 3.2b, Figure 3.2). *Brassica fruticulosa* is a short-lived perennial plant forming vigorous rosettes during pre-flowering initial growth. The mature plant may reach a height up to 120 cm. Its leaves are alternate, medium green in color, sparsely haired, and pinnately lobed. The lower leaves are larger in size, with serrated mar-gins. Flowers are medium in size and yellow in color. Siliques are erect or semi-erect, cylindrical,

TABLE 3.1
Chromosome Checklist of the Species of Wild Crucifers Discussed in This Chapter

S. No.	Species	Ch. No. (*n*)	Ref.
1.	*Brassica berrelieri* (L.) Janka	10	Herberd (1972)
2.	*Brassica doesnotesii* Emb. & Maire	10	Takahata and Hinata (1978)
3.	*Brassica fruticulosa* Cyr.	8	Mizushima and Tsunoda (1967)
4.	*Brassica fruticulosa* ssp. *cossoniana* (Boiss & Rent.) Maire	16	Herberd (1972)
5.	*Brassica oxyrrhina* (Coss.) Wilk. & Lange	9	Herberd (1972)
6.	*Brassica tournefortii* Gouan	10	Sikka (1940)
7.	*Cakile maritima* Scop.	9	Manton (1932)
8.	*Crambe abyssinica* Hochst. ex O.E. Schulz	45	Manton (1932)
9.	*Diplotaxis berthautii* Braun-Blanq. & Maire	9	Herberd (1972) Takahata and Hinata (1978)
10.	*Diplotaxis catholica* (L.) DC.	9	Manton (1932)
11.	*Diplotaxis cossoniana* (Rent. Ex Boiss.) O.E. Schulz	7	Gómez Campo (1978)
12.	*Diplotaxis cretacea* Kotov	11	Herberd (1972)
13.	*Diplotaxis erucoides* (L.) DC	7	Jaretzky (1932)
14.	*Diplotaxis harra* (Forsk.) Boiss	13	Amin (1972)
15.	*Diplotaxis harra* ssp. *crassifolia* (Raf.) Maire	13	Herberd (1972)
16.	*Diplotaxis muralis* (L.) DC	11	Jaretzky (1932), Herberd (1972)
17.	*Diplotaxis siettiana* Maire	8	Takahata and Hinata (1978)
18.	*Diplotaxis siifolia* G. Kunze	10	Herberd (1972)
19.	*Diplotaxis tenuifolia* (L.) DC	11	Manton (1932)
20.	*Diplotaxis tenuisiliqua* Del.	9	Herberd (1972)
21.	*Diplotaxis virgata* (Cav.) DC	10	Reese (1957)
22.	*Eruca sativa* (Mill.) Thell.	11	Jaretzky (1932)
23.	*Erucastrum abyssinicum* (A.Rich) O.E. Schulz	16	Jaretzky (1932)
24.	*Erucastrum arabicum* Fischer & C.A. Meyer	8	Jonsell (1976)
25.	*Erucastrum gallicum* (Willd.) Schulz	15	Manton (1932)
26.	*Erucsatrum littoreum* (Pau & Font Quer) Maire	16	Gómez -Campo(1978)
27.	*Erucastrum nasturtifolium* (Poiret) O.E. Schulz	8	Coutinho and Lor.-Andreu (1948) Manton (1932)
28.	*Erucastrum virgatum* (J.C. Presl) C. Presl	7	Herberd (1972)
29.	*Hirschfeldia incana* (L.) Lagréze-Fossat	7	Baez-Major (1934)
30.	*Hirschfeldia incana* ssp. *consobrina* (Pomel ex Batt.) Maire	7	Herberd (1972)
31.	*Moricandia arvensis* (L.) DC	14	Jaretzky (1932)
32.	*Sinapis alba* L.	12	Karpechanko (1924)
33.	*Sinapis arvensis* L.	9	Karpechanko (1924)
34.	*Sinapis pubescens* L.	9	Manton (1932)
35.	*Sisymbrium irio* L.	7,14,21,28	Halvorson, 2003
36.	*Sisymbrium altissimum* L.	7	Halvorson, 2003
37.	*Sisymbrium officinale* (L.) Scop.	7	Halvorson, 2003
38.	*Sisymbrium orientale* (L.) DC	7	Halvorson, 2003
40.	*Trachystoma balli* O.E. Schulz	8	Herberd (1972)

TABLE 3.2A
Morpho-physiological Variation in *Diplotaxis* under Subtropical Conditions in India

Character	D. tenuifolia	D. muralis	D. cretacea	D. harra ssp. crassifolia	D. siettiana	D. tenuisiliqua
Plant Characters						
Plant form	Perennial	Annual	Annual	Biennial/perennial	Annual	Annual
Pre-flowering growth habit	Erect	Semi erect	Semi-erect	Semi erect	Rosette	Semi-erect
Plant height (cm)	64–75	66–71	39–50	76–100	20–40	Tall
Primary branches (No.)	7–9	15–23	5–9	41–60	13–25	Medium
Secondary branches (No.)	12–18	18–35	12–18	7–12	3–5	4–8
Leaf Characters						
Arrangement	Alternate	Alternate	Alternate	Alternate	Alternate	Alternate
Hairyness	Absent	Absent	Absent	Sparse	Absent/Sparse	Absent
Angle	Open	Semi-erect	Semi-erect	Semi-erect	Semi-erect	Semi-erect
Color	Medium green	Medium green	Light green	Dark green	Medium green	Medium green
No. of lobes	2–5	3–5	3–5	Medium	Many	Medium
Waxiness	Absent	Absent	Absent	Absent	Absent	Absent
Flower Characters						
Color of petals	Bright yellow	Yellow	Yellow	Deep yellow	Bright yellow	Yellow
Petal length (cm)	0.9–1.6 cm	0.7–0.8	1.1–1.3	0.5–0.6	0.9–1.0	Long
Petal width (cm)	0.9–1.00	0.5–0.6	0.4–1.1	0.4–0.5	0.6–0.7	Broad
Silique Characters						
Length (cm)	2.5–4.6	2.6–3.2	2.6–7	4.5–5.5	3.0	2.0–3.0
Length of beak (cm)	0.2–0.5, seedless	0.1–0.2, seedless	0.1–0.2	0.8	0.1–0.2, seedless	0.1–0.2
Angle with main shoot	Semi-appressed	Erecto-patent	Erecto-patent	Pendent	Erecto-patent	Erecto-patent
Surface texture	Intermediate	Smooth	Smooth	Intermediate	Smooth, flattened	Smooth
Number on main shoots	29–35	Many	16–23	Medium	Few	Few
Seeds per silique	40–86	39–51	58–74	32–42	30–60	Few
Seed Characters						
Color	Orange-brown	Orange-brown	Beige to brown	Light brown	Light brown-yellow	Orange-yellow
Shape	Ovoid to widely elliptical	Elliptical	Spherical	Elliptical	Elliptical	Elliptical

TABLE 3.2B

Morpho-physiological Variation in *Brassica* under Subtropical Conditions in India

Character	B. fruticulosa	B. fruticulosa ssp. cossoniana	B. tournefortii	B. desnottesii	Moricandia arvensis
Plant Characters					
Plant form	Short-lived perennial	Annual	Annual	Annual	Perennial
Pre-flowering growth habit	Rosette	Rosette	Semi-erect	Semi-erect	Erect
Plant height (cm)	51–98	40–120	80–100	35–95	25–65
Primary branches (No.)	12–17	8–16	10–25	5–11	6–10
Secondary branches (No.)	38–42	25–40	15–40	8–16	9–17
Leaf Characters					
Arrangement	Alternate	Alternate	Alternate	Alternate	Alternate
Hairyness	Sparse	Sparse to dense	Absent/Sparse	Sparse	Absent
Angle	Open	Open	Open	Open	
Color	Medium green	Dark green	Medium green	Dull green	Green
No. of lobes	4–7	Medium	Many (8–14)	Medium	Many
Waxiness	Absent	Present	Absent	Absent	Present
Length (c.)	3.0–25.0	Medium	6.0–25.0	6.0–18.0	4–11
Width (cm)	2.0–19.0	Medium	2.5–8.0	3.0–8.0	2.5–4.6
Flower Characters					
Color of petals	Yellow	Pale yellow	Pale yellow	Yellow	White, cream, purple
Petal length (cm)	0.98–1.02	0.8–1.1	0.4–0.8	Medium	1.5–2.0
Petal width (cm)	0.40	0.4–0.5	1.5–2.0	Medium	0.5–0.6
Silique Characters					
Length (cm)	1.9–2.3	1.7–2.5	3–7	1.5–1.8	2.0–3.0
Length of beak (cm)	0.1–0.2, seeded	0.2–0.3, seeded	1.0–1.5	0.2–0.4	0.1–0.3
Angle with main shoot	Semi-appressed	Open	Open	Open	Semi-appressed
Surface texture	Intermediate	Constricted	Constricted	Smooth	Intermediate
Number on main shoots	22–57	17–49	Few	Medium	Medium
Seeds per silique	15–26	15–17	Many	Many	Many
Seed Characters					
Color	Pale to dark brown	Pale to dark brown	Brown-purplish brown/beige	Brown	Brown
Shape	Oval/spherical	Oval/spherical	Ovate/Oblong	Ovate to obovate	Spherical

FIGURE 3.2 Morpho-physiological variation in genus *Brassica*: (A) pre-flowering plants of *B. fruticulosa;* (B) leaf structure in *B. fruticulosa;* (C) fruiting plants of *B. tournefortii;* (D) flowers of *B. fruticulosa ssp cosoniana;* (E) constricted siliques of *B. fruticulosa* ssp. *Cosoniana;* (F) a young plant of *B. oxyrrhina;* (G) silique pattern in *B. desnottessi;* (H) flowers of *B. desnottesii;* and (I) leaf structure in *B. desnottesii.*

and intermediate in surface texture, and may be 1.9- to 2.3-cm long with a small beak. Seeds are very small and brown in color.

Another *Brassica* species, *B. tournefortii*, commonly known as Asian mustard, Sahara mustard, African mustard, or Mediterranean mustard, is native to northern Africa, the Middle East, eastern Pakistan, and the Mediterranean region (GRIN, 2000; Tutin et al., 1964; Zohary, 1966). It is well adapted to the drier habitats of central Asia and the Middle East (Prakash, 1974), being native to the habitats in semi-arid and arid deserts (Minnich and Sanders, 2000). It has further spread tremendously into lowland desert regions, especially in places with sandy soils (Felger, 1990). In Morocco and Algeria, it grows on the alluvial sand and on inland sand dunes (Tsunoda, 1980).

Brassica tournefortii is an annual herbaceous plant that is erect and can reach a height up to 1 m or more. However, the size of the plants can vary considerably, depending on the available soil moisture (Felger, 1990; Van Devender et al., 1997). The plants have a good number of primary branches and a large number of secondary branches that may be up to 40 in number. The roots are well developed in a sturdy taproot system. Leaves are medium green in color, pinnately lobed with 8 to 14 lobes per leaf. Lower leaves are quite large in size (10- to 40-cm long) and have serrate-dentate margins. Upper leaves are much reduced, oblong to linear, and glabrous. Inflorescence is a terminal raceme with 6 to 20 or more perfect flowers. The flowers are of medium size (Halvorson and Guertin, 2003) and pale yellow in color with a purplish tinge toward the base. Siliques are linear, constricted, glabrous, and 3.5- to 7-cm long. Beaks are 1- to 1.5-cm long and cylindrical with one or two seeds. Seeds are globose and brown to purplish-brown in color. Some plants may produce up to 9000 seeds (Minnich and Sanders, 2000).

Brassica tournefortii is an autogamous species (Hinata et al., 1974; Minnich and Sanders, 2000) with almost 100% fruit set. Seeds require dark conditions for germination to occur with an optimum temperature range of 15 to 20°C (Thanos et al., 1991). However, its germination can be totally inhibited when exposed to light. Early and quickly growing habit enable faster establishment of *B. tournefortii* in cereal crops before the crop becomes competitive (Moore and Williams, 1983). It is reported to be a drought-tolerant crop (Salisbury, 1989; West and Nabhan, 2002) and also to have allelopathic properties (Patterson, 1983), in addition to being fairly resistant to pod shattering (Salisbury, 1989). Pradhan et al. (1991) reported it as serving as a source of CMS (*cytoplasmic* male sterility).

ERUCA

Eruca is a native of the southwestern Mediterranean region but has long been cultivated in southern Europe, and eastern and central Asia where it is commonly known as rocket, rocket salad, and white pepper. The genus has a unique species *Eruca vesicaria* (Cav.) DC. with three subspecies that show their maximum genetic variability in Spain and Morocco. *E. vesicaria* subsp. *sativa* is widely cultivated by its pungent leaves, as a source of oil or as a vegetable.

Cultivated *Eruca* is found distributed throughout the Mediterranean, central Europe, Afghanistan, northern India, North and South America, South Africa, and Australia. This species has been known since antiquity and is listed in the Greek Herbal of Dioscorides (*Materia Medica*) written in the first century, as well as in the English Herbal of John Gerard (Morales and Janick, 2002). *Eruca* is a very hardy plant and requires little care. Although it grows well in warm temperatures, it can also tolerate temperatures down to −4°C. In Portugal, it is grown on a commercial scale for export to the United Kingdom (Silva Dias, 1997). In various Mediterranean countries, it is cultivated as a green salad or for cooked vegetables; and in Asia, it is grown as an oilseed crop (Morales and Janick, 2002). However, in some other countries such as Turkey, Spain, France, Great Britain, and Greece, it is rarely found in cultivation. *E. vesicaria* (syn. *E. sativa*) is a minor oilseed crop in India. Here, the major *Eruca* growing areas are the states of Punjab, Haryana, Rajasthan, Madhya Pradesh, and Uttar Pradesh.

Eruca vesicaria is a fast-growing annual crop and it flowers under long-day and high-temperature conditions. It is a low-growing herbaceous plant that can grow up to a height of 80 cm. (Table 3.2c). It has a slender tap root with an erect, stiff, little-branched, and hairy stem. The basal leaves in a pre-flowering plant occur in a rosette and are deeply lyrate-pinnatifid although not quite pinnatisect. Upper leaves are more or less sessile and pinnatifid with a long-oblong or obovate terminal lobe, which may be entire to coarsely toothed or lobed. Trichomes are either absent or simple. All leaves are rather fleshy, dull green in color, deeply cut and compound, and have a characteristic pungent flavor (Palada and Crossmann, 1999). Flowers are large in size and few in number, borne on small terminal racemes. Petals are four in number, white to pale yellow in color with prominent brown or purple veins, and may be 1.2- to 2.2-cm long and 0.7- to 0.9-cm wide. Stamens are six in number and strongly tetradynamous. Siliques are erect, can measure up to 4 cm, and have a subcylindrical valvar portion. These are almost parallel to the stem or sub-erect on more or less erect stalks. The terminal portion is a flattened beak that may be 5- to 10-mm long, indehiscent, and seedless. Seeds are ovate to widely elliptical, wingless, 1.5- to 2.0-mm long, light brown to olive green in color, and

FIGURE 3.3 Morpho-physiological variation in *Enarthrocarpus*: (A) silique structure of *E. arcuatus*; (B) flowers of *E. arcuatus*; (C) leaf and silique pattern in *E. lyratus*; (D) plants of *E. lyratus*; (E) silique structure of *E. lyratus*; (F) leaf structure of *E. lyratus*.

are arranged into two or three rows on each side. Seeds also exhibit a characteristic mucilaginous cover when wet.

Eruca oil is mainly characterized by a high content of sulfur- and nitrogen-containing compounds (Miyazava et al., 2001). It is pungent and inedible; and because it is high in erucic acid, it is primarily used for industrial purposes (Bhandari and Chandel, 1997). It is mainly used in lubricant, soap making, and illuminating industries, as well as in pharmaceuticals in addition to its (not recommended) use as salad oil. The young *Eruca* plants are used as salad, vegetables, and as green fodder. *Eruca* leaves are the essential ingredient of the "misticanza" (mixed salad) consumed in Italy. Tender leaves are reported to have stimulant, stomachic, diuretic, and antiscorbutic activities (Bhandari and Chandel, 1997). This plant is considered an excellent aphrodisiac. Its seeds exhibit a strong antioxidant property (Alam et al., 2007). Like other brassicas, it contains glucosides such as allyl sulfonocyanates, while the seed oil contains erucic acid (Nuez and Hernández-Bermejo, 1994).

ERUCASTRUM

The genus *Erucastrum* consists of 25 species (Warwick et al., 2006) bearing the basic chromosome number $n = 7$, 8, and 9. The plants are herbaceous and can be annual, biennial, or perennial species that may sometimes be subshrubs and shrubs also (Table 3.2d, Figure 3.3). Leaves in this species are

TABLE 3.2C
Morpho-physiological Variation in *Erucastrum* under Subtropical Conditions in India

Character	E. virgatum (Italian subsp.)	E. laevigatum (Spanish subsp.)	E. gallicum	E. abyssinicum	E. nasturtiifolium
Plant Characters					
Plant form	—	—	Annual/biennial	Annual	Biennial/perennial
Pre-flowering growth habit	Erect	Rosette	Semi-erect/rosette	Erect	Rosette
Plant height (cm)	40–80	35–75	60–101	100	25–70
Primary branches (No.)	5–12	4–12	6–8	5–9	4–12
Secondary branches (No.)	7–13	8–17	12–29	6–15	9–23
Leaf Characters					
Arrangement	Alternate	Alternate	Alternate	Alternate	Alternate
Hairyness	Absent	Dense	Absent/Sparse	Absent	Sparse
Angle	Open	Semi-erect	Semi-erect	Open	Open
Color	Green	Dark green	Green	Dark green	Dark green
No. of lobes	Medium	Present	Many	Absent	Many
Waxiness	Present	Absent	Absent	Absent	Absent
Flower Characters					
Color of petals	Yellow	Yellow	Cream to pale yellow	Yellow/Orange	Pale-bright yellow
Petal length (cm)	0.5–0.6	0.3–0.4	0.8–1.2	0.4–0.6	0.7–0.9
Petal width (cm)	0.3–0.4	0.5–0.6	0.5–0.8	0.2–0.3	0.4–0.5
Silique Characters					
Length (cm)	3.0–4.0	1.8–2.6	2.9–4.2	4.0–5.0	3.0–4.0
Length of beak (cm)	0.2–0.3	0.1–0.2	0.1–0.3	0.1–0.2	0.2–0.3
Angle with main shoot	Semi-appressed	Semi-appressed	Erecto-patent	Erecto-patent	Erecto-patent
Surface texture	Intermediate	Intermediate	Intermediate	Smooth	Intermediate
Number on main shoots	Medium	Medium	Medium	Medium	Medium
Seeds per silique	Many	Many	46–55	Many	Many
Seed Characters					
Color	Brown/dark brown	Pale to dark brown	Golden to reddish-brown	Reddish-brown to beige	Reddish-brown
Shape	Ovate to elongate	Ovate-elongate, flattened	Ovate to obovate	Ovate to obovate	Ovate-elongate, flattened

simple, entire or pinnately divided, and dark green in color. Lower leaves may arrange in a rosette. The cauline leaves are almost always entire and alternate, rarely opposite or whorled. Inflorescences in *Erucastrum* are prominently racemes — more or less elongated. Flowers are small to medium in size and petals may vary from cream to pale orange in color in the different species. Fruits are segmented, typically a two-valved dehiscent capsule. Most *Erucastrum* species seem to possess small amounts of mucilage when wet — with the exceptions of *E. nasturtiifolium* and *E. gallicum* (Gómez-Campo, 1980).

Commonly known as dogmustard, *E. gallicum* is an annual or biennial plant with an erect, upright growth habit, with the initial pre-flowering growth witnessing a rosette or semi-erect structure. The fully grown plant can reach a height of up to 60 to 100 cm. The plant is native to Eurasia but is also found widely distributed around the world. It is a weedy species with pungent watery juice. Leaves are alternate, simple, generally entire, sometimes pinnately dissected, and green in color. Hairs on leaves and stem are simple, sometimes absent. Flowers are medium in size with cream to pale yellow petals. Siliques are segmented, two-valved, erect, and can vary in length between 3 and 4.5 cm. Each silique can contain a large number of seeds that are ovate to obovate in outline and golden to reddish-brown in color with the hilum end darker in color. The seed surface is reticulated with an irregular net pattern and granulation is visible in the interspaces.

Another species of *Erucastrum*, *E. nasturtiifolium*, is commonly known as watercress-leaved rocket and is a biennial or perennial plant species. Mature plants can vary between 25- and 70-cm in height and have a well-developed root system. The pre-flowering plant, which has quite large and conspicuous leaves, forms a rosette, which may later become erect during the reproductive phase. The stems have short and whitish hairs on the lower portions. Leaves are large in size, dark green in color, and may be sparsely hairy or glabrous. The lower leaves are deeply pinnate-dissected and have a large number of lobes on both sides of the mid-veins. However, cauline leaves are alternate and their lower lobes characteristically clasp the stem. Flowers are medium in size and pale to bright yellow in color. They later develop into erect siliques. The siliques can sometimes be slightly sickle shaped toward the end and can vary from 3 to 4 cm in length. Seeds are ovate-elongate and a little flattened, smooth, and small in size with reddish-brown color.

HIRSCHFELDIA

Hirschfeldia Moench is small genus with only one species (*H. incana*) (Warwick and Al-Shehbaz, 2006; Warwick et al., 2006) and is taxonomically very similar to *Erucastrum* C. Presl. Previously, however, three species were described in the genus *Hirschfeldia*, of which two are reportedly much localized in northern Africa and Socotra (Gómez-Campo, 1993; Mabberly, 1998). Commonly known as Buchan weed, *H. incana* is native to the Mediterranean and Irano-Turanian regions. It is well distributed in western, central, and southeastern Europe and the Mediterranean Sea, extending to southwestern Asia (Davis, 1965; Greuter et al., 1986). It often forms big colonies in the shelter of roads and in other places with relatively fine soil texture (Tsunoda, 1980). In Australia, it occurs in large numbers along railways and roadsides in canola growing regions (Dignam, 2000) and is also found in Queensland, New South Wales, and Tasmania. In general, it grows in wastelands, abandoned cultivations at the fringes of the cultivated habitat, and along roadsides and railroad tracks.

Hirschfeldia incana is an annual to biennial crucifer, although perennial forms are also found. The plants of *H. incana* may reach up to 150 cm in height and are divariately branched with a potential to turn in the mature stage into a tumbleweed, with a high individual seed production (Chronopoulos et al., 2005) (Table 3.2e, Figure 3.4). Plants flower from January to May.

Hirschfeldia is a nitrophilus species. It is not a major weed problem in cultivated crops but has been reported as a minor problem in the agricultural areas of Queensland and New South Wales (Groves et al., 2000). Due to its common appearance in canola growing areas, spontaneous hybridization with canola is also known to occur in the species. When canola is pollinated with *H. incana*, up to 70% hybridization is reported (Lefol et al., 1996).

TABLE 3.2D

Morpho-physiological Variation in *Hirschfeldia*, *Crambe*, *Sinapis*, *Eruca*, and *Cakile* under Subtropical Conditions in India

Character Studied	H. incana	C. abyssinica	S. pubescens	S. alba	E. sativa	C. maritima
Plant Characters						
Plant form	Annual/biennial	Annual	Annual	Annual	Annual	Annual/biennial
Pre-flowering growth habit	Rosette	Erect	Semi-erect	Erect	Semi-spreading/rosette	Erect
Plant height (cm)	60–150	68–105	60–170	75–200	30–80	20–50
Primary branches (No.)	9–12	22–30	9–15	8–15	6–12	6–12
Secondary branches (No.)	11–19	19–42	12–25	10–19	10–22	10–22
Leaf Ch aracters						
Arrangement	Alternate	Alternate		Alternate	Alternate	Alternate
Hairyness	Sparse/absent	Absent	Sparse/dense	Sparse/dense	Absent/sparse	Absent
Angle	Open	Open	Open	Open	Open	Open
Color	Med. Green to purplish	Light dull green	Dark green	Yellowish-green	Dull green	Bright green
No. of lobes	Many	Few	Medium	Many	Many	Many
Waxiness	Absent	Absent	Absent	Absent	Absent	Absent
Length (cm)	6–16	4–8	Long	6–11	5.0–10.0	Long
Width (cm)	5–13	2.5–4.5	Medium	5–8	3.0–6.0	Medium
Flower Characters						
Color of petals	Yellow	White	Bright yellow	Yellow	Pale yellow, brown/purple veins	Purplish white to lilac
Petal length (cm)	0.3–0.5	0.4	0.9–1.0	0.5–0.6	1.2–2.2	0.6–0.8
Petal width (cm)	0.2–0.4	0.2	0.6–0.7	0.4–0.5	0.7–0.9	0.3–0.4
Silique Characters						
Length (cm)	2.0–3.0	0.7–1.1	1.8–2.2	2.0–3.5	1.5–3.0	1.1–2.0
Length of beak (cm)	0.1–0.2	Absent	0.1–0.2	0.2	0.4–0.8	Absent
Angle with main shoot	Erect	Open	Erect	Erecto-patent	Variable	Semi-appressed
Surface texture	Intermediate	Smooth	Intermediate	Intermediate	Intermediate	Smooth
Number on main shoot	Many	Many	Medium	Medium	Medium	Few (usually 1)
Seeds per silique	5–7	1	16–18	44–58	Many	Few (usually >2)
Seed Characters						
Seed color	Brown	Olive green to beige	Yellow to dark brown	Creamy yellow to yellow	Light brown to olive green	Light to dark brown (Upper seeds larger)
Seed shape	Ovate/elongate	Spherical	Spherical	spherical, sometimes flattened	Ovate-elongate, flattened	Ovate to spherical

FIGURE 3.4 (See color insert following page 128.) Morpho-physiological variation in *Erucastrum*: (A) flowers of *E. abyssinicum*; (B) siliques of *E. abyssinicum*; (C) a young plant of *E. arabicum*; (D) pre-flowering plant of *E. virgatum*; (E) flowers of *E. gallicum*; (F) pod arrangement in *E. gallicum*; (G) leaf structure of *E. gallicum*; (H) flowers of *E. laevigatum*; (I) pre-flowering plant of *E. laevigatum*; (J) leaf structure in *E. laevigatum*; (K) pre-flowering plant of *E. laevigatum* ssp. *Elongatum*; (L) pre-flowering plant of *E. nasturtifolium*; (M) flowers of *E. nasturtifolium*; and (N) a young plant of *E. littoreum*

SINAPIS

Genus *Sinapis* consists of four species (Warwick et al., 2006; Warwick and Al-Shehbaz, 2006) and is typically characterized by open patent sepals (Gómez-Campo, 1980). *Sinapis alba*, commonly known as yellow mustard or white mustard, is found wild in the Mediterranean region. It prefers sites with an abundant moisture supply and a high level of soil fertility (Tsunoda, 1980) where it witnesses vigorous growth. It can also coexist with *S. arvensis* at sites with moderate soil moisture levels (Tsunoda, 1980). The plants of *S. alba* under favorable conditions may become quite vigorous, reaching heights of up to 2 m and bearing numerous seeds (see Table 3.2e, Figure 3.4). The petiole and lamina, as well as the juvenile leaves, have very dense hairs (Gómez-Campo and Tortosa, 1974).

TABLE 3.2E

Morpho-physiological Variation in *Sisymbrium*, *Camelina*, and *Thlaspi* under Subtropical Conditions in India

Character Studied	S. officinale	S. orientale	S. irio	S. altissimum	Camelina sativa	T. arvense
Plant Characters						
Plant form	Annual/Biennial	Annual/Biennial	Annual	Annual/Biennial	Annual/winter annual	Annual
Pre-flowering growth habit	Erect	Erect	Semi-erect	Erect	Erect	Erect
Plant height (cm)	60–90	15–60	50–70	60–150	30–90	84–97
Primary branches (No.)	6–8	6–10	Few	4–6	6–18	4–5
Secondary branches (No.)	10–12	10–18	Medium	8–12	10–24	12–24
Leaf Characters						
Arrangement	Alternate	Alternate	Alternate	Alternate	Alternate	Alternate
Hairyness	Sparse/absent	Sparse	Sparse/absent	Sparse/absent	Absent/sparse	Absent
Angle	Semi-erect	Semi-erect	Semi-erect	Semi-erect	Semi-erect	Open
Color	Green	Green	Green	Green	Bright green	Dull green
No. of lobes	Many	Many	Many	Many	Few	Few
Waxiness	Absent	Absent	Absent	Absent	Absent	Absent
Length (cm)	8.0–18.0	4.0–12.0	8.0–13.0	7.0–15.0	5.0–8.0	4–11
Width (cm)	6.0–8.0	3.0–5.0	2.0–5.0	3.0–5.0	3.0–4.0	3–6
Flower Characters						
Color of petals	Pale yellow	Pale yellow-yellow	Pale cream to yellow	Cream to white	Pale greenish-yellow	White
Petal length (cm)	0.4–0.5	0.8–1.0	0.2–0.4	0.5–0.9	0.4–0.6	0.34
Petal width (cm)	0.2–0.3	0.3–0.5	0.2–0.3	02–03	0.2–0.3	0.14
Silique Characters						
Length (cm)	0.8–1.5	5–8	2.0–5.0	0.5–1.2	0.6–1.4	1.1–1.4
Length of beak (cm)	0.1–0.2	0.2–0.3	0.2–0.3	0.1–0.2	0.1–0.2	Absent
Angle with main shoot	Appressed	Patent	Erecto-patent	Erecto-patent	Erecto-patent	Erecto-patent
Surface texture	Intermediate	Intermediate	Intermediate	Intermediate	Intermediate	Smooth
Seed Characters						
Seed color	Light to dark brown	Brown	Yellowish-orange to brown	Yellow to dark brown	Pale yellow to brown	Dark reddish-brown to black
Seed shape	Ovoid to Oblong	Oval	Ovoid to ellipsoid	Ovoid	Oblong	Oval, ovate, obovate

FIGURE 3.5 (See color insert following page 128.) Morpho-physiological variation in *Trachystoma, Moricandia, Crambe, Hirschfeldia,* and *Sinapis*: (A) flowers of *T. ballii*; (B) pre-flowering plants of *T. ballii*; (C) leaf structure of *T. ballii*; (D) a flowering plant of *M. arvensis*; (E) plants of *Crambe abyssinica* at reproductive phase; (F) inflorescence *of C. abyssinica*; (G) spherical fruits of *C. abyssinica*; (H) leaf structure of *C. abyssinica*; (I) pre-flowering plants of *H. incana*; (J) flowers of *H. incana*; (K) leaf structure in *H. incana*; (L) pre-flowering plants of *H. incanca* ssp. *cosobrina*; (M) a young plant of *S. alba*; (N) pre-flowering plants of *S. pubescens*; (O) flowers of *S. pubescens;* and (P) leaf structure of *S. pubescens*.

Hairs are also found on the siliques (Lamb, 1980). The seeds are spherical in outline although they may sometimes be flattened longitudinally from the hilum to the apex. The seed surface is faintly reticulated and its diameter is about 0.2 to 0.3 mm; seed color varies from light creamy yellow to yellow. Seeds are mucilaginous when wet. Seed cultivation has gained momentum during the past quarter century.

Sinapis arvensis is an annual, non-succulent species, commonly known as Wild mustard or Charlock. It is also of Mediterranean origin but occurs widely in many other parts of the world, including Australia and Tasmania. In Australia, it is reported to occur in the disturbed sites along the roadsides and railways in canola growing areas (Dignam, 2001). Its seeds are spherical, although immature seeds may be oval in outline. Immature seeds may be much shrunken, and a fuzzy seed pod material may keep adhering to the surface. The color of seeds of *S. arvensis* is highly variable and may occur in all shades of orange brown to black. The seeds are smooth in appearance although they are actually finely reticulated and their diameter ranges from approximately 1.5 to 2.0 mm. Seeds do not have mucilage when wet. It is a particularly serious weed in the cropping regions of New South Wales (Groves et al., 2000). However, chances of gene transfer from *S. arvensis* to cultivated rapeseed are considered unlikely as they are not considered sexually compatible.

Sinapis alba is widely used as a mustard condiment. It has also been used to treat tuberculosis (Arnason et al., 1981). Because the erucic content is very high (>45%) in the oil of *S. alba* (Yaniv et al., 1995), it has tremendous potential as an industrial oil. *S. alba* is reported to have a fairly high degree of resistance to *Alternaria* blight (Kolte, 1985; Brun et al., 1987; Sharma and Singh, 1992). *S. arvensis* has been reported to have allelopathic properties (Bowers et al., 1997) and its oil is being used in mosquito repellant formulations.

MORICANDIA

Originating in the Mediterranean, stems of *Moricandia* are erect, ascending or prostrate. Leaves are simple, exstipulate, entire, or variously pinnately lobed. Cauline leaves are almost always alternate. The pre-flowering plant may sometimes form a rosette although they generally have an erect growing habit. Flowers are hypogynous, actinomorphic, small, and white to cream and also purple in color. Fruits are typically a two-valved dehiscent silique.

Moricandia arvensis is a short-lived glaucous perennial herb and its plants may reach up to 65 cm (see Table 3.2e, Figure 3.5). Stems are branched. Lower leaves are obovate; obtuse at the apex, and narrower at the base. Cauline leaves are cordate, widened, and entire. Inflorescence is raceme with more than 15 large-sized, loosely borne flowers. Petals may be up to 2-cm long and color varies from cream to white to purple-violet. Siliques are compressed, four-angled, and vary between 2 and 3 cm in length. Seeds are small and brown in color.

Moricandia has C_3-C_4 intermediate plants (McVetty et al., 1989; Apel, 1997) and it can be used to improve water use efficiency in several cultivated crucifers because it increases the assimilation of carbon under conditions where water availability limits the rate of photosynthesis (McVetty et al., 1989). It has been shown that *Moricandia* × *Brassica* hybrids possess intermediate C_3-C_4 photosynthetic activity (Apel et al., 1984).

CRAMBE

The genus *Crambe* contains a total of 34 species (Warwick et al., 2006; Warwick and Al-Shehbaz, 2006) that are mostly perennial herbs, although a few are annuals or shrubs. Originating in the Mediterranean region, it is prevalent across Asia and western Europe, Euro-Siberia, and the Turko-Iranian region. In Ethiopia, two species of *Crambe* (i.e., *C. abyssinica* and *C. sinuatodentata*) are found together, indicating that these have evolved from a common related species, *C. hispanica,* as it migrated down from the Nile Valley and Red Sea in ancient times (Weiss, 2002). Among the different species of *Crambe*, *C. abyssinica* is the only cultivated species (Weiss, 2000). Endemic to East Africa and the Red Sea (Warwick and Francis, 1993), it is found widely distributed in the areas of moderate rainfall and warm-temperate climates. Although it is sensitive to frost and drought, it has also developed adaptability to the cooler and drier areas of the world. Its deep tap root system (about 15 cm) makes it relatively drought resistant. Another closely related species is *C. hispanica*. *C. hispanica* and *C.*

abyssinica are mainly distinguished on the basis of their basal leaves (White, 1975) and both can cross readily.

In recent years, particularly in Europe and the United States, there has been increasing interest in the use of crops as biorenewable industrial feedstocks, for uses such as fuels, lubricants, and bioplastics. *Crambe abyssinica* is one such crop with huge industrial potential (Knights, 2002).

Commonly referred to as Abyssinian mustard, Ethiopian kale, and crambe, *Crambe abyssinica* is an erect herbaceous annual herb with plant height varying between 70 and 150 cm. (see Table 3.2e, Figure 3.4). The plants have a straight stem, branching mostly on the upper half. Leaves are large, ovate in outline, and pinnately lobed (White and Higgins, 1966). They are light to dark green in color and progressively smaller upward. The plants may become quite vigorous, depending on the season and planting density (Figure 3.6). Flowering is of indeterminate type and continues for 3 to 4 weeks. The flowers are small, white in color, borne on long recemes, and numerous in number. Flowers are primarily self-pollinated with some natural out-crossing (Beck et al., 1975). Later, they turn into spherical-shaped, small, smooth, and pale green capsules of about 0.5 to 0.7 cm in diameter, which are beige colored at maturity. The siliques are indehiscent and bear a single spherical seed that is greenish-brown in color and up to 2.5 mm in diameter. The fruits remain attached to the plant even after maturity, unless heavy rain or wind causes extensive shedding (Hanzel et al., 1993). Crop is ready for harvesting 85 to 100 days after planting, and a single plant can produce 500 to 1800 fruits. The plant is quite attractive in appearance and some of the *Crambe* species (*C. strigosa* and *C. gigantea*) have been seen cultivated as ornamentals in the Canary Islands (Gómez-Campo, 1980). Different species of *Crambe* have been grown in Europe, Africa, the Near East, central and western Asia, as well as North and South America (Mazanni, 1954).

Crambe seed is highly desirable for industrial applications and its cultivation has reached commercial status in North America and Europe. Erucic acid represents the major portion of fatty acids in *Crambe* oil. Due to its high smoke point and viscosity, it is potentially useful in the lubricating,

FIGURE 3.6 Morpho-physiological variation in *Sisymbrium*: (A) pre-flowering plant of *S. irio;* (B) pre-flowering plants of *S. officinale;* (C) leaf structure of *S. officinale;* (D) mature pods of *S. orientale;* (E) silique pattern in *T. arvense;* (F) mature siliques of *T. arvense;* and (G) leaf structure of *T. arvense.*

emulsifying, and refrigeration industries (Lazzeri et al., 1994), and also as plasticizers and foam suppressants (Princen, 1983). In addition, the oil is highly suitable for other industries, such as pharmaceuticals, surfactants, cosmetics, perfumery, waxes, and flavors. It has been suggested to be a better potential domestic crop than rapeseed (Lessman and Anderson, 1981). *Crambe* seed has an oil content of 35% to 60% and a protein content of 20% to 40%. The oil content can be raised from 35% to 46% by removing the silique wall or hull from the seeds during harvest, which otherwise remains attached to the seed (Erickson and Bassin, 1990). The meal from dehulled seeds contains about 50% crude protein (Steg et al., 1994). Meal is also suitable as animal feed, a green manure (Bondioli et al., 1997), and insecticide (Peterson et al., 2000). In the United States and Canada, *C. abyssinica* is considered a potential industrial crop because of its high level of erucic acid (Hirsinger et al., 1989). Due to this high erucic acid content, *Crambe* oil is in direct competition with high erucic acid rape (HEAR). The seeds are usually processed in a two-step operation with half of the oil pressed out and the remainder extracted with a solvent (Duke, 1983).

In *Crambe* species, epiprogoitrin is the quantitatively dominating glucosinolate, accompanied by minor amounts of progoitrin (Bellostas et al., 2007). The whole seed of *Crambe* contains 3% to 4%, or about 90 moles/g, of glucosinolates [(S)-2-hydroxy-3-butenyl glucosinolate] (Lazzeri et al., 1994). This concentration is about two times that in high glucosinolate rapeseed (Carlson et al., 1985). Although glucosinolates as such do not cause harm as a part of mammalian diets (McMillan et al., 1986; Vermorel et al., 1986), their breakdown products — thiocyanates, isothiocyanates, and/ or nitriles — are undesirable in animal feeds (Agnihotri et al., 2007). They adversely affect iodine uptake in non-ruminants and reduce palatability and feed efficiency in terms of development and weight gain (Bille et al., 1983; Fenwick et al., 1983). Glucosinolates also reduce rumen microflora activity for 6 days after the animals are fed diets containing them (Duncan and Milue, 1991). It has been reported that 1-cyano-2-hydroxy-3-butene, a breakdown product in *Crambe* meal, is toxic to both plants (Vaughn and Berhow, 1998) and animals (Carlson, 1985). Therefore, high glucosinolate levels could limit the use of *Crambe* meal, particularly in pig or poultry diets, unless they are properly removed. Because of this, a major breeding goal in *Crambe* breeding should be a reduction in or elimination of toxic glucosinolates in *Crambe* seed meal. However, its cultivation and development remains in a juvenile phase and requires attention as a biorenewable source of energy.

CAKILE

Commonly known as sea rocket, *Cakile maritima* Scopoli is an annual or biennial species that is specifically restricted in habitat to the coastal strandline (Cordazzo et al., 2006). Its perennial forms are rarely found. Native to the shores of the Mediterranean Sea, this succulent species is widely distributed throughout the world (Barbour, 1972). It occurs sporadically along the south-eastern Brazilian coast although it is restricted to more protected sites at the base of frontal ridges (Cordazzo, 1985, 2006; Seelinger, 1992; Cordazzo and Seelinger, 1993).

The plants of *Cakile maritima* are small, ranging from 20 to 50 cm in height (see Table 3.2d). They prefer sandy and well-drained soils although they may also survive well on nutritionally poor soils. Leaves are pinnate-pinnatifid with dentate or serrate margins, and alternately arranged along the stem. However, simple types of leaves may also appear in *Cakile*. The leaves and stems are bright green, hairless, and fleshy, the water in the cells helping to dilute the salt that it comes in contact with either from inundation or sea sprays by violent waves. Flowers are medium-sized and white to purplish-white in color. Pollination by bees is prominent although other insects such as beetles, flies, and moths also frequently visit the plants. The species is distinguished by a marked heteroarthrocarpy and seed polymorphism (Baskin and Baskin, 1976; Fenner, 1985). The dimorphism is represented by two morphologically distinct types of fruit segments, upper and lower (Cordazzo et al., 2006). The upper fruit segment is brittle and easily separated, whereas the lower segment remains attached to the plant. The lower segment is thought to contain a larger number of aborted seeds in relation to the upper segment (Barbour, 1970; Rodman, 1974; Maun et al., 1990).

Upper seeds are slightly larger than the lower seeds and this is a typical characteristic of all *Cakile* taxa (Rodman, 1974). Fruit production per plant is highly variable, with a range of 15 to more than 3500 fruits per plant (Cordazzo, 2006).

Fruits are well adapted for water dispersal because of the presence of spongy tissue that makes them buoyant in seawater for a longer time (Maun et al., 1990). The upper segments can be dispersed over long distances by water waves, whereas the tumbling caused by wind can disperse it over shorter distances. The lower fruit segments are buried along with the plants in the wind-blown sand and germinate as a club of seedlings in the next season (Barbour, 1972; Keddy, 1982). Most of the seedlings that germinate are restricted to within a 1-m radius of the mother plant and mainly originate from the lower fruit segment. The leaves, stems, flower buds, and immature pods of *Cakile* are rich in vitamin C but have a very bitter taste (Freethy, 1985). *Cakile* plants strongly compete with the growth of nearby plants and this depresses their growth. As a result, it has become a noxious weed of cultivated fields in some coastal areas into which it has been introduced (Hedrick, 1972).

SISYMBRIUM

Sisymbrium belongs to the tribe Sisymbrieae and is native to southern Europe, Africa, and temperate and tropical Asia (Halvorson, 2003). It has several species of interest (see Table 3.2e, Figure 3.5). Many species have passed to North America, where *S. irio* was reported from Los Angeles, California, in the early 1900s. It is usually found in abandoned fields, orchards, roadsides, unattended pastures, and deserts (Rollins, 1993; Wilkins and Hannah, 1998; DeFalco and Brooks, 1999). *S. irio* is an annual nitrophilous species with $n = 7$. With a coarse tap root system, the plant is erect and herbaceous, and can reach a height of 70 cm. The stem is much branched from near the base and may be glabrous to sparsely hairy. Leaves are alternate along the stem, dissected, green in color, thin, and large sized. Basal leaves are bigger and may be 8- to 13-cm long and 2- to 5-cm or more wide. Leaf blades are ovate to lanceolate, pinnatifid with lateral lobes lanceolate with entire to slightly dentate or serrate margins. Spinescence is absent. Inflorescence is terminal raceme with a large number of hermaphrodite flowers. Petals are pale cream to yellow in color and may be 0.25- to 0.4-cm long. Siliques are dehiscent, 2- to 5-cm long and 0.8- to 1.5-mm in diameter, slightly curved toward the inner side, glabrous, and three-veined. Seeds are yellowish-orange to brown in color and ovoid to ellipsoid in shape.

The *Sisymbrium* species is a complex of diploid to octaploid races (Khooshoo, 1955). *S. irio* cannot survive in hot temperatures (Parker, 1972). It is self-compatible and autogamous (Wilkins and Hannah, 1998) and behaves like other mustards. It is a prolific seed producer with several thousand seeds per plant (Khooshoo, 1955). A large plant was estimated to produce up to 30,000 seeds (Halvorson, 2003). The seeds are primarily dispersed by animals (Drezner et al., 2001). It could also spread as a contaminant in crop seeds (Wester, 1992) and by cows via ingestion and their manure (Cudney et al., 1992). *S. irio* competes well with native annual plant species and may displace them (Marshall et al., 2000).

Another species, *Sisymbrium officinale* is also very prominent and is commonly known as hedge mustard. It is an erect annual to biennial herbaceous plant that can reach up to 90 cm in height (see Table 3.2c). It is reproduced by seeds, which are many in number on a plant. Stems are rigid, simple or much branched from above the base. Herbage on the lower portion of the plant is hirsute-hispid (Halvorson, 2003). Leaf arrangement is alternate along the stem, and leaves are lyrate-pinnatifid to irregularly pinnately lobed. As in *S. irio*, lower leaves in *S. officinale* are also bigger in size, ranging from 8 to 15 cm in length and 6 to 8 cm or more in width. They are glabrous to sparsely haired and petioled. Upper leaves are reduced, lanceolate, and dentate. Flowers are pale yellow borne on long and narrow racemes. Siliques are erect, 0.8- to 1.5-cm long and appressed along the stem, linear, and tapering toward the terminal apex (acuminate). They may be sparsely pubescent to glabrous. The silique beak is small and seedless. Seeds are small, ovoid to oblong, and brown in color with various shades.

Sisymbrium orientale is commonly known as Indian hedge mustard or Oriental hedge mustard (Hickman, 1993; Rollins, 1993). This is an annual to biennial species with an erect stem, much branched, and the plant may be 15- to 60-cm tall. Leaves are alternate, petioled, pinnately lobed to compound, and hirsute-pubescent with dentate or serrate margins. The lower leaves may be 4- to 12-cm long and 2.5- to 7-cm wide with entire to toothed margins. Inflorescence is raceme with pale yellow to yellow flowers of medium size. Petals are 0.8- to 1.0-cm long and 0.3- to 0.5-cm wide. Siliques can be very long (5- to 6-cm but up to 10- to 12-cm in var. *macroloma*), linear, glabrous to sparsely hairy, having a small stylar portion. Seeds are oval in shape and brown in color.

Sisymbrium altissimum, commonly known as tumble mustard, is an erect, annual to biennial plant with a well-developed tap root system. Stems are erect and much branched from above, giving it a bushy appearance (Halvorson, 2003). The plants are tall and may reach heights of 150 cm. Leaves are alternate and glabrous to sparsely hirsute. As in other *Sisymbrium* species, the lower leaves are larger in size, being 7- to 15-cm long, runcinate-pinnatifid with lanceolate lateral lobes. Upper leaves are reduced and deeply pinnatifid. Flowers are medium in size and light cream to nearly white in color, petals being 0.5- to 0.9-cm long. Anthers are often sagittate. Siliques are linear and 0.5- to 1.2-cm long with a prominent mid-vein evident. Seeds are ovoid and yellow to dark brown in color.

CAMELINA

Camelina sativa is native to the Mediterranean region and central Asia (Putnam et al., 1993; McVay, 2008). Its cultivation probably began in Neolithic times and by the Iron age, *Camelina* became a common oil-supplying plant (Knorzer, 1978). It was cultivated in antiquity from Rome to southeastern Europe and the southwestern Asian Steppes (Knorzer, 1978; Putnam et al., 1993). In Europe, it has been reported to be cultivated since the Bronze age (Pilgeram, 2007). In modern times, its limited cultivation exists in Germany, Poland, and the erstwhile USSR (Seehuber and Dambroth, 1984; Euge and Olsson, 1986), in addition to its cultivation in Canada (Downey, 1971; Robinson, 1987). In general, it is a spring or winter annual crop adapted to cooler areas (Fröhlich and Rice, 2005; Straton et al., 2007).

Camelina sativa, commonly known as linseed dodder or falseflax, is an annual or a winter annual with branched and erect stems. The plants can reach heights of 30 to 90 cm (see Table 3.2e). Leaves are medium in size, arrow shaped, pointed toward the apex, and green in color. They may be 5- to 8-cm long with smooth edges (Putnam et al., 1993). The flowers are small, prolific, and pale yellow to greenish-yellow in color, which later turn into siliques of 0.6 to 1.4 cm length resembling flax bolls. Seeds are very small, oblong, and pale yellow to dark brown in color.

The oil content of *Camelina sativa* ranges between 29 and 39% in the studies conducted by Putnam et al. (1993), although it may vary considerably under different growing conditions. Marquard and Kuhlamann (1986) reported an oil content between 37 and 41% in the seed samples of *C. sativa* in Germany. Its meal consists of 45% to 47% crude protein and 10% to 11% fiber (Korsrud et al., 1978), and in terms of its nutritional profile, it is comparable to soybean meal.

Camelina oil has good potential for industrial use (Vollmann et al., 1996) and it can be used for nutritional, health, biodiesel, biolubricant, and seed amendment purposes. Its oil can be effectively used for the production of biodiesel although the omega-3-fatty acid (α-linolenic acid) and γ-tocopherol content of the seed may limit its use for biofuel because of its high value in food and feed (Pilgeram et al., 2007). Approximately 35% to 39% of the total oil content of *Camelina* is linolenic acid (C18:3), while the remaining fatty acids are oleic (15% to 20%), linoleic (20% to 25%), gondoic (5% to 10%), and erucic (4% to 5%) acid. Putnam et al. (1993) found the fatty acid composition of *C. sativa* to be generally similar to that reported for *C. rumelica* (Umarov et al., 1972) and other reports on *C. sativa* (Seehuber and Dambroth, 1983). Its oil has also been used as a replacement for petroleum oil in pesticidal sprays (Robinson and Nelson, 1975).

Camelina meal may contain up to 10% to 15% oil by weight, with a protein content of about 40%, thereby making it competitive with soybean meal as animal feed (Pilgeram et al., 2007). *Camelina* meal could be a good source of nutrition for poultry, dairy, and beef. *Camelina* has also been shown to have allelopathic properties (Grummer, 1961; Lovelt and Duffield, 1981). In addition, there are also reports of its seeds being fed to birds (Fogelfors, 1984). In general, the research reports prove that *Camelina* has a tremendous potential for use in animal and human food, edible and industrial oil, and also in other applications. Furthermore, interest in *Camelina* as a crop is high, partly due to its low level of required inputs (Vollmann et al., 2005). Keeping this in mind, a more dedicated strategy should be framed for breeding and research of this crop to fully use its potential.

THLASPI

Thlaspi arvense, commonly known as stinkweed, fanweed, field pennycress, or Frenchweed, is an annual temperate species. Although widely distributed throughout the world, particularly in Asia, Europe, Africa, North and South America, and Oceania, its major occurrence is in Africa. It behaves as a weed in most temperate crops but is rarely a problem in tropical crops (Holm et al., 1997). The plants may grow in a range of wet or dry soils but prefer fertile areas (Best and McIntyre, 1975).

The *Thlaspi* species has a very high degree of ecotypic and morphological variation. In shallow, dry, and infertile soils, the plants may remain very tiny and unbranched, whereas, at favorable sites and under suitable climates, the plants may produce several lateral flowering branches and their height may reach up to 80 cm or more (Best and McIntyre, 1975). The plant is glabrous with bright green leaves and erect stems, and reaches up to 1 m in length (see Table 3.2e, Figure 3.5). When bruised, the leaves and stem produce an unpleasant odor. The stems may branch above and may produce more than 20 secondary branches under favorable conditions. Leaves are alternate along the stem, the lower being narrow, obovate, and petioled, and the upper leaves oblong, entire or irregularly serrated. Flowers are small, white, perfect and regular, and 3- to 5-mm long. Siliques are ovate to almost circular, dehiscent, 1- to 1.4-cm across, flattened, and winged. Seeds are oval to obovate in shape, flattened in cross-section, and are slightly elongated at the hilum. These are winged and their color varies from dark reddish-brown to black with a glossy look. Seeds are covered with distinct concentric rings like a typical fingerprint, giving them a roughened appearance. The siliques in general, have a lower degree of resistance to shattering, thus allowing release and dispersal of seeds prior to harvest (McIntyre and Best, 1975).

Thlaspi arvense is self-compatible and autogamous. It produces a good quantity of viable seeds (Mulligan, 1997) and is a prolific seed producer with an average number of seeds per plant being approximately 7000 seeds (Stevans, 1954), which may increase to as high as 20,000 seeds per plant (Holm et al., 1997). Fresh seeds are non-dormant and exhibit about 100% germination when exposed to light and a temperature regime of 10 to 25°C. (Chepil, 1946; Best and McIntyre, 1975). Salisbury (1964) recorded good germination of fresh seeds, although a proportion of them were dormant. Holm et al. (1997) reported that the seeds produced by winter annuals were non-dormant in autumn, and those produced by summer plants were dormant in autumn and became non-dormant during the winter. However, germination could be stimulated by treatment with gibberelic acid (Corns, 1960a, 1960b), sodium hypochlorite (Hsiao, 1980), and scarification (Pelton, 1956; Salisbury, 1964). *T. arvense* seeds have an ability to survive in the soil seed bank for periods of between 20 and 30 years (Duvel, 1905; Smith, 1917).

Thlaspi arvense is reported to be a weed of 30 crops in 45 countries (Holm et al., 1997), being a serious weed in 12 countries (Holm et al., 1991). In addition, it also acts as an alternate host to a number of crop pests and diseases, including the diamondback moth (*Plutella xylostella*) (Kmec and Weiss, 1997), *Leptosphaeria maculans* (Pedras et al., 1996), cabbage root fly (*Delia radicum*) (Finch and Ackley, 1977), and *Meladogyne hapla* (Belair and Benoit, 1996). Davis et al. (1996) reported the seeds of *T. arvense* to be the contaminants of oilseed rape stocks in the United States. Zero tillage (Blackshaw et al., 1994) and seed bed preparation at night (Kuhbaugh et al., 1992;

Pellezynski et al., 1996) are reported to effectively control the emergence and spread of *T. arvense*. Among the chemical control agents, 2,4-D (Muzik, 1970; Best and McIntyre, 1975), tribenuron-methyl and MCPA-tribenuron-methyl (Stratil, 1987), and pre-emergence application of chlorsulfuron (Kang, 1983) are suggested to be effective.

Some *Thlaspi* species accumulate toxic metals from polluted soils and are being used in modern techniques of phytoremediation.

ACKNOWLEDGMENTS

Our sincere thanks to the Department of Science & Technology, Government of India, New Delhi, for providing financial assistance in the form of ad hoc project funds (Grant No. SR/FT/L-122/2005, dated July 12, 2006), which enabled us to carry out the experiments on crucifers. We also thank Professor Cesar Gómez-Campo, Universidad Politécnica de Madrid, Madrid, Spain, for providing seeds of wild crucifers and technical guidance for their evaluation and also for a critical review of this manuscript.

REFERENCES

Agnihotri, A., Prem, D., and Gupta, K. 2007. The chronicles of oil and meal quality improvement in rapeseed. In: Gupta, S.K. (Ed.), *Advances in Botanical Research-Rapeseed Breeding*, Academic Press, CA, p. 50–99.

Alam, S.M., Kaur, G., Jabbar, Z., and Athar, M. 2007. *Eruca sativa* seeds possess antioxidant activity and exert a protective effect on mercuric chloride induced renal toxicity. *Food Chem. Toxicol.*, 45:910–920.

Al-Shehbaz, I.A. 1978. Chromosome number reports in certain Cruciferae from Iraq. *Iraqi J. Biol. Sci.*, 6:26–31.

Al-Shehbaz, I.A. 1984. The tribes of Cruciferae (*Brassicaceae*) in the southeastern United States. *J. Arnold Arboretum*, 65:343–373.

Al-Shehbaz, I.A. and Al-Omar, M.M. 1982. In Love, A. (Ed.), IOPB Chromosome Number Reports LXXVI. *Taxon*, 31:587–589.

Al-Shehbaz, I.A. and Al-Omar, M.M. 1983. In Love, A. (Ed.), IOPB Chromosome Number Reports LXXX. *Taxon*, 32:508–509.

Al-Shehbaz, I.A. and Al-Shammary, K. 1987. Distribution and chemotaxonomic significance of glucosinolates in certain Middle-Eastern Cruciferae. *Bioch. Syst. Ecol.*, 15:559–569.

Anonymous. 1991. *Horta e Saude*. Editoria Abril, Sao Paulo, Brazil.

Apel, P., Bauve, H., and Ohle, H. 1984. Hybrids between *Brassiceae alboglabra* and *Moricandia arvensis* and their photosynthetic properties. *Biochem. Physiol. Pflanzen*, 179:793–797.

Apel, P., Hilmer, M., Pfeffer, M., and Muhle, K. 1996. Carbon metabolism type of *Diplotaxis tenuifolia* (L.) DC. (*Brassicaceae*), *Photosynthetica*, 32:237–243.

Apel, P., Horstmann, C., and Pfeffer, M. 1997. The *Moricandia* syndrome in species of the Brasicaceae — evolutionary aspects. *Photosynthetica,*. 33:2005–215.

Arietti, N. 1965. *Flora Medica ed Erboristica nel Territorio Bresciano,* 1st ed. Commentari Ateneo di Brescia, Tip. Fratelli Geroldi, Brescia.

Arnason, T., Hebda, R.J., and Johns, T. 181. Use of plants for food and medicine by native peoples of eastern Canada. *Can. J. Bot.,* 59:2189–2325.

Baillargeon, G. 1986. Eine Taxonomische Revision der Gattung *Sinapis* (Cruciferae: Brassiceae). Ph.D. thesis, Universitat Berlin, Berlin, Germany.

Baldrati, I. 1950. *Trattato delle Coltivazioni Tropicali e Subtropicali,* 1st ed. Hoepli, Milano.

Bang, S.W., Mizuno, Y., Kaneko, Y., Matsuzawa, Y., and Bang, K.S. 2003. Production on intergeneric hybrids between the C3-C4 intermediate species *Diplotaxis tenuifolia* (L.) DC. and *Raphanus sativus* L. 2003. *Breeding Sci.*, 53:231–236.

Barbour M.G. 1972. Seedling establishment of *Cakile maritima* at Bodega Head, California. *Bull. Torrey Botanical Club*, 99:11–16.

Barbour, M.G. 1970. Germination and early growth on the strand plant *Cakile maritima*. *Bull. Torrey Botanical Club*, 97:13–22.

Baskin, J.K. and Baskin, C.C. 1976. Germination dimorphism in *Heterotheca subaxillaris* var. *subaxillaris*. *Bull. Torrey Botanical Club,* 103:201–206.

Beck, L.C., Lessman, K.J., and Buker. R.J. 1975. Inheritance of pubescence and its use in out crossing measurements between a *Crambe hispanica* type and *C. abyssinica* Hochst. ex. R.E. Fries. *Crop Sci.* 15:221–224.

Beckman, C. 2005. Vegetable oils: competition in a changing market. *Bi-weekly Bulletin. Agriculture and Agri-Food, Canada* 18(11). (Accessed at http://www.agr.gc.ca/mad-dam/e/bulletine/v18e/v18n11_e.htm)

Belair, G. and Benoit, D.L. 1996. Host suitability of 36 common weeds to *Meloidogyne hapla* in organic soils of southwestern Quebec. *J. Nematol.,* 38:643–647.

Best, K.F. and McIntyre, G.I. 1975. The biology of Canadian weeds. 9. *Thlaspi arvense* L. *Can. J. Plant Sci.,* 55:279–292.

Bhandari, D.C. and Chandel, K.P.S. 1997. Status of rocket germplasm in India: research accomplishments and priorities. In: Padulosi, S. and Pignone, D. (Eds.), *Rocket: A Mediterranean Crop for the World.* Report of a workshop, 13–14 December, 1996, Legnaro, (Padova), Italy. International Plant Genetic Resources Institute, Rome, Italy, p. 67–75.

Bhat, S.R., Kumar, P., and Prakash, S. 2008. An improved cytoplasmic male sterile (*Diplotaxis berthautii*) *Brassica juncea*: identification of restorer and molecular characterization. *Euphytica,* 159:145–152.

Bianco, V.V. 1995. Rocket, an ancient underutilized vegetable crop and its potential.. In: Padulosi, S. (Ed.), *Rocket Genetic Resources Network, Report of the First Meeting,* 13–15 November 1994, Lisbon, Portugal. International Plant Genetic Resources, Rome, Italy, p. 35–57.

Bianco, V.V. and Boari, F. 1997. Up-to-date developments in wild rocket cultivation. In: Padulosi, S. and Pignone, D. (Eds.), *Rocket: A Mediterranean Crop for the World.* Report of a workshop, 13–14 December, 1996, Legnaro, (Padova), Italy. International Plant Genetic Resources Institute, Rome, Italy, p. 41–43.

Bille, N. Eggum, B.O., Jacobson, I., Olseno, O., and Sorensen, N. 1983. Anti- nutritional and toxic effects in rats of individual glucosinolates (+) myrosinases added to a standard diet. I. Effects on protein utilization and organ weight. Tierphysiol. *Tierer. nahar Futter-mittellkd,* 49:195–210.

Blackshaw, R.E., Larney, F.O., Lindwall, C.W., and Kozub, G.C. 1994. Crop rotation and tillage effects on weed populations on the semi-arid Canadian prairies. *Weed Technol.,* 8:231–237.

Blangiforti, S. and Venora, G. 1997. Cytological study on rocket species by means of image analysis system. In: Padulosi, S. and Pignone, D. (Eds.), *Rocket: A Mediterranean crop for the World.* Report of a workshop, 13–14 December, 1996, Legnaro, (Padova), Italy. International Plant Genetic Resources Institute, Rome, Italy, p. 38.

Bondioli, P., Inzaghi, L., Postorino, G., and Quarticcio, P. 1997. *Crambe abyssinica* oil and its derivatives as renewable lubricants: synthesis and characterization of different esters based on Crambe fatty acids. *La Revista Italiana Delle Sostanze Grasse. LXXIV:* 137–141.

Boulos, L. 1977. Studies on the flora of Jordal. 5. On the flora of El Jafr-Bayir Desert. *Candollea,* 32:99–110.

Brun, H., Pleiss, J., and Renard, M. 1987. Resistance to some crucifers to *Alternaria Brassicae* (Berk.) Sacc. 7th Int. Rapeseed Congr. Poznan, 11–14 May, 1987. p. 1222-1227.

Carlson, K.D., Baker, E.C,. and Mustakas, G.C. 1985. Processing of *Crambe abyssinica* seeds in commercial extraction facilities. *J. Am. Oil Chemists Soc.,* 62:897–905.

Caso, O. 1972. Fisilogia de la regeneracion de *Diplotaxis tenuifolia* (L.) DC. *Bol. Soc. Argent. Bot.,* 14:335–346.

Chepil, W. 1946. Germination of weed seeds. II. The influence of tillage treatments on germination. *Scientif. Agric.,* 26:347–357.

Chronopoulas, G., Theocharopoulas, M., and Christodoulakis, D. 2005. Phytosociological study of *Hirschfeldia incana* (L.) Lagraze-Fossat (Cruciferae) communities in mainland Greece. *Acta Bot. Croat.,* 64:75–114.

Cordazzo, C.V. 1985. Taxonomia e Ecologia da Vegetacao das Dunas Costeiras ao sul do Cassino (RS). Dissertacao de mestrado, Universidade do Rio Grande, Rio Grande. (Cited in Cordazzo, 2006.)

Cordazzo, C.V. 2006. Seed characteristics and dispersal of dimorphic fruit segments of *Cakile maritima* Scopoli (*Brassicaceae*) population of southern Brazilian coastal dunes. *Revista Brasil. Bot.,* 29:259–265.

Cordazzo, C.V. and Seelinger, U. 1993. Zoned habitats of southern Brazilian coastal foredunes. *J. Coastal Res.* 9:317–323.

Corns, W. 1960a. Combined effects of gibberlelin and 2,4-D on dormant seeds of *Thlaspi arvense. Can. J. Bot.,* 38:871–874.

Corns, W. 1960b. Effects of gibberelin treatments on germination of various species of weed seeds. *Can. J. Pl. Sci.,* 40:47–51.

Cudney, D.W., Wright, S.D., Schultz, T.A., and Reints, J.S. 1992. Weed seed in dairy manure depends on collection site. *California Agric.,* 46:31–32.

Davis, J.B., Brown, J., Brennan, J.S., and Thill, D.C. 1996. Potential effect of weed seed contamination on the quality of canola produced in the Pacific Northwest of the USA. *Cruciferae Newslett.,* 18:136–137.

Davis, P.H. 1965. *Flora of Turkey and the East Aegean Islands, 1.* Edinburgh University Press, Edinburgh.

De Feo, D., Senatore, F,. and De Feo, V. 1993. Medicinal plants and phytotherapy in the Amalfitan coast, Salerno, Province, Campania, southern Italy. *J. Ethnopharmacol.,* 39:343–385.

DeFalco, L.A. and Brooks, M.L. 1999. Ecology and management of exotic annual plant species. In: *Presentation Abstracts, Mojave Desert Science Symposium,* February 25–27, 1999. (Available at: http://www.werc. usgs.gov/mojave-symposium/abstracts.html).

Dignam, M. 2001. Bush, parks, road and rail weed management survey. Monsanto Australia Limited. In: *The Biology and Ecology of Canola,* July 2002, p. 35.

Diosdado, J.C., Ojeda, F., and Pastor, J. 1993. In: Stace, C.A. (Ed.,) IOPB chromosome data 5. *IOPB Newslett.,* 20:6.

Downey, R.K. 1971. Agricultural and genetic potential of cruciferous oilseed crops. *J. Am. Oil Chem. Soc.,* 48:718–722.

Drezner, T.D., Fall, P.L., and Stromberg, J.C. 2001. Plant distribution and dispersal mechanisms at the Hassayampa River Reserve, Arizona, USA. *Global Ecol. and Biogeogr.,* 10:205–217.

Duncan, A.J. and Milne, J.A. 1991. Rumen microbial degradation of allyl cyanide as a possible explanation for the tolerance of sheep to *Brassica*-derived glucosionolates. *J. Sci. Food. Agric.,* 58:15.

Duvel, J. 1905. Vitality of Buried Seeds. Bureau of Plant Industries, USDA Bulletin No. 83.

Ellison, J.A., Hylands, P., Paterson, A., Pick, C., Sanecki, K., and Stuart, M. 1980. *Enciclopedia delle erbe. 1st ed.* A. Mondadori, Verona.

Erickson, D.B. and Bassin, P. 1990. Rapeseed and *Crambe*: Alternative Crops with Potential Industrial Uses. Agricultural Experimental Station, Kansas State University, Manhattan, Bulletin 656.

Facciola, S. 1990. *Cornucopia. A Source of Edible Plants. 1st ed.* Kampong Publication, Vista, California.

Felger, R.S. 1990. Non-native plants of Organ Pipe Cactus National Monument, Arizona. Technical Report No. 31. U.S. Geological Survey, Cooperative Park Studies Unit, The University of Arizona and National Park Service, Oregon Pipe Cactus Monument.

Fenner, M. 1985. *Seed Ecology.* Chapman & Hall, New York.

Fenwick, G.R., Heaney, R.K., and Mullin, W.J. 1983. Glocosinolates and their breakdown products in food and food plants. *Crit. Rev. Food Nutrit.,* 18:123–201.

Fernald, M.L. 1993. *Gray's Manual of Botany. 1st ed.* 2. Dioscorides Press, Portland, OR.

Finch, S. and Ackley, C.M. 1977. Cultivated and wild host plants supporting populations of the cabbage root fly. *Ann. Appl. Biol.,* 85:13–22.

Fogelfors, H. 1984. Useful weeds? Part 5. *Lantmannen (Sweden),* 105:28.

Freethy, R. 1985. *From Agar to Zenery,* The Crowood Press, Marlborough, UK.

Fröhlich, A. and Rice, B. 2005. Evaluation of *Camelina sativa* oil as a feedstock for biodiesel production. *Industr. Crops Products,* 21:25–31.

Gómez-Campo, C. 1978. Studies on Cruciferae. IV. Chronological notes. *Anales Inst. Bot. Cavanilles,* 34:485–496.

Gómez-Campo, C. 1980. Morphology and morpho-taxonomy of the tribe Brassiceae. In: Tsunoda, S., Hinata, K., and Gómez-Campo, C. (Eds.), *Brassica Crops and Wild Allies.* Japan Scientific Society Press, Tokyo, p. 1–31.

Gómez-Campo, C. 1983. Studies on Cruciferae. X. Concerning some west Mediterranean species of Erucastrum. *Annales Jard. Bot. Madrid,* 40:63–72.

Gómez-Campo, C. 1993. *Hirschfeldia* Moench. In: Castroviejo, S, Aedo, C., Gómez-Campo, C., Lainz, M., Montserrat, P., Morales, R., Munoz-Garmendia, F., Nieto Feliner, G., Rico, E., Talavera, S., and Villar, L. (Eds.), *Flora Iberica, 4. Cruciferae-Monotropaceae.* Real Jardin Botanico, Madrid, Spain, p. 398–400.

Gómez-Campo, C. 1999. Taxonomy. In: C. Gómez-Campo, C. (Ed.), *Biology of Brassica Coenospecies.* Elsevier, Amsterdam, p. 3–32.

Gómez-Campo, C. and Hinata, K. 1980. A check-list of chromosome numbers in the tribe Brassiceae. In: Tsunoda, S., Hinata, K., and Gómez-Campo, C. (Eds.), *Brassica Crops and Wild Allies, Biology and Breeding,* Japan Science Societies Press, Tokyo, p. 51–63.

Gómez-Campo, C. and Prakash, S. 1999. Origin and domestication. In: Gómez-Campo, C. (Ed.), *Biology of Brassica Coenospecies.* Elsevier, Amsterdam, p. 33–58.

Gómez-Campo, C. and Tortosa, M.E. 1974. The taxonomic and evolutionary significance of some juvenile characters in the Brassiceae. *Bot. J. Linn. Soc.,* 69:105–124.

Greuter, W., Burdet, H.M., and Long, G. (Eds.), 1986. Med-Checklist: a critical inventory of vascular plants of the circum-Mediterranean countries, 3. Conservatoire et Jardin Botaniques de la Ville de Geneve, Geneve.

GRIN. 2000. Grin Taxonomy. United States Department of Agriculture. Agricultural Research Service, The Germplasm Resources Information Network (GRIN). (Available at: http://www.ars-grin.gov/npgs/tax/index.html)

Groves, R.H., Hosking, J.R., Batianoff, D.A., Cooke, D.A., Cowie, I.D., Keighery, B.J., Rozefelds, A.C., and Welsh, N.G. 2000. The naturalized non-native flora of Australia: its categorization and threat to native plant biodiversity. Unpublished. Cited in: *The Biology and Ecology of Canola*, Office of the Gene Technology Regulator, July 2002, p. 35.

Grummer, G. 1961. The role of toxic substances in the interrelationships between higher plants. In: *Mechanisms in Biological Competition*. Academic Press, New York, p. 226–227.

Gupta, S.K. and Pratap, A. 2007. History, origin and evolution.In: Gupta, S.K. (Ed.), *Advances in Botanical Research-Rapeseed Breeding,* Vol. 45, Academic Press, London, p. 1–20.

Gupta, S.K. and Pratap, A. 2008. Evaluation of wild crucifers under subtropical conditions of Jammu & Kashmir (India). *Cruciferae Newslett.,* 27:97–98.

Halvorson, W.L. 2003. Fact Sheet for *Brassica tournefortii* Gouan, U.S. Geological Survey, South west Biological Science Centre, University of Arizona, Tucson, AZ.

Halvorson, W.L. 2003. Fact Sheet for *Sisymbrium irio*. U.S. Geological Survey National Park Service, p. 24.

Hanzel, J.J., Mitchell-Fetch, J.W., Montgomery, M.L., Schatz, B.G., and Hanson, B.K. 1993. Breeding of *Crambe* and *Brassica* oilseeds. In: *Alternative Crop Production Research,* North Dakota State University, May 1993, p. 38–44.

Harberd, D.J. and McArthur, E.D. 1980. Meiotic analysis of some species and genus hybrids in the *Brassicacae*. In: Tsunoda, S., Hinata, K., and Gómez-Campo, C. (Eds.), Brassica Crops and Wild Allies, Biology and Breeding. Japan Scientific Society Press, Tokyo, p. 65–87.

Hedge, I.C., Lamond, J.M., and Townsend, C.C. 1980. Cruciferae. In: Townsend, C.C. and Guest, E. (Eds.), *Flora of Iraq 4(2)*. Ministry of Agriculture and Agrarian Reform, Baghdad, p.–1085.

Hedric, U.P. 1972. *Sturtvant's Edible Plants of the World*. Dover Publications, New York, NY.

Hickman, J.C. 1993. *The Jepson Manual: Higher Plants of California*. University of California Press, Berkley and Los Angeles, CA.

Hinata, K., and Konno, N. 1979. Studies on a male-sterile strain having the *Brassica campestris* nucleus and the *Diplotaxis muralis* cytoplasm. *Japan J. Breed.,* 29:305.

Hinata, K., Konno, N., and Mizushima, U. 1974. Interspecific crossability in the tribe Brassiceae with special reference to self-incompatibility. *Tohoku J. Agricultural Res.,* 25:58–66.

Holm, L.G., Pancho, J.V., Herberger, J.P., and Plucknett, D.L. 1991. A Geographic Atlas of World Weeds. Krieger Publishing, Malabar, FL.

Holm, L.G., Doll, J. Holm, E., Pancho, J.V. and Herberger, J.P. 1997. *World Weeds: Natural Histories and Distribution*. John Wiley & Sons, New York.

Hsiao, A. 1980. The effect of sodium hypochlorite, gibberelic acid and light on seed dormancy and germination of stinkweed, *Thlaspi arvense* and wild mustard, *Brassica kaber*. Can. J. Pl. Sci., 60:643–650.

Ibarra, F. and La Porte, J. 1947. Las Cruciferas del genero *Diplotaxis adventicias* en la Argentina. *Rev. Argent. Agron,.* 14:261–272.

Jaretzky, R. 1932. Beziehungen zwischen Chromosomenzahl und Systematik bei den Cruciferen. *Jaharb. Wiss. Bot.,* 4:485–527.

Kang, B.H. 1983. Behaviour, persistence and selectivity of chlorsulphuron in crop plants and weeds. *Verhalten und Verbleb sowie Ursachen fur die selective Wirkung von Chlorsulpfuron in Kulturpflanzen und Unkrautern*. Universitat Hohenheim German Federal Republic.

Karpechanko, G.D. 1924. Hybrids of *Raphanus sativus* L. X *Brassica oleracea* L. *J. Genet..* 14:375–396.

Keddy, P.A. 1982. Population ecology on an environmental gradient: *Cakile edentula* on sand. *Oecologia,* 52:348–355.

Khatri, R., Sethi, V., and Kaushik, A. 1991. Inter-population variation of *Kochia indica* during germination under different stresses. *Ann. Bot.,* 67:413–415.

Khooshoo, T.N. 1955. Biosystematics of *Sisymbrium irio* complex. *Nature,* 176:608.

Kmec, P. and Weiss, M.J. 1997. Seasonal abundance of diamondback moth (Lepidoptera: Yponomeutidae) on *Crambe abyssinica. Environ. Entomol.,* 26:483–488.

Knights, K.E. 2002. *Crambe*: A North Dakotan Case Study. A Report for the Rural Industries Research and Development Corporation. p. 25

Knorzer, K.H. 1978. Evolution and spread of gold of pleasure (*Camelina sativa* S.L.). *Ber. Deutsch. Botan.n Gesellschaft,* 91:187–195.

Kolte, S.J. 1985. Diseases of Annual Oilseed Crops. Vol. II. Rapeseed-Mustard and Sesame Diseases. CRC Press, Boca Raton, FL.

Korsud, G.O., Keith, M.O., and Bell, J.M. 1978. A comparison of the nutritional value of *Crambe* and *Camelina* seed meals with egg and casein. *Can. J. Anim. Sci.,* 58:493–499.

Kuhbaugh, W., Gerhards, R., and Klumper, H. 1992. Weed control by soil cultivation at night?. *PSP Pflanzenschulz Praxiz,* 1:13–15.

Laghetti, G., Piergiovanni, A.R., and Perrino, P. 1995. Yield and oil quality in selected lines of *Crambe abyssinica* grown in Italy. *Industr. Crops Products,* 4:205–-12.

Lamb, R.J. 1980. Hairs protect pods of mustard (*Brassica hirta* 'Gisilba') from flea beetle feeding damage. *Can. J. Plant Sci.,* 60:1439–1440.

Lazarides, M., Cowley, K., and Hohnen, P. 1997. *CSIRO Handbook of Weeds.* CSIRO, Melbourne, p. 64.

Lazzeri, L. Leoni, O., Conte, L.S., and Palmieri, S. 1994. Some technological characteristics and potential uses of *Crambe abyssinica* products. *Industr. Crops and Products,* 3:103–112.

Lefol, E., Fleury, A., and Darmency, H. 1996. Gene dispersal from transgenic crops. II. Hybridization between oilseed rape and wild hoary mustard. *Sexual Plant Reproduc.,* 9:189–196.

Lessman, K.J. and Anderson, W.P. 1981. Crambe. In: Pryde, E.H., Princen, L.H., and Mukherjee, K.D. (Eds.), *New Sources of Fats and Oils.* American Oil Chemists Society, Champaign, IL, p. 223–246.

Lovett, J.V. and Duffield, A.M. 1981. Allelochemicals of *Camelina sativa. J. Appl. Ecol.,* 18:283–290.

Mabberley, D.J. 1998. *The Plant-Book: A Pocket Dictionary of the Vascular Plants, 2nd ed.* Cambridge University Press, Cambridge, U.K.

Malik, M., Vyas, P., Rangaswamy, N.S., and Shivanna, K.R. 1999. Development of two new cytoplasmic male-sterile lines of *Brassica juncea* through wide hybridization. *Plant Breed.,* 118:75–78.

Manton, I. 1932. Introduction to the general cytology of the Cruciferae. *Ann. Bot.,* XLVI:509–555.

Marquard, R. and Kuhlmann, H. 1986. Investigations of productive capacity and seed quality of linseed dodder (*Camelina sativa* Crtz.). *Fette Seifen Anstrichmittel,* 88:245–249.

Marshall, R.M., Anderson, S., Batcher, M., Comer, P., Cornelius, S. Cox., R., Gondor, A., Gori, D., Humke, J., Paredes Aquilar, R., Parra, I.E., and Schwartz, S. 2000. An Ecological Analysis of Conservation Priorities in the Sonoran Desert Ecoregion. Prepared by The Nature Conservancy Arizona Chapter, Sonoran Institute, and Instituto del University. 97 pp. United States Geological Survey, Northern Prairie Wildlife Research Centre Homepage (Available at: http://www.npwrc.usgs.gov/resource.othrdata/Explant/explant.htm).

Martin Ciudad, A. 1990. Indice de recuentos publicados en la serie "Numeros cromosomaticos de plantas occidentals," 1–638. *Annales Jard. Bot. Madrid,* 47:445–460.

Martínez-Laborde, J.B. 1988. Estudio sistematico del genero *Diplotaxis* DC. (Cruciferae, Brassiceae). Unpublished Ph.D. thesis, Universidad Polytechnica de Madrid, Spain.

Martínez-Laborde, J.B. 1991. Two additional species of *Diplotaxis* (Cruciferae, Brassiceae) with n = 8 chromosomes. *Willdenowia,* 21:63–68.

Martínez-Laborde, J.B. 1997. A brief account of the genus *Diplotaxis.* In: Padulosi, S. and Pignone, D. (Eds.), *Rocket: A Mediterranean Crop for the World.* Report of a workshop, 13–14 December, 1996, Legnaro, (Padova), Italy. International Plant Genetic Resources Institute, Rome, Italy, p. 13–22.

Maun, M.A., Boyd, R.S., and Olson, L. 1990. The biological flora of coastal dunes and wetlands. 1. *Cakile edentula* (Bigel.) Hook. *J. Coastal Res.,* 6:137–156.

Mazzani, B. 1954. Introduction de plantas oleoginosas nuevas para Venezuela, *Crambe abyssinica* Hochst. *Agron. Trop.,* 4:101–104.

McIntyre, G.I. and Best, K. 1975. Studies on the flowering of *Thlaspi arvense.* A competitive study of early and late flowering strains. *Bot. Gaz.,* 136:151–158.

McMillan, M., Spinks, E.A., and Fenwick, G.R. 1986. Preliminary observations on the effect of dietary Brussels sprouts on thyroid function. *Hum. Toxicol.,* 5:15–19.

McVay, K.A. 2008. *Camelina* Production in Montana: A self learning resource from MSU Extension (MT200701AG). Available at: http://www. msuextension.org

McVetty, P.B.E., Austin, R.B., and Morgan, C.L. 1989. A comparison of the growth, photosynthesis, stomatal conductance and water use efficiency in *Moricandia* and *Brassica* species. *Ann. Bot.,* 64:87–94.

Minnich, R.A. and Sanders, A.C. 2000. *Brassica tournefortii* Gouan. In: Bossard, C.C., Randall, J.M., and Hoshovsky, M.C. (Eds.), *Invasive Plants of California's Wild Lands.* University of California Press, Berkeley and Los Angeles, CA.

Miyazawa, M., Maehara, T,. and Kurose, K. 2001. Composition of the essential oil from the leaves of *Eruca sativa. Flavor Fragrance J.,* 17:187–190.

Moore, R.M. and Williams, J.D. 1983. Competition among weedy species: diallel experiments. *Austral. J. Agric. Res.,* 34:119–131.

Morales, M. and Janick, J. 2002. Argula: a promising specialty leaf vegetable. In: Janick, J. and Whipkey, A. (Eds.), *Trends in New Crops and New Uses,* ASHS Press, Alexandria, VA, p. 418–423.

Mulligan, 1959. Chromosome numbers of Canadian weeds. III. *Can. J. Bot.,* 37:81–92.

Mulligan, G.A. 1972. Autogamy, allogamy and pollination in some Canadian weeds. In: Best and McIntyre, *Can. J. Bot.,* 50:1767–1771.

Muzik, T.J. 1970. *Weed Biology and Control.* McGraw-Hill, New York.

Negre, R. 1961. Petite Flore des Regions Arides du Maroc Occidental. Vol. I, CNRS, Paris.

Nuez, F. and Hernandez-Bermejo, J.E. 1994. Neglected horticultural crops. In: Hernandez-Bermejo, J.E. and Leon, J. (Eds.), Neglected Crops: 1492 From a Different Perspective. Plant Production and Protection Series 26. FAP, Rome Italy, p. 303–332.

Palada, M.C. and Crossman. S.M.A. 1999. Evaluation of tropical leaf vegetables in the Virgin Islands. In: Janick, J. (Ed.), *Perspectives on New Crops and New Uses.* ASHS Press, Alexandria, VA, p. 388–393.

Pallczynski, J., Dobrzanski, A., and Anyszka, Z. 1996. The influence of seed bed preparation at night on weed infestation and herbicide efficacy in carrots. In: *Proc. 2nd Int. Weed Control Congr.,* Copenhagen, Denmark. Denmark: Department of Weed Control and Pesticide Ecology, p. 1267–1271.

Parker, K.F. 1972. *An Illustrated Guide to Arizona Weeds.* The University of Arizona Press, Tucson, AZ. (Available at : http://www.uapress.arizona.edu/online.bks/weeds/)

Parsons, W.T. and Cuhbertson, E.G. 1992. *Noxious Weeds of Australia,* Inkata Press, Melbourne.

Pathania, A., Bhat, S.R., Dinesh Kumar, V., Asutosh, Kirti, P.B., Prakash, S., and Chopra, V.L. 2003. Cytoplasmic male sterility in alloplasmic *Brassica juncea* carrying *Diplotaxis catholica* cytoplasm: molecular characterization and genetics of fertility restoration. *Theor. Appl. Genet.,* 107:455–461.

Pathania, A., Kumar, R., Dinesh Kumar, V., Ashutosh, Dwivedi, K.K., Kirti, P.B., Prakash, S., Chopra, V.L., and Bhat, S.R. 2007. A duplication of *CoxI* gene is associated with CMS (*Diplotaxis catholica*) *Brassica juncea* derived from somatic hybridization with *Diplotaxis catholica. J. Genet.,* 86:93–101.

Patterson, D.T. 1983. Research on exotic weeds. In: Wilson, C.L. and Graham, C.L. (Eds.), *Exotic Plant Pests and North American Agriculture.* Academic Press, New York.

Pedras, M.S.C., Taylor, J.L., and Morales, V.M. 1996. The blackleg fungus of rapeseed: How many species? In: Dias, J.S., Crute, I., and Monteiro, A.A. (Eds.), International Symposium on *Brassicas.* Ninth Crucifers Genetics Workshop, 1994, Lisbon, Portugal. *Acta Horticulturae,* 407:411–446.

Pellan-Delourme, R. and M. Renard. 1987. Identification of maintainer genes in *Brassica napus* L. for male sterility inducing cytoplasm of *Diplotaxis muralis* L. *Plant Breed.,* 99:89–97.

Pelton, J. 1956. A study of seed dormancy in eighteen species of high-altitude Colorado species. *Butler University Botanical Studies ,Indiana, USA,* 13:74–84.

Peterson, C.J., Cosse, A., and Coats, J.R. 2000. Insecticidal components in the meals of *Crambe abyssinica. J. Agric. Urban Entomol.,* 17:27–36.

Pilgeram, A.L., Sanda, D.C., Boss, D., Dale, N., Wichman, D., Lamb, P., Lu, C., Barrows, R., Kirkpatrick, M., Thompson, B., and Johnson, D.L. 2007. Camelina sativa, A Montana Omega-3- and Fuel Crop. In: Janick, J. and Whipkey, A. (Eds.), *Issues in New Crops and New Uses,* ASHS Press, Alexandria, VA, p. 129–131.

Pradhan, A.K., Mukhopadhyay, A., and Pental, D. 1991. Identification of putative cytoplasmic donor of CMS system in *Brassica juncea. Plant Breed.,* 106:204–208.

Prakash, S. 1974. Haploid meiosis and origin of *Brassica tournefortii* Goaun. *Euphytica,* 23:591–595.

Praksh, S. and Hinata, K. 1980. Taxonomy, cytogenetics and origin of crop *Brassicas,* A review. *Opera Botanica,* 55:1–57.

Pratap, A., Gupta, S.K., and Sharma, M. 2008. Genetic amelioration of crop *Brassicas* through intergeneric hybridization using *Diplotaxis.* Proc. Natl. Semin. Physiolog. Biotechnolog. Approaches to Improve Plant Productivity, March 15–17, 2008. Centre for Plant Biotechnology, Hisar, India. p. 85–86.

Princen, L.H. 1983. New oilseed crops on the horizon. *Econ. Bot.,.* 37:478–492.

Putnam, D.H., Budin, J.T., Field, L.A., and Breene, W.M. 1993. *Camelina*: A promising low-input oilseed. In: Janick, J. and Simon, J.E. (Eds.), *New Crops.* Wiley, New York, p. 314–322.

Rai, B., Gupta, S.K., and Pratap, A. 2007. Breeding methods. In: Gupta, S.K. (Ed.), *Advances in Botanical Research-Rapeseed Breeding,* Vol. 45, Academic Press, London, U.K., p. 22–48.

Rao, G.U. and Shivanna, K.R. 1996. Development of a new alloplasmic CMS *Brassica napus* in the cytoplasmic background of *Diplotaxis siifolia. Cruciferae Newslett.,* 18:68–69.

Rao, G.U., Batra, V.S, Prakash, S., and Shivanna, K.R. 1994. Development of a new cytoplasmic male sterile system in *Brassica juncea* through wide hybridization. *Plant Breed.,* 112:171–174.

Robinson, R.G. 1987. *Camelina*: A Useful Research Crop and a Potential Oilseed Crop. *Minnesota Agr. Expt. Sta. Bull.* 579 (AD-SB-3275).

Robinson, R.G. and Nelson, W.W. 1975. Vegetable oil replacements for petroleum oil adjuvants in herbicide sprays. *Econ. Bot.,* 29:146–151.

Rodman, J.E. 1974. Systematics and evolution of the genus Cakile (Cruciferae) in Australia. *J. Biogeogr.,* 13:159–171.

Romano, S., Mazzola, P., and Raimondo, F.M. 1986. Chromosome numbers of Italian flora, 1070-1081. *Inf. Bot. Ital.,* 18:159–167.

Salisbury, E. 1964. *Weeds and Aliens, 2nd ed.* Collins, London, U.K.

Salisbury, P. 2006. Biology of *Brassica juncea* and gene transfer from *B. juncea* to Other *Brassicacae* Species in Australia. Faculty of Land and Food Resources, University of Melbourne, Miscellaneous Report PAS 2006/1.

Salisbury, P.A. 1989. Potential utilization of wild crucifer germplasm in oilseed *Brassica* breeding. *Proc. ARAB 7th Workshop,* Toowoombu, Queensland, Australia, p. 51–53.

Sánchez-Yelamo, M.D. 1994. A chemosystematic survey of flavanoids in the Brassicinae: Diplotaxis. *Bot. J. Linn.* Soc. 115:9–18.

Seehuber, R. and Dambroth, M. 1983. Studies on genetic variability of yield components in linseed (*Linum usitatissimum* L.), poppy (*Papaver somniferum* L.) and *Camelina sativa* Crtz. *Landbauforschung Volkenrode.,* 33:183–188.

Seelinger, U. 1992. Coastal foredunes of southern Brazil: physiography, habitats, and vegetation. In: Seelinger, U. (Ed.), *Coastal Plant Communities of Latin America,* Academic Press, San Diego, p. 367–381.

Sharma, T.R. and Singh, B.M. 1992. Transfer of resistance of *Alternaria Brassicae* in *Brassica juncea* through interspecific hybridization among *Brassica. J. Genet. Breed.,* 46:373–378.

Sikka, K. and Sharma, A.K. 1979. Chromosome evaluation in certain genera of *Brassicaceae. Cytologia,* 44:467–477.

Sikka, S.M. 1940. Cytogenetics of *Brassica* hybrids and species. *J. Genet.,* 40:441–509

Silva D., Joao C.1996. *Crops of Britain and Europe.* Collins.

Smith, J. 1917. Weeds of Alberta, Canada. Alberta Department of Agriculture, Bulletin No. 2.

Snogerup, B. 1985. In: Love, A. (Ed.), IOPB Chromosome Number Reports. LXXXIX. *Taxon,* 34:727.

Steg, A., Hindle, V.A., and Liu, Y.G. 1994. By-products of some novel oilseeds from feeding: laboratory evaluation. *Anim. Feed Sci. Technol.,* 50: 87-99.

Stevans, O.A. 1964. (In: Best and McIntyre) *Weed Seed Facts.* N.D. Agri. Coll. Circ. p. A218.

Stock, R., Britton, T., Klopfenstein, T., Katges, K., Krehbiel, C., and Huffman, R. 1993. Feeding value of *Crambe* meal. Nebraska Beef Cattle Report, p. 51–53.

Stratil, J. 1987. Results of pilot trial studies on herbicides in winter rape. *Agrochemia,* 27:169–171.

Straton, A., Kleinschmidt, J. and D. Keeley. 2007. *Camelina.* Institute for Agriculture and Trade Policy. (Available at: http://www.iatp.org/iatp/publications.cfm?accountID=258&refID=97279).

Takahata, Y. and Hinata, K. 1986. Consideration of the species relationships in subtribe Brassicinae (Cruciferae) in view of cluster analysis of morphological characters. *Plant Sp. Biol.,* 1:79–88.

Takahata, Y. and Hinata, K. 1983. Studies in cytodemes in the subtribe Brassicinae. *Tohoku J. Agric. Res.,* 33:111–124.

Takahata, Y. and Hinata, K. 1986. Consideration of the species relationship in subtribe Brassicenae (Cruciferae) in view of cluster analysis of morphological characters. *Plant Sp. Biol.,* 1:79–88.

Thanos, C.A., Georghiou, K., Douma, D.J., and Marangaki, C.J. 1991. Photoinhibition of seed germination in Mediterranean maritime plants. *Ann. Bot.,* 68:469–475.

Tsunoda, S. 1980. Eco-physiology of wild and cultivated forms in *Brassica* and allies genera. In: Tsunoda, S., Hinata, K., and Gómez-Campo, C.(Eds.), *Brassica Crops and Wild Allies,* Japan Scientific Societies Press, Tokyo, p. 109–120.

Tutin. T.G., Heywood, V.H., Burges, N.A., Velentine, D.H., Walters, S.M., and Webb, D.A. (Eds.). 1964. *Flora Europea.* Cambridge University Press, Cambridge, U.K..

Umarov, A.U., Chernenko, T.V., and Markman, A.L. 1972. The oils of some plants of the family Cruciferae. *Khimiya Prirodnykh Soedinenii (USSR),* 1:24–27.

Vachova, A. and Franzen, R. 1981. In: Love, A. (Ed.) IOPB Chromosome Number Reports. LXXIII. *Taxon,* 30:834.

Vachova, M. and Ferakova, V. 1978. In: Love, A. (Ed.), IOPB Chromosome Number Reports. LXI. *Taxon,* 27:382.

Van Devender, T.R., Felger, R.S., and Burquez, A.M. 1997. Exotic plants in the Sonoran Desert region, Arizona and Sonora. *1997 Symposium Proceedings,* California , Exotic Pest Plant Council (available at: http://www.caleppc.org/symposia/97/symposium/VanDevender.pdf)

Van Loon, J.C. and De Jong, H. 1978. In: Love, A. (Ed.) IOPB Chromosome Number Reports. LIX. *Taxon,* 27:57.

Vaughn, S.F. and Berhow, M.A. 1998. 1-Cyano-2-hydroxy-3-butene, a phytotoxin from *Crambe (Crame abyssinica* L.) seedmeal. *J. Chem. Ecol.,* 24:1117–1126.

Vermorel, M., Heaney, R.K,. and Fenwick, G.R. 1986. Nutritive value of rapeseed meals: effects of individual glucosinolates. *J. Agric. Food Sci.,* 37:1197–1202.

Vollmann, J., Damboeck, A., Eckl, A., Schrems, H., and Ruckenbauer, P. 1996. Improvement of *Camelina sativa,* an underexploited oilseed. In: Janick (Ed.), Progress in New Crops. ASHS Press. Alexandria, VA, p. 357–362.

Vollmann, J., Grausgruber, H., Stift, G., Dryzhruk, V., and Lelley, T. 2005. Genetic diversity in *Camelina* germplasm as revealed by seed quality characteristics and RAPD polymorphism. *Plant Breeding,* 124:446–453.

Warwick, S.I. and Al-Shehbaz, I.A. 2006. Chromosome Number Index and Database on CD-ROM. *Pl. Syst. Evol.,* 259:249–258.

Warwick, S.I. and Francis, A. 1994. Guide to Wild Germplasm of *Brassica* and Allied Crops. Part V. Life History and Geographical Data for Wild Species in the Tribe Brassiceae (Cruciferae), Centre for Land and Biological Resources Research, Agriculture Canada, March, 1994.

Warwick, S.I., Francis, A., and Al-Shehbaz, I.A. 2006. *Brassicaceae*: Species Checklist and Database on CD-ROM. *Pl. Syst. Evol.,* 259:249–258.

Weiss, E.A. 2000. *Oilseed Crops, 2nd ed.* Blackwell Science, New York.

West, P. and Nabhan, G.P. 2002. Invasive plants: their occurrence and possible impact on the central Gulf Coast of Sonora and the Midriff islands in the Sea of Cortes, p. In: Tellman, B. (Ed.), *Invasive Exotic Species in the Sonoran Desert Region.* The University of Arizona Press and the Arizona-Sonora Desert Museum, Tucson, Arizona, p. 91–111.

Wester, L. 1992. Origin and distribution of adventive alien flowering plants in Hawaii. In: Stone, C.P., Smith, C.W., and Tunison, J.T. (Eds.), *Alien Plant Invasions in Native Ecosystems of Hawaii: Management and Research.* University of Hawaii Cooperation National Park Resources Study Unit, University of Hawaii, Honolulu, Hawaii.

White, G.A. 1975. Distinguishing characters of *Crambe abyssinica* and *C. hispanica. Crop Science,* 15:91.

White, G.A. and Higgins, J.J. 1966. Culture of *Crambe,* A New Industrial Oilseed Crop. USDA Production Research Report No. 95.

Wilkins, D. and Hannah, L. 1998. *Sisymbrium irio (Brassicaceae)* London rocket. Santa Barbara Botanic Garden, for Channel Islands National Park. (Cited in Halvorson, 2003.)

Yaniv, Z. 1995. Activities conducted in Israel.. In: Padulosi, S. (Ed.), *Rocket Genetic Resources Network.* Report of the first meeting, 13–15 November, 1994. Lisbon, Portugal. International Plant Genetic Resources Institute, Rome, Italy, p. 2–6.

Zohari, M. 1966. *Flora Palaestina. I Equisetaceae to Moringaceae.* The Israeli Academy of Sciences and Humanities, Jerusalem.

4 Floral Variation in the Subtribe Brassicinae with Special Reference to Pollination Strategies and Pollen-Ovule Ratios

Yoshihito Takahata

CONTENTS

INTRODUCTION

It is generally recognized that variations in overall floral shape and floral components represent adaptations to various modes of pollination, and it has been reported that the characteristics of floral morphology are associated with reproduction systems.

Ornduff (1969) listed 32 floral characters that distinguish xenogamous plants from their autogamous derivatives for the cruciferous genus *Laevenworthia*. He indicated that the switch from xenogamy to autogamy was accompanied by alterations in many morphological characters — mostly floral ones. Cruden (1977) mentioned that decreasing flower size and alterations in floral morphology, which mediated the evolutionary shift in breeding system, reduced the energy cost per flower and facilitated self-pollination. Further, he emphasized that the pollen-ovule ratios (P/Os) were a better predictor of a plant's breeding system than other morphological characters. Since Cruden's work (1977), many researchers have reported the relationships between floral components including P/Os and reproductive systems in each plant group (summarized by Cruden, 2000)

In the Cruciferae, it was reported that of 182 species representing 12 tribes, 80 were self-sterile and 102 were self-fertile (Bateman, 1955). In subtribe Brassicinae, (genera *Brassica, Diplotaxis, Eruca, Erucastrum, Hutera* (syn. *Coincya*), *Hirschfeldia, Sinapidendron, Sinapis,* and *Trachystoma*), which is important as a potential genetic resource for improving *Brassica* crops (Harberd, 1976; Takahata and Hinata, 1983; Gomez-Campo, 1999), it was reported that 50 species are self-incompatible and 9 are self-compatible (Hinata and Nishio, 1980). Several studies have been performed using the species of Brassicinae, which have dealt with interspecific floral diversity related with reproductive system or intraspecific floral variation for revealing the traits concerning the increase of the production of F_1 hybrid seeds.

This chapter describes the floral variation in the subtribe Brassicinae with respect to (1) the differentiation of various floral component characters associated with floral diversity, (2) the overall floral shape and its relationships to pollen production and pollination mechanism, and (3) the correlation of P/Os with energy costs per flower and breeding systems.

RELATIONSHIPS AMONG FLORAL CHARACTERISTICS

Cruden (2000) insisted that a number of floral traits involved in animal-pollination may interact evolutionally and the change of one trait might influence a change in another one. The correlation coefficients between 15 floral characteristics are shown in Table 4.1, and their interrelationships are summarized schematically in Figure 4.1. These values are calculated based on the data from previous reports (Takahata and Hinata, 1980, 1986; Takahata et al., 2008) using 119 strains of 53 species in subtribe Brassicinae.

Based on the correlation coefficients, the 15 characters were divided into two major groups (Figure 4.1). The first group contained the characteristics that were primarily concerned with overall flower size — that is vegetative floral structure such as sepal length, petal length, pistil length, and anther length. The number of pollen grains per flower belongs in this group. Such close relationships between flower size and pollen production have been reported in several species, including *Brassica napus* and *Raphanus sativus* (Stanton and Preston, 1988; Damgaard and Loeschcke, 1994). The second group consisted of characteristics such as number of ovules per flower and stigma width that were concerned chiefly with reproductive potential.

On the other hand, certain inter-relationships between the characteristics that belonged to different groups respectively were also observed. A positive correlation between the ovary length and the number of placentas per ovary ($r = 0.429$), and a negative correlation between the seed length and the number of ovules in flower ($r = -0.398$), were good examples. Cruden (2000) insisted that pollen grain number was positively related to ovule number, based on the several observations; however, such a relationship was not observed in Brassicinae (for which $r = 0.180$)

Furthermore, there were two other characteristics — pollen grain diameter and the number of seeds per beak — that could not be classified into either of the groups. The pollen grain diameter and the number of pollen grains per flower showed the highest negative correlation coefficient ($r = -0.548$), while the number of seeds per beak and style length had a positive correlation ($r = 0.469$). Such a

Table 4.1 Correlation coefficient among 15 floral characters in 119 strains of 53 species in Brassicinae

	SPL	PTL	ANL	PIL	OVL	STL	SGW	NPG	POD	VAP	BKS	NOV	P/O	OUV
PTL	0.946													
ANL	0.803	0.783												
PIL	0.895	0.940	0.739											
OVL	0.523	0.481	0.697	0.527										
STL	0.657	0.740	0.335	0.778	−0.122									
SGW	0.059	−0.053	0.302	−0.010	0.429	−0.351								
NPG	0.780	0.828	0.708	0.768	0.480	0.549	−0.041							
POD	−0.115	−0.220	0.034	−0.140	−0.143	−0.082	0.410	−0.548						
VAP	−0.020	−0.042	0.185	−0.042	0.431	−0.375	0.605	0.175	−0.007					
BKS	0.014	0.107	−0.196	0.194	−0.314	0.464	−0.283	0.096	−0.203	−0.221				
NOV	−0.019	−0.038	0.179	−0.034	0.422	−0.359	0.599	0.180	−0.012	0.999	−0.182			
P/O	0.584	0.611	0.388	0.535	−0.085	0.688	−0.325	0.585	−0.284	−0.432	0.244	−0.426		
OUV	0.143	0.108	0.497	0.140	0.489	−0.216	0.646	0.153	0.019	0.459	−0.267	0.452	−0.177	
SEL	0.629	0.510	0.568	0.490	0.256	0.377	−0.030	0.297	0.138	−0.392	−0.103	−0.399	0.517	0.164

SPL: Sepal length, PTL: Petal length, ANL: Anther length, PIL: Pistil length, OVL: Ovary length, STL: Style length, SGW: Stigma length, NPG: No. of pollen grains, POD: Pollen diameter, VAP: No. of placentas in valve, BKS: No. of seeds in beak, NOV: No. of ovules per flower, P/O: pollen-ovule ratio, OUV: No. of veins reached to the fringe in petal.
Correlation coefficient of POD with other characters were based on 72 strains of 38 species.

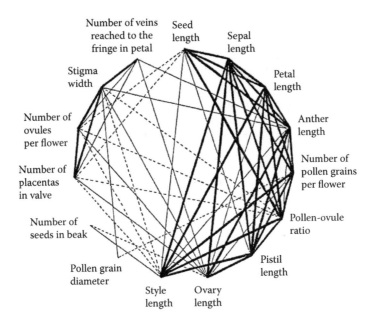

FIGURE 4.1 Schematic representation of correlations among 15 floral characteristics in 119 strains of 53 species in Brassicinae ($-$: $0.5 \leq r < 1.0$; $-$: $0.3 \leq r < 0.5$, ... $1.0 \leq r < -0.3$).

negative relationship between pollen grain size and number is well documented in other plants, most of which were attributed to a simple trade-off between size and number; however, a number of selection pressures may influence both the pollen grain size and/or number (summarized by Cruden, 2000).

Plitmann and Levin (1983) mentioned that much more attention had been given to floral structure in relation to pollination than to the integrated evolution of floral components. They continued further that it was apparent that most evolutionary alterations in one component must be compensated for by change elsewhere if the system was to remain functional. In the subtribe Brassicinae, the floral components showed a wide range of variation among the taxa and they seem to keep close relationships with each other. However, it is possible to divide the floral characteristics roughly into two groups by degrees of correlation among them. The division into two character groups— (1) one related to the floral size useful to attract pollinators, and (2) the other related to the reproductive structure, concerning seed production — seems to suggest that they have been differentiating independently, retaining a rather loose relationship between them. That is, it suggests that two trends of differentiation — one resulting from the adaptation for pollinators and the other from the adaptation for habitats — have been proceeding in the evolution of the floral components in Brassicinae.

The close relationship between the number of ovules per flower and the stigma width in Brassicinae is likely to endorse the idea that the stigma size responds evolutionary to the ovule number. However, Cruden and Miller-Ward (1981) remarked that in typical distylous species, the number of ovules in the two morphs were equivalent, having no relation to their stigmatic areas. Based on the above observation, they argued that the ovule number did not appear to be the major selective force contributing to stigmatic area.

In Brassicinae, the number of pollen grains that could occupy the stigmatic area per ovule did not vary much between the species with large vs. small stigmas, when the pollen sizes were similar. Furthermore, there were significant relationships between the stigmatic area per ovule and the pollen size ($r = 0.329$, $P < 0.001$). From these facts it may follow that the number of pollen grains occupying the stigmatic area per ovule remains unvaried, regardless of the variation in pollen size. Therefore, it is plausible to conclude that not only the ovule number, but also the pollen size could be the candidate for selection pressure for the stigmatic area. This idea may be supported by interpreting the above-mentioned case in distylous species as follows. In typical distylous species, although the ovule number in both morphs is constant, a thrum stigma with small area is pollinated with small pin pollen grains, while a pin stigma with large area is always pollinated with large thrum pollen grains, thus keeping constant the number of pollen grains accommodating the stigmatic area per ovule.

Whether or not the correlation between style length and pollen grain size is close has been disputed since Darwin (1896) by many authors (Taylor and Levin, 1975; Cruden and Miller-Ward, 1981; Plitmann and Levin, 1983) in many species. Although the strong correlation between them was reported in related species (Taylor and Levin, 1975; Baker and Baker, 1979; Plitmann and Levin, 1983), no relation was found in the subtribe Brassicinae ($r = -0.082$).

Intraspecific variation in floral traits has been reported in *Brassica* crops such as *B. napus* and *B. rapa*, in order to determine the floral traits influencing pollination for improving the production of F_1 hybrid seeds. Syafaruddin et al. (2006b) found variations of several floral characteristics on 30 strains of 8 varieties in *B. rapa*, and they (2006c) also reported that the effect of the spatial position of pistil to stamens on F_1 seed production was low. The genotypic variability in nectar secretion among 71 genotypes of rapeseeds (*B. napus*) were examined by Pierre et al. (1999), who reported that (1) double low [00] varieties secreted much more nectar than single low [0] and double high varieties; (2) male-sterile lines secreted one-half to one-third the amount of their isogenic lines; and (3) their nectar was predominantly composed of hexose (such as glucose and fructose). Cruciferous flowers produce a nectar guide, which consists of colored pattern (UV-absorbing center of the flower) invisible to humans but visible to insects (Horovitz and Cohen, 1972). Yoshioka et al. (2005) found intraspecific variation in UV color proportion (ratio of UV-absorbing area to flower area) when 24 genotypes of 8 varieties of *B. rapa* were surveyed. Furthermore, Syafaruddin et al. (2006a) reported that the genetic variation in the nectar guide was mainly due to dominance effects.

RELATIONSHIPS AMONG OVERALL FLORAL SHAPES, POLLEN PRODUCTION, AND POLLINATION STRATEGY

Based on overall floral shapes, the species belonging to Brassicinae were classified into three types: *Sinapis-*, *Brassica-*, and *Eruca*-type (Figure 4.2). The *Sinapis*-type flower was characteristic of having an open dish-like calyx, petals and stamens, and, accordingly, displaying exposed pistils and nectaries. Three species of *Sinapis*, *B. nigra*, *B. juncea*, and *Diplotaxis tenuisiliqua* were assigned to the *Sinapis* type. On the contrary, the *Eruca*-type flower had a closed cylinder-like calyx. The basal parts of petals and stamens were wrapped with sepals tightly around the pistil, so that the anthers were located very close to the stigma. All species of *Eruca* and *Hutera*, *B. oleracea*, *B. oxyrrhina*, and *B. tournefortii* were of this type. The *Brassica*-type flower was intermediate between the two in its overall floral shape, showing a bowl-like calyx. Many species were included in this type.

These flower types were closely related to the number of pollen grains per flower in xenogamous species. The *Sinapis*-type showed a small number of pollen grains — that is, from 4.6×10^4 (*Diplotaxis tenuisiliqua*) to 6.4×10^4 (*Sinapis turgida*), while those of *Eruca* type, on the other hand, revealed a large number of pollen grains — that is from 14.6×10^4 (*Hutera longirostra*) to 23.8×10^4 (*Eruca sativa*). The amount of pollen grains produced in *Brassica*-type species seemed intermediate between those of *Sinapis-* and *Eruca*-type species, as indicated by the number of pollen grains per flower, which ranged from 2.4×10^4 (*Brassica souliei*) to 13.9×10^4 (*Erucastrum nasturtiifolium*) (Figure 4.2). However, some of the species, such as *B. fruticulosa*, *D. assurgens*, *Er. varium*, and *Hirschfeldia incana* produced similar amounts of pollen grains to those of *Sinapis*-type species. Further, in *Brassica soulieri*, *D. catholica*, and *Erucastrum elatum*, the number of pollen grains per flower was found to be smaller than those of *Sinapis*-type species. Autogamous species had a smaller number of pollen grains than xenogamous species. Exceptional cases were found in *Brassica* amphidiploids such as *B. juncea*, *B. napus*, and *B. carinata*, which were not distinguished from xenogamy. These results are consistent with those of Hinata and Konno (1975).

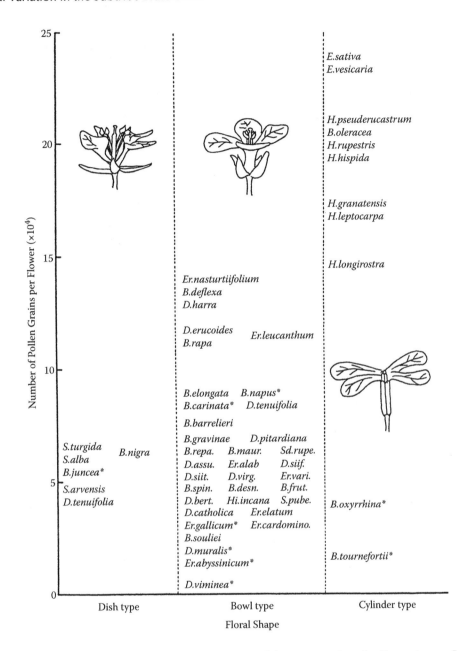

FIGURE 4.2 Relationships between three floral shapes and the mean number of pollen grains per flower in 53 species in Brassicinae.(entogamous species).

Most species of Brassicinae are known to be outcrossing through entomorphily. Although nectar secretion was not measured in these species, valuable data were obtained from observations made by others studying members of the Brassicinae. The *Sinapis*-type flower, which was characteristic of forming a dish-like open calyx and producing a comparatively small number of pollen grains, is likely to secrete plenty of nectar with high sugar concentration. For example, the *Brassica alba* (= *Sinapis alba*) flower produced between 0.2 and 0.6 mg nectar, the sugar concentration of which may reach 60% (Ermakova, 1959; Haragassimova-Neprasova, 1960), and *B. juncea* nectar shows an average of 52% sugar content (Sharma, 1969). Concerning the behavior of bees visiting *B. juncea* flowers, an excellent observation was made by Howard et al. (1916), who observed that a bee reaches its tongue to nectaries. From the data on nectar secretion, the behavior of pollinators, and pollen production, it seems reasonable to say that the *Sinapis*-type flower mainly attracts pollinators by large volume of nectar with high sugar concentration, and low pollen production remains only as a subsidiary means to attract pollinators.

On the other hand, an *Eruca*-type flower was characteristic of having a closed cylinder-like calyx and producing a large amount of pollen grains, and is known to have poor nectar development (Goukon, 1979). Pearson (1932) reported that the inner nectaries of *Brassica oleracea* secrete 0.1 ml each day for 3 days, and Butler (1945) found that its nectar had an average sugar content of 39%. Jablonsky (1961) observed that newly open flowers of *Eruca sativa* were visited and pollinated by pollen-gathering honeybees. Therefore, it may be inferred that to proceed with pollination, the main attraction provided for pollinators is abundant pollen grains, while nectar secretion is complementary in *Eruca*-type flowers.

The *Brassica*-type flower, to which many species of Brassicinae belong, varies widely in terms of pollen production and shows a bowl-like calyx of intermediate shape are between the other two. The nectaries of most species belonging to this flower type are similar to those of the *Sinapis*-type, which are characterized by the large amount of nectar secretion, and are designated as T-1 type by Goukon (1979). *B. napus* flowers, for example, yield means of 9.3 and 11.7 mg nectar with a sugar content of 32% to 39% (Maksymiuk, 1958). Most of the honeybees that visit *B. napus* flowers collect nectar and only rarely does a honeybee deliberately scrabble for pollen grains (Free and Nuttal, 1968). Further observation revealed that some honeybees packed the pollen grains into their corbiculae, whereas others discarded them (Free and Nuttal, 1968). As described above, the genotypic variability of nectar secretion in *B. napus* was reported by Pierre et al. (1999). It is supposed that most species of this flower type use their large amounts nectar production as their primary means to attract pollinators. However, in some cases, pollination proceeded mainly by pollen collecting honeybees that are attracted by abundant pollen production.

Concerning the pollination strategy of plants in Brassicinae, therefore, it is plausible to say that nectar secretion and pollen production are both major elements working complementarily to attract pollinators. The overall floral shape seems to be a subsidiary element that makes pollination more effective. Conner and Rush (1996) have indicated that flower size and number as floral attractants affected the pollinator visitation on wild radish (*Raphanus raphanistrum*). In Brassicinae, those few studies that have demonstrated the relationships between variation of such floral attractants and pollination.

RELATIONSHIPS AMONG POLLEN-OVULE RATIO, ENERGY COST PER FLOWER, BREEDING SYSTEM, AND ADAPTATION STRATEGY

P/Os have been used as an indicator to examine the breeding system, sexual system, and pollen vector (Cruden, 2000). Takahata et al. (2008) reported the P/Os in 119 strains of 53 species in Brassicinae and their relationships with the breeding system. The value of P/O varies widely, ranging from 19,800 for *Hutera rupestris* (H09) to 100 for *Diplotaxis viminea* (D32). The authors indicated the close relationships ($r = 0.578$, $P < 0.001$) between P/Os and the energy cost per flower, which is the first component score by principal component analysis based on six floral characteristics (Figure 4.3). The plants with high P/Os tend to have larger energy cost per flower, while those expending only small amounts of energy per floral unit have lower P/Os.

The P/Os and the energy cost per flower are related to the breeding system. In Brassicinae, most of which are self-incompatible xenogamous species, nine self-compatible autogamous species are included (Hinata and Nishio, 1980). Of these, three obligate autogamous ones — *Brassica tournefortii* (B51-B53), *Diplotaxis viminea* (D32), and *Erucastrum abyssinicum* (M01, M02) — had the lowest P/Os and the smallest energy costs per flower. *D. muralis* (D15, D16) and *Er. gallicum* (M06), which are autogamous allotetraploids, show the lowest P/Os but slightly larger energy costs per flower than

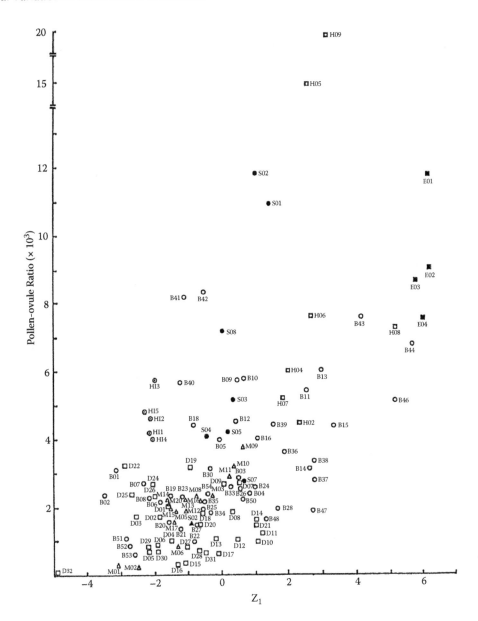

FIGURE 4.3 119 strains scattered according to the pollen-ovule ratios and the first component scores (Z1) by principal component analysis based on six floral characteristics. The first component score could be considered the indicator of energetic cost per flower, because it expresses the variation in general size factor of the flower. ○: *Brassica*, □ : *Diplotaxis*, ■: *Eruca*, △ : *Erucastrum*, ⊙ : *Hirschfeldia*, ▣ : *Hutera*, ▲: *Sinapidendron*, ●: *Sinapis*. (*Source:* From Takahata et al., 2008.)

the obligate ones. Slightly larger costs per flower seem to be due to their xenogamous parental species (Prakash et al., 1999), and they may retain a partial outcrossing mechanism by their slightly attractive larger flowers. The autogamous diploid, *B. oxyrrhina* (B06-B08), was characteristic of much reduced energy cost per flower and a rather high P/O. From these results, Takahata et al. (2008) considered that *B. oxyrrhina* has been in the process of alteration from facultative to obligate autogamy.

Three amphidiploid *Brassica* crop species (*B. carinata*, *B. juncea*, and *B. napus*) are self-compatible and facultative autogamous partially because of outcrossing (Rivers, 1957; Olsson and Persson,

1958; Ohsawa and Namai, 1987), although their parental species are self-incompatible xenogamous species (Okamoto et al., 2007). Both the P/O and energy cost per flower of these amphidiploid *Brassica* crops — *B. carinata* (B13, B14), *B. juncea* (B30), and *B. napus* (B36-B39) — were similar to those of xenogamous species (Figure 4.3). When compared with parental species, the amphidiploid *Brassica* crops showed intermediate energy cost per flower between their xenogamous parental species and lower P/Os than their parents (Figure 4.4). From these results, these facultative autogamous amphidiploids are considered more recently differentiated than other autogamous species, and Takahata et al. (2008) suggest that they have obtained a more efficient device for pollination than their xenogamous parental species, although no reduction in energy cost per flower has occurred.

Although the P/O and the energy cost per flower in xenogamous species of Brassicinae varied widely, each genus seems to have a characteristic distribution (Figure 4.3). *Eruca* and *Hutera* species had the high P/O and the large energy cost per flower. In *Sinapis* species, the P/Os varied widely in range, from 11,900 of *S. alba* (S02) to 2800 of *S. pubescens* (S07), despite the rather invariable energy cost per flower. *Hirschfeldia incana* (HI1-HI5) had relatively high P/O values (4000 to 5800). *Brassica* species showed large variations in both P/O and energy cost per flower. The P/O of *B. fruticulosa* ssp. *cossoneana* (B22) was similar to that of the autogamous species, while *B. nigra* (B40-B42) and *B. oleracea* (B43, B44, B46) were closer to *Eruca* species. The energy cost per flower of *B. soulieri* (B01, B02) was similar to that of autogamous species, while *B. oleracea* was close to *Eruca* and *Hutera* species. *B. nigra* showed the highest P/O among *Brassica* species, although it had relatively small flower. A similar case was also shown in *S. alba* (S01, S02). The distributions of *Diplotaxis*, *Erucastrum,* and *Sinapidendron* species were overlapping with that of *Brassica* species. In *Diplotaxis*, however, *D. harra* (D10-D14), *D. tenuifolia* cytodeme (D17, D21), and *D. sittiana* (D31) were increasing the energy cost per flower without accompanying a distinct increase in P/O.

As described above, in Brassicinae the shift from xenogamy to autogamy is generally accompanied by a reduction in floral size and a decrease in P/O. This agrees with the results of previous workers (Ornduff, 1969; Cruden, 1977; Spira, 1980; Preston, 1986). Some xenogamous species had low P/Os, similar to those of autogamous ones. Cruden (2000) indicated that many xenogamous

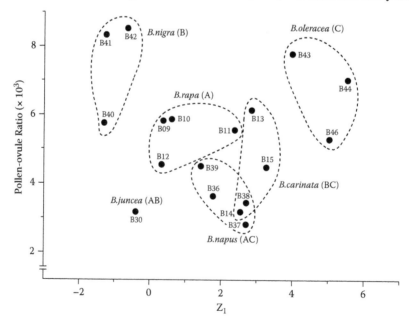

FIGURE 4.4 Three autogamous amphidiploid species and their xenogamous parental ones in *Brassica* crops scattered according to the pollen-ovule ratios and the first component scores (Z1) as described in Figure 4.3. Genome symbols are in parenthesis. (*Source:* From Takahata et al. 2008.)

species with low P/Os have large pollen grains and/or large stigma areas, and both the P/O and pollen grain size and/or other traits should be used to determine the plant's breeding system.

Cruden (2000) suggests that in animal pollination plants, the P/O is influenced negatively by pollen grain size and ovule number, and positively by pollen grain number. In Brassicinae, a negative relationship between P/O and pollen grain size, although not very significant statistically, was also found ($r = -0.284$). In addition, it became clear that the P/Os were closely related to energy cost per flower ($r = 0.561$, $P < 0.001$). Similar positive relationships were found between flower diameter and P/O in *Trichostema* of Labitatae (Spira, 1980), and between the weight of the flower and P/O in *Brassica napus* (Damgaard and Loeschcke, 1994). However, in *B. nigra*, *Sinapis alba*, and *Hirschfeldia incana*, the high P/Os were not always accompanied by the large floral energetics, which is contrary to our expectation. In a previous study (Takahata et al., 2008), however, the indicators of the energy cost per flower were estimated without determining nectar production. Therefore, the energy cost per flower of *Sinapis*-type and some *Brassica*-type species, as shown in Figure 4.3, may have been underestimated. *Sinapis*-type species and some *Brassica*-type species must have been investing more floral energy than was actually estimated.

The species in Brassicinae are not only dissimilar in their habitats, but also very different in their lifestyle forms, consisting of many annuals or biennials and a few perennials (Tsunoda, 1980). Takahata et al. (2008) have indicated the relationships of P/Os with natural selection in this subtribe. Concerning theories of natural selection, Grime (1977) proposed three primary strategies — competition (C), stress (S), and ruderal (R) — that are closely related with r-K strategy (Pianka, 1970). Taking into consideration the habitats and lifestyles of each species, it is speculated that in the subtribe Brassicinae, as a whole, the species with low P/Os are R-selected plants and with increasing P/O, they seem to shift from R- to C- or/and S-selected plants.

REFERENCES

Baker, H.G. and I. Baker. 1979. Starch in angiosperm pollen grains and its evolutionary significance. *Am. J. Bot.*, 66:591–600.

Bateman, A.J. 1955. Self-incompatibility system in angiosperms. III. Cruciferae. *Heredity*, 9:53–68.

Butler, C.G. 1945. The influence of various physical and biological factors of the environment on honeybee activity. An examination of the relationship between activity and nectar concentration and abundance. *J. Exp. Biol.*, 21:5–12.

Conner, J.K. and S. Rush. 1996. Effects of flower size and number of pollinator visitation to wild radish, *Raphanus raphanistrum*. *Oecologia*, 105:509–516.

Cruden, R.W. 1977. Pollen-ovule ratios: a conservative indicator of breeding system in flowering plants. *Evolution*, 31:32–46.

Cruden, R.W. 2000. Pollen grains: why so many? *Plant Syst. Evol.*, 222:143–165.

Cruden, R.W. and S. Miller-Ward. 1981. Pollen-ovule ratio, pollen size, and the ratio of atigmatic area to the pollen-bearing area of the pollination: an hypothesis. *Evolution*, 35:964–974.

Damgaard, C. and V. Loeschcke. 1994. Genotypic variation for reproductive characters, and the influence of pollen-ovule ratio on selfing rate in rape seed (*Brassica napus*). *J. Evol. Biol.*, 7:599–607.

Darwin, C. 1896. The Difference Forms of Flowers on Plants of the Same Species. New York: Appleton.

Ermakova, I. A. 1959. Nectar productivity of white mustard. *Pchelovodstvo Mosk.*, 36:29–31. From: *Insect Pollination of Crop*, Ed. J.B. Free. 1970. London & New York: Academic Press.

Free, J.B. and P.M. Nuttall. 1968. The pollination of oilseed rape (*Brassica napus*) and the behaviour on the crops. *J. Agric. Sci. Camb*, 71:91–94. From: *Insect Pollination of Crop*, Ed. J.B. Free. 1970. London & New York: Academic Press.

Gomez-Campo, C. 1999. Taxonomy. In *Biology of Brassica Coenospecies*, Ed. C. Gomez-Campo, p. 3-32. Amsterdam: Elsevier.

Goukon, K. 1979. The nectary of Cruciferae. *Saishu to shiiku*, 41:428–431.

Grime, J.P. 1977. Evidence for the existence of three primary strategies in plants and its relevance to ecological and evolutionary theory. *Am. Natur.*, 111:1169–1194.

Haragusimova-Neprasova, L. 1960. Zjistovani nektarodarnosti rostlin. Ved Pr vyzk Ust veelar CSAZV 2, p. 63–79. From: *Insect Pollination of Crop*, Ed. J.B. Free. 1970. London & New York: Academic Press.

Harberd, D.J. 1976. Cytotaxonomic studies of *Brassica* and related genera. In *The Biology and Chemistry of the Cruciferae*, Eds. J.G. Vaughan, A.J. Macleod, and B.M.J. Jones, p. 47–68. New York: Academic Press.

Hinata, K. and N. Konno. 1975. Number of pollen grains in *Brassica* and allied genera. *Tohoku J. Agri. Res.,* 25:58–66.

Hinata, K. and T. Nishio. 1980. Self-incompatiblity in Crucifers. In *Brassica* Crops and Wild Allies, Eds. Tsunoda, S., K. Hinata, and C. Gomez-Campo, p. 223–234. Tokyo: Japan Scientific Societies Press.

Horovitz, A. and Y. Cohen. 1972. Ultraviolet reflectance characteristic in flower of Cruciferae. *Am. J. Bot.,* 59:706–713.

Howard, A., L.C. Gabrielle, and K. Abduur Rahman. 1916. Studies in Indian oil-seeds. I. safflower and mustard. *Mem Dep Agriculture Indian Botany Ser 7,* p. 237–272. From: *Insect Pollination of Crop,* Ed. J.B. Free. 1970. London & New York: Academic Press.

Jablonsky, B. 1961. Wyniki badan nad wartoscia pszczelarska rukwi siewnej. *Pszczelnicze Zeszyty Naukowe 5,* p. 33–51. From: *Insect Pollination of Crop,* Ed. J.B. Free. 1970. London & New York: Academic Press.

Maksymiuk, K. 1958 Nectar secretion in winter rape. *Pszczelicze Zeszyty Naukowe 2,* p. 49–54. From: *Insect Pollination of Crop,* Ed. J.B. Free. 1970. London & New York: Academic Press.

Ohsawa, R. and H. Namai. 1987. The effect of insect pollinators on pollination and seed setting in *Brassica campestris* cv. Nozawana and *Brassica juncea* cv. Kikarashina. *Jpn. J. Breed.,* 37:453–463.

Okamoto, S., M. Odashima, R. Fujimoto, Y. Sato, H. Kitashiba, and T. Nishio. 2007. Self-compatibility in *Brassica napus* is caused by independent mutations in S-locus genes. *Plant J.,* 50:391–400.

Olsson, G. and B. Persson. 1958. In korsingsgrad och sjalvsteriliter hos raps. *Sveriges Utsadeforeninge Tidskrift 68,* p. 74–78. From: *Insect Pollination of Crop,* Ed. J.B. Free. 1970. London & New York: Academic Press.

Ornduff, R. 1969. Reproductive biology in relation to systematics. *Taxon,* 18:121–133.

Pearson, O.P. 1932. Incompatibility in broccoli and the production of seed under cages. *Proc. Am. Soc. Hort. Sci.,* 29:468 471.

Pianka, E.R. 1970. On *r*- and *K*-selection. *Am. Natur.,* 104:592–597.

Pierre, J., J. Mesquida, R. Marilleau, M.M. Pham-Delegue, and M. Renard. 1999. Nectar secretion in winter oilseed rape, *Brassica napus* — quantitative and qualitative variability among 71 genotypes. *Plant Breed.,* 118:471–476.

Plitmann, U. and D.A. Levin. 1983. Pollen-pistil relationships in the Polemoniaceae. *Evolution,* 37:957–967.

Prakash, S., Y. Takahata, P.B. Kirta, and V.L. Chopra 1999. Cyotogenetics. In *Biology of Brassica Coenospecies,* Ed. C. Gomez-Campo, p. 59–106. Amsterdam: Elsevier.

Preston, R.E. 1986. Pollen-ovule ratios in the Cruciferae. *Am. J. Bot.,* 73:1732–1740.

Rives, M. 1957. Etudes sur la selection du colza d'hive. *Ann. Inst. Natn Rech Agron, Paris, Ser B (Annls Amel),* 7:61–107. From: *Insect Pollination of Crop,* Ed. J.B. Free. 1970. London & New York: Academic Press.

Sharma, P.L. 1958. Sugar concentration of nectar of some Punjab honey plants. *Indian Bee J.,* 20:86–91.

Spira, T.P. 1980. Floral parameter, breeding system and pollinator type in *Trichostema* (Labiatae). *Am. J. Bot.,* 67:278–284.

Stanton, M.L. and R.E. Preston. 1988. Ecological consequences and phenotypic correlates of petal size variation in wild radish, *Raphnus sativus* (Brassicaceae). *Am. J. Bot.,* 75:528–539.

Syafaruddin, K. Kobayashi, Y. Yoshioka, A. Horisaki, S. Niikura, and R. Ohsawa. 2006a. Estimation of heritability of the nectar guide of flowers in *Brassica rapa* L. *Breed. Sci.,* 56:75–79.

Syafaruddin, Y. Yoshioka, A. Horisaki, S. Niikura, and R. Ohsawa. 2006b. Intraspecific variation in floral organs and structure in *Brassica rapa* L. analyzed by principal component analysis. *Breed. Sci.,* 56:189–194.

Syafaruddin, A. Horisaki, S. Niikura, Y. Yoshioka, and R. Ohsawa. 2006c. Effect of floral morphologyon pollination in *Brassica rapa* L. *Euphytica,* 149:267-272.

Takahata, Y. and K. Hinata. 1980. A variation study of subtribe *Brassicinae* by principal component analysis. In *Brassica Crops and Wild Allies,* Eds. Tsunoda, S., K. Hinata, and C. Gomez-Campo, p. 33–49. Tokyo: Japan Scientific Societies Press.

Takahata, Y. and K. Hinata. 1983. Studies on cytodemes in subtribe *Brassicinae* (Cruciferae). *Tohoku J. Agr. Res.,* 33:111–124.

Takahata, Y. and K. Hinata. 1986. A consideration of the species relationships in subtribe *Brassicinae* (Cruciferae) in view of cluster analysis of morphological character. *Plant Sp. Biol.,* 1:79–88.

Takahata, Y., N. Konno, and K. Hinata. 2008. Genotypic variation for floral characters in *Brassica* and allied genera with special reference to breeding system. *Breed Sci.* 58:385–392.

Taylor, T.N. and D.A. Levin. 1975. Pollen morphology of Polemoniaceae in relationship to systematics and pollination system: scanning electron microscopy. *Grana,* 15:91–112.

Tsunoda, S. 1980. Eco-physiology of wild and cultivated forms in *Brassica* and allied genera. In *Brassica Crops and Wild Allies,* Eds. Tsunoda, S., K. Hinata, and C. Gomez-Campo, p.109–120. Tokyo: Japan Scientific Societies Press.

Yoshioka, Y., A. Horisaki, K. Kobayashi, Syafaruddin, S. Niikura, S. Ninoyama, and R. Ohsawa. 2005. Intraspecific variation in the ultraviolet colour proportion of flowers in *Brassica rapa* L. *Plant Breed.,* 124:551–556.

5 Breeding Methods

S.K. Gupta and Aditya Pratap

CONTENTS

INTRODUCTION

The Cruciferae contains a number of species and diversity of crop plants that have an amalgam of breeding systems ranging from complete cross-pollination to a high level of self-pollination (Rai, 1997; Rai et al., 2007). The different species in this group are mainly known throughout the world for their oil-yielding properties. However, a few of them are also cultivated as salad, vegetable, and condiment crops. This chapter deals principally with species comprising the economically important crucifers that require agronomic and genetic improvement.

All the important crucifers are propagated from seed, but a few minor crops such as horseradish (*Armoracia rusticana* Gaertnia, Meyer & Scherb, syn. *Cochlearia armoracia* L.), sea-kale (*Crambe maritima*), and watercress (*Nasturtium* spp.) are vegetatively propagated. Because the mode of reproduction and the breeding objectives differ in different species, breeding methods may be quite different within each of them.

Brassica juncea is of much importance in Asia and *B. napus* in Europe and Canada. Under European and Canadian conditions, both winter and summer (spring planted) forms of *B. campestris* (syn. *B. rapa*) and *B. napus* are being grown; but in *B. juncea,* only the spring form has evolved much. Winter types of *B. napus* are largely grown under northern-European, Chinese, and Canadian conditions (Rai et al., 2007). However, spring types of *B. campestris*, are usually preferred and are largely grown in Sweden, Finland, and some parts of Canada and northwestern China. On the Indian sub-continent, two spring types of *B. juncea* and *B. campestris* cultivars are largely grown. In addition to these crops, small acreages of *Sinapis* are grown in Sweden and *Camelina sativa* is grown as a traditional crop in various parts of Europe.

On the Indian sub-continent, genetic improvement of seed yield is the prime objective, while in the Western world, breeding for quality receives greater attention. Erucic acid-free types of both

Brassica napus (Stefansson et al., 1961) and *B. campestris* (Downey, 1964, 1966) have been found in Canada in fodder forms of these species, and this trait has been incorporated into both annual and biannual oilseed forms in Canada and Europe (Jonsson, 1973).

BREEDING OBJECTIVES

On the Indian sub-continent, genetic improvement in seed yield is the primary objective, whereas in Europe and Canada, breeding for oil and meal quality receives greater attention in *Brassica* breeding. Development of high-yielding and early maturing varieties is a major objective in central China and western Canada where frost days during the growing season are usually less than 100 because the early maturing varieties complete their life cycle during this period and escape frost injury. On the Indian sub-continent, early maturing varieties (80 to 90 days) are required for fitting in relay-, multiple-, and inter-cropping systems.

Breeding for resistance to diseases and insect-pests is also a major objective in *Brassicas*. On the Indian sub-continent, *Alternaria* blight, white rust, and downy and powdery mildews are the major diseases, while in Western countries, blackleg (*Leptosphaeria maculans* Desm.) is important in Canada and Australia (Rai et. al., 2007). In addition, clubroot (*Plasmodiophora brassicae*) and root rot (*Rhizoctinina solani*) are other important diseases. Among the major insects, mustard aphid (*Lipaphis erysimi* Kalt.), mustard sawfly (*Athalia proxmia*), and leaf minor (*Bagrada cruciferarum*) are the important insect pests that cause considerable economic losses.

In recent years, specific emphasis is being placed on the development of double-low (low erucic acid and glucosinolates) cultivars. In Europe and Canada, breeding for oil and cake better suited to human nutrition and livestock feeding has received higher research priority than anywhere in the Asian countries. Currently, breeding for double-low quality is also gaining importance in China. While high erucic acid is liked by industry for specific purposes, zero or low glucosinolate oil is usually required for human consumption. The rapeseed and mustard oil with zero erucic acid content is more or less parallel to groundnut or sesame oil in its fatty acid composition. To further improve the oil quality, the major objective is to raise the level of stearic acid in order to reduce the need for hydrogenation in the manufacture of margarine and frying oil. Also, reduced levels of palmitic acid and the development of a commercial source of high oleic canola with more than 80% oleic acid to improve the stability of frying oils as well as suitability for industrial applications remains an important objective.

A number of CMS (cytoplasmic male sterility) systems are available in crucifers and are being utilized for the production of hybrids. Moreover, both self-compatible and self-incompatible forms occur in the two main *Brassica campestris* variety Toria and *B. juncea,* which strongly suggests that hybrid cultivars based on self-incompatible material should be produced.

GENETIC RESOURCES

To strengthen the genetic resources of crucifers, germplasm is being continuously augmented through plant explorations and introduction activities worldwide. Globally, there are more than 74,000 accessions of oilseed and vegetable *Brassica* germplasm lines, available from five countries — namely, China, India, the United Kingdom, the United States, and Germany. Together these countries are holding more than 60% of the world's total rapeseed and mustard germplasm (Singh and Sharma, 2007). The genetic stocks/wild crucifers are being utilized in crossing programs in India and other parts of the world in interspecific crosses to create new generic variability, and some of these are also being utilized as a base population for breeding work. At the international level, the Cruciferae germplasm is being maintained by IBPGR/IPGRI, Rome, Italy; Universidad Politecnica de Madrid, Spain; Tohoku University, Sendei, Japan; the Banco Nationale de Germoplasma in Bari (Italy); the Kew Gardens and Horticulture Research Institute, Willesbourne in the United Kingdom; and Nordic Gene Bank, Sweden. In Australia, the cultivated and wild crucifers are being maintained by the Australian Temperate Field Crop Collections, and Victoria Institute for Dryland Agriculture, Horsham. In India, crucifer genetic resources are being maintained by the National Bureau of Plant Genetic Resources, New Delhi, which during the past three decades has introduced more than 3950 accessions of rapeseed and mustard from more than 25 countries (Singh and Sharma, 2007).

CREATION OF NEW GENETIC VARIABILITY

In crucifers, enough variability is available because of the cross-pollinating nature of its primary species. However, in searching new desired genes or gene-complexes for resistance to diseases, insect pests, male sterility and fertility restoration, etc., it is necessary to resort to purposeful inter-varietal or distant hybridization. The success rate of interspecific hybridization greatly depends on the genetic relationship and genomic constitution of the parents used, and also on the direction of the cross.

In general, interspecific hybridization is more successful if an amphidiploid species is used as the female parent, a species that has one genome in common with the pollen parent (Zhang et al., 2003, 2006a). However, hybrids between monogenomic primary species are somewhat more difficult, with a success rate of 0.002 hybrids per pollinated flower (Downey et al. 1980; Mohapatra and Bajaj, 1987; Quazi, 1988). To rescue these interspecific hybrids, ovaries containing embryos are cultured using various culture media, including plant growth regulators, with successful plant regeneration from the interspecific hybridization between *Brassica rapa* and *B. oleracea* through ovary culture (Zhang et al., 2004; Zhang and Zhou, 2006).

Many of the wild relatives of oilseed rape are abundant in the cultivated fields, and therefore these are potential donors to oilseed rape (Chèvre et al., 2004; Scheffer and Dale, 1994). However, among the different species belonging to the *Brassicaceae* tribe, only four wild species are reported to hybridize spontaneously with *Brassica napus*. These are *B. rapa* (syn. *B. campestris*), *Raphanus raphanistrum*, *Sinapis arvensis,* and *Hirschfeldia incana* (Jorgensen, 2007). In addition, *B. napus* can also hybridize spontaneously with the cultivated species brown mustard (*B. juncea*) and cultivated radish (*Raphanus sativus*). Successful artificial hybridizations with *B. napus* have been

TABLE 5.1
Crossability Relationship among Various Oilseed *Brassicas*

Brassica Species with Their Genomic Constitution (as female parents)	*Brassica* Species with Their Genomic Constitution (as male parents)						
	Digenomic			Monogenomic			
	B. juncea (BJN)	*B. napus* (BNP)	*B. carinata* (BCN)	*B. campestris*	*B. nigra* (BNG)	*B. oleracea* (BOL)	*B. tournefortii* (BTF)
Digenomic	(AABB)	(AACC)	(BBCC)	(AA)	(BB)	(CC)	(DD)
BJN (AABB)	S	++	=	+	+	−	−
BNP (AACC)	+	S	+	+	−	+	−
BCN (BBCC)	=	+	S	−	+	+	−
Monogenomic BCM (AA)	−	−	−	S	−	−	−
BNG (BB)	−	−	−	−	S	−	−
BOL(CC)	−	−	−	−	−	S	−
BTF(DD)-	−	−	−	−	−	−	S

Note: S Selfed, + Successful, ++ Easily Crossed, − Crossed failed or Unsuccessful, = Made with great difficulty utilizing technique to remove crossability barriers like non-synchrony of flowering.

achieved using *B. elongata, B. fruticulosa, B. souliei, Diplotaxis tenuifolia, Hirschfeldia incana, Coincya monensis, Sinapis arvensis* (Plumper, 1995), and *Sinapis alba* (Gupta, 1993). All of these hybrids showed resistance to *Leptosphaeria maculans* in a cotyledon test, but resistance to *Alternaria* was lost in the hybrids *B. napus* × *B. souliei* and reduced in *B. napus* × *B. elongata* as compared to the wild species involved. The hybrids between *B. napus* and wild species (*B. napus* × *Hirschfeldia incana, B. napus* × *Sinapis arvensis*, and *B. napus* × *Coincya monensis*) have been successfully backcrossed to *B. napus* (Plumper, 1995). Successful artificial hybridizations have also been achieved in the crosses involving *Diplotaxis cretacea, D. harra* ssp. *Crassifolia*, and *D. muralis* with *B. campestris* var. Toria (RSPT-1), when aided by post-fertilization *in vivo* treatment of pollinated buds with 0.01% gibberelic acid (GA), followed by embryo rescue on modified MS (Murashige and Skoog, 1962) and Gamborg's media (Gamborg, et al., 1968) after 18 to 20 days of pollination (Pratap et al., 2008).

One of the most effective approaches to genetic improvement of oilseed brassicas is to seek the morphological ideotype, which appears with erect leaf posture and/or splitting leaf and waxy leaf surface, etc. Unusual floral structures such as apetalous genotypes are supposed to have advantages such as increased photosynthetic efficiency and less severe disease infection spread by petals (Jiang and Becker, 2001). Bijral et al. (2004), in F_5 generation of 53 intergeneric crosses between *Brassica napus* and *Eruca sativa*, observed a plant with abnormal floral characteristics. On furthering the generations of this plant, Pratap and Gupta (2007) also observed multipetalous and apetalous flowers with seven or eight stamens in many of its progenies.

It has now been possible to transfer the blackleg resistant gene from *Brassica juncea* to *B. napus* because of the possible recombination between A and C genomes in *B. juncea* crosses and A and B genomes in *B. carinata* crosses (Sacristan and Gerdemann, 1986). The intrageneric hybridization with *B. napus* has been used to transfer the resistance gene to *Leptosphaeria maculans* into the gene pool of oilseed rape (Roy, 1984; Sacristan and Gerdemann, 1986; Zhu et. al., 1993), a technique widely used by breeders.

A number of potential sources of resistance are available among crucifers for various pathogens (see Table 5.2). The potential sources of resistance to pathogen *Plasmodiopora brassicae* and the possibility for use in the interspecific hybridization with *B. napus* were reviewed by Siemens (2002). Triazine resistance has also been transferred from *B. napus* to *B. oleracea* (Ayotte et. al., 1986, 1987).

Cytoplasmic male sterility (CMS) was transferred from radish to *Brassica oleracea* (Bannerot et al., 1974; McCollum, 1988). Various wild species — that is, *B. oxyrrhina* (Prakash and Chopra, 1988); *B. tournefortii* (Pradhan et al., 1991); *D. catholica; D. erucoides* (Malik et al., 1999); *D. berthautii* (Malik et al., 1999); *D. harra; D. muralis* (Hinata and Konno, 1979); *D. siifolia* (Rao et al., 1994; Rao and Shivanna 1996); *Hirschfeldia incana; Moricandia arvensis* (Prakash et al., 1998); and *Raphanus sativus* (Ogura, 1968) — have been used as donor species for the development of new CMS lines. Similarly, genes for earliness have been introgressed from *Erucastrum gallicum* into cultivated species. The genes for high linoleic acid have been transferred from *B. juncea* to *B. napus* through selection in the F_2 generation (Roy and Tarr, 1985, 1986).

There are good possibilities of incorporating resistance genes from *Brassica juncea* and *B. carinata* to *B. napus* cultivars (Prakash and Chopra, 1988; Rao, 1990). *B. macrocarpa; B. juncea* (Prakash and Chopra, 1988a); *B. tournefortii* (Salisbury, 1989; Ahuja-Ishita et al., 1999a); *Hirschfeldia incana* (Salisbury, 1989); and *Raphanus species* (Agnihotri et al., 1991) have been identified as potential donors for the development of shattering resistance varieties in rapeseed. Also, wide hybridization has been reported with some degree of success in the crosses of *B. spinescens* ($2n = 16$) × *B. campestris* ($2n = 20$), *Eruca sativa* ($2n = 22$EE) × *B. campestris* ($2n = 20$ AA), and for the production of *Brassica napus* × *Raphanobrassica* hybrids by embryo rescue and ovary-culture technique(s) (Agnihotri et al., 1990a, b, c). Four F_1s of *Eruca sativa* × *B. campestris* ($2n = 21$ EA) showed a maximum of twelve bivalents, of which five were attributed to allosyndetic pairing between *Eruca* and *Brassica* genomes. This suggests that the possibility exists for the flow of

TABLE 5.2
Potential Sources of Resistance among *Brassicas* to Pathogens
Leptosphaeria maculans, Alternaria brassicola. A. brassicae, A. raphani, and
Plasmodiophora brassicae

Source	Resistance To	Ref.
B. elongata	*L. maculans*	Plumper (1995)
	A. brassicola, A. brassicae	Klewer et al. (2002)
	A. raphani	Klewer and Sacristan 2002
B. souliei	*A. brassicola, A. brassicae, A. raphani*	Plumper (1995)
B. fruticulosa	*L. maculans A. brassicola, A. brassicae,*	Plumper (1995)
	A. raphani	Klewer et al. (2002)
	L. maculans	Zhu and Spanier (1991)
	A. brassicola, A. brassicae	Plumper (1995)
Sinapis alba	*A. raphani*	Klewer et al. (2002)
Sinapis arvensis	*L. maculans*	Plumper (1995)
	A. brassicola, A. brassicae	Snowdon et al. (2000)
Raphanus sativus	*L. maculans*	Plumper (1995)
	Plasmodiophora brassicae	Xing et al. (1988)
Eruca vericaria	*L. maculans*	Plumper (1995)
Diplotaxis erucoides	*A. brassicola, A. brassicae*	Klewer et al. (2002)
	A. raphani	Klewer and Sacristan (2002)
Diplotaxis tenuifolia	*A. brassicola, A. brassicae*	Plumper (1995)
	A. raphani	Klewer et al. (2002)
Coincya monensis	*L. maculans*	Plumper (1995); Winter et al. (2002)
Capsella bursa-pastoris	*A. brassicola, A. brassicae, A. raphani,* *Plasmodiophora brassicae*	Plumper (1995); Siemens (unpubl.)
Camelina sativa	*A. brassicola, A. brassicae, A. raphani*	Plumper (1995)
Hemicrambe fruticulosa	*A. brassicola, A. brassicae, A. raphani*	Plumper (1995)
Hirschfeldia incana	*L. maculans*	Plumper (1995); Klewer et al.(2002)
Hesperis matronalis	*A. brassicola, A. brassicae, A. raphani*	Plumper (1995)
Neslia paniculata	*A. brassicola, A. brassicae, A. raphani*	Plumper (1995)

Source: From Plumper, 2006.

useful genes from *Eruca* to rapeseed for resistance to aphids and drought. Wild allies of *Brassica* were evaluated under natural drought conditions and identified as potential donors for drought resistance (Gupta et al., 1995). Wild crucifers (namely, *B. tournefortii* (Salisbury, 1989), *Diplotaxis acris,* and *D. harra*) have also been identified as potential donors for drought resistance genes and are being used in the interspecific hybridization program for the transfer of resistance genes into cultivated crucifers.

BREEDING METHODS

Crucifers include a number of cultivated crops and wild species that have a breeding system ranging from complete cross-pollination to a high level of self-pollination. Therefore, these are quite interesting materials from the breeding point of view. Selection procedures in cross-pollinated species vary from mass selection to recurrent selection; and in predominantly self-pollinated ones, the desirable plants are usually selected from a broad-based population such as land races, segregated population, germplasm complexes, gene pools, etc. and are bulked. This bulked seed is repeatedly grown cycle after cycle. One cycle of mass selection in Toria *B. campestris* is reported to have given a yield improvement of 8.2% (Chaubey, 1979).

Segregating populations or the progenies from the crosses could also make a good base population for initiating recurrent selection programs. In this method, the desirable individual open-pollinated plants (around 3000) are harvested and threshed separately. A part of this seed is saved and another part of it is planted in progeny rows, evaluated visually, and then superior rows are selected, tagged, and harvested separately. After harvesting and threshing, the seeds are analyzed for their 1000-seed weight, oil content, glucosinolates and protein content, etc. Thereafter, equal quantities of the reserved seed from the selected plants are composited. In this way, the first recurrent selection cycle is completed and this composited seed is grown again in the field in isolation, where intercrossing takes place among the plants within the composited populations. The second cycle of recurrent selection starts with harvesting the single plants (around 1000) from this population. A bulk seed sample is harvested from the remaining plants of the population for use in replicated yield trials to determine response to selection in each recurrent cycle for the characteristic under improvement (e.g., oil content, seed yield, or tolerance to a disease). Recurrent cycle selection is continued until a reasonable level of improvement is achieved.

In self-pollinated crucifers, pure line selection is usually followed in India; this involves the isolation of superior-performing lines from a genetically broad-based population based upon their progeny performance. Various improved varieties such as Varuna, Krishna, Kranti, Shekhar, Sita, RH-30, and Durgamani have all been developed from such simple breeding efforts (Rai, 1983a, b).

PEDIGREE BREEDING METHOD

This method can be effectively utilized for concentrating favorable genes for various economic traits and has been used to produce many cultivars in *Brassica napus* and *B. juncea*. In India, various high-yielding varieties were developed following pedigree selection. In this method, five to ten F_1 plants are grown to obtain F_2 seed and 1000 to 3000 F_2 plants are grown and harvested individually, from which F_3 progenies are secured. In the F_4 generation, the selection is practiced. The variation among F_4 families is a good indication of the effectiveness of further selection. This method has been utilized to develop a low erucic acid, high yielding, and winter-hardy *B. napus* variety from a cross between high erucic winter *B. napus* variety 'Rapol' and the low erucic acid spring *B. napus* variety 'Oro'.

BACKCROSS BREEDING METHOD

When the desirable gene is available from an unadapted or wild population, backcrossing would be the right choice; but if the favorable gene is available in an adapted or cultivated material's background, then the pedigree method of selection would be the most appropriate procedure. The spring *Brassica napus* variety 'Wester' had been developed through a combination of backcross and pedigree breeding. Backcross breeding has been used to transfer the low glucosinolate content of *B. napus* variety 'Bronowski', into a number of commercial cultivars of Gobhi Sarson (*B. napus*) in various parts of the world. This method is also used to transfer new traits such as fatty acid composition, seed color, and herbicide and insect-pest resistance.

DEVELOPMENT OF SYNTHETICS AND COMPOSITES

On the Indian subcontinent, the development of composite varieties is being viewed as a possible way to increase the average yield production of *Brassicas* as these are primarily grown under rain-fed conditions, and are subjected to all sorts of biotic and abiotic stresses (Rai, 1979). Although synthetic *Brassica napus* cultivars were also marketed in Europe, they were often not uniform and therefore this method of breeding is no longer used in *B. napus*. In Canada, efforts to develop synthetic spring *B. napus* were not very encouraging in terms of successful commercial cultivation. The composite breeding program in *B. rapa* and other Brassicas usually involves the production of a number of intervarietal hybrids or by making their blends. This is followed by evaluating the inbreeding depression in seed yield from F_2 and later generations, and evaluating the performance of the experimental checks against the ruling checks (Rai, 1982).

The development of synthetic varieties requires the development of inbred lines, their testing for general combining ability (GCA) by making all possible cross combinations, predicting F_2 performance constituting a number of experimental synthetic lines, testing the yield levels in gene trails over location, and finally releasing those that excel over the standard choice.

DEVELOPMENT OF HYBRIDS

At present, the ultimate goal of breeding cross-pollinated crops to exploit the nonadditive gene action is the high heterosis. A number of initial studies have demonstrated that there is considerable heterosis for yield in *Brassicas* (Schuster and Michael, 1976; Lefort-Buson and Dattee, 1985; Brandle and McVetty, 1990); *B. rapa* (Sernyk and Stefansson, 1983; Schuler et al., 1992); and *B. juncea* (Singh, 1973; Larik and Hussain, 1990; Pradhan et al., 1993). In India, 11% to 82% check parent yield heterosis has been reported in mustard (*B. juncea*), 10% to 72% in *Gobhi Sarson*, and 20% to 107% in *B. campestris* (Das and Rai, 1972; Labana et al., 1975; Yadava et al., 1974; Doloi, 1977; Srivastava and Rai, 1993), which is sufficiently high for its exploitation in hybrid cultivars. A range of 14% to 30% natural outcrossing is usually observed in these crops. This is sufficient to justify the efforts to develop cytoplasmic male-sterile (CMS) lines and search for usable fertility restorer lines to produce the hybrids.

In oilseed *Brassicas*, a number of CMS sources (e.g., *Brassica carinata* CMS, *B. juncea* CMS, *B. oxyrrhina* CMS, *B. tournefortii* CMS, *Raphanus*-based Ogura CMS, *B. napus*-based Polima CMS, *Siettiana* CMS, *Siifolia* CMS, etc.) are now well known and some of them are being studied rather intensively. Of these CMS sources, fertility restoration has been identified in *Raphanus*-based Ogura CMS and Polima CMS in Western countries, and it has been detected in CMS-based crosses in *B. tournefortii, B. juncea* CMS, Polima CMS, and *Siifolia* CMS in India. Fortunately, Punjab Agricultural University, Ludhiana, in India has recommended the release of the first CMS-based Gobhi Sarson hybrid PGSH-51 for cultivation in Punjab state in India.

DH BREEDING AND *IN VITRO* MUTAGENESIS

Doubled-haploidy (DH) breeding through microspore culture is quite well developed in *Brassicas* (Maluszynski et al., 2003; Xu et al., 2007). The DH technology in *Brassicas* aims at developing fully homozygous plants in a single generation, which could be further used in mutation breeding, genetic engineering, *in vitro* screening for complex traits like drought, cold, and salinity tolerance, and for developing mapping populations for linkage maps using molecular markers (Pratap et al., 2007). Several methods are available for DH production in *Brassicas*, such as microspore culture, anther culture, and ovary/ovule culture. The possibility to produce haploids in *Brassica napus* from anther culture (Keller and Armstrong, 1978) and microspore culture (Lichter, 1982) has provided breeders with a new tool for breeding improved cultivars of rapeseed and mustard (Zhou et al., 2002a, b).

The initiation of microspore culture experiments was followed by extensive investigations into various aspects of embryogenesis in anther- and microspore culture, and as a result, DH technology has developed to its present form in *Brassica*s (Charne and Beversdorf, 1988; Yu and Liu, 1995; Wang et al., 1999, 2002; Shi et al., 2002). The microspore culture technique has widespread applications in *Brassica* breeding due to its relative simplicity, efficiency in haploid and doubled haploid production, mutation and germplasm regeneration, and gene transformation (Xu et al., 2007). Also, microspore cultures provide the best material for mutation induction in haploid cells (Szarejko and Forster, 2006). Microspore embryogenesis is affected by a number of factors, including donor plant genotype and conditions, pretreatment, growth stage of the anther/microspore to be cultured, culture media and environment, diploidization process, etc. (Dunwell, 1996; Gu et al., 2003, 2004; Zhang et al., 2006b; Pratap et al., 2007). These factors are discussed in detail in Chapter 15 in this book.

Mutagenic treatments may have significant effects on the efficiency of DH breeding. McDonald et al. (1991) reported that UV light had harmful effects on embryo formation in rapeseed although regeneration remained unaffected and, at the same time, γ-irradiation decreased the frequency of embryos and plants. The induction of mutation in haploid cells involves isolating the developing microspores at the late uninucleate stage, followed by pretreatment and culturing on specialized media, which lead to direct embryogenesis rather than formation of pollen (Szarejko and Forster, 2006). Mutagenic treatment is given shortly after the isolation of microspores or after pretreatment, before the first nuclear division. Due to direct embryogenesis, the uninucleate microspore is the ideal target for *in vitro* mutagenesis. Also, microspores are far more sensitive to mutagenic treatments than other explants and therefore yield better results.

DHs also provide an efficient screening material for the desired mutants and other material for complex traits. Because through microspore-derived DHs one can obtain a very large number of synchronously developing embryos, one can modify the system to screen them *in vitro* for various desirable traits. For example, for development of herbicide-resistant *Brassica*s, the active chemical is introduced in the culture medium after mutation treatment (Beversdorf and Kott, 1987), and the surviving plants after chromosome doubling could be raised under controlled conditions and later screened for this trait. Similarly, effective selection could also be done for drought, cold, and salinity tolerance. Using this technique, several herbicide-resistant mutants have been developed in rapeseed (Kott, 1995, 1998; Swanson et al., 1988, 1989).

Although embryogenic microspores are the prime targets for mutagenic treatment, other haploid tissues and cells have also been treated with mutagens in *Brassica*s. In *Brassica napus*, isolated microspores have been treated with chemicals such as EMS (Beversdorf and Kott, 1987), NaN_3 (Polsoni et al., 1988), MNU (Cegielska-Taras et al., 1999) and ENU (Swanson et al., 1988; 1989), and physical mutagens such as γ-rays (Beversdorf and Kott, 1987; McDonald et al., 1991), x-rays (McDonald et al., 1991), and UV rays (Ahmad et al., 1991; McDonald et al., 1991). *B. napus* anthers have also been treated with γ-rays and fast neutrons by Jedrzejaszek et al. (1997). Similarly, microspores of *B. carinata* have been treated with EMS and UV rays (Barro et al., 2001, 2002), and *B. campestris* with UV rays (Zhang and Takahata, 1999; Ferrie and Keller, 2002). In *B. juncea*, isolated microspores as well as haploid embryos have been treated with chemical mutagens.

Despite great promise, the use of DH technology as a routine breeding tool for *Brassica* improvement is yet to be seen, mainly due to problems associated with anther/microspore culture (Pratap et al., 2007). These include low regeneration rate, highly genotype-specific response, and high frequency of callogenesis, but low recovery of DH plants. The focus of rapeseed breeders has lately shifted toward more specific and practical goals such as the development of herbicide-tolerant varieties, development of male-sterile lines for hybrid seed production, oil and meal quality improvement, and also drug production (Gupta and Pratap, 2008). For this, DH breeding must be adopted in conjugation with newer ideas such as directed *in vitro* mutagenesis, *in vitro* screening for desirable traits, and incorporation of molecular markers.

Genetic Transformation

Biotechnology has opened up new horizons for novel agronomic and quality traits in responsive crops such as the brassicas by providing access to novel molecules, the ability to change the level and pattern of gene expression, and the development of transgenics with insecticidal genes. With the development of genetic transformation techniques, it has become possible to bring about quick and dramatic improvements in the tolerance to many Lepidopteran and other insect pests, and herbicides, improvement in oil quality for industrial and domestic use, and the development of pharmaceuticals and industrial products. Much emphasis is currently placed on transgenic technology with the aim of improving cultivated *Brassica*s. As a result, the global area of biotech canola reached to an estimated 5.5 million ha in 2007 (James, 2007), the majority of it being herbicide-resistant canola.

Successful genetic transformation systems have been developed in many economically important *Brassica*s, such as *Brassica napus* (Moloney et al., 1989), *B. oleracea* (De Block et al., 1989), *B. juncea* (Barfield and Pua, 1991), *B. carinata* (Narasimhulu et al., 1992), *B. rapa* (Radke et al., 1992), and *B. nigra* (Gupta et al., 1993). However, among all the systems, *Agrobacterium tumefaciens*-mediated gene transfer is most widely used in *Brassica* and it is also quite efficient and practical in most of the species in the genus (Cardoza and Stewart, 2004).

Rapeseed cultivars tolerant to herbicides such as imidazoline, glyphosate, and glufosinate are available commercially in the United States and Canada (Cardoza and Stewart, 2004). For insect-resistance, the gene from *Bacillus thuriengiensis* has been introduced in canola cultivars (Stewart et al., 1996; Halfhill et al., 2001), which leads to overproduction of δ-endotoxins in the insects feeding on transgenic canola. This crystalline prototoxin gets inserted into the midgut plasma membrane of the insect, leading to lesion formulation and production of pores that disturb the osmotic balance. These cause swelling and lysis of the cells and, as a result, the larvae stop feeding and die (Hofte and Whiteley, 1989; Schnepf et al., 1998; Shelton et al., 2002).

Canola varieties with increased linolenic acid (Liu et al., 2001), stearate (Hawkins and Kridl, 1998), laurate (Knutzon et al., 1999), and increased enzyme activity (Facciotti et al., 1999) have been developed through genetic transformation. Further, *Brassica*s have been transformed to develop various industrial and pharmacological products. For example, *Brassica carinata* has been transformed for the production of hirudin, a blood anticoagulant protein (Chaudhary et al., 1998), while *B. napus* has been used for the production of carotenoids (Shewmaker et al., 1999). Development of male-sterile lines and fertility restoration systems has also been achieved through genetic transformation in *B. napus* (Jagannath et al., 2001, 2002), which could be of tremendous potential for the development of commercial hybrid cultivars. Similarly, salt- and cold-tolerant lines have also been developed in *B. juncea* by engineering the bacterial *codA* gene (Prasad et al., 2000). Transgenic lines of Wester variety having high palmitic and stearic acid contents have been developed by Hitz et al. (1995). High oleic acid containing *B. napus* and *B. juncea* lines with an increased shelf life have also been obtained through transgenic technology (Stoutjesdijk et al., 1999).

Quality Improvement

Brassica oil is nutritionally superior to most other edible oils due to its containing the lowest amounts of harmful saturated fatty acids and a good proportion of mono- and polyunsaturated fatty acids (Agnihotri et al., 2007). However, the value of its oil and meal becomes restricted due to the presence of two major anti-nutritional substances: (1) erucic acid, a long carbon chain unsaturated fatty acid; and (2) glucosinolate, the sulfur-containing compounds.

Oil quality mainly relates to the fatty acid composition of the seed. High contents of erucic and eicosenoic acids in *Brassica* oils decrease the profile of oleic, linoleic, and linolenic acids, rendering them inferior in quality to those from other oilseeds (Gupta and Pratap, 2007). Therefore, one of the most important breeding objectives in *Brassica* breeding has been genetic modification of the seed quality by changing the proportion of fatty acids suitable for nutritional as well as industrial purposes. Modifications in the compositions of fatty acids have been achieved in the past through various conventional breeding methods coupled with biotechnological techniques such as induced mutation, *in vitro* embryo rescue, DH techniques, and genetic engineering, especially post-transcriptional gene-silencing (Agnihotri et al., 2007).

Dietary recommendations in many countries focus attention on limiting total fat intake to 30% of energy, and saturated fat intake to 10% of energy. Breeding approaches in reducing the saturates include interspecific crosses followed by selection, reconstitution of *Brassica napus* from *B. rapa* and *B. oleracea* strains with reduced saturated levels, and mutagenesis in both *B. rapa* and *B. napus*. For a reduction in linolenic acid, both mutagenic sources and genetic transformation can be used. Gas liquid chromatography (GLC) (Craig and Murty, 1959), and the technique of using only half of the cotyledon to test the erucic acid content together provide a quick means to screen very large populations, which is necessary to identify genetically changed *Brassica* strains with low or zero erucic acid contents. Using this technique, the desirable strains with half cotyledon intact have been grown and carried forward, and with this, low erucic acid strains of *B. napus* (Stefansson et al., 1961; Downey and Harvey, 1963) and *B. campestris* (Downey, 1964) were developed in early 1960s. Later, such strains were developed in *B. juncea* (Kirk and Oram, 1981) and *B. carinata* (Alonso et al., 1991). Gupta et al. (1994, 1998) identified low erucic acid genetic stocks among the Indian accessions of *B. juncea*. Several low erucic acid *B. juncea* genotypes were developed in India through interspecific hybridization (Khalatkar, 1991; Malode et al., 1995), and transgressive segregation through interspecific/intergeneric hybridization followed by the pedigree method (Agnihotri et al., 1995; Agnihotri and Kaushik, 1998; 1999). Similarly, other fatty acids have also been modified in oleiferous brassicas, and high oleic and low linolenic acid *B. juncea* genotypes have been developed (Oram et al., 1999; Potts et al., 1999).

Brassica seed meal is an important source of nutrition for animals. However, undesirable components in the meal, such as glucosinolates, render them unfit for animal and human consumption. In high concentrations, in nonruminants such as swine and poultry, glucosinolates hydrolyze to form thiocyanates, isothiocyanates, or nitriles, which can adversely affect iodine uptake by the thyroid gland, and can reduce weight gains (Fenwick et al., 1983). High-performance quantitative GLC techniques (McGregor et al., 1983; Spinks et al., 1984; Brzezinski et al., 1986) have made it possible to obtain the profiles of glucosinolates and also measure their absolute levels. In addition to glucosinolates, other anti-nutritional factors such as sinapine, phenolic acid, tannins, and phytic acid also interfere with digestive enzymes, especially those affecting protein hydrolysis. To improve the quality of *Brassica* seed meal, the glucosinolate content should either be decreased or altogether eliminated from the meal through appropriate breeding techniques. Unfortunately, however, the genes controlling glucosinolate content in rapeseed are either pleiotropic or in linkage with the seed-filling stage and have a positive correlation with 1000-seed weight (Oliveri and Parrini, 1986). This renders strict selection difficult for quality traits in early segregating generations, lest genotypes for high seed yield could be lost. Therefore, it is advocated to keep the population heterozygous for quality characters and select the plants for these characters in advanced generations.

To date, the Bronowski gene is the only known source for low glucosinolate content and no natural germplasm source for stable low-glucosinolate genes has been reported (Agnihotri et al., 2007). Bronowski is a Polish *Brassica napus* cultivar that has a glucosinolate content of about 12 μmol/g oil free meal and 7% to −10% erucic acid in the oil. Considerable success has been achieved in Australia in the development of low-glucosinolate genotypes using mutagenesis, interspecific hybridization, and tissue culture coupled with pedigree selection (Oram et al., 1999). In India, two transgressive segregants (TERI 5 and TERI 6) with low glucosinolate and a Canadian accession BJ-1058 have been used to develop low-glucosinolate genotypes in the background of *B. juncea* var. Pusa Bold (Agnihotri and Kaushik, 2003; Agnihotri et al., 2007).

Breeding of a canola type of cultivar in *Brassica* with less than 2% erucic acid in the oil and less than 30 μmol/g glucosinolate in defatted meal (commonly known as '00') has been achieved, and several such varieties (e.g., Cyclone in Denmark, Shiralee in Australia, and AC Excel in Canada) are now available (Rakow, 1995). In Australia, several double-low cultivars of *Brassica juncea* have shown promising yield potential (Burton et al., 2003). In India, Agnihotri and Kaushik (2003) have reported successful introgressions of double-low traits in *B. juncea* cultivar Varuna using low erucic acid donors TERI (OE) M21 and Zem-1, and low glucosinolate line BJ-1058. The double-low *B. napus* varieties GSC-865 and TERI-Uttam-Jawahar have been released for commercial cultivation in the Indian states of Punjab and Madhya Pradesh, respectively (Agnihotri et al., 2007).

The introduction of double-low (i.e., low erucic acid-low glucosinolates) genotypes of *Brassica napus* followed their extensive cultivation in many countries around the world, and experimental work toward development and improvement of low erucic acid germplasm for other species is being pursued on a global level (Rakow and Raney, 2003). At present, the breeding efforts in the development of canola-quality, double-low *B. napus* cultivars in improving the oil composition, and enhancing vitamin levels are underway in many countries, including Germany (Luhs et al., 2003), Canada (Raney et al., 2003a, b), the United States (Corbett and Sernyk, 2003), Australia (Gororo et al., 2003), France (Carre et al., 2003), and Poland (Spasibionek et al., 2003).

Yellow seed coat color also adds to high oil content and therefore this could also be another breeding objective for improved *Brassica*s.

FUTURE CRUCIFEROUS CROPS AND POSSIBLE RESEARCH DEVELOPMENT

Crucifer breeders have explored and domesticated a few wild oleiferous *Brassica*s that hold tremendous potential for oil and meal in the future. However, owing to zero or little suitability of these crops to local agronomic practices, anti-nutritional compounds present in their oil and meal, and also little value of the byproducts and meal after oil extraction make their commercialization a difficult task. Still, a few species have proven their worth and could be potential industrial and domestic oil producers.

1. *Brassica carinata*. This species of *Brassica* is grown on a small scale in Ethiopia. The seed is large and predominantly dark, although some yellow seeded forms are also available. Apart from being resistant to various major diseases (e.g., *Alternaria* blight and white rust*)* and pests, it also revealed high yield potential in research and adapted trials (Gupta et al., 1999). *B. carinata* has an edge over other domesticated species, especially under rainfed and natural aphid infestation conditions (Anand and Rawat, 1984). Pure line selection is being used to develop high-yielding varieties in this crop. Because the available germplasm is late maturing, efforts have been made to induce earliness through mutation breeding or using interspecific/intergeneric hybridization. In Canada, at the Saskatoon Research Centre, attempts are being made to develop high-yielding, early-maturing varieties.

2. *Camelina sativa*. *Camelina* oil is used as a crude oil in Canada and many European countries. It is a major oilseed crop in Siberia (Francis and Campbell, 2003). *Camelina* oil has very good potential for industrial use (Vollmann et al., 1996). The oil can also be used for the production of biodiesel (Pilgeram et al., 2007). Vollmann et al., (2005) reported that there was enough genetic variation in the material studied for the various traits (i.e., oil content, seed yield, maturity, drought resistance, etc.). The variability inherent in *Camelina* has yet to be fully exploited. Therefore, the conventional breeding techniques may be some of the options that can be utilized for the development of suitable ideotypes. Somatic embryogenesis and microspore-derived embryogenesis in combination with mutagenesis could speed up the breeding efforts for particular fatty acids.

3. *Lesquerella species*. A species common in drier parts of the United States, Canada, and Mexico (Dierig et al., 2006), it is being used as a domestic replacement for a single hydroxy fatty acid in castor oil, ricinoleic acid. *Lesquerella* is a highly cross-pollinated crop with 86% to -90% outcrossing. Inter- and intra-population improvement can be used to develop high-yielding genotypes in this crop. Hybrid cultivar development may also be another

possibility because of the resistance of the male sterility system for *L. fendleri* (Dierig et al., 1996).

4. *Crambe species.* These species have been evaluated as a potential oilseed crop in Canada, Denmark, Germany, Poland, Russia, and Sweden since 1932 (White and Higgins, 1966). Due to a high erucic acid content and high smoke point and viscosity, *Crambe* oil is potentially useful in lubricating, emulsifying, and refrigeration industries (Lazzeri et al., 1994). The main objective in *Crambe* research is to increase seed yield, oil production, and protein meal quality. Mutagenesis and interspecific hybridization have been used as a breeding approach to produce new cultivars with improved agronomic and seed quality traits (Lessman and Meier, 1972).

5. *Eruca sativa.* This crop is mainly in cultivation in Iran, Pakistan, and India. It is highly cross-pollinated with more than 50% to 70% outcrossing, and is grown in the drier parts of India (Punjab, Haryana, Uttar Pradesh, and Rajasthan) and some parts of Pakistan. *Eruca* oil is characterized by a high content of sulfur and nitrogen compounds, and finds main use in industry and pharmaceuticals. It is also highly resistant to aphids and thrives well under rain-fed and drought conditions. As a result, this species is being used for the transfer of drought/aphid resistance genes to cultivated species using interspecific and intergeneric hybridization. Very little work has been done in terms of genetic improvement in this crop. In India, variety type-27 has been developed by following mass selection either from local land races or from genetic variable germplasm entries. This variety has been released for commercial cultivation in India. Crucifer breeders are using new breeding methods, polycross methods, synthetics, and composites to develop high-yielding varieties in this crop.

ACKNOWLEDGMENT

The authors thank Prof. Wei Jun Zhou, Institute of Crop Science, College of Agriculture & Biotechnology, Zhejiang University, China, for a critical review of this manuscript.

REFERENCES

Agnihotri, A. and Kaushik, N. 1998. Transgressive segregation and selection of zero erucic acid strains from intergeneric crosses of *Brassica. Ind. J. Plant Genet. Resources,* 11:251–255.

Agnihotri, A. and Kaushik, N. 1999a. Genetic enhancement for double low characteristics in Indian rapeseed mustard In: *Proc. 10th Int. Rapeseed Congr.,* 26–29 September 1999, Canberra, Australia.

Agnihotri, A. and Kaushik, N. 1999b. Transfer of double low characteristics in Indian *B. napus. J. Oilseeds Res.,* 16:227–229.

Agnihotri, A. and Kaushik, N. 2003a. Towards nutritional quality improvement in Indian mustard (*Brassica juncea* [L]. Czern and Coss) var. Pusa Bold In: *Proc. 11th Int. Rapeseed Congr.,* Copenhagen, Denmark, 2:501–503, (Eds.) H. Sorensen, J.C. Sorensen, S. Sorensen, N.B. Muguerza, C. Bjergegaard, et al. The Royal Veterinary and Agricultural University, Copenhagen, Denmark, 6–10 July.

Agnihotri, A. and Kaushik, N. 2003b. Combining canola quality, early maturity and shattering tolerance in *B. napus* for Indian growing conditions. In *Proc. 11th Int. Rapeseed Congr.,* Copenhagen, Denmark, 2:436–439 (Eds.) H. Sorensen, J.C. Sorensen, S. Sorensen, N.B. Muguerza, C. Bjergegaard, et al. The Royal Veterinary and Agricultural University, Copenhagen, Denmark, 6–10 July.

Agnihotri, A., Gupta, V., Lakhmikumaran, M., Shivanna, K.R., Prakash, S., and Jagannathan, V. 1990b. Production of *Eruca-Brassica* hybrids by embryo rescue. *Plant Breed.,* 104:281–289.

Agnihotri, A., Lakshmikumaran, M., Shivanna, K.R. and Jagannathan, V. 1990a. Embryo rescue of interspecific hybrids of *Brassica spinescens* × *B. campestris* and DNA analysis. *Curr. Plant Sci. Biotechnol. Agric., Progr.Plant Cell. Mol. Biol.,* 1990:270–274.

Agnihotri, A., Prem, D., and Gupta, K. 2007. The chronicles of oil and meal quality improvement in rapeseed, p. 50–99. In: Gupta, S.K. (Ed.) *Advances in Botanical Research-Rapeseed Breeding*, Academic Press/ Elsevier Ltd., San Diego, CA.

Agnihotri, A., Raney, J.P., Kaushik, N., Singh, N.K., and Downey, R.K. 1995. Selection for better agronomical and nutritional characteristics in Indian rapeseed-mustard In *Proc. 9th Int. Rapeseed Congr.*, 4–7 July, Cambridge, U.K., 2:425–427.

Agrihotri, A., Shivanna, K.R., Raina, S.N., Lakshmikumaran, M., Prakash, S., and Jagannathan, V. 1990c. Production of *Brassica napus × Raphanobrassica* hybrids by embryo rescue. *Plant Breed.*, 105:292–299.

Ahmad, I., Day, J.P., MacDonald, M.V., and Ingram, D.S. 1991. Haploid culture and UV mutagenesis in rapid cycling *Brassica napus* for the generation of resistance to chlorsulfuron and *Alternaria brassicola*. *Ann. Bot.*, 67:521–525.

Alonso, L.C., Fernandez-Serrano, O., and Fernandez-Escobar, J. 1991. The outset of a new oilseed crop: *Brassica carinata* with low erucic acid. In: *Proc. 8th Int. Rapeseed Conf.*, Saskatoon, SK, Canada, p. 170–176.

Anand, I.J. and Rawat, D.S. 1984. Recent plant breeding efforts towards productivity breakthrough in rapeseed-mustard oilseed. *Proc. Symp. Oilseed Production and Utilization — Constraints and Opportunities,* New Delhi, India.

Ayotte, R., Harney, P.M., and Machado, V.S. 1986. The transfer of triazine resistance from *B. napus* to *B. oleracea*. *Cruciferae Newslett.*, 11:95–96.

Ayotte, R., Harney, P.M., and Machado, V.S. 1987. Transfer of triazine resistance from *Brassica napus* to *B. oleracea*. I. Production of F1 hybrids through embryo rescue. *Euphytica*, 36: 615–624.

Bannerot, H., Boulidar, L., Canderon, Y., and Tompe J. 1974. Transfer of cytoplasmic male sterility from *Raphanus sativus* to *B. oleracea*. In: *Proc. Eucarpia Meeting on Cruciferae, Crop. Sci.*, 25:52–54.

Barfield, D.G. and Pua, E.C. 1991. Gene transfer in plants of *Brassica juncea* using *Agrobacterium tumefaciens* mediated transformation. *Plant Cell Rep.*, 10:308–314.

Barro, F., Fernandez-Escobar, J., De La Vega, M., and Martin, A. 2001 Doubled haploid lines of *Brassica carinata* with modified erucic acid content through mutagenesis by EMS treatment of isolated microsporas. *Plant Breed.*, 120:262–264.

Barro, F., Fernandez-Escobar, J., De la Vega, M., and Martin, A. 2002. Modification of glucosinolate and erucic in doubled haploid lines of *Brassica carinata* by UV treatment of isolated microspores. *Euphytica*, 129:1–6.

Beversdorf, W.D. and Kott, L.S. 1987. An *in vitro* mutagenesis/selection system for B*rassica napus*. *Iowa State J. Res.*, 61:435–443.

Bijral, J.S., Gupta, S.K., and Dey, T. 2004. Unusual floral morphology in an advanced generation of an intergeneric cross of *Brassica napus* X *Eruca sativa*. *J. Res. SKUAST-J*, 3:61–64.

Brandle, J. and McVetty, P. 1990. Geographic diversity, parental selection and heterosis in oilseed rape. *Can. J. Plant. Sci.* 40:35–94.

Brzezinski, W., Mendelewski, P., and Musse, B.G. 1986. Comparative study on determination of glucosinolates in rapeseed *Cruciferae Newslett.*, 11:128–129.

Burton, W., Salisbury, P., and Potts, D. 2003a. Inheritance of allyl glucosinolate in the development of canola quality *Brassica juncea* for Australia In: *Proc. 11th Int. Rapeseed Congr.*, p 280. The Royal Veterinary and Agricultural University, Copenhagen, Denmark, 6–10 July.

Burton, W., Salisbury, P., and Potts, D. 2003b. The potential of canola quality *Brassica juncea* as an oilseed crop for Australia In: *Proc. 11th Int. Rapeseed Congr.*, p. 5–7. The Royal Veterinary and Agricultural University, Copenhagen, Denmark, 6–10 July.

Cardoza, V. and Stewart, N.C. Jr. 2004. *Brassica* biotechnology: progress in cellular and molecular biology. *In Vitro Cell Dev. Biol.- Plant*, 40:542–551.

Carre, P., Dartenuc, C., Evrard, J., Judde, A., Labalette, F., Raoux, R., and Renard, M. 2003. Frying stability of rapeseed oils with modified fatty acid composition In *Proc. 11th Int. Rapeseed Congr.*, Copenhagen, Denmark, 2:540–543 The Royal Veterinary and Agricultural University, Copenhagen, Denmark, 6—10 July.

Cegielska-Taras, T., Szala, L., and Krzymanski, J. 1999. An *in vitro* mutagenesis-selection system for *Brassica napus* L. New horizons for an old crop. *Proc. 10th Int. Rapeseed Congr.*, Canberra, Australia, p. 1–4.

Charne, D.G. and Beversdorf, W.D. 1988. Improving microspore culture as a rapeseed breeding tool: the use of auxins and cytokinins in an induction medium. *Can. J. Bot.*, 66:1671–1675.

Chaubey, C.N. 1979. Mass selection in Toria Brassica-Campestris-Var -Toria.. *Indian J. Genet.*, 39:194–201.

Chaudhary, S., Parmenter, D.L., and Moleney, M.M. 1998. Transgenic *Brassica carinata* as a vehicle for the production of recombinant proteins in seeds. *Plant Cell Rep.*, 17:195–200.

Chèvre, A.M., Ammitzbøll, H., Breckling, B., Dietz-Pfeilstetter, A., Eber. F., Fargue, A., Gomez-Campo, C., Jenczewsk, E., Jørgensen, R.B., Lavigne, C., Meier, M.S., Nijs, H., Pascher, K., Seguin-Swartz, G., Sweet, J., Steward, C.N., and Warwick, S. 2004. A review on interspecific gene flow from oilseed rape to wild relatives. In: *Introgression from Genetically Modified Plants into Wild Relatives and its Consequences* (Eds. Nijs, H., Bartsch, D., and Sweet, p. 235–251, CABI Publishing, UK.

Corbett, P. and Sernyk, L. 2003. Global opportunities for naturally stable canola/rapeseed oils. In: *Proc. 11th Int. Rapeseed Congr.*, Copenhagen, Denmark, 2:524–527. The Royal Veterinary and Agricultural University, Copenhagen, Denmark, 6–10 July.

Craig, B.M. and Murty, N.L. 1959. Quantitative fatty acids analysis of vegetable oils by gas chromatography. *J. Am. Oil Chem. Soc.*, 36:549–552.

Das, B. and Rai, B. 1972. Heterosis in intervarietal crosses of Toria. *Ind. J. Genet.*, 32:197–202.

De Block, M., De Brower, D., and Tenning, P. 1989. Transformation of *Brassica napus* and *Brassica oleracea* using *Agrobacterium tumefaciens* and the expression of bar and neo genes in the transgenic plants. *Plant Physiol.*, 91:694–701.

Dierig, D.A., Coffelt, T.A., Nakayama, F.S., and Thompson, A.E. 1996. *Lesquerella* and *Vernonia*: oilseeds for arid land. In: J. Janick (Ed.), *Progress in New Crops*. ASHS Press, Alexandria, VA, p. 347–354.

Dierig, D.A. Dahlquist, G.H., and Tomasi, P.M. 2006a. Registration of WCL-LO3 high oil *Lesquerella fendleri* germplasm. *Crop Sci.*, 46:1832–1833.

Dierig, D.A., Salywon, A.M,. and De Rodriquez, D.J. 2006b. Registration of a mutant *Lesquerella* genetic stock with cream flower color. *Crop Sci.*, 46:1836–1837.

Doloi, P.C. 1977. Levels of Self-incompatibility: Heterosis Inbreeding Depression in *Brassica campestris*. Unpublished Ph.D. thesis. Govind Ballabh Pant University of Agriculture and Technology, Pant Nagar (Nainital) U.P., India.

Downey, R.K. 1964. A selection of *Brassica campestris* L. containing no erucic acid in its seed oil. *Can. J. Plant Sci.*, 44:295.

Downey, R.K. 1966. Breeding for fatty acid composition in oils of *Brassica napus* L. and *B. campestris* L. Qualitas. *Plant Mater. Veget.*, 13:171–180.

Downey, R.K. and Harvey, BL. 1963. Method of breeding for oil quality in rape *Can. J. Plant Sci.*, 43:271–275.

Downey, R.K., Klassen, A.J., and Stringan, G.S. .1980. Rapeseed-mustard. In: Fehr, W.R. Hadley, H.H. (Eds.) *Hybridization of Crop Plant*. Aust. Soc. Agron. Inc. Madison, WI, p. 495–509.

Dunwell, J.M. 1996. Microspore culture. In: Jain S.M., Sopory S.K., and Veilleux R.E. (Eds.), *In Vitro Haploid Production in Higher Plants*, 1:205–216. Kluwer Academic Publishers, Dordrecht.

Facciotti, M.T., Bertain, P.B., and Yuan, L. 1999. Improved stearate phenotype in transgenic canola expressing a modified acyl-acyl carrier protein thioesterase. *Nat. Biotechnol.*, 17:593–597.

Fenwick, G.R., Heaney, R.K., and Mullin, W.J. 1983. Glucosinolates and their breakdown products in food and food plants. *Crit. Rev. Food Nutr.*, 18:123–201.

Ferrie, A.M.R. and Keller, W.A. 2002. Application of double haploidy and mutagenesis in *Brassica*. *13th Cruciferae Genetics Workshop*, March 23–26, University of California, Davis.

Francis, C.M. and Campbell, M.C. 2003. New High Quality Oil Seed Crops for Temperate and Tropical Australia. Rural Industries Research and Development Corp. Publication No. 03/045, RIRDC Project No. UWA-47A.

Gamborg, O.L., Miller, R.A., and Ojima, K. 1968. Nutrient requirements of suspension cultures of soybean root cells. *Exp. Cell Res.* 50:151–158.

Gororo, N., Salisbury, P., Rebetzke, G., Burton, W., and Bell, C. 2003. Genotypic variation for saturated fatty acid content of Victorian canola In: *Proc. 11th Int. Rapeseed Congr.*, Copenhagen, Denmark, 1:215–217 The Royal Veterinary and Agricultural University, Copenhagen, Denmark, 6–10th July.

Gu, H.H., Hagberg, P., and Zhou, W.J. 2003. Cold pretreatment enhances microspore embryogenesis in oilseed rape (*Brassica napus* L.). *Plant Growth Regulation*, 2004, 42:137–143.

Gu, H.H., Zhou, W.J., and Hagberg, P. 2004. High frequency spontaneous production of doubled haploid plants in microspore cultures of *Brassica rapa* ssp. *chinensis*. *Euphytica*, 134:239–245.

Gupta, S.K., 1993. Wild hybridization in *Brassica*. *Nucleus*, 36:149–151.

Gupta, S.K., 1997. Production of interspecific and intergeneric hybrids in *Brassica* and *Raphanus*. *Cruciferae Newslett.*, 19:21–22.

Gupta, M.L., Ahuja, K.L., Raheja, R.K., and Labana, K.S. 1998. Variation for biochemical quality traits in promising genotypes of Indian mustard. *J. Res.*, 25:1–5.

Gupta, M.L., Banga, S.K. Banga, S.S., Sandha, GS., Ahuja, K.L., and Raheja, R.K. 1994. A new genetic stock for low erucic acid in Indian mustard. *Cruciferae Newslett.*, 16:104–105.

Gupta, S.K. and Pratap, A. 2007a. History, origin and evolution. In: Gupta, S.K. (Ed.), *Advances in Botanical Research-Rapeseed Breeding*, Vol. 45, Academic Press, London, p. 120.

Gupta, S.K. and Pratap, A. 2008. Recent trends in oilseed *Brassicas*. In: Nayyar, H. (Ed.), *Crop Improvement: Challenges and Strategies*. I.K. International, New Delhi, India, p. 284–299.

Gupta, S.K., Bhagat, K.L., Khanna, Y.P., and Gupta, S.C., 1999. Performance of Abyssinian mustard under rainfed conditions of Jammu. *J. Oilseed Res.,* 16(2):285–88.

Gupta, S.K., Sharma, D.R., and Chib, H.S., 1995. Evaluation of wild allies of *Brassica* under natural conditions. *Cruciferae Newslett.,* 17:10–11.

Gupta, V., Sita, G.L., Shaila, M.S., and Jagannathan, V.1993. Genetic transformation in *Brassica nigra* by *Agrobacterium* based vector and direct plasmid uptake. *Plant Cell Rep.,* 12:418–421.

Halfhill, M.D., Richards, H.A., Mabon, S.A., and Stewart, N.C. Jr. 2001. Expression of GFP and Bt transgenes in *Brassica napus* and hybridization with *Brassica rapa*. *Theor. Appl. Genet.,* 103:151–156.

Hawkins, D. and Kridl, L. 1998. Characterization of acyl-ACP thioesterase of mangosteen (*Garcinia mangosteena*) seed and high levels of state production in transgenic canola. *Plant J.,* 13:743–752.

Hinata, K. and Konno, N. 1979. Studies on a male sterile strain having the *Brassica campestris* nucleus and the *Diplotaxis muralis* cytoplasm. I. On the breeding procedure and some characteristics of the male sterile strain. *Jap. J. Breed.,* 29:305–311.

Hitz, W.D., Mauvis, C.J., Ripp, K.G., Reiter, R.J., DeBonte, L., and Chen, Z. 1995. The Use of Cloned Rapeseed Genes for the Cytoplasmic Fatty Acid Desaturases and the Plastid Acyl-ACP Thioesterases to Alter Relative Levels of Polyunsaturated And Saturated Fatty Acids in Rapeseed. D5-Breeding Oil Quality. GCIRC, Cambridge, UK., p. 470–478.

Hofte, H. and Whiteley, H.R. 1989. Insecticidal crystal proteins of *Bacillus thuringiensis. Microbiolog. Rev.,* 53:242–255.

Jagannath, A., Arumugam, N., Guipta, V., Pradhan, A., Burma, P.K., and Pental, D. 2002. Development of transgenic *barstar* lines and identification of a male sterile (*barnase*/restorer (*barstar*) combination for heterosis breeding in Indian oilseed mustard (*Brassica juncea*). *Curr. Sci.,* 82:46–52.

Jagannath, A., Bandhopadhyay, P., Arumugam, N., Burma, P.K., and Pental, D. 2001. The use of a spacer DNA fragment insulates the tissue specific expression of a cytotoxic gene (*barnase*) and allows high frequency generation of transgenic male sterile lines in *Brassica juncea* L. *Mol. Breed.,* 8:11–23.

Jedrzejaszek, K., Kruczkowska, H., Pawlowska, H., and Skucinska, B. 1997. Simulating effect of mutagens on *in vitro* plant regeneration. *MBNL,* 43:10–11.

Jiang, L. and Becker, H.C. 2001. Effect of apetalous flowers on crop physiology in winter oilseed rape (*Brassica napus*). *Pflanzenbauwissenchaft,* 5:58–63.

Jonsson, 1973. Breeding for low erucic acid contents in summer turnip rape (*Brassica campestris* L. var. *annua*). *Z. Pflan Zen Zu Chtz.,* 69:1–18.

Jørgensen, R.B. 2007. Gene flow from wild species into rapeseed. In: Gupta, S.K. (Ed.), *Advances in Botanical Research-Rapeseed Breeding*, Academic Press/Elsevier Ltd., San Diego, CA, Vol. 45, p. 452–554.

Kalia, H.R. and Gupta, S.K. 1997. Breeding Methods. In: *Recent Advances in Oilseed Brassicas*, Kalia, H.R. and Gupta, S.K. (Eds.), Kalyani Publishers.

Keller, W.A. and Armstrong, K.C. 1978. High frequency production of microspore derived plants from *Brassica napus* anther cultures. *Z. Pflanzenzchtg.,* 80:100–108.

Khalatkar, A.S., Rakow, G., and Downey, R.K. 1991. Selection for quality and disease resistance in *Brassica juncea* cv. Pusa Bold. In: *Proc. 8th Int. Rapeseed Congr.,* Saskatoon, SK, Canada, 9–11 July.

Kirk, J.T.O. and Oram, R.N. 1981 Isolation of erucic acid free lines of *Brassica juncea*: Indian mustard now a potential oilseed crop in Australia. *J. Australian Inst. Agric. Sci.,* 47:51–52

Klewer, A. and Sacristán, M.D. 2002. Erschließung neur Resistenzquellen bei raps-Schwerpunkt *Alternaria*-Resistenz. *Vortr. Pfl.-Zücht.*

Klewer, A., Mewes, S., Mai, J., and Sacristán, M.D. 2002. *Alternaria*-Resistenz in interspezifischen Hybriden und deren Rückkreuzungsnachkommenschaften in Tribus *Alternaria*-Resistenz. *Vortr. Pfl.-Zücht.*

Knutzon, D.S., Hayes, T.R., Wyrick, A., Xiong, H., Davies, H.M., and Voelker, T.A. 1999. Lysophosphatidic acid acyltransferase from coconut endosperm mediates the insertion of laurate at the sn-2 position of triglycerols in lauric rapeseed oil and can increase total laurate levels. *Plant Physiol.,* 120:739–746.

Kott, L. 1995. Production of mutants using the rapeseed doubled haploid system. In: *Induced Mutations and Molecular Techniques for Crop Improvement*. IAEA, Vienna, p. 505–515.

Kott, L. 1998. Application of doubled haploid technology in breeding of oilseed *Brassica napus. AgBiotech News Inform.,* 10:69N–74N.

Labana, K.S., Badwal, S.S., and Chaurasia, B.D. 1975. Heterosis and combining ability in *B. juncea. Crop Improv.,* 2:46–51.

Larik, A.S. and Hussain, M. 1990. Heterosis in Indian mustard *Brassica juncea* (L.) Coss. *Pakistan J. Bot.,* 22(2):168–171.

Lazzeri, L. Leoni, O., Conte, L.S., and Palmieri, S. 1994. Some technological characteristics and potential uses of *Crambe abyssinica* products. *Industr. Crops Prod.,* 3:103–112.

Lefort-Buson, M. and Datte, Y. 1982. Genetic study of some agronomic characters in winter oilseed rape (*Brassica napus*). I. Heterosis. *Agronomie,* 2:315–332.

Lessman, K.J., and Meir, V.D. 1972. Agronomic evaluation of *Crambe* as a source of oil. *Crop Sci.,* 12:224–227.

Lichter, R. 1982. Induction of haploid plants from isolated pollen of *Brassica napus. Z. Pflanzenphysiol.,* 103:229–237.

Liu, J., W., DeMichele, S., Bergana, M., Bobik, E., Hastilow, C., Chuong, L.T., Mukerji, P., and Huang, Y.S. 2001. Characterization of oil exhibiting high gamma-linolenic acid from a genetically transformed canola strain. *J. Am. Oil Chem. Soc.,* 78:489–493.

Luhs, W., Weier, D., Marwede, V., Frauen, M., Lekband, G., Becker, H.C., Frentzen, M., and Friedt, W. 2003. Breeding of oilseed rape (*Brassica napus* L.) for modified tocopherol composition — synergy of conventional and modern approaches In *Proc. 11th Int. Rapeseed Congr.,* Copenhagen, Denmark, 1:194–197, The Royal Veterinary and Agricultural University, Copenhagen, Denmark, 6–10 July.

MacDonald, M.V., Ahgmad, I., Menten, J.O.M., and Ingram, D.S. 1991. Haploid culture and *in vitro* mutagenesis (UV light, x-rays, and gamma rays) of rapid cycling *Brassica napus* for improved resistance to disease. In *Plant Mutation Breeding for Crop Improvement,* Vol. 2. IAEA, Vienna, p. 129–138.

Malik, M., Vyas, P., Rangaswamy, N.S., and Shivanna, K.R.. 1999. Development of two new cytoplasmic male-sterile lines of *Brassica juncea* through wide hybridization. *Plant Breed.,* 118:75–78.

Malode, S.N., Swamy, R.V., and Khalatkar, A.S. 1995. Introgression of 'OO' quality characters in *Brassica juncea* cv. Pusa bold. In *Proc. 9th Int. Rapeseed Congr.,* Cambridge, UK. 4–7 July 1995, p. 431–438.

Maluszynski, M., Kasha, K.J., Forster, B.P., and Szarejko, I. 2003. *Doubled Haploid Production in Crop Plants: A Manual.* Kluwer Academic Publisher, Dordrecht, The Netherlands.

McCollum, G.D. 1988. CMS (ESG-508) and CMS (FSG-512) cytoplasmic male sterile cabbage germplasm with radish cytoplasm. *Hort. Sci.,* 23:227–228.

McGregor, D.I., Mullim, W.J., and Fenwick, G.R. 1983. Review of analysis of glucosinolate analytical methodology for determining glucosinolate composition and content. *J. Assoc. Official Anal. Chemists,* 66:825–849.

Mohapatra, D. and Bajaj, Y.P.S. 1987. Interspecific hybridization of *B. juncea* × *B. hirta* using embryo rescue. *Euphytica,* 36:321–326.

Moloney, M.M., Walker, J.M., and Sharma, K.K. 1989. High efficiency transformation of *Brassica napus* using *Agrobacterium* vectors, *Plant Cell Rep.,* 8:238–242.

Muralhige, T. and Skoog, F. 1962. A revised medium for rapid growth and bioassays with tobacco tissue cultures. *Physiol. Plant,* 15:473–479.

Narasimhulu, S.B., Kirti, P.B., Mohapatra, T., Prakash, S., and Chopra, V.L. 1992. Shoot regeneration in stem explants and its amenability to *Agrobacterium tumefaciens* mediated gene transfer in *Brassica carinata. Plant Cell Rep.,* 11:359–362.

Ogura, H. 1968. Studies on a new male-sterility in Japanese radish, with special reference to utilization of this sterility towards the practical raising of hybrid seeds. *Mem. Fac. Agr. Kogoshima Univ.,* 6:39–78.

Oliveri, A.M. and Parrini, P. 1986. Relationship between glucosinolate content and yield component in rapeseed. *Cruciferae Newslett.,* 11:126–127.

Oram, R.N., Salisbury. P.A., Krick, J.T.O., and Burton, W.A. 1999. Development of early flowering, canola grade *Brassica juncea* germplasm In *Proc. 10th Int. Rapeseed Congr.,* 26–29 September 1999, Canberra, Australia.

Pilgeram, A.L., Sanda, D.C., Boss, D., Dale, N., Wichman, D., Lamb, P., Lu, C., Barrows, R., Kirkpatrick, M., Thompson, B., and Johnson, D.L. 2007. *Camelina sativa,* A Montana Omega-3- and Fuel Crop. In: Janick, J. and Whipkey, A. (Eds.), *Issues in New Crops and New Uses,* ASHS Press, Alexandria, VA, p. 129–131.

Plumper, B. 1995. Somatische and sexual hybridisierung fiir den Transfer von Krankheitsreristenzen and *Brassica napus,* Ph.D. thesis, Free Univ. of Berlin, Berlin.

Polsoni, L., Kott, L.S., and Beversdorf, W.D. 1988. Large-scale microspore culture technique for mutation-selection studies in *Brassica napus. Can. J. Bot.,* 66:1681–1685.

Potts, D.A., Rakow, G.W., and Males, D.R. 1999. Canola-quality *Brassica juncea,* a new oilseed crop for the Canadian prairies. In *Proc. Xth GCIRC Int. Rapeseed Congr.,* 26–29 September 1999, Canberra, Australia.

Pradhan, A.K., Mukhopadhyay, A., and Pental, D. 1991. Identification of putative cytoplasmic donor of CMS system in *Brassica juncea*. *Plant Breed.,* 106:204–208.

Pradhan, A.K., Sodhi, Y.S., Mukhopadhyay, A., and Pental, D. 1993. Heterosis breeding in Indian mustard (*Brassica juncea* L. Czern & Coss): analysis of component characters contributing to heterosis for yield. *Euphytica,* 69, 219–229.

Prakash, S. and Chopra, V.L. 1988. Introgression of resistance to shattering in *Brassica napus* from *Brassica juncea* through non-homologous recombination. *Plant Breed.,* 101:167–168.

Prakash, S., Kirti, P.B., Bhat, S.R., Gaikwad, K.V., Dinesh Kumar, V., and Chopra, V.L. 1998. A *Moricandia arvensis* based cytoplasmic male sterility and fertility restoration system in *Brassica juncea*. *Theor. Appl. Genet.,* 97:488–492.

Prasad, K.V.S.K., Sharmila, P., Kumar, P.A., and Saradhi, P.P. 2000. Transformation of *Brassica juncea* (L.) Czern with bacterial coda gene enhances its tolerance to salt and cold stress. *Mol. Breed.,* 6:489–499.

Pratap, A. and Gupta, S.K. 2007. Unusual floral morphology in advanced generations of intergeneric hybrids between *Brassica napus* and *Eruca sativa*. ISOR 2007. Extended summaries: *National Seminar on Changing Vegetable Oils Scenario: Issues and Challenges before India*, Indian Society of Oilseeds Research, Hyderabad, India, p. 50–51.

Pratap, A., Gupta, S.K., and Sharma, M. 2008. Genetic amelioration of crop brassicas through intergeneric hybridization using *Diplotaxis. National Seminar on Physiological and Biotechnological Approaches to Improve Plant Productivity,* March 15–17, 2008, Center for Plant Biotechnology, DST, Haryana, Hisar, India, p. 85–86.

Pratap, A., Gupta, S.K., and Vikas. 2007. Advances in doubled haploid technology of oilseed rape. *Indian J. Crop Sci.,* 2:267–271.

Quazi, M.H. 1988. Interspecific hybrids between *B. napus* and *B. oleracea* developed by embryo rescue. *Theor. Application Genet.,* 75:309–318.

Radke, S.E., Turner, J.C., and Facciotti, D. 1992. Transformation and regeneration of *Brassica rapa* using *Agrobacterium tumefaciens*. *Plant Cell Rep.,* 11:499–505.

Rai, B. 1976. Considerations in the genetic improvement of oil quality in rapeseed. *Oilseed J.,* 6:13–15.

Rai, B. 1979. *Heterosis Breeding*. Agro-Biological Publications, Azad Nagar, New Delhi, India, p. 183.

Rai, B. 1982. Breeding strategy for developing high yielding varieties of toria (*Brassica campestris* var. toria). *Research and Development Strategies for Oilseeds Production in India.* ICAR, New Delhi, p. 131–135.

Rai, B. 1983a. Genetic improvement of seed yield and disease resistance in rapeseed and mustard oilcrops. *Oilseeds J.,* 13:6–13.

Rai, B. 1983b. Advances in rapeseed and mustard breeding research. *Ind. Fmg.,* 33:3–8.

Rai, B. 1987. PT 303 the first national variety of toria. Ind. Fmg., 37:16–17.

Rai, B. and Kumar, A. 1978. Rapeseed and mustard production programme. *Ind. Fmg.,* 28:27–30.

Rai, B. and Sehgal, V.K. 1975. Field resistance of *Brassica* germplasm to mustard aphids (*Lipaphis Erysimi* kalt). *Sci. and Cult.,* 41:444–445.

Rai, B. and Singh, A. 1976. Commercial seed production in rapeseed. *Ind. Fmg.,* 26:15–17.

Rai, B., Gupta, S.K., and Pratap, A. 2007. Breeding methods. In: Gupta, S.K. (Ed.), *Advances in Botanical Research-Rapeseed Breeding*, 45:21–48. Academic Press/Elsevier, San Diego, CA.

Rakow, G. 1995. Developments in the breeding of oil in other *Brassica* species. In: *Proc. 9th Int. Rapeseed Congr.,* Cambridge, UK., 2:401-406.

Rakow, G. and Raney, J.P. 2003. Present status and future perspectives of breeding for seed quality in *Brassica* oilseed crops In: *Proc. 11th Int. Rapeseed Congr.,* The Royal Veterinary and Agricultural University, Copenhagen, Denmark, 6–10th July, p. 181–185.

Raney, J.P., Olson, T.V., Rakow, G., and Ripley, V.L. 2003a. *Brassica juncea* with a canola fatty acid composition from an interspecific cross with *Brassica napus* In: *Proc. 11th Int. Rapeseed Congr.,* The Royal Veterinary and Agricultural University Copenhagen, Denmark, 6–10 July, p. 281–283.

Raney, J.P., Olson, T.V., Rakow, G., and Ripley, V.L. 2003b. Selection of near zero aliphatic glucosinolate *Brassica juncea* from an interspecific cross with *B. napus* In: *Proc. of 11th Int. Rapeseed Congr.,* The Royal Veterinary and Agricultural University, Copenhagen, Denmark, 6–10 July, p. 284–286.

Rao, G.U. and Shivanna, K.R. 1996. Development of a new alloplasmic CMS *Brassica napus* in the cytoplasmic background of *Diplotaxis siifolia. Cruciferae Newslett.,* 18:68–69.

Rao, G.U., Batra, V.S., Prakash, S., and Shivanna, K.R. 1994. Development of a new cytoplasmic male sterile system in *Brassica juncea* through wide hybridization. *Plant Breed.,* 112:171–174.

Rao, M.V.B. 1990. Widening Variability in Cultivated Digenomic *Brassica* through Interspecific Hybridization. Ph.D. thesis, IARI, Library, New Delhi, India.

Roy, N.N. 1984. Interspecific transfer of *Brassica juncea* type high blackleg resistance to *Brassica napus*, *Euphytica*, 33:295–303.

Roy, N.N. and Tarr, A.W. 1985. IXLIN — an interspecific source for high linoleic acid content in rapeseed. *Plant Breed.*, 95:201–209.

Roy, N.N. and Tarr, A.W. 1986. Development of new zero linolenic acid (18:3) lines of rapeseed (*Brassica napus* L.). *Z. Planzenzuchtg.*, 96:218–233.

Sacristan, N.D. and Gerdemann, M. 1986. Different behavior of *Brassica juncea* and *B. carinata* as sources of interspecific transfer to *B. napus*. *Plant Breed.*, 97:304–314.

Salisbury, P. 1989. Potenital utilization of wild crucifer germplasm in oilseed Brassica breeding. *Proc. ARAB 7th Workshop*, Toowomba, Queensland, Australia, pp. 51–53.

Scheffler, J.A. and Dale, P.J. 1994. Opportunities for gene transfer from transgeneric oilseed rape (*Brassica napus*) to related species. *Transgen. Res.*, 3:263–278.

Schnepf, E., Crickmore, N., Van Rie., J., Lereclus, D., Baum, J., Feitelson, J., Ziegler, D.R., and Dean, D.H. 1998. *Bacillus thuringiensis* and its pesticidal crystal proteins. *Microbiol. Molec. Biol. Rev.*, 62:775–806.

Schuler, T.J., Hutcheson, D.S,. and Downey, R.K. 1992. Heterosis in intervarietal hybrids of summer turnip rape in western Canada. *Canad. J. Plant Sci.*, 72:127–136.

Schuster, W. and Michael, J. 1976. Untersuchungen uber Inzuchtdepressionen und Heterosis effekte bei Raps (*Brassica napus* oleifera). *Z. Pflanzenzuchtung.*, 77:56–66.

Sernyk, J.L. and Stefansson, B.R. 1983. Heterosis in summer rape (*Brassica napus* L.). *Canad. J. Plant Sci.*, 63:407–413.

Shelton, A.M., Zhao, J.Z., and Roush, R.T. 2002. Economic, ecological, food safety and social consequences of the deployment of Bt transgenic plants. *Annu. Rev. Entomol.*, 47:845–881.

Shewmaker, C.K., Sheehy, J.A., Daley, M., Colburn, S., and Ke, D.Y. 1999. Seed-specific over expression of phytoene synthase: increase in carotenoids and other metabolic effects. *Plant J.*, 20:401–412.

Shi, S.W., Wu, J.S., Zhou, Y.M., and Liu, H.L. 2002. Diploidization techniques of haploids from *in vitro* culture microspores of rapeseed (*Brassica napus* L.). *Chinese J. Oil Crop Sci.*, 24:1–5.

Siemens, J. 2002. Interspecific hybridization between wild relatives and *Brassica napus* to introduce new resistance traits into the oilseed rape gene pool. *Czech. J. Genet. Plant Breed.*, 38:155–157.

Singh, S.P. 1973. Heterosis and combining ability estimates in Indian mustard, *Brassica juncea* (L.) Czern and Coss. *Crop Sci.*, 13:497–499.

Singh, R. and Sharma, S.K. 2007. Evaluation, maintenance and conservation of germplasm. In: Gupta, S.K. (Ed.), *Advances in Botanical Research — Rapeseed Breeding*, Academic Press, London, 45:465–481.

Snowdon, R.J., Winter, H., Diestel, A., and Sacristán, M.D. 2000. Development and characterization of *Brassica napus-Sinapis arvensis* addition lines exhibiting resistance to *Leptosphaeria maculans*. *Theor. Appl. Genet.*, 101:1008–1014.

Spasibionek, S., Krzymanski, J., and Bartkowiak-Broda, I. 2003. Mutants of *Brassica napus* with changed fatty acid composition In: *Proc. 11th Int. Rapeseed Congr.*, Copenhagen, Denmark, 1:221–224 The Royal Veterinary and Agricultural University, Copenhagen, Denmark, 6–10 July.

Spinks, E.A., Sones, K., and Fenwick, G.R. 1984. The quantitative analysis of glucosinolates in cruciferous vegetables, oilseeds and forage crops using high performance liquid chromatography. *Fette Seifen Anstrichmittel*, 86:228–231.

Srivastava, K. and Rai, B. 1993. Expression of heterosis for yield and its attributes in rapeseed. *Ind. J. Agric. Sci.*, 63:243–245.

Stefansson, B.R., Hougen, F.W., and Downey, R.K. 1961. Note on the isolation of rape plants with seed oil free from erucic acid. *Canad. J. Plant Sci.*, 41:218–219.

Stewart, C.N., Adang, M. J., All, J.N., Raymer, P.L., Ramachandran, S., and Parrott, W.A. 1996. Insect control and dosage effects in transgenic canola containing a synthetic *Bacillus thuringiensis* cry1Ac gene. *Plant Physiol.*, 112:115–120.

Stoutjesdijk, P.A., Hurlestone, C., Singh, S.P., and Green, A.G. 2000. High oleic-acid Australian *Brassica napus* and *B. juncea* varieties produced by co-suppression of endogenous delta 12-desaturases. *Biochem. Soc. Trans.*, 28:938–940.

Swanson, E.B., Coumans, M.P., Brown, G.L., Patel, J.D., and Beversdorf, W.D. 1988. The characterization of herbicide tolerant plants in *Brassica napus* L. after *in vitro* selection of microspores and protoplasts. *Plant Cell Rep.*, 7:83–87.

Swanson, E.B., Herrgesell, M.J., Arnoldo, M., Sippell, D.W., and Wong, R.S.C. 1989. Microspore mutagenesis and selection: canola plants with field tolerance to the imidazolinones. *Theor. Appl. Genet.*, 78:525–530.

Szarejko, I. and Forster, B.P. 2007. Doubled haploidy and induced mutation. *Euphytica,* 158:359–370.

Vollmann, J., Damboeck, A., Eckl, A., Schrems, H., and Ruckenbauer, P. 1996. Improvement of *Camelina sativa*, an underexploited oilseed. In: Janick (Ed.), *Progress in New Crops*. ASHS Press, Alexandria, VA, p. 357–362.

Vollmann, J., Grausgruber, H., Stift, G., Dryzhruk, V., and Lelley, T. 2005. Genetic diversity in *Camelina* germplasm as revealed by seed quality characteristics and RAPD polymorphism. *Plant Breed.,* 124:446–453.

Wang, H.Z., Liu, G.H., Zheng, Y.B., Wang, X.F., and Yang, Q. 2002. Breeding of *Brassica napus* cultivar Zhongshuang No. 9 with resistance to *Sclerotinia sclerotiorum*. *Chinese J. Oil Crop Sci.,* 24:71–73.

Wang, M., Farnham, M.W., and Nannes, J.S.P. 1999. Ploidy of broccoli regenerated from microspore culture versus anther culture. *Plant Breed.,* 118:249–252.

White, G.A. and J.J. Higgins. 1966. Culture of Crambe: A New Industrial Oilseed Crop. Agr. Res. Serv., USDA, ARS Production Research Report 95.

Winter, H., Diestel, A., Gärtig, S., Krone, N., Sterenberg, K., and Sacristán, M.D. 2002. Transfer von Resistenzen gegen *Leptosphaeria maculans* aus Wilderuciferen in den Raps. *Vortr. Pflanzensuchtg.,* 56:51–62.

Xing, G.M., Long, M.H., Tanaka, S., and Fujieda, K. 1988. Clubroot resistance in *Brassico-raphanus*. *J. Fac. Agric. Kyushu Univ.,* 33:189–194.

Xu, L., Najeeb, U., Tang, G.X., Gu, H.H., Zhang, G.Q., He, Y., and Zhou, W.J. 2007. Haploid and doubled haploid technology. In: *Advances in Botanical Research, Vol. 45: Rapeseed Breeding*, S.K. Gupta (Ed.). Academic Press/Elsevier Ltd., San Diego, CA, p. 181–216.

Yadava, T.P., Singh, H., Gupta, V.P. and Rana, R.K. 1974. Heterosis and combining ability in raya for yield and its component. *Ind. J. Genet.,* 34A:648–695.

Yu, F.Q. and Liu, H.L. 1995. Effects of donor materials and media on microspore embryoid yield of *Brassica napus*. *J. Huazhong Agric. Univ.,* 14:327–332.

Zhang, F. and Takahata, Y. 1999. Microspore mutagenesis and *in vitro* selection for resistance to soft rot disease in Chinese cabbage (*Brassica campestris* L. ssp. pekinensis). *Breed. Sci.,* 49:161–166.

Zhang G.Q., Tang, G.X., Song, W.J., and Zhou, W.J. 2004. Resynthesizing *Brassica napus* from interspecific hybridization between *Brassica rapa* and *B. oleracea* through ovary culture. *Euphytica,* 140:181–187.

Zhang G.Q., Zhang, D.Q. Tang, G.X., He, Y., and Zhou, W.J. 2006b. Plant development from microspore-derived embryos in oilseed rape as affected by chilling, desiccation and cotyledon excision. *Biologia Plantarum,* 50:180–186.

Zhang G.Q., Zhou W.J., Gu H.H., Song W.J., and Momoh E.J.J. 2003. Plant regeneration from the hybridization of *Brassica juncea* and *B. napus* through embryo culture. *J. Agron. Crop Sci.,* 189:347–350.

Zhang, G.Q. and Zhou, W.J. 2006a. Genetic analyses of agronomic and seed quality traits of synthetic oilseed *Brassica napus* produced from interspecific hybridization of *B. campestris* and *B. oleracea*. *J. Genet.,* 85:45–51.

Zhang, G.Q., He, Y., Xu, L., Tang, G.X., and Zhou, W.J. 2006b. Genetic analyses of agronomic and seed quality traits of doubled haploid population in *Brassica napus* through microspore culture. *Euphytica,* 149:169–177.

Zhou, W.J., Hagberg, P., and Tang, G.X. 2002a. Increasing embryogenesis and doubling efficiency by immediate colchicine treatment of isolated microspores in spring *Brassica napus*. *Euphytica,* 128:27–34

Zhou, W.J., Tang, G.X., and Hagberg, P. 2002b. Efficient production of doubled haploid plants by immediate colchicine treatment of isolated microspores in winter *Brassica napus*. *Plant Growth Reg.,* 37:185–192.

Zhu, J., Struss D., and Röbbelen, G. 1993. Studies on resistance to *Phoma lingam* in *Brassica napus-Brassica nigra* addition lines. *Plant Breed.,* 111:192–197.

6 Self-Incompatibility

Hiroyasu Kitashiba and Takeshi Nishio

CONTENTS

INTRODUCTION

Many flowering plants have a self-incompatibility (SI) system to prevent self-fertilization for avoidance of inbreeding depression and for maintenance of genetic variations in populations. Many species of cruciferous plants such as *Brassica rapa* and *Brassica oleracea* also have the SI system. Genetic studies have indicated that this trait is controlled by a single locus, the *S* locus, with multiple alleles (Bateman, 1955). The SI phenotype of the pollen side is determined not gametophytically by the haploid genotype, but sporophytically by the diploid genotype of the parent plant. Co-dominance and/or dominance relationships between *S* alleles are observed on both the pollen and stigma sides and influence the SI phenotype (Thompson and Taylor, 1966). That is, when an *S* phenotype of pollen is identical to that of a stigma, pollen grains are rejected by the inhibition of pollen germination and/or pollen tube penetration into papillar cells on the stigma. Three genes located at the *S* locus — namely, *S*-receptor kinase (SRK) as a female determinant, *S*-locus protein 11/ *S*-locus cysteine-rich protein (*SP11/SCR*) as a male determinant, and *S*-locus glycoprotein (SLG) highly similar to the extracellular domain (*S*-domain) of SRK — have been characterized (Nasrallah et al., 1988; Stein et al., 1991; Schopfer et al., 1999; Suzuki et al., 1999). SRK, *SP11/SCR,* and SLG are closely linked to each other at the *S*-locus (Figure 6.1), and the alleles of these three genes are inherited by progeny as one set. Therefore, a set of alleles of these genes is termed *S* haplotype (Nasrallah and Nasrallah, 1993). Due to dominance relationships between *S* haplotypes, the *S* phenotypes in the respective pollen and stigma are determined by a dominant *S* haplotype (Thompson and Taylor, 1966; Hatakeyama et al., 1998a). This chapter reviews the molecular genetics of self-incompatibility in *Brassicaceae* species.

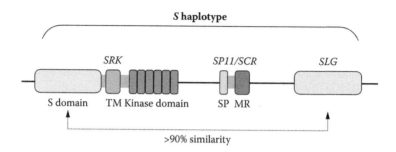

FIGURE 6.1   (See color insert following page xxx.) Linkage of SRK, SLG, and *SP11/SCR* at the *S* locus. (TM = transmembrane domain; SP = signal peptide; MR = region of mature protein.)

GENES CONTROLLING SELF-INCOMPATIBILITY

SLG AND SRK

In *Brassica* species, molecular analysis of self-incompatibility (SI) started with the finding of *S*-specific glycoproteins in the stigma (Nasrallah and Wallence, 1967; Nishio and Hinata, 1977), which was later designated the *S* locus glycoprotein (SLG). SLG proteins of respective SLG alleles have different pI values and co-segregate with *S* haplotypes (Hinata and Nishio, 1978; Nou et al., 1993). In addition, SLG proteins are abundant in the mature papilla cell wall, where SI reaction occurs (Nasrallah et al., 1988). Amino acid sequences of SLG proteins have been determined (Takayama, 1987) and, to date, many alleles of SLG in *Brassicaceae* species such as *Brassica rapa*, *B. oleracea*, and *Raphanus sativus* have been characterized (Nasrallah et al., 1987, 1988; Trick and Flavell, 1989; Chen and Nasrallah, 1990; Kusaba et al., 1997; Sakamoto et al., 1998). SLG has a hydrophobic signal peptide at the N-terminus for secretion to the outside of cells, several *N*-glycosylation sites, three hypervariable regions, and twelve conserved cysteine residues.

The study of SLG led to the identification of the second *S*-linked gene encoding for *S* receptor kinase (SRK), which has an extracellular domain (*S*-domain) highly similar to SLG, a transmembrane domain, and a seine/threonine kinase domain toward the C-terminus (Stein et al., 1991). Transcripts of the SRK gene accumulate in papillar cells before flower opening, coinciding with the timing of SI acquisition in the stigma as in the case of SLG (Watanabe et al., 1994; Delorme et al., 1995). SLG and the *S*-domain of SRK of the same *S* haplotype exhibit high similarity, more than 90% (Stein et al., 1991; Watanabe et al., 1994; Hatakeyama et al., 1998b).

ROLE OF SLG AND SRK AS A FEMALE DETERMINANT

Gain-of-function studies have suggested the involvement of SLGs and SRKs in the SI response (Takasaki et al., 1999, 2000; Silva et al., 2001). Expression of an SRK-9 allele transgene in *Brassica rapa* plants conferred the ability to reject pollen grains of *S-9* homozygotes, whereas SLG-9 did not, suggesting that SRK is the sole determinant of the *S*-haplotype specificity of the stigma (Takasaki et al., 2000). A transgenic *B. rapa* plant harboring the SRK-9 transgene along with the SLG-9 transgene produced fewer seeds by *S-9* pollen than a transgenic plant expressing the SRK-9 transgene alone, indicating that SLG acts to enhance the SI response (Takasaki et al., 2000). In contrast, another gain-of-function study revealed no evidence for the enhancing role of SLG for the SI reaction, although the role of SRK was confirmed (Silva et al., 2001). *Brassica S* haplotypes and *Arabidopsis lyrata* plants that harbor defects in or lack the SLG gene exhibit strong self-incompatibility (Kusaba et al., 2001; Sato et al., 2002; Suzuki et al., 2000a). Furthermore, an SRK allele having the sequence of SLG, which may be generated by gene conversion, has no function of self-pollen rejection, indicating that SLG cannot recognize self-pollen (Fujimoto et al., 2006a). The SI enhancing role of SLG might be limited to some *S* haplotypes having extremely high similarity between SLG and SRK, for example, *S-29* in *B. rapa*.

SP11/SCR AS A MALE DETERMINANT

In 1999, two groups — Suzuki et al. (1999) and Schopfer et al. (1999) — succeeded in isolating the male determinant gene named *SP11* (*S*-locus protein 11) or *SCR* (*S*-locus cysteine rich) of *Brassica rapa* by cloning and sequencing the *S*-locus region. *SP11/SCR* encodes a basic cysteine-rich protein. Eight cysteine residues (C1 to C8), a glycine residue, and an aromatic amino acid residue, located between C1 and C2 and between C3 and C4, respectively, are conserved among SP11/SCR proteins encoded by different alleles (Schopfer et al., 1999; Takayama et al., 2000; Watanabe et al., 2000). The locus of *SP11/SCR* is adjacent to those of SLG and SRK within the *S* locus (Suzuki et al., 1999), and the sequences of mature proteins exhibit high polymorphism between alleles within species (below 50% at amino acid identity), although the signal peptide sequence is highly conserved between alleles. The transcript is specifically detected in the anthers, especially in tapetum cells and microspores (Suzuki et al., 1999; Schopfer et al., 1999; Takayama et al., 2000; Shiba et al., 2002).

Takayama et al. (2001) and Kachroo et al. (2001) have revealed that SP11/SCR interacts directly with its self-SRK and that the binding induces autophosphorylation of SRK in an *S*-haplotype-specific manner. The membrane-anchored form of SRK exhibits high-affinity binding to SP11/SCR, whereas eSRK, which is another form of SRK protein lacking the transmembrane domain and the kinase domain, does not (Shimosato et al., 2007). Although the extracellular domain of SRK interacts with SP11/SCR, SRK tends to form a dimer (or oligomer) *in vivo* in the absence of the ligand (Giranton et al., 2000; Shimosato et al., 2007; Naithani et al., 2007). SRK interacts more strongly with an identical SRK molecule derived from the same SRK allele than with a variant SRK derived from another SRK allele (Naithani et al., 2007). These results suggest that a homodimer of SRK tends to predominantly bind an SP11/SCR protein from the same *S* haplotype. Furthermore, the stigmas reject pollen grains from transgenic plants expressing the *SP11/SCR* alleles of the same *S* haplotype as the stigmas (Schopfer et al., 1999; Shiba et al., 2001). Thus, *SP11/SCR* is considered the male determinant.

DOMINANCE RELATIONSHIPS BETWEEN S HAPLOTYPES

On the basis of nucleotide sequences, *S* haplotypes have been classified into two groups: class I and class II (Nasrallah et al., 1991). The sequence similarities among SLGs or SRKs within each class are about 80% to 90%, whereas those between classes are only about 65%. Class I *S* haplotypes are generally dominant over class II *S* haplotypes on the pollen side (Hatakeyama et al., 1998a, c). In class I *S* haplotypes, nonlinear dominance relationships among *SP11/SCR* alleles have been observed, whereas in class II *S* haplotypes, linear dominance relationships have been reported (Thompson and Taylor, 1966; Kakizaki et al., 2003). *In situ* hybridization of developing anthers in *Brassica rapa* has suggested that a dominant class I *S* haplotype suppresses transcription of a class II *SP11/SCR* gene at the RNA level (Shiba et al., 2002). Between four class-II *S* haplotypes, similar control of gene expression has been observed at the RNA level (Kakizaki et al., 2003). Bisulfite sequencing analysis has suggested that transcriptional suppression of recessive *SP11/SCR* alleles is caused epigenetically by a *de novo* methylation of 5′ promoter sequences in tapetum cells before the initiation of transcription of the recessive *SP11/SCR* alleles (Shiba et al., 2006). Interestingly, the suppression of recessive *SP11/SCR* alleles requires no expression of dominant *SP11/SCR* alleles (Fujimoto et al., 2006b; Okamoto et al., 2007). However, the precise epigenetic mechanism of interaction between dominant and recessive *SP11/SCR* alleles is still unclear, as is the regulation of dominance or co-dominance relationships between class I *S* haplotypes.

On the stigma side, the dominance relationships between *S* haplotypes have been inferred to be determined by SRK protein itself, not by the difference of their relative expression levels (Hatakeyama et al., 2001). Although the kinase domain of BrSRK-54 is highly similar to those of BrSRK-8 and BrSRK-46, different dominance relationships are observed; that is, *BrS-54* is co-dominant and recessive to *BrS-8* and *BrS-46*, respectively (Takuno et al., 2007; Sugimura et al., unpublished). These results suggest the possibility that amino acid sequences of the *S* domain play a significant role in the dominance relationships of SRK. Recently, yeast two-hybrid assays between eSRKs have revealed a preference for SRK homodimerization, suggesting that its strong affinity might cause co-dominance relationships, which are exhibited by a majority of the SRK alleles (Naithani et al., 2007). This result also supports the important role of the *S* domain of SRK. However, to address the mechanism resulting in the dominance relationships in the stigma, further genetic and biochemical studies would be necessary.

OTHER GENES FOR SI REACTION

Many classical and molecular genetic studies have clarified a recognition mechanism of sporophytic SI, beginning with interaction between SP11/SCR as a male factor and SRK as a female factor, finally resulting in inhibition of germination or tube growth of self-pollen grains on the stigma. Since the late 1990s, explorations of genes coding for proteins interacting with SRK have been carried out to reveal the signaling pathway to rejection of self-pollen, and three genes participating in the pathway but not linked to the *S* locus have been reported.

ARC1 (Arm Repeat Containing 1) was isolated by yeast two-hybrid screen for SRK kinase domain-interacting proteins (Gu et al., 1998). Suppression of *ARC1* at the RNA level in a self-incompatible *Brassica napus* line conferred a partial breakdown of SI, resulting in seed production after self-pollination (Stone et al., 1999). This result suggests that ARC1 is one of the positive effectors of the *Brassica* SI response. Furthermore, ARC1 has been inferred to promote the ubiquitination and subsequent proteasomal degradation of compatibility factors in the stigma by E3 ubiquitin ligase activity of ARC1, causing pollen rejection (Stone et al., 2003).

THL1 and *THL2* encoding thioredoxin *h* proteins have also been identified as genes for SRK kinase domain-interacting proteins (Bower et al., 1996). *In vivo* autophosphorylation of SRK is prevented by THL1 in the absence of a ligand, but this inhibition is released in the presence of SP11/SCR in an *S*-haplotype-specific manner (Cabrillac et al., 2001). This result indicates that THL1 plays an important role in SI signaling as a negative regulator of SRK in the absence of the pollen ligand. Furthermore, transgenic *Brassica napus* cv. Westar plants expressing antisense *THL1/2* transgenes showed a decline in *THL1/2* mRNA levels along with a reduction in seed production by compatible pollen (Haffani et al., 2004).

Hinata and Okazaki (1986) described a self-compatible *B. rapa* variety, Yellow sarson, cultivated in India. Classical genetic analysis indicated that this self-compatible trait is regulated by two independent loci: *S* and *M* (Hinata and Okazaki, 1986). The identified mutation in SRK is insertion of a retrotransposon-like sequence and that in *SP11/SCR* is deletion in the promoter region (Fujimoto et al., 2006b). On the other hand, the *M* gene identified by positional cloning has been characterized as being a gene encoding a membrane-anchored cytoplasmic protein kinase, which is designated *M* locus protein kinase (MLPK) (Murase et al., 2004). The kinase region of the mutated MLPK has a single base nonsynonymous substitution, resulting in no autophosphorylation activity and an absence of detectable MLPK protein in the mutant (Murase et al., 2004). More recently, two different *MLPK* transcripts from the same allele by alternative splicing were identified, both proteins being found to interact directly with SRK (Kakita et al., 2007). Because the both isoforms can complement mutation caused by the *mlpk/mlpk* genotype, resulting in self-incompatibility, MLPK is also one of the factors involved in the signaling pathway of SI reaction downstream from SRK.

POLYMORPHISM OF *S* HAPLOTYPES IN *BRASSICACEAE*

INTERSPECIFIC PAIRS OF *S* HAPLOTYPES BETWEEN *B. RAPA* AND *B. OLERACEA*

The number of *S* haplotypes in *Brassica oleracea* is deemed to be 50 (Ockendon, 2000) and estimated to be greater than 100 in *B. rapa* (Nou et al., 1993). Sequence analysis of more than 30 class-I

SLG alleles from *B. rapa* and *B. oleracea* has revealed a high degree of intraspecific variation and high interspecific similarity (Kusaba et al., 1997); and subsequently, similar results in SRK alleles and *SP11/SCR* alleles were also indicated (Sato et al., 2002). In *Raphanus,* which has SI and is closely related to *B. rapa* and *B. oleracea,* about ten alleles of respective SLG, SRK, and *SP11/SCR* have been reported (Sakamoto et al., 1998; Okamoto et al., 2004). Even between different genera (i.e., *Raphanus* and *Brassica*) several interspecific *S* haplotype pairs with high similarity between the SRK alleles and the *SP11/SCR* alleles have been found (Okamoto et al., 2004). These results suggest that the interspecific pair of *S* haplotypes with high similarity in SI-related genes might have originated from a common ancestral *S* haplotype.

An amphidiploid plant produced by interspecific hybridization between a homozygote of the class-I *S* haplotype of *Brassica rapa* and that of the class-II *S* haplotype of *B. oleracea* rejected pollen grains of an *S* homozygote, whose *S* haplotype is an interspecific pair with the *B. rapa S* haplotype of the amphidiploid (Kimura et al., 2002). In addition, experiments of plant transformation and bioassay have also revealed the same recognition specificity between *S* haplotypes in the interspecific pairs (Sato et al., 2003). These results demonstrate that *S* haplotypes of interspecific pairs have maintained the same recognition specificity as their ancestral *S* haplotype.

By analysis of sequence polymorphism in some interspecific pairs of *S* haplotypes, a key motif for recognition specificity in the SI reaction was inferred. Six regions have been assigned to SP11/SCR (Regions I to VI) on the basis of conserved cysteine residues. Domain swapping experiments have revealed that region III and V of SP11/SCR are necessary to confer the recognition of self-SRK, followed by SI reaction, and are key sequences that maintain the recognition specificity between the interspecific pairs (Sato et al., 2004).

In class II *S* haplotypes, three pairs with highly similar SRK and *SP11/SCR* alleles (*BoS-2b/BrS-44, BoS-5/BrS-40,* and *BoS-15/BrS-60*) have been found (Sato et al., 2006). In these pairs, the recognition specificities between *BoS-2b* and *BoS-5* are slightly and completely different from *BrS-44* and *BrS-40*, respectively, possibly because of several amino acid substitutions within region V of SP11/SCR (Sato et al., 2006). The significance of region V has also been supported by Chookajorn et al. (2004), who used a domain swapping experiment. Based on these studies concerning the interspecific pairs of *S* haplotypes, further identification of the recognition site of SP11/SCR and SRK can be advanced in the future.

GENOMIC STRUCTURE OF THE *S* LOCUS

In the early 1990s, genomic structural analysis of the *S*-locus began with an estimation of the physical distance between SLG and SRK at the *S* locus of *Brassica oleracea* using the PFGE (pulse field gel electrophoresis) technique (Boyes and Nasrallah, 1993). Subsequently, sequencing of a 76-kb fragment of the *S* locus region containing SLG, SRK, and *SP11/SCR* of *S-9* haplotype was carried out (Suzuki et al., 1999). In the flanking regions of the *S* core region containing the three SI-related genes (i.e., SRK, *SP11/SCR,* and SLG), several genes of unknown function, such as *SP6* (*S*-locus protein 6) and *SLL2* (*S*-locus-linked gene 2), were found (Suzuki et al., 1999). Structural and transcriptional comparative analysis of *S* haplotypes in *B. rapa* and *B. napus* (Cui et al., 1999; Shiba et al., 2003) has suggested that the region outside the *SP6* and *SLL2* genes shares high synteny between different *S* haplotypes in the genomic structure, whereas the region between the two genes is highly polymorphic and rich with *S*-haplotype-specific intergenic sequences. This diversity might contribute to the suppression of recombination at the *S* locus between *S* haplotypes. In class-II *S* haplotypes, the order of *SP11/SCR,* SRK, and SLG has been found to be the reverse of that in the class-I *S* haplotype (Fukai et al., 2003). Comparative analysis has revealed that the direction of transcription of the SI-related genes of four class-II *S* haplotypes in *B. rapa* is completely conserved and that the region between SRK and *SP11/SCR* shares high synteny (Kakizaki et al., 2006). Recent study has indicated that SI-related genes disfavor recombination, which results in the breakdown of SI, but other genes not involved in SI in the *S* locus complex as well as the kinase domain of SRK have experienced recombination between *S* haplotypes (Takuno et al., 2007).

Comparison of the genomic structure of the *S* locus between *Brassica rapa* and *B. oleracea* has been carried out. PFGE analysis has revealed that the physical size of the *S* locus complex in *B. rapa* is significantly shorter and less variable among *S* haplotypes than that in *B. oleracea* (Boyes and Nasrallah, 1993; Suzuki et al., 2000b). However, it is still unknown whether these differences are due to the difference between species or to variation among different *S* haplotypes in a species. Comparative genomic structural analysis by PFGE (Kimura et al., 2002) and by nucleotide sequence analysis (Fujimoto et al., 2006c) between the interspecific pairs of *S* haplotypes has also indicated that the distance between SRK and *SP11/SCR* in *B. oleracea* is larger than that in *B. rapa* and has revealed several retrotransposon-like sequences in the *S* locus complex (Fujimoto et al., 2006c). These results suggest that several insertions into the *S* locus region of *B. oleracea* might have occurred independently in respective *S* haplotypes. However, the possibility that extension in the size of the *S* locus in *B. oleracea* occurred by chance only in some limited *S* haplotypes remains.

IDENTIFICATION OF *S* HAPLOTYPES

Discrimination of *S* haplotypes is necessary not only for the study of self-incompatibility, but also for seed production of F_1 hybrid cultivars in cruciferous vegetables such as cabbage, Chinese cabbage, broccoli, cauliflower, and radish. The standard method for *S* haplotype identification is pollination with tester lines with defined *S* haplotypes, followed by observation of pollen tube growth by fluorescent microscopy or seed set. However, this work requires significant time, labor, and maintenance of *S* tester lines. Identification of SLG protein has enabled development of a method using isoelectric focusing analysis of stigma proteins and detection of SLG with concanavalin A or SLG antibody (Hinata and Nishio, 1981; Nou et al., 1991; Okazaki et al., 1999). Sequence polymorphism of the SLG alleles has been revealed, and such information has led to the development of a PCR-RFLP technique (Brace et al., 1993; Nishio et al., 1994, 1996) that is widely used for discrimination of *S* genotypes (Sakamoto et al., 2000; Lim et al., 2002). Both the standard method as well as electrophoretic analysis of SLG protein requires flowers, whereas a leaf of a seedling or a single seed can be used as materials for DNA polymorphism analysis of the SLG alleles (Sakamoto et al., 2000). In addition to SLG, polymorphism of the kinase domain of the SRK gene can also be utilized for discrimination of *S* haplotypes lacking SLG, such as *BrS-32*, *BrS-33*, and *BrS-36* (Nishio et al., 1997; Suzuki et al., 2003). Progress in accuracy of *S* genotyping by these PCR-RFLP methods has contributed to efficient production of F_1 hybrid seeds using the SI system.

Because nucleotide sequence polymorphism of the *SP11/SCR* alleles (about 40 %) is higher than that of the SRK and SLG alleles (about 10%), an alternative method using dot-blot analysis of the *SP11/SCR* alleles to identify the *S* haplotypes has been examined (Fujimoto and Nishio, 2003). This method consists of two techniques: (1) dot-blotting of plant genomic DNA as samples probed with labeled *SP11m* (mature protein region of *SP11/SCR* cDNA), and (2) dot-blotting of *SP11m* DNA fragments amplified from each allele hybridized with the *SP11/SCR* coding region labeled by PCR using a template of plant genomic DNA. The former technique is considered helpful for the screening of desirable *S* haplotypes, and the latter technique for *S* genotyping of plants for which there is no information on the *S* haplotypes. Discrimination of *S* genotypes is possible by PCR-RFLP analysis of SLG or SRK alleles; but for identification of *S* haplotypes, nucleotide sequencing is necessary because of the complicated banding patterns, especially in the case of *S* heterozygotes (Sakamoto et al., 2000). The improved dot-blot technique (Fujimoto and Nishio, 2003) has overcome this problem and enabled us to identify *S* haplotypes simply and distinctly. Although there are still many *S* haplotypes whose *SP11/SCRs* have not been identified, in the near future, this dot-blot method should prove to be a powerful tool for many researchers and breeders.

EVOLUTION OF *S* HAPLOTYPE IN *BRASSICACEAE*

S HAPLOTYPES IN GENUS *ARABIDOPSIS*

Many studies on self-incompatibility have been carried out in the genus *Arabidopsis. A. lyrata* and *A. halleri* are outcrossing (self-incompatible) species, whereas *A. thaliana* has an inbreeding (self-compatibility) trait. At the *S* locus of *A. lyrata* and *A. halleri*, several *S* haplotypes with SRK and *SPI1/SCR* orthologs have been identified (Schierup et al., 2001; Bechsgaard et al., 2006). As described previously for *Brassica*, the *S* haplotypes of *A. lyrata* differ in the order and orientation of the genes as well as in their physical size (Kusaba et al., 2001). Dominance relationships at the pollen and stigma sides have been observed between *S* haplotypes (Schierup et al., 2001). Comparison of *S*-locus maps between *A. lyrata* and *Brassica* indicates that the *S* locus of crucifers has undergone several duplication events since separation of *Brassica* from *Arabidopsis* (Kusaba et al., 2001). On the other hand, *A. thaliana* has nonfunctional alleles of the SRK (ΨSRK) and *SCR* (Ψ*SCR*) genes (Kusaba et al., 2001), and their sequences from respective *A. thaliana* accessions show polymorphisms (Nasrallah et al., 2004). Transformation with just two transgenes, which are the *SRK-b* and *SCR-b* genes isolated from the *S-b* haplotype of *A. lyrata* (Kusaba et al., 2001, 2002), conferred a self-incompatible phenotype on several ecotypes in *A. thaliana* (Nasrallah et al., 2002) in the short period from mature bud to young flower in floral development (flower stage 13 and early stage 14 as referred to in Smyth et al., 1990). This reverted phenotype indicates that the self-compatibility of *A. thaliana* is caused by defects in both the *SRK* and *SCR* genes. However, in one ecotype, C24, the transformant recovered a stable self-incompatibility phenotype almost identical to that of naturally self-incompatible *Arabidopsis* species (Nasrallah et al., 2004), implying that the signal transduction pathway downstream from SRK is variable, depending on ecotype. Recent studies on the difference of SI reaction among transformants of these ecotypes identified a candidate factor, PUB8, for modification of the degree of SI reaction (Liu et al., 2007). This *PUB8* gene is tightly linked to the *S* locus and encodes an uncharacterized ARM repeat and U box-containing protein that regulates *SRK* transcript levels (Liu et al., 2007).

EVOLUTIONAL DIVERSITY OF *S* HAPLOTYPES IN *BRASSICACEAE*

Gene diversity analysis in *Arabidopsis lyrata* has revealed that nucleotide diversity in the *S* domain of *SRK* alleles is much higher in *A. lyrata* than in *Brassica* species (Schierup et al., 2001). This implies that divergence of *S* haplotypes in *Brassica* has occurred more recently than that in *Arabidopsis*. Ancestral *S* haplotypes of *Arabidopsis* and *Brassica* may have been present in their ancestral species and are thought to have been carried over to the present *Arabidopsis* species after separation of both genera (Figure 6.2). In *Brassica*, it is inferred that a small number of *S* haplotypes derived from the ancestral species have been carried over and that a bottleneck effect may have resulted in a remarkable decrease in *S* haplotypes, followed by rediversification of *S* haplotypes (Schierup et al., 2001; Castric and Vekemans, 2007).

More recently, using sequence information of the *S* locus containing *SRK* and *SCR* in *Arabidopsis lyrata* and *A. halleri,* together with the respective pseudogenes (Ψ*SCR*, Ψ*SRK*) from *A. thaliana,* the coalescent time of diversity in *S* haplotype has been estimated by genetic analysis such as nucleotide diversity analysis and linkage disequilibrium analysis (Charlesworth and Vekemans, 2005; Bechsgaard et al., 2006; Tang et al., 2007). According to these studies, the coalescent time of the speciation event between *A. lyrata* and *A. halleri* is estimated at 2 million years ago, and that between the ancestor of those two species and *A. thaliana* at 5 million years ago (Charlesworth et al., 2005). In addition, the time of transition to inbreeding in *A. thaliana* has been estimated at about 1 to 3 million years ago by several researchers (Charlesworth, 2005; Bechsgaard, 2006; Tang et al., 2007). Future genetic and evolutionary studies should yield more precise knowledge, such as when

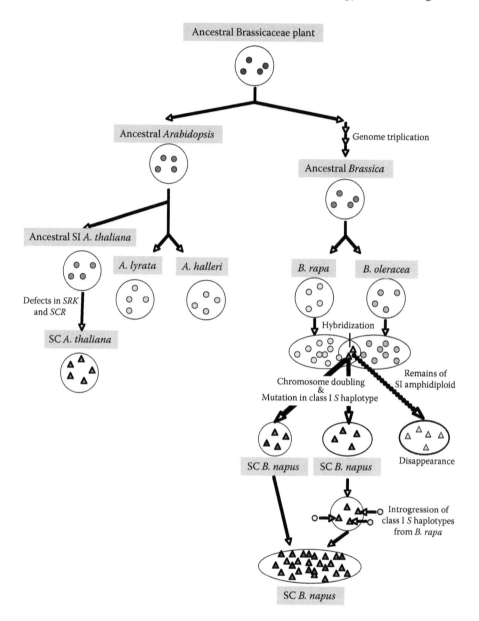

FIGURE 6.2   (See color insert following page 128.) A model of evolution of *S* haplotypes and establishment of SC species in *Arabidopsis* and *Brassica*. (SC = self-compatibility; SI = self-incompatibility.)

S haplotypes arose in *Brassicaceae*, as well as the increase and decrease in *S* haplotypes during speciation in *Brassicaceae*.

SELF-COMPATIBLE AMPHIDIPLOID SPECIES IN *BRASSICA* AND *ARABIDOPSIS*

Most monogenomic *Brassica* species are self-incompatible, whereas digenomic species in this genus (i.e., amphidiploid species with two different genomes) exhibit self-compatibility (SC) (Takahata and Hinata, 1980). *Brassica napus*, *B. juncea*, and *B. carinata* are amphidiploid species having AC, AB, and BC genomes, in which A, B, and C are the genomes of *B. rapa*, *B. nigra*, and *B. oleracea*, respectively (U, 1935). However, artificially synthesized amphidiploids with the AC genome are self-incompatible (Hinata and Nishio, 1980). This discrepancy between natural amphidiploid species and artificial amphidiploids has remained a great question for a long time.

Analyzing *S* haplotypes in 45 lines of *Brassica napus*, six *S* genotypes, mainly combinations of a class I dominant *S* haplotype and a class II recessive one, have been identified (Okamoto et al., 2007). Self-compatibility of *B. napus* cv. Westar has been shown to be caused by suppression of transcription from a recessive *SP11/SCR* allele by a dominant nonfunctional *SP11/SCR* allele (Okamoto et al., 2007), as observed in monogenomic *Brassica* species (Fujimoto et al., 2006b). In two other genotypes, SRK alleles of dominant *S* haplotypes have been revealed to have knockout mutations. Because generation of an interspecific hybrid plant between *B. rapa* and *B. oleracea* followed by chromosome doubling of the hybrid is a rare event due to reproductive barriers, original hybrid plants are considered to have emerged by chance as a single plant at different times or in different populations. A knockout mutation in a dominant *S* haplotype can lead to the generation of an SC plant by a single mutation event, and is therefore likely to have contributed to the development of a new amphidiploid species (i.e., *B. napus*) (Figure 6.2). The presence of several *S* haplotypes in *B. napus* may suggest independent generation of the original *B. napus* plants. Another possibility is introgression of several *S* haplotypes from *B. rapa* into *B. napus* after establishment of the amphidiploid species, because interspecific hybrids can be easily crossed with *B. rapa*. This possibility is supported by the presence of only one *S* haplotype derived from *B. oleracea* in all surveyed *S* genotypes. Independent interspecific hybridization, introgression, or both may have contributed to the evolutional development of many varieties of SC *B. napus* (Figure 6.2).

Arabidopsis suecica, which is an amphidiploid species derived from spontaneous hybridization between *A. thaliana*, a self-compatible species, and *A. arenosa*, a self-incompatible species, is self-compatible (Mummenhoff and Hurka, 1995; O'Kane et al., 1996). In artificial interspecific *Arabidopsis* hybrids and artificial amphidiploids (allotetraploids) between self-compatible plants and self-incompatible plants, a breakdown of self-incompatibility is observed. *A. thaliana-lyrata* hybrids obtained by a cross between self-compatible *A. thaliana* as female and self-incompatible *A. lyrata* as male exhibit compatibility to paternal pollen from *A. lyrata*. Artificial amphidiploids generated from *A. thaliana-lyrata* hybrids also showed compatibility to the pollen from *A. lyrata* as well as self-compatibility (Nasrallah et al., 2007). In the *A. thaliana-lyrata* hybrids and amphidiploids, suppression of *SRK* transcripts from *A. lyrata* was observed, whereas backcrossing of the hybrids to *A. lyrata* restored the normal expression of the SRK gene along with the stigma SI response (Nasrallah et al., 2007). Therefore, it has been inferred that self-compatibility in the hybrid and in the amphidiploids might have resulted from epigenetic changes in expression of a functional *SRK* allele from *A. lyrata*.

Knowledge about SI in *Brassicaceae* has been accumulated primarily from vegetable crops such as tribe Brassiceae, including genus *Brassica* and *Raphanus*. Recent molecular genetic analysis in the genus *Arabidopsis* has advanced remarkably and contributed particularly to the understanding of evolution of SI, such as the origin and dynamics of SI-related genes. However, the family *Brassicaceae* has many genera (approximately 350), and there has been little progress in the study of SI in these genera other than *Brassica* and *Arabidopsis*. Many questions on SI have not yet been clarified. For example, when did SLG, which is harbored generally in the genus *Brassica* but not in the genus *Arabidopsis*, break out? What is the significance of the existence of SLG? How many *S* haplotypes are there in respective species, and how did the SC species occur? In addition, the mechanism of dominance relationships between *S* haplotypes on both the pollen and the stigma sides has not been sufficiently elucidated. Given the above, it is likely that genetic resources and collection of *S* haplotypes will become more important for these studies, surely providing us with new insights into SI in *Brassicaceae*.

REFERENCES

Bateman, A.J. 1955. Self-incompatibility systems in angiosperms. III. Cruciferae. *Heredity,* 9:52–68.

Bechsgaard, J.S., V. Castric, D. Charlesworth, X. Vekemans, and M.H. Schierup. 2006. The transition to self-compatibility in *Arabidopsis thaliana* and evolution within *S*-haplotypes over 10 Myr. *Mol. Biol. Evol.,* 23:1741–1750.

Bower, M.S., D.D. Matias, E. Fernandes-Carvalho, M. Mazzurco, T. Gu, S.J. Rothstein, and D.R. Goring. 1996. Two members of the thioredoxin-h family interact with the kinase domain of a *Brassica S* locus receptor kinase. *Plant Cell,* 8:1641–1650.

Boyes, D.C. and J.B. Nasrallah. 1993. Physical linkage of the SLG and SRK genes at the self-incompatibility locus of *Brassica oleracea. Mol. Gen. Genet.,* 236:369–373.

Brace, J., D.J. Ockendon, and G.J. King. 1993. Development of a method for the identification of *S* alleles in *Brassica oleracea* based on digestion of PCR-amplified DNA with restriction endonucleases. *Sex. Plant Reprod.,* 6:133–138.

Cabrillac, D., J.M. Cock, C. Dumas, and T. Gaude. 2001. The *S*-locus receptor kinase is inhibited by thioredoxins and activated by pollen coat proteins. *Nature,* 410:220–223.

Castric, V. and X. Vekemans. 2007. Evolution under strong balancing selection: how many codons determine specificity at the female self-incompatibility gene SRK in *Brassicaceae? BMC Evol. Biol.,* 7:132.

Charlesworth, D. and X. Vekemans. 2005. How and when did *Arabidopsis thaliana* become highly self-fertilising? *BioEssays,* 27:472–476.

Chen, C.H. and J. B. Nasrallah. 1990. A new class of *S* sequences defined by a pollen recessive self-incompatibility allele of *Brassica oleracea. Mol. Gen. Genet.,* 222:241–248.

Chookajorn, T., A. Kachroo, D.R. Ripoll, A.G. Clark, and J.B. Nasrallah. 2004. Specificity determinants and diversification of the *Brassica* self-incompatibility pollen ligand. *Proc. Natl. Acad. Sci. U.S.A.,* 101:911–917.

Cui, Y., N. Brugière, L. Jackman, Y.-M. Bi, and S.J. Rothstein. 1999. Structural and transcriptional comparative analysis of the *S* locus regions in two self-incompatible *Brassica napus* lines. *Plant Cell,* 11:2217–2231.

Delorme, V., J.-L. Giranton, Y. Hetzfeld, A. Friry, P. Heizmann, M.J. Ariza, C. Dumas, T. Gaude, and J.M. Cock. 1995. Characterization of the *S* locus genes, SLG and SRK, of the *Brassica S_3* haplotype: identification of a membrane-localized protein encoded by the *S* locus receptor kinase gene. *Plant J.,* 7:429–440.

Fujimoto, R. and T. Nishio. 2003. Identification of *S* haplotypes in *Brassica* by dot-blot analysis of *SP11* alleles. *Theor. Appl. Genet.,* 106:1433–1437.

Fujimoto, R., T. Sugimura, and T. Nishio. 2006a. Gene conversion from SLG to SRK resulting in self-compatibility in *Brassica rapa. FEBS Lett.,* 580:425–430.

Fujimoto, R., T. Sugimura, E. Fukai, and T. Nishio. 2006b. Suppression of gene expression of a recessive *SP11/SCR* allele by an untranscribed *SP11/SCR* allele in *Brassica* self-incompatibility. *Plant Mol. Biol.,* 61:577–587.

Fujimoto, R., K. Okazaki, E. Fukai, M. Kusaba, and T. Nishio. 2006c. Comparison of the genome structure of the self-incompatibility (*S*) locus in interspecific pairs of *S* haplotypes. *Genetics,* 173:1157–1167.

Fukai, E., R. Fujimoto, and T. Nishio. 2003. Genomic organization of the *S* core region and the *S* flanking regions of a class-II *S* haplotype in *Brassica rapa. Mol. Gen. Genom.,* 269:361–369.

Giranton, J.-L., D. Dumas, J.M. Cock, and T. Gaude. 2000. The integral membrane *S*-locus receptor kinase of *Brassica* has serine/threonine kinase activity in a membranous environment and spontaneously forms oligomers *in planta. Proc. Natl. Acad. Sci. U.S.A.,* 97:3759–3764.

Gu, T., M. Mazzurco, W. Sulaman, D.D. Matias, and D.R. Goring. 1998. Binding of an arm repeat protein to the kinase domain of the *S*-locus receptor kinase. *Proc. Natl. Acad. Sci . U.S.A.,* 95:382–387.

Haffani, Y.Z., T. Gaude, J.M. Cock, and D.R. Goring. 2004. Antisense suppression of thioredoxin h mRNA in *Brassica napus* cv. Westar pistils causes a low level constitutive pollen rejection response. *Plant Mol. Biol.,* 55:619–630.

Hatakeyama, K., M. Watanabe, T. Takasaki, K. Ojima, and K. Hinata. 1998a. Dominance relationships between *S*-allele in self-incompatible *Brassica campestris* L. *Heredity,* 80:241–247.

Hatakeyama, K., T. Takasaki, M. Watanabe, and K. Hinata. 1998b. High sequence similarity between SLG and the receptor domain of SRK is not necessarily involved in higher dominance relationships in stigma in self-incompatible *Brassica rapa* L. *Sex. Plant Reprod.,* 11:292–294.

Hatakeyama, K., T. Takasaki, M. Watanabe, and K. Hinata. 1998c. Molecular characterization of *S* locus genes, SLG and SRK, in a pollen-recessive self-incompatibility haplotype of *Brassica rapa* L. *Genetics,* 149:1587–1597.

Hatakeyama, K., T. Takasaki, G. Suzuki, T. Nishio, M. Watanabe, A. Isogai, and K. Hinata. 2001. The *S* receptor kinase gene determines dominance relationships in stigma expression of self-incompatibility in *Brassica*. *Plant J.,* 26:69–76.

Hinata, K. and T. Nishio. 1978. *S*-allele specificity of stigma proteins in *Brassica oleracea* and *B. campestris*. *Heredity,* 41:93–100.

Hinata, K. and T. Nishio. 1980. Self-incompatibility in crucifers. In *Brassica Crops and Wild Allies,* Eds. S. Tunoda, K. Hinata, and C. Gomez-Campo, C. p. 223–234. Tokyo: Japan Scientific Societies Press.

Hinata, K. and T. Nishio. 1981. Con A-peroxidase method: an improved procedure for staining *S*-glycoproteins in cellulose-acetate electrofocusing in crucifers. *Theor. Appl. Genet.,* 60:281–283.

Hinata, K. and K. Okazaki. 1986. Role of stigma in the expression of self-incompatibility in crucifers in view of genetic analysis. In *Biotechnology and Ecology of Pollen,* Eds. D.L. Mulcahy, G.B. Mulcahy, and E. Ottaviano, p. 185–190. New York: Springer-Verlag.

Kachroo, A., C.R. Schopfer, M.E. Nasrallah, and J.B. Nasrallah. 2001. Allele-specific receptor-ligand interactions in *Brassica* self-incompatibility. *Science,* 293:1824–1826.

Kakita, M., K. Murase, M. Iwano, T. Matsumoto, M. Watanabe, H. Shiba, A. Isogai, and S. Takayama. 2007. Two distinct forms of *M*-locus protein kinase localize to the plasma membrane and interact directly with *S*-locus receptor kinase to transduce self-incompatibility signaling in *Brassica rapa*. *Plant Cell,* 19:3961–3973.

Kakizaki, T., Y. Takada, A. Ito, G. Suzuki, H. Shiba, S. Takayama, A. Isogai, and M. Watanabe. 2003. Linear dominance relationship among four class-II *S* haplotypes in pollen is determined by the expression of *SP11* in *Brassica* self-incompatibility. *Plant Cell Physiol.,* 44:70–75.

Kakizaki, T., Y. Takada, T. Fujioka, G. Suzuki, Y. Satta, H. Shiba, A. Isogai, S. Takayama, and M. Watanabe. 2006. Comparative analysis of the *S*-intergenic region in class-II *S* haplotypes of self-incompatible *Brassica rapa* (syn. *campestris*). *Genes Genet. Syst.,* 81:63–67.

Kimura, R., K. Sato, R. Fujimoto, and T. Nishio. 2002. Recognition specificity of self-incompatibility maintained after the divergence of *Brassica oleracea* and *Brassica rapa*. *Plant J.,* 29:215–223.

Kusaba, M., T. Nishio, Y. Satta, K. Hinata, and D.J. Ockendon. 1997. Striking sequence similarity in inter- and intra-specific comparisons of class I SLG alleles from *Brassica oleracea* and *Brassica campestris*: implications for the evolution and recognition mechanism. *Proc. Natl. Acad. Sci. U.S.A.,* 94:7673–7678.

Kusaba, M., K. Dwyer, J. Hendershot, J. Vrebalov, J.B. Nasrallah, and M.E. Nasrallah. 2001. Self-incompatibility in the genus *Arabidopsis*: characterization of the *S* locus in the outcrossing *A. lyrata* and its autogamous relative *A. thaliana*. *Plant Cell,* 13:627–643.

Kusaba, M., C.-W. Tung, M.E. Nasrallah, and J.B. Nasrallah. 2002. Monoallelic expression and dominance interactions in anthers of self-incompatible *Arabidopsis lyrata*. *Plant Physiol.,* 128:17–20.

Lim, S.H., H.J. Cho, S.J. Lee, Y.H. Cho, and B.D. Kim. 2002. Identification and classification of *S* haplotypes in *Raphanus sativus* by PCR-RFLP of the *S* locus glycoprotein (SLG) gene and the *S* locus receptor kinase (*SRK*) gene. *Theor. Appl. Genet.,* 104:1253–1263.

Liu, P., S. Sherman-Broyles, M.E. Nasrallah, and J.B. Nasrallah. 2007. A cryptic modifier causing transient self-incompatibility in *Arabidopsis thaliana*. *Current Biol.,* 17:734–740.

Mummenhoff, K. and H. Hurka. 1995. Allopolyploid origin of *Arabidopsis suecica* (Fries) Norrlin: evidence from chloroplast and nuclear genome markers. *Bot. Acta,* 108:449–456.

Murase, K., H. Shiba, M. Iwano, F.-S. Che, M. Watanabe, A. Isogai, and S. Takayama. 2004. A membrane-anchored protein kinase involved in *Brassica* self-incompatibility signaling. *Science,* 303:1516–1519.

Naithani, S., T. Chookajorn, D.R. Ripoll, and J.B. Nasrallah. 2007. Structural modules for receptor dimerization in the *S*-locus receptor kinase extracellular domain. *Proc. Natl. Acad. Sci. U.S.A.,* 104:12211–12216.

Nasrallah, J.B., T.H. Kao, C.H. Chen, M.L. Goldberg, and M.E. Nasrallah. 1987. Amino-acid sequence of glycoproteins encoded by three alleles of the *S* locus of *Brassica oleracea*. *Nature,* 326:617–619.

Nasrallah, J.B., S.-M. Yu, and M.E. Nasrallah. 1988. Self-incompatibility genes of *Brassica oleracea* expression, isolation, and structure. *Proc. Natl. Acad. Sci. U.S.A.,* 85:5551–5555.

Nasrallah, J.B., T. Nishio, and M.E. Nasrallah. 1991. The self-incompatibility genes of *Brassica*: expression and use in genetic ablation of floral tissues. *Annu. Rev. Plant Physiol. Plant Mol. Biol.,* 42:393–422.

Nasrallah, J.B. and M.E. Nasrallah. 1993. Pollen-stigma signaling in the sporophytic self-incompatibility response. *Plant Cell,* 5:1325–1335.

Nasrallah, J.B., P. Liu, S. Sherman-Broyles, R. Schmidt, and M.E. Nasrallah. 2007. Epigenetic mechanisms for breakdown of self-incompatibility in interspecific hybrids. *Genetics,* 175:1965–1973.

Nasrallah, M.E. and D.H. Wallence. 1967. Immunogenetics of self-incompatibility in *Brassica oleracea* L. *Heredity*, 22:519–527.

Nasrallah, M.E., P. Liu, and J.B. Nasrallah. 2002. Generation of self-incompatible *Arabidopsis thaliana* by transfer of two *S* locus genes from *A. lyrata. Science*, 297:247–249.

Nasrallah, M.E., P. Liu, S. Sherman-Broyles, N.A. Boggs, and J.B. Nasrallah. 2004. Natural variation in expression of self-incompatibility in *Arabidopsis thaliana*: implications for the evolution of selfing. *Proc. Natl. Acad. Sci. U.S.A.*, 101:16070–16074.

Nishio, T. and K. Hinata. 1977. Analysis of *S* specific proteins in stigma of *Brassica oleracea* L. by isoelectric focusing. *Heredity*, 38:391–396.

Nishio, T., K. Sakamoto, and J. Yamaguchi. 1994. PCR-RFLP of *S* locus for identification of breeding lines in cruciferous vegetables. *Plant Cell Rep.*, 13:546–550.

Nishio, T., M. Kusaba, M. Watanabe, and K. Hinata. 1996. Registration of *S* alleles in *Brassica campestris* L. by the restriction fragment sizes of SLGs. *Theor. Appl. Genet.*, 92:388–394.

Nishio, T., M. Kusaba, K. Sakamoto, and D.J. Ockendon. 1997. Polymorphism of the kinase domain of the *S*-locus receptor kinase gene (SRK) in *Brassica oleracea* L. *Theor. Appl. Genet.*, 95:335–342.

Nou, I.S., M. Watanabe, A. Isogai, H. Shiozawa, A. Suzuki, and K. Hinata. 1991. Variation of *S*-alleles and *S*-glycoproteins in a naturalized population of self-incompatible *Brassica campestris* L. *Jpn. J. Genet.*, 66:227–239.

Nou, I.S., M. Watanabe, K. Isuzugawa, A. Isogai, and K. Hinata. 1993. Isolation of *S*-alleles from a wild population of *Brassica campestris* L. at Balcesme, Turkey and their characterization by *S*-glycoproteins. *Sex. Plant Reprod.*, 6:71–78.

Ockendon, D.J. 2000. The *S*-allele collection of *Brassica oleracea*. *Acta Horticulturae*, 539:25–30.

Okamoto, S., Y. Sato, K. Sakamoto, and T. Nishio. 2004. Distribution of similar self-incompatibility (*S*) haplotypes in different genera, *Raphanus* and *Brassica*. *Sex. Plant Reprod.*, 17:33–39.

Okamoto, S., M. Odashima, R. Fujimoto, Y. Sato, H. Kitashiba, and T. Nishio. 2007. Self-compatibility in *Brassica napus* is caused by independent mutations in *S*-locus genes. *Plant J.*, 50:391–400.

O'Kane, S.L., B.A. Schaal, and I.A. Al-Shehbaz. 1996. The origins of *Arabidopsis suecica* (*Brassicaceae*) as indicated by nuclear rDNA sequences. *Syst. Bot.*, 21:559–566.

Okazaki, K., M. Kusaba, D.J. Ockendon, and T. Nishio. 1999. Characterization of *S* tester lines in *Brassica oleracea*: polymorphism of restriction fragment length of SLG homologues and isoelectric points of *S*-locus glycoproteins. *Theor. Appl. Genet.*, 98:1329–1334.

Sakamoto, K., M. Kusaba, and T. Nishio. 1998. Polymorphism of the *S*-locus glycoprotein gene (SLG) and the *S*-locus related gene (*SLR1*) in *Raphanus sativus* L. and self-incompatible ornamental plants in the Brassicaceae. *Mol. Gen. Genet.*, 258:397–403.

Sakamoto, K., M. Kusaba, and T. Nishio. 2000. Single-seed PCR-RFLP analysis for the identification of *S* haplotypes in commercial F₁ hybrid cultivars of broccoli and cabbage. *Plant Cell Rep.*, 19:400–406.

Sato, K., T. Nishio, R. Kimura, M. Kusaba, T. Suzuki, K. Hatakeyama, D.J. Ockendon, and Y. Satta. 2002. Coevolution of the *S*-Locus genes SRK, SLG and *SP11/SCR* in *Brassica oleracea* and *B. rapa*. *Genetics*, 162:931–940.

Sato, Y., R. Fujimoto, K. Toriyama, T. Nishio. 2003. Commonality of self recognition specificity of S haplotypes between *B. oleracea* and *B. rapa*. *Plant Mol. Biol.* 52:619–626.

Sato, Y., S. Okamoto, and T. Nishio. 2004. Diversification and alteration of recognition specificity of the pollen ligand SP11/SCR in self-incompatibility of *Brassica* and *Raphanus*. *Plant Cell*, 16:3230–3241.

Sato, Y., K. Sato, and T. Nishio. 2006. Interspecific pairs of class II *S* haplotypes having different recognition specificities between *Brassica oleracea* and *Brassica rapa*. *Plant Cell Physiol.*, 47:340–345.

Schierup, M.H., B.K. Mable, P. Awadalla, and D. Charlesworth. 2001. Identification and characterization of a polymorphic receptor kinase gene linked to the self-incompatibility locus of *Arabidopsis lyrata*. *Genetics*, 158:387–399.

Schopfer, C.R., M.E. Nasrallah, and J.B. Nasrallah. 1999. The male determinant of self-incompatibility in *Brassica*. *Science*, 286:1697–1700.

Shiba, H., S. Takayama, M. Iwano, H. Shimosato, M. Funato, T. Nakagawa, F.-S. Che, G. Suzuki, M. Watanabe, K. Hinata, and A. Isogai. 2001. A pollen coat protein, SP11/SCR, determines the pollen *S* specificity in the self-incompatibility of *Brassica* species. *Plant Physiol.*, 125:2095–2103.

Shiba, H., T. Kakizaki, M. Iwano, T. Entani, K. Ishimoto, H. Shimosato, F.-S. Che, Y. Satta, A. Ito, Y. Takada, M. Watanabe, A. Isogai, and S. Takayama. 2002. The dominance of alleles controlling self-incompatibility in *Brassica* pollen is regulated at the RNA level. *Plant Cell*, 14:491–504.

Shiba, H., M. Kenmochi, M. Sugihara, M. Iwano, S. Kawasaki, G. Suzuki, M. Watanabe, A. Isogai, and S. Takayama. 2003. Genomic organization of the *S*-locus region of *Brassica*. *Biosci. Biotechnol. Biochem.*, 67:622–626.

Shiba, H., T. Kakizaki, M. Iwano, Y. Tarutani, M. Watanabe, A. Isogai, and S. Takayama. 2006. Dominance relationships between self-incompatibility alleles controlled by DNA methylation. *Nature Genet.*, 38:297–299.

Shimosato, H., N. Yokota, H. Shiba, M. Iwano, T. Entani, F.-S. Che, M. Watanabe, A. Isogai, and S. Takayama. 2007. Characterization of the SP11/SCR high-affinity binding site involved in self/nonself recognition in *Brassica* self-incompatibility. *Plant Cell*, 19:107–117.

Silva, N.F., S.L. Stone, L.N. Christie, W. Sulaman, K.A.P. Nazarian, L.A. Burnett, M.A. Arnoldo, S.J. Rothstein, and D.R. Goring. 2001. Expression of the *S* receptor kinase in self-incompatible *Brassica napus* cv. Westar leads to the allele-specific rejection of self-incompatible *Brassica napus* pollen. *Mol. Gent. Genom.*, 265:552–559.

Smyth, D.R., J.L. Bowman, and E.M. Meyerowitz. 1990. Early flower development in Arabidopsis. *Plant Cell*, 2:755–767.

Stein. J.C., B. Howlett, D.C. Boyes, M.E. Nasrallah, and J.B. Nasrallah. 1991. Molecular cloning of a putative receptor protein kinase gene encoded at the self-incompatibility locus of *Brassica oleracea*. *Proc. Natl. Acad. Sci. U.S.A.*, 88:8816–8820.

Stone, S.L., M.A. Arnoldo, and D.R. Goring. 1999. A breakdown of *Brassica* self-incompatibility in ARC1 antisense transgenic plants. *Science*, 286:1729–1731.

Stone, S.L., E.M. Anderson, R.T. Mullen, and D.R. Goring. 2003. ARC1 is an E3 ubiquitin ligase and promotes the ubiquitination of proteins during the rejection of self-incompatible *Brassica* pollen. *Plant Cell*, 15:885–889.

Suzuki, G., N. Kai, T. Hirose, K. Fukui, T. Nishio, S. Takayama, A. Isogai, M. Watanabe, and K. Hinata. 1999. Genomic organization of the *S* locus: identification and characterization of genes in SLG/SRK region of *S9* haplotype of *Brassica campestris* (syn. *rapa*). *Genetics*, 153:391–400.

Suzuki, G., M. Watanabe, and T. Nishio. 2000b. Physical distances between *S*-locus genes in various *S* haplotypes of *Brassica rapa* and *B. oleracea*. *Theor. Appl. Genet.*, 101:80–85.

Suzuki, G., T. Kakizaki, Y. Takada, H. Shiba, S. Takayama, A. Isogai, and M. Watanabe. 2003. The *S* haplotypes lacking SLG in the genome of *Brassica rapa*. *Plant Cell Rep.*, 21:911–915.

Suzuki, T., M. Kusaba, M. Matsushita, K. Okazaki, and T. Nishio. 2000a. Characterization of *Brassica S*-haplotypes lacking *S*-locus glycoprotein. *FEBS Lett.*, 482:102–108.

Takahata, Y. and K. Hinata. 1980. A variation study of subtribe Brassicinae by principal component analysis. In *Brassica Crops and Wild Allies*, Eds. S. Tunoda, K. Hinata, and C. Gomez-Campo, p. 33–49. Tokyo: Japan Scientific Societies Press.

Takasaki. T., K. Hatakeyama, M. Watanabe, K. Toriyama, A. Isogai, and K. Hinata. 1999. Introduction of SLG (*S* locus glycoprotein) alters the phenotype of endogenous *S* haplotype, but confers no new *S* haplotype specificity in *Brassica rapa* L. *Plant Mol. Biol.*, 40:659–668.

Takasaki, T., K. Hatakeyama, G. Suzuki, M. Watanabe, A. Isogai, and K. Hinata. 2000. The *S* receptor kinase determines self-incompatibility in *Brassica* stigma. *Nature*, 403:913–916.

Takayama, S., A. Isogai, C. Tsukamoto, Ueda, Y.K. Hinata, K. Okazaki, and A. Suzuki. 1987. Sequences of *S*-glycoproteins, products of the *Brassica campestris* self-incompatibility locus. *Nature*, 326:102–105.

Takayama, S., H. Shiba, M. Iwano, H. Shimosato, F.-K. Che, N. Kai, M. Watanabe, G. Suzuki, K. Hinata, and A. Isogai. 2000. The pollen determinant of self-incompatibility in *Brassica campestris*. *Proc. Natl. Acad. Sci. U.S.A.*, 97:1920–1925.

Takayama, S., H. Shimosato, H. Shiba, M. Funato, F.-K. Che, M. Watanabe, M. Iwano, and A. Isogai. 2001. Direct ligand-receptor complex interaction controls *Brassica* self-incompatibility. *Nature*, 413:534–538.

Takuno, S., R. Fujimoto, T. Sugimura, K. Sato, S. Okamoto, S.-L. Zhang, and T. Nishio. 2007. Effects of recombination on hitchhiking diversity in the *Brassica* self-incompatibility locus complex. *Genetics*, 177:949–958.

Tang, C., C. Toomajian, S. Sherman-Broyles, V. Plagnol, Y.-L. Guo, T.T. Hu, R. M. Clark, J. B. Nasrallah, D. Weigel, and M. Nordborg. 2007. The evolution of selfing in *Arabidopsis thaliana*. *Science*, 317:1070–1072.

Thompson, K.F. and J.P. Taylor. 1966. Non-linear dominance relationships between *S* alleles. *Heredity*, 21:345–362.

Trick, M. and R.B. Flavell. 1989. A homozygous *S* genotype of *Brassica oleracea* expresses two *S*-like genes. *Mol Gen. Genet.*, 218:112–117.

U, N. 1935. Genome analysis in *Brassica* with special reference to the experimental formation of *B. napus* and peculiar mode of fertilization. *Jpn. J. Bot.*, 7:389–452.

Watanabe, M., T. Takasaki, K. Toriyama, S. Yamakawa, A. Isogai, A. Suzuki, and K. Hinata. 1994. A high degree of homology exists between the protein encoded by SLG and the *S* receptor domain encoded by *SRK* in self-incompatible *Brassica campestris* L. *Plant Cell Physiol.*, 35:1221–1229.

Watanabe, M., A. Ito, Y. Takada, C. Ninomiya, Y. Kakizaki, Y. Takahata, K. Hatakeyama, K. Hinata, G. Suzuki, T. Takasaki, Y. Satta, H. Shiba, S. Takayama, and A. Isogai. 2000. Highly divergent sequences of the pollen self-incompatibility (*S*) gene in class-I *S* haplotypes of *Brassica campestris* (syn. *rapa*) L. *FEBS Lett.*, 473:139–144.

7 Wild Germplasm and Male Sterility

Shyam Prakash, S.R. Bhat, and Ting-Dong Fu

CONTENTS

INTRODUCTION

Male sterility, the inability of the plant to produce fertile pollen, is widespread in angiosperms. It provides one of the most efficient means of directed pollination for large-scale production of hybrid seeds in crops. Of two types of male sterility — that is, genic and cytoplasmic — the latter is more exploitable as it is maternally inherited. Cytoplasmic male sterility (CMS) is encoded in mitochondrial genomes as evidenced by detailed molecular analyses of CMS systems in different crop species such as wheat, rice, maize, sunflower, *Petunia* and *Brassica,* etc. A well-coordinated expression of mitochondrial and nuclear genes is required for plant development and function (Mackenzie and McIntosh, 1999). Floral development, in particular anther development, is very vulnerable to nuclear-mitochondrial incompatibilities and manifests as male sterility. Such incompatible interactions are encountered in alloplasmic lines derived from wide hybridization, which carry nuclear and mitochondrial genomes from different species. CMS also arises through mutations in the mitochondrial genome and is referred to as autoplasmy. A well-known example of autoplasmic CMS is the Polima system discovered in China in a Polish variety of *Brassica napus* by Fu in 1972 (Fu et al., 1990) that is extensively utilized for hybrid development, primarily in China but also in Canada.

Alloplasmics are generally obtained by repeated backcrossings of sexual hybrids or somatic hybrids with the cultivated species. Such wide hybridizations may include interspecific, intergeneric, or even intertribal combinations. CMS becomes evident in BC_3–BC_4 generations when a normal nuclear genome is reconstituted by backcrossing. The first alloplasmic CMS *Brassica* was reported in cabbage (*B. oleracea* var. *capitata*), which was obtained by repeated backcrossings of the interspecific hybrid *B. nigra* × *B. oleracea* (Pearson, 1972).

Because mitochondrial (mt) genomes encode the CMS trait, knowledge of the availability of mitochondrial genome variability is of considerable importance in developing CMS systems of diverse origin. *Brassica*-related wild germplasm provides wide variability for the mt-genome as revealed by mt-DNA RFLP studies (Pradhan et al., 1992). Thus, there is an opportunity to generate a spectrum of novel CMS systems of alloplasmic origin in *Brassica*. *In vitro* techniques of embryo rescue and protoplast fusion have enabled production of a large number of sexual and somatic hybrids — a prerequisite for the introgression of cytoplasm. The first examples of CMS of alloplasmic origin were *Brassica oleracea* and *B. napus,* which carried sterility-inducing *Raphanus*/Ogura cytoplasm (Bannerot et al., 1974). Subsequently, Hinata and Konno (1979) developed CMS *B. rapa* utilizing *Diplotaxis muralis* cytoplasm. Since then, a large number of CMS systems have been reported. In this chapter, the development, characterization, and utilization of CMS systems that have originated from the use of wild germplasm, are discussed. In recent years, several informative reviews dealing with different aspects of cytoplasmic male sterility in *Brassica* have appeared, notably those by Delourme and Budar (1999), Prakash (2001), Budar et al. (2004), Carlsson et al. (2008), and Prakash et al. (2009).

Brassica-related wild germplasm has a wide distribution, from the Iberian Peninsula to north-western India with a concentration in Mediterranean phytochorea. Referred to as *Brassica* coeno-species, taxonomically it comprises subtribe Brassicinae and part of subtribes Raphaninae and Moricandinii of the tribe Brassiceae (Harberd, 1972). The boundaries of this group have further expanded in recent years, thus creating a large germplasm pool for *Brassica* improvement.

CROP *BRASSICAS*

Six species of the genus *Brassica* are important for human use. Of these, *B. oleracea, B. rapa,* and *B. juncea* are highly polymorphic. Several varieties of *B. oleracea* provide vegetables [e.g., cabbage (var. *capitata*), cauliflower (var. *botrytis*), and broccoli (var. *italica*)]; fodder viz. (var. *acephala*); etc. Forms of *B. rapa* serve as sources of oilseed (var. *oleifera*), vegetables [e.g., Chinese cabbage (var. *pekinensis*), var. *chinensis, narinosa,* etc.]; and fodder, viz. turnip (var. *rapifera*). *B. juncea* (Indian mustard) is the predominant oilseed species of the Indian subcontinent. It is also widely grown as a vegetable in China, in addition to oilseed forms. Rapeseed (*B. napus*) is extensively cultivated in Canada, Europe, China, and Australia for its oil. Seeds of black mustard (*B. nigra*) are used as condiments, while cultivation of Ethiopian mustard (*B. carinata*) is very limited and is a dual-purpose crop used both as oilseed and fodder.

HETEROSIS

Conventional breeding programs have been pursued vigorously to increase seed yield in major oilseed *Brassica* species in many countries during the past 50 years. Many investigations into *Brassica* species have revealed significant heterosis for seed yield in intervarietal crosses. Studies on heterosis in *B. napus* have been carried out mostly in Canada, China, and Europe, while those with *B. juncea* are confined to India. Seed heterosis has been reported in the range of 24% to 60% in *B. rapa* (Falk et al., 1994), 30% to 90% in *B. juncea* (Pradhan et al., 1993; Ghosh et al., 2002), and 30% to 80% in *B. napus* (Brandle and McVetty, 1990). These studies suggest that heterosis can be exploited profitably in *B. napus* and *B. juncea*. In recent years, a relationship has been suggested between the level of heterosis and genetic distance sometimes reflected in geographical diversity. For example, heterosis was greater in hybrids between Asiatic and European forms than within Asiatic or European types (Brandle and McVetty, 1990; Jain et al., 1994). Also, it is more pronounced in hybrids between natural and synthetic strains (Pradhan et al., 1993; Becker et al., 1995; Seyis et al., 2006). These studies strongly suggest the potential for developing hybrid cultivars and have generated considerable commercial interest (Renard et al., 1998). Morphological and isozyme markers were used to assess genetic diversity (Ali et al., 1995) in earlier studies. However, in recent years, various molecular markers that show a higher level of polymorphism than isozymes are currently employed — for example, RFLPs in *B. napus* (Becker et al., 1995; Diers et al., 1995) and *B. rapa* (McGrath and Quiros, 1992); RAPDs in *B. juncea* (Jain et al., 1994); AFLPs in *B. napus* (Shen et al., 2003); SRAP in *B. napus* (Riaz et al., 2001); and SSR in *B. napus* (Plieske and Struss, 2001). These techniques may be of considerable practical value to predict potential parents for heterotic combinations.

FLORAL BIOLOGY

Typical of the *Brassicaceae* family, a *Brassica* flower has four petals in crucifer cross-form and these are mostly yellow in color, but occasionally also light yellow to white. The inflorescence is a corymbose-raceme with indeterminate flowering habit. The flower has six stamens: 2 lateral ones with short filaments, and 4 median stamens with long filaments. The stigma is receptive for 5-6 days, 3 days prior to and 3 days after anthesis. Two functional and two nonfunctional nectaries are present at the base of short and long stamens, respectively. Anthers are introrse in self-compatible *B. carinata*, *B. juncea*, *B. napus* and *B. rapa* var. yellow sarson, while self-incompatible *B. oleracea*, *B. nigra*, and *B. rapa* possess extrorse anthers. Alloploid species *B. juncea* and *B. napus* are partially allogamous with variable range of cross pollination. Related species *Raphanus sativus*, *Eruca sativa*, and *Sinapis alba* are self-incompatible and show extrorse anthers.

HISTORY OF CMS IN *BRASSICA*

Although cytoplasmic male sterility was observed in intervarietal crosses of *Brassica napus* (Shiga and Baba, 1973), serious investigations into CMS in *Brassica* for its utilization started with the discovery of male sterility in Japanese radish (*Raphanus sativus*) (Ogura, 1968). This species has the distinction of being one of the most extensively investigated CMS systems among crop plants.

A spontaneous cytoplasmic male sterile *Brassica juncea* plant was isolated by Rawat and Anand (1979) and later found to have the cytoplasm of a related wild species *B. tournefortii* (Pradhan et al., 1991). This CMS probably originated from outcrossings of synthetic alloploid *B. tounefortii* × *B. nigra* synthesized earlier by Narain and Prakash (1972). Around that time, Hinata and Konno (1979) demonstrated the importance of wild germplasm in obtaining male-sterile lines of alloplasmic origin in a designed experiment by placing the nucleus of *B. rapa* var. *chinensis* in the cytoplasm of a related wild species (*Diplotaxis muralis*), followed by the repeated backcrossings of the intergeneric hybrid *D. muralis* × *B. rapa*.

SYNTHESIS

CMS lines are obtained following repeated backcrossings of sexual or somatic hybrids. In sexual hybridizations, the wild species is used as the female parent and contributes unaltered chloroplast (cp) and mt genomes. In the case of somatic hybrids, the initial hybrid must contain the mt genome of wild species. As sexual hybrids carry the entire cytoplasm from the maternal parent, only one type of CMS is obtained. However, frequent intergenomic mitochondrial recombinations in somatic hybrids may yield several different alloplasmic lines with different mt-genome constitutions and associated floral morphologies (Leino et al. 2003). To date, a spectrum of diverse CMS systems has been developed in *Brassica* species, primarily in *B. juncea* and *B. napus,* by combining an array of wild species (Table 7.1).

TABLE 7.1
Cytoplasmic Male Sterility Systems in *Brassica*

CMS Species	Cytoplasmic Donor	Origin	Ref.
Raphanus sativus	*Raphanus sativus*	Spontaneous mutation	Ogura (1968)
B. oleracea	*Raphanus sativus*	Intergeneric cross	McCollum (1981)
	CMS (Ogura) *Raphanus sativus*	Intergeneric cross	Bannerot et al. (1974)
		Protoplast fusion	Pelletier et al. (1989)
			Kao et al. (1992)
			Walters et al. (1992)
	Raphanus sativus (Shougoin)	Protoplast fusion	Kameya et al. (1989)
B. rapa	*Diplotaxis muralis*	Intergeneric cross	Hinata and Konno (1979)
	B. oxyrrhina	Interspecific cross	Prakash and Chopra (1988)
	Modified CMS (Ogura) *B. napus*	Interspecific cross	Delourme et al. (1994b)
	Eruca sativa	Intergeneric cross	Matsuzawa et al. (1999)
	Enarthrocarpus lyratus	Intergeneric cross	Deol et al. (2003)
B. juncea	*B. tournefortii*	Spontaneous	Rawat and Anand (1979)
			Pradhan et al. (1991)
	B. oxyrrhina	Interspecific cross	Prakash and Chopra (1990)
	Refined CMS (Ogura) *B. napus*	Interspecific cross	Delourme et al. (1994b)
			Kirti et al. (1995a)
	Diplotaxis siifolia	Intergeneric cross	Rao et al. (1994)
	Trachystoma ballii	Protoplast fusion	Kirti et al. (1995b)
	Moricandia arvensis	Protoplast fusion	Prakash et al. (1998)
	Diplotaxis erucoides	Intergeneric cross	Malik et al. (1999)
			Bhat et al. (2006)
	Diplotaxis berthautii	Intergeneric cross	Malik et al. (1999)
			Bhat et al. (2008)
	Erucastrum canariense	Intergeneric cross	Prakash et al. (2001)
	Diplotaxis catholica	Intergeneric cross	Pathania et al. (2003)
		Protoplast fusion	Pathania et al. (2007)
	Enarthrocarpus lyratus	Intergeneric cross	Banga et al. (2003)
B. napus	*Raphanus sativus*/Ogura	Intergeneric cross, protoplast fusion	Bannerot et al. (1974) Pelletier et al. (1983)
	Diplotaxis muralis	Intergeneric cross	Pellan-Delourme and Renard (1987)
	Raphanus sativus	Intergeneric cross	Paulmann and Röbbelen (1988)
	Raphanus sativus/Kosena	Protoplast fusion	Sakai and Inamura (1990)
	B. tournefortii	Interspecific cross	Mathias (1985)
			Bartkowiak-Broda (1991)
		Protoplast fusion	Stiewe and Röbbelen (1994)
			Liu et al. (1996)
	Diplotaxis siifolia	Interspecific cross	Rao and Shivanna (1996)
	Enarthrocarpus lyratus	Intergeneric cross	Janeja et al. (2003)
	Arabidopsis thaliana	Protoplast fusion	Leino et al. (2003)
	Orychophragmus violaceus	Protoplast fusion	Mei et al. (2003)
	Sinapis arvensis	Protoplast fusion	Hu et al. (2004)

GENERAL FEATURES OF CMS

CMS plants, in general, closely resemble euplasmic plants in morphology and important agronomic traits. However, several developmental and floral abnormalities are commonly observed. These include late flowering and varying degrees of leaf chlorosis ranging from mild (*Oxyrrhina*) to severe (*Raphanus*/Ogu, *Moricandia*) — a trait leading to late flowering and poor productivity. Floral abnormalities are quite common in several of them. *Raphanus*/Ogu induces a range of abnormalities, including petaloid and carpeloid stamens in *Brassica oleracea* (McCollum, 1981), and petaloid anthers, crooked style, and reduced nectaries in *B. napus* (Bannerot et al., 1974) and *B. juncea* (Kirti et al., 1995a; Meur et al., 2006). CMS (*Trachystoma*) *B. juncea* flowers have narrow petals; stamens are replaced by petal-like structures with a slender filament and contain one or two sacs similar to anther locules full of sterile pollen. Sometimes, rudimentary anther-like protuberances at the base are also present (Kirti et al., 1995b). CMS (*Eruca*) *B. rapa* lines have abnormal flowers with petaloid, partial petaloid, or slender anthers and have poorly developed nectaries (Matsuzawa et al., 1999). In CMS (*Diplotaxis catholica*) *B. juncea* derived from sexual hybridization, the stamens form tubular structures. Occasionally, spherical, ovule-like developments on modified stamens are also observed. Carpel deformities viz. curved style, bi- or multilobed stigma and poorly developed nectaries are common (Pathania et al., 2003). Several lines (*Arabidopsis thaliana*) of *B. napus* have stamen-resembling carpels with stigmatoid tissues and ovule-like structures (Leino et al., 2003). However, flowers in several CMS systems — for example, (*Oxyrrhina*) *B. rapa*, *B. juncea*, and *B. napus* (Prakash and Chopra, 1988, 1990); (*Moricandia*) *B. juncea* (Prakash et al., 1998); (*Erucastrum canariense*) *B. juncea* (Prakash et al., 2001); (*Diplotaxis erucoides*) *B. juncea* (Bhat at al., 2006); (*Diplotaxis catholica*) *B. juncea* of somatic hybrid origin (Pathania et al., 2007) and (*Diplotaxis berthautii*) *B. juncea* (Bhat et al., 2008) — are normal except for slender anthers that make them almost indistinguishable from normal euplasmic flowers. Female fertility is low in CMS lines carrying mitochondrial genomes of *Raphanus*/Ogu, *Catholica*, *Trachystoma*, *Eruca*, and *Enarthrocarpus*. Characteristic features of some important CMS systems are given in Table 7.2.

FERTILITY RESTORATION

Fertility-restoring nuclear gene(s) for alloplasmic CMS systems are generally not available in natural populations and thus must be introgressed from the cytoplasm-donor species. Homoeologous chromosome pairing between the chromosomes of an alien genome carrying the fertility restoration gene(s) and crop species leads to introgression. *Raphanus*/Ogu was the first CMS to be exploited for commercial purposes. No restorer was found for this system in any *Brassica* spp. (Rousselle, 1982). However, restorer genes were identified in European radish cultivars (Bonnet, 1975) and were introgressed through synthetic *Raphanobrassica* (*Raphanus sativus* × *B. napus*) and (*R. sativus* × *B. oleracea*) by Heyn (1976) and Rouselle and Dosba (1985), respectively. Restorers were successfully obtained for (*Raphanus*) *B. napus* (Paulman and Röbbelen, 1988), (*Tournefortii*) *B. napus* (Stiewe and Röbbelen, 1994), (*Moricandia*) *B. juncea* (Prakash et al., 1998), (*Trachystoma*) *B. juncea* (Kirti et al., 1997), (*Canariense*) *B. juncea* (Prakash et al., 2001), and (*Enarthrocarpus*) *B. rapa*, *B. juncea*, and *B. napus* (Banga et al., 2003; Deol et al., 2003). These restorer lines possess normal phenotypes and exhibit regular meiosis with normal pollen and seed fertility, indicating that introgressed genetic material is fully integrated with the recipient species' chromosomes. *Rf* genes for (*Diplotaxis muralis*) *B. rapa* were frequently found in *B. rapa* (Hinata and Konno, 1979). *Moricandia arvensis* restorer is unique. It restores fertility in four CMS systems: (*Moricandia*) *B. juncea*, (*Catholica*) *B. juncea* (sexual), (*Erucoides*) *B. juncea*, and (*Berthautii*) *B. juncea* (Pathania et al., 2003; Bhat et al., 2005, 2006, 2008). In (*Catholica*) *B. juncea* derived from sexual hybridization, it is sporophytic (Bhat et al., 2005) while in the other three systems, the restorer function is required at the

TABLE 7.2
Characteristics of Some Male Sterility Systems

Cytoplasmic Donor	Recipient Species	Characteristics	Ref.
Raphanus sativus (Ogura)	*Brassica oleracea, B. rapa, B. juncea, B. napus*	Leaves highly chlorotic; flowers with petaloid anthers or absent nectaries; female fertility poor	Bannerot et al. (1974) Mathias (1985)
		Abnormalities removed following protoplast fusion	Pelletier et al. (1983) Jarl and Bornman (1988) Kao et al. (1992) Kirti et al. (1995a)
Diplotaxis muralis (Muralis)	*B. rapa, B. napus*	Flowers with narrow petals and occasionally with petaloid anthers; nectaries very poorly developed; female fertility low; cp and mt: *D. muralis*	Hinata and Konno (1979) Pellan-Delourme and Renard (1987)
Brassica oxyrrhina (Oxyrrhina)	*B. rapa, B. juncea, B. napus*	Flowers normal with non-dehiscent anthers and excellent nectaries; female fertility 96%; cp and mt: *B. oxyrrhina*; leaf chlorosis in *B. juncea* rectified following protoplast fusion	Prakash and Chopra (1988, 1990) Kirti et al. (1993)
Raphanus sativus	*B. oleracea var. capitata*	Plants similar to *B. oleracea* in floral morphology and female fertility; cp: *R. sativus*	Kameya et al. (1989)
Raphanus sativus (Kosena)	*B. napus*	Leaves green; flowers with stunted filaments and aborted anthers; female fertility normal; mt: recombined	Sakai and Inamura (1990)
Brassica tournefortii (Tournefortii)	*B. napus*	Leaves green; flowers with normal or without petals and reduced empty anthers; female fertility normal; cp: *B. napus*, mt: recombined	Stiewe and Röbbelen (1994) Liu et al. (1996)
Diplotaxis siifolia (Siifolia)	*B. juncea, B. napus*	Leaves green; flowers with reduced anthers and well-developed nectaries; female fertility 95%; cp and mt: *D. siifolia*	Rao et al. (1994) Rao and Shivanna (1996)
Trachystoma ballii (Trachystoma)	*B. juncea*	Leaves green; flowers with 2 types of anthers: petaloid and slender non-dehiscent; female fertility 30–40%; cp: *B. juncea*, mt: recombined	Kirti et al. (1995b)
Moricandia arvensis (Moricandia)	*B. juncea, B. napus*	Leaves highly chlorotic, almost yellowish; delayed flowering; normal flowers with slender anthers and excellent nectaries; female fertility 96 %; cp and mt: *M. arvensis*; chlorosis rectified following protoplast fusion	Prakash et al. (1998) Kirti et al. (1998)

— continued

TABLE 7.2
Characteristics of Some Male Sterility Systems

Cytoplasmic Donor	Recipient Species	Characteristics	Ref.
Diplotaxis erucoides (*Erucoides*)	*B. juncea*	Leaves normal green; flowers with smaller slender anthers and excellent nectaries; female fertility 96%; cp and mt: *D. erucoides*	Malik et al. (1999) Prakash (2001) Bhat et al. (2006)
Diplotaxis berthautii (*Berthautii*)	*B. juncea*	Leaves green; flowers with short indehiscent or petaloid anthers; female fertility 95%; cp and mt: *D. berthautii*	Malik et al. (1999) Bhat et al. (2008)
Erucastrum canariense (*Canariense*)	*B. juncea,*	Leaves normal green; flowers with slender anthers and excellent nectaries; female fertility 95 %; cp and mt: *E. canariense*	Prakash et al. (2001)
Diplotaxis catholica (*Catholica-*somatic)	*B. juncea*	Leaves green; flowers with small rudimentary anthers and excellent nectaries; female fertility 96%; cp: *B. juncea*, mt: recombined	Prakash (2001) Pathania et al. (2007)
Diplotaxis catholica (*Catholica*-sexual)	*B. juncea*	Leaves green; flowers with petaloid/tubular structures; nectaries poorly developed; abnormal style; female fertility 40%; cp and mt: *D. catholica*	Pathania et al. (2003)
Arabidopsis thaliana (*Thaliana*)	*B. napus*	Plants normal green; 2 types of flowers: (1) with feminized anthers and abnormal pistils, and (2) with reduced shrunken anthers; normal nectaries; female fertility normal; mt: recombined	Leino et al. (2003)

gametophytic stage. Because *Moricandia* restorer is an introgression from *M. arvensis* into *B. juncea*, it may be that more than one restorer gene is present at this locus. The restorer genes in *Raphanus*/Ogu and Kosena are identical. In *Raphanus*/Ogu, genetic studies indicate the involvement of two genes; however, one dominant allele of the restorer gene *Rfo* can fully restore the cybrids (Pellan-Delourme, 1986). Delourme et al. (1994b) subsequently introgressed this gene to cybrid lines of *B. rapa* and *B. juncea*. Two nuclear genes *Rfk1* and *Rfk2* are required to restore Kosena CMS in radish and were introgressed into *B. napus* following protoplast fusion (Sakai et al., 1996). However, in *B. napus* cybrids, only one gene (*Rfk2*) can restore fertility (Koizuka et al., 2000). Sources and mode of fertility restoration for various CMS systems are given in Table 7.3.

THE *RAPHANUS*/OGURA SYSTEM

This system was discovered by Ogura (1968) in a population of *Raphanus sativus* in Japan. Subsequently, it was introgressed through conventional backcross breeding to *Brassica oleracea*

TABLE 7.3
Summary of Fertility Restoration

CMS System	Source and Mode of Restoration	Ref.
Raphanus/Ogura	*Rf* gene introgressed from *Raphanus sativus* to *B. napus*; monogenic dominant, sporophytic restoration	Heyn (1976)
Raphanus/Kosena	*Rf* gene introgressed from *Raphanus sativus* to *B. napus*; monogenic dominant, sporophytic restoration	Sakai et al. (1996)
Moricandia arvensis	*Rf* gene introgressed from *Moricandia* to *B. juncea*, single dominant, gametophytic restoration	Prakash et al. (1998) Bhat et al. (2005)
Diplotaxis erucoides	*Moricandia* restorer, single dominant, gametophytic restoration	Bhat et al. (2006)
Diplotaxis berthautii	*Moricandia* restorer, single dominant, gametophytic restoration	Bhat et al. (2008)
Diplotaxis catholica (sexual*)*	*Rf* gene in the progeny of *D. catholica* + *B. juncea;* single dominant, sporophytic restoration	Pathania et al. (2003)
Trachystoma ballii	*Rf* gene introgressed from *Trachystoma* to *B. juncea;* single dominant	Kirti et al. (1997)
Enarthrocarpus lyratus	*Rf* gene introgressed from *E. lyratus* to *B. rapa, B. juncea, B. napus*	Deol et al. (2003) Banga et al. (2003) Janeja et al. (2003)
Brassica tournefortii -Stiewe	*Rf* gene introgressed from *B. tournefortii*	Stiewe and Röbbelen (1994)
Brassica tournefortii	*Rf* gene in natural accessions of *B. napus*; single dominant	Sodhi et al. (1994)

(Bannerot et al., 1974), *B. napus* (Bannerot et al., 1974), *B. rapa* (Delourme et al., 1994b), and *B. juncea* (Kirti et al., 1995a). The leaves appear yellow at low temperatures (less than 12°C) due to chlorophyll deficiency, thus making this system unsuitable for hybrid development (Rouselle, 1982). Although the male sterility is stable, the flowers have several abnormalities, including petaloid anthers and poorly developed or absent nectaries. Female fertility was poor (Rouselle, 1982). These defects have now been corrected in *B. napus,* leading to the recovery of green plants with normal flowers having well-developed nectaries (Pelletier et al., 1983; Jarl and Bornman, 1988), in *B. oleracea* (Kao et al., 1992), and in *B. juncea* (Kirti et al., 1995a). Fertility restorer nuclear genes were located in several European *Raphanus* cultivars (Bonnet, 1975) and introduced *B. napus* through conventional breeding (Heyn, 1976; Pellan-Delourme and Renard, 1988). Difficulties were encountered while attempting to transfer *Rf* genes, primarily due to the low frequency of viable hybrids between *B. napus* and *Raphanus*, the sterility of F$_1$ hybrids, and also the low seed productivity of introgressed progeny. Restoration was achieved in the progeny of CMS (Ogu) *B. napus* × *Raphanobrassica* by Heyn (1976), who obtained fertile white or yellow flowered *B. napus* plants carrying *Raphanus*/Ogu cytoplasm. However, female fertility in the restored plants was very poor due to high degree of ovule abortion, which was linked to male fertility restoration. It was caused by meiotic irregularities such as formation of up to three multivalents per cell (Pellan-Delourme and Renard, 1988). Also, it was tightly linked with high glucosinolate content. The *Rfo* gene has close linkage with an isozyme marker phosphoglucomutage PGI-2 (Delourme and Eber, 1992) and four RAPD markers (Delourme et al., 1994a). The fertility restorer gene was also introgressed into *B. rapa* and *B. juncea* (Delourme et al., 1994b). Genetical studies revealed that the original Ogu CMS required several genes for restoration. However, the improved CMS obtained through protoplast fusion containing recombinant mitochondrial genomes requires only one dominant gene. Extensive investigations at INRA (France) resulted in double low R-lines with good female fertility and regular meiosis (Primard et al., 2005). Bartkowiak-Broda et al. (2003) also obtained double low winter rapeseed restorer lines.

The Ogu cytoplasm confers male sterility in both the *Raphanus* and *Brassica* species. Makaroff and Palmer (1988) detected novel rearrangements in the mitochondrial genome of CMS as compared to normal radish. Analysis of fertile and sterile cybrids (containing the nuclear genome from *Brassica* but radish cytoplasm) revealed a 2.5-kb *NcoI* fragment present in male sterile cybrids but absent in revertant fertile counterparts. The *NcoI* 2.5-kb fragment of Ogu origin is transcribed in male sterile cybrids to give a specific 1.4-kb transcript (Bonhomme et al., 1991). Male fertiles show a 1.1-kb transcript with the same *NcoI* 2.5-kb probe. Sequence analysis of this fragment revealed a t-RNA^{fmet} sequence, a putative 138 amino acid ORF (*orf138*), and a 158 amino acid ORF (*orf158*) previously observed in several other plant species (Bonhomme et al., 1992). Based on physical mapping and transcript analysis studies, *orf138* was considered the best candidate gene responsible for inducing CMS. *orf138* encodes a 19-kD polypeptide that is membrane associated and has been detected in sterile lines (Grelon et al., 1994; Krishnasamy and Makaroff, 1994). However, the transcript profile of *orf138* was not affected in fertile versus CMS lines (Krishnasamy and Makaroff, 1993; Bellaoui et al., 1999). Upon restoration, ORF138 protein was reduced in flowers, indicating post-transcriptional regulation of *orf138*. In recent years, this *Raphanus* restorer gene has been cloned independently by three different groups (Brown et al., 2003; Desloire et al., 2003; Koizuka et al., 2003).

THE *MORICANDIA ARVENSIS* SYSTEM

This system was obtained in *Brassica juncea* following backcrossings of somatic hybrid *Moricandia arvensis* + *B. juncea* ($2n = 64$, MMAABB; cp and mt: *M. arvensis*; Prakash et al., 1998). CMS plants have delayed flowering and the leaves are highly chlorotic and appear yellowish. Flowers are normal with slender anthers and excellent nectaries, and thus are indistinguishable from euplasmic *B. juncea* flowers. Pollen sterility is absolute, while female fertility is normal (>96%). Molecular analysis of the CMS line revealed that *Moricandia* has contributed both mt and cp genomes. Incongruous interactions between *Moricandia* chloroplasts and a *B. juncea* nucleus result in a high degree of leaf chlorosis and poor productivity. Protoplast fusion between sterile and fertile *B. juncea* was undertaken to rectify chlorosis. The resulting green plants have *Moricandia* chloroplasts substituted with those of *B. juncea* (Kirti et al., 1998). Transfer of fertility restoring gene(s) from *Moricandia* was facilitated by having a *Moricandia* chromosome addition line in *B. juncea*. The additional chromosome-carrying fertility restorer gene(s) formed a rare trivalent arising from homoeologous association between a *B. juncea* bivalent and a *Moricandia* univalent that allowed introgression (Prakash et al., 1998). The *Rf* gene introgressed plants show normal pollen fertility.

Ashutosh et al. (2007) tagged the fertility restorer gene for its potential cloning and identified closely linked markers. Using AFLP markers in conjunction with the BSA method, the restorer locus was mapped and one of the AFLP markers linked to the *Rf* locus was converted to a SCAR marker. The genetic distance between the markers and the *Rf* locus ranged from 0.6 to 2.9 cM. The SCAR 3 marker was observed to be tightly linked to the *Rf* locus with a map distance of 0.6 cM. The 200-bp SCAR 3 amplicon shows homology with the *Brassica rapa* genome sequence.

The mitochondrial genome of this CMS was characterized by Gaikwad et al. (2006). A Northern profile of flower buds and leaf mtRNA of *Brassica juncea*, *Moricandia arvensis* was done, and the CMS line revealed altered expression of *atpA* in flower buds. A 1900 nt-long transcript was detected in *B. juncea*, the fertility restored, fertility restorer, and *M. arvensis* lines in both bud and leaf. However, in the CMS line, a 2800 nt-long transcript was observed in the bud tissue, whereas a normal 1900 nt-long *atpA* transcript was present in the leaf. Thus, it was concluded that floral-specific alterations in mitochondrial *atpA* expression is associated with male sterility. Subsequently, Ashutosh et al. (2008) identified a novel *orf108* cotranscribed with the *atpA*. In fertility-restored plants, the bi-cistronic *orf108-atpA* transcripts are specifically cleaved within the *orf108* to yield mono-cistronic *atpA* transcripts. As genetic studies revealed the gametophytic fertility restoration, transcript patterns of *atpA* in sepals, petals, and anthers of CMS; and fertility restorer F_1 plants were studied to know whether the *Rf* gene was active in all floral tissues or confined to just the anthers. Although the anthers of fertility restored plants expressed both 2800 nt and 1900 nt transcripts, the intensity of the longer transcript was weak. Comparative analysis of the *atpA* transcripts of the CMS and the fertility restorer lines revealed a novel *orf108* upstream to the *atpA* region in the male-sterile line. In the fertility restorer line, the *orf108* was abolished as the transcript was cleaved at 12 nt downstream to the *orf108* initiation codon. Thus, *atpA* transcript of the male-sterile line was bi-cistronic, whereas the transcript of the male fertile line was mono-cistronic, capable of coding for only the *atpA* protein. These studies also revealed a perfect hairpin loop structure at the 5′ end of the *atpA* transcript of the CMS line and absent in the *atpA* transcript of the fertility restorer.

THE *DIPLOTAXIS CATHOLICA* SYSTEM

This system, available in *Brassica juncea*, was developed through both sexual and somatic hybridizations, the former from *Diplotaxis catholica* × *B. juncea* allopolyploid, while the latter ones from *D. catholica* + *B. juncea* somatic hybrid (2n = 54, DDAABB; cp: *D. catholica*, mt: recombined; Kirti et al., 1995c). CMS plants are green and similar to *B. juncea* in growth and development. However, those of sexual origin have altered floral morphology, particularly with respect to stamen and carpel. Stamens are transformed into petals in the majority of flowers, while in some they form a tubular structure. Spherical ovule-like developments are also occasionally observed on petaloid stamens. Carpel deformities are also common; for example, the style is curved and the stigma is bi- or multi-lobed. Sometimes, the ovary splits open, exposing the ovules. Nectaries are also poorly developed. These factors lead to poor female fertility (<50%). An interesting feature is the occurrence of trilocular siliques (Pathania et al., 2003). CMS plants of somatic origin are green and vigorous, and late in flowering. Flowers are normal with slender anthers and needle-like short filaments. Female fertility is normal, centering around 97% (Pathania et al., 2007).

A molecular comparison of mt genomes of CMS of somatic origin and parental lines through RAPD and RFLP revealed that the mt genome is largely derived from *Brassica juncea*. However, a recombination of the mt genome was observed, leading to duplication of the *coxI* gene. Further, Northern hybridization suggested a strong association between the duplicated *coxI* gene and the CMS trait. The origin of this gene duplication is not clear; however, the occurrence of long repeats at the 5′ region of *coxI*-2 suggests the recombination event at this region (Pathania et al., 2007). In CMS (sexual), the transcript pattern of *coxI* is similar to *B. juncea* and an altered transcription pattern of the *atpA* gene is associated with the CMS trait (Pathania et al., 2003). However, in CMS (somatic), this region is not at all involved, suggesting the creation of a new CMS-inducing locus involving the *coxI* gene. Also, as no major difference between the two *coxI* genes in the coding region was observed, and no novel ORFs could be identified in the duplicated *coxI* gene region, male sterility may be due to defects in post-transcriptional processing or translation of the novel transcript (Pathania et al., 2007).

The fertility restorer gene for CMS (sexual) was found in the BC_4–BC_5 progenies of the somatic hybrid *Diplotaxis catholica* + *Brassica juncea*. Restored plants have normal stamen development and are without any ovary abnormalities. Restored plants always have bilocular siliques, thus suggesting that the number of locules in the silique is governed by nuclear cytoplasmic interaction and is linked to CMS. The pollen and seed fertility of fertility restored lines match euplasmic lines. Genetic studies revealed that a single dominant gene controls male fertility restoration and is sporophytic in nature (Pathania et al., 2003).

SOMATIC HYBRIDIZATION AND CMS

Somatic hybridization is of relevance for the development of novel CMS lines. In particular, protoplast fusion is employed (1) to create new alloplasmic combinations of wild cytoplasm and crop nucleus; (2) for direct transfer of male sterility-inducing cytoplasm to a new nuclear background; (3) to improve existing CMS systems by rectifying developmental and floral abnormalities; and finally (4) to combine organellar encoded traits. The first point was discussed previously (see Table 7.1). As described , CMS lines based on *Raphanus*/Ogu, *Oxyrrhina,* and *Moricandia* exhibit mild to severe leaf chlorosis in the early stages of plant development. This is due to incompatibilities between wild species chloroplasts and crop nuclei. In *Raphanus*/Ogu, the chlorosis is temperature conditioned and expresses below 20°C. *Moricandia* chloroplasts cause very severe chlorosis in *Brassica juncea,* where the leaves become almost yellowish. *B. oxyrrhina* chloroplasts induce severe chlorosis in *B. rapa* and *B. juncea,* and mild chlorosis in *B. napus.* Experiments have been performed to rectify this undesirable trait using protoplast fusion methodology. Pelletier et al. (1983), for the first time, elegantly demonstrated it when they removed the chlorosis in (*Raphanus*/Ogu) *B. napus* through fusion of the protoplasts of a *B. napus* cultivar and the CMS line. Similar experiments were carried out to restore normal chlorophyll levels in CMS (*Raphanus*/Ogu) *B. napus* (Jarl and Bornman, 1988), *B. oleracea* (Kao et al., 1992), *B. juncea* (Kirti et al., 1995a); *(Oxyrrhina) B. juncea* (Kirti et al., 1993; Arumugam et al., 2000), and (*Moricandia*) *B. juncea* (Kirti et al., 1998). Substitution of alien chloroplasts by those of crop chloroplasts resulted in normal green leaves. It would be desirable to combine a cytoplasm-encoded trait (e.g., atrazine resistance (ATR)) with male sterility. This can be accomplished only through protoplast fusion. CMS (*Raphanus*/Ogu) *B. napus* and *B. oleracea* var. *botrytis* have been combined with ATR (Jourdan et al., 1989a, b).

COMMERCIAL UTILIZATION OF CMS

With the development of different male sterility–fertility restorer systems, breeding hybrid varieties has become feasible in oilseed *Brassica.* A number of hybrid varieties in *Brassica napus* based on an improved Ogura-INRA CMS system have been marketed in Europe, Canada, and Australia since 1994, and presently occupy a large area. These were developed by various seed companies, including Serasam, Monsanto, Syngenta, Adventa, Pacific Seeds, Pioneer, Cargil, and Rustica. This improved CMS sterility has also been widely used to produce commercial F_1 hybrids in vegetable forms of *B. oleracea* viz. cauliflower, cabbage, and Savoy cabbage (Pelletier et al., 1989). CMS *Moricandia*-based *B. juncea* hybrid varieties are in the final stages of evaluation in India and show 20% to 30% yield advantage. Advanta India has recently marketed a *B. juncea* hybrid based on improved Ogura CMS. Because efficient restorers are available for various CMS systems such as the Ogu INRA system in *B. napus, Moricandia, Berthautii, Erucoides, Canariense,* and Ogu IARI in *B. juncea,* these developments will accelerate the use of CMS systems in developing heterotic hybrid cultivars.

REFERENCES

Ali, M., L.O. Copeland, S.G. Elias, and J.D. Kelly. 1995. Relationship between genetic distance and heterosis for yield and morphological traits in winter canola (*Brassica napus*). *Theor. Appl. Genet.,* 91:118–121.

Arumugam, N., A. Mukhopadhyay, V. Gupta, Y.S. Sodhi, J.K. Verma, D. Pental, and A.K. Pradhan. 2000. Somatic cell hybridization of '*oxy*' CMS *Brassica juncea* (AABB) with *B. oleracea* (CC) for correction of chlorosis and transfer of novel organelle combinations to allotetraploid brassicas. *Theor. Appl. Genet.,* 100:1043–1049.

Ashutosh, P.C. Sharma, Prakash, S. and S.R. Bhat. 2007. Identification of AFLP markers linked to the male fertility restorer gene of CMS (*Moricandia arvensis*) *Brassica juncea* and conversion to SCAR marker. *Theor. Appl. Genet.,* 114:385–392.

Ashutosh, P. Kumar, P.C. Sharma, S. Prakash, and S.R. Bhat, 2008. A novel *orf108* co- transcribed with atpA gene is associated with cytoplasmic male sterility in *Brassica juncea* carrying *Moricandia arvensis* cytoplasm. *Plant Cell Physiol.*, 49:284–289.

Banga, S.S., J.S. Deol, and S.K. Banga. 2003. Alloplasmic male sterile *Brassica juncea* with *Enarthrocarpus lyratus* cytoplasm and the introgression of gene(s) for fertility restoration from cytoplasmic donor species. *Theor. Appl. Genet.*, 106:1390–1395.

Bannerot, T., L. Boulidard, Y. Cauderon, and J. Tempe. 1974. Transfer of cytoplasmic male sterility from *Raphanus sativus* to *Brassica oleracea*, *Eucarpia Meeting Cruciferae*, Dundee, Scotland. p. 52–54.

Bartkowiak-Broda, I., W. Poplawska, and M. Gorska-Paukszta. 1991. Transfer of CMS *juncea* to double low winter rape (*Brassica napus* L.). *Proc. 8th Int. Rapeseed Congress*, Saskatoon, Canada, 5:1502–1505.

Bartkowiak-Broda, I., W. Poplawska, and A. Furguth. 2003. Characteristic of winter rapeseed double low restorer line for CMS Ogura system. In: *Proc 11th Int. Rapeseed Congr.* Copenhagen, Denmark. 6-10 July 2003. p. 1:303–305

Becker, H.C., G.M. Engqvist, and B. Karlsson. 1995. Comparison of rapeseed cultivars and resynthesized lines based on allozyme and RFLP markers. *Theor. Appl. Genet.*, 91:62–67.

Belllaoui, M., M. Grelon, G. Pelletier, and F. Budar. 1999. The restorer *Rfo* gene acts post-translationally on the stability of the ORF138 Ogura CMS-associated protein in reproductive tissues of rapeseed cybrids. *Plant Mol. Biol.*, 40:893–902.

Bhat, S.R., S. Prakash, P.B. Kirti, V. Dineshkumar, and V.L. Chopra. 2005. A unique introgression from *Moricandia arvensis* confers male fertility to two different cytoplasmic male sterile lines of *Brassica juncea*. *Plant Breed.*, 124:117–120.

Bhat, S.R., V. Priya, Ashutosh, K.K. Dwivedi, and S. Prakash. 2006. *Diplotaxis erucoides* induced cytoplasmic male sterility in *Brassica juncea* is rescued by the *Moricandia arvensis* restorer: genetic and molecular analyses. *Plant Breed.*, 125:150–155.

Bhat, S.R., P. Kumar, and S. Prakash. 2008. An improved cytoplasmic male sterile (*Diplotaxis berthautii*) *Brassica juncea*: identification of restorer and molecular characterization. *Euphytica*, 159:145–152.

Bonhomme, S., F. Budar, M. Ferault, and G. Pelletier. 1991. A 2.5 kb NcoI fragment of Ogura radish mitochondrial DNA is correlated cytoplasmic male sterility in *Brassica* cybrids. *Curr. Genet.*, 19:121–127.

Bonhomme, S., F. Budar, D. Lancelin, I. Small, M.C. Defrance, and G. Pelletier. 1992. Sequence and transcript analysis of the Nco2.5 Ogura-specific fragment correlated with cytoplasmic male sterility in *Brassica* cybrids. *Mol. Gen. Genet.*, 235:340–348.

Bonnet, A. 1975. Introduction et utilization d'une stérilité male cytoplasmique dans des variétés précoces européennes de radis *Raphanus sativus* L. *Ann. Amélior Plantes*, 25:381–397.

Brandle, J. and P. McVetty. 1990. Geographical diversity, parental selection and heterosis in oilseed rape. *Can. J. Plant Sci.*, 70:35–40.

Brown, G.G., N. Formanová, H. Jin, et al. 2003. The radish *Rfo* restorer gene in Ogura cytoplasmic male sterility encodes a protein with multiple pentatricopeptide repeats. *Plant J.*, 35:262–272.

Budar, F., R. Delourme, and G. Pelletier. 2004. Male sterility. In: E.C. Pua and C.J. Douglas (Eds.), *Biotechnology in Agriculture and Forestry*. Springer, NewYork. Vol. 54, p. 43–64.

Carlsson, J., M. Leino, J. Sohlberg, J.F. Sundström, and K. Glimelius. 2008. Mitochondrial regulation of flower development. *Mitochondrion*, 8:74–86.

Delourme, R. and F. Budar. 1999. Male Sterility. In: C. Gómez-Campo (Ed.), *Biology of Brassica coenospecies*. Elsevier Science, Amsterdam. p. 185–216.

Delourme, R. and F. Eber. 1992. Linkage between an isozyme marker and a restorer gene in radish cytoplasmic male sterility of rapeseed (*Brassica napus* L.). *Theor. Appl. Genet.*, 85:222–228.

Delourme, R., A. Bouchereau, M. Renard, and B.S. Landry. 1994a. Identification of RAPD markers linked to a fertility restorer gene for the Ogura radish cytoplasmic male sterility of rapeseed (*Brassica napus* L.). *Theor. Appl. Genet.*, 88:741–748.

Delourme, R., F. Eber, and M. Renard. 1994b. Transfer of radish cytoplasmic male sterility from *Brassica napus* to *B. juncea* and *B. rapa*. *Cruciferae Newslett.*, 16:79.

Deol J.S., K.R. Shivanna, S. Prakash, and S.S. Banga. 2003. *Enarthrocarpus lyratus*-based cytoplasmic male sterility and fertility restorer system in *Brassica rapa*. *Plant Breed.*, 122:438–440.

Desloire, S., H. Gherbi, W. Laloui, et al. 2003. Identification of the fertility restoration locus, *Rfo*, in radish, as a member of the pentatricopeptide repeat protein family. *EMBO Rep.*, 4:588– 594.

Diers, B.W., P.B.E McVetty, and T.C. Osborn. 1995. Relationship between hybrid performance and genetic diversity based on restriction fragment length polymorphism markers in oilseed rape (*Brassica napus* L.). *Crop Sci.*, 36:79–83.

Falk, K., G. Rakow, R. Downey, and D. Spurr. 1994. Performance of inter-cultivar summer turnip rape hybrids in Saskatchewan. *Can. J. Plant Sci.*, 74:441–445.

Fu, T., G. Yang, and X. Yang. 1990. Studies on " three line" Polima cytoplasmic male sterility developed in *Brassica napus. Plant Breed.*, 104:115–120.

Gaikwad, K., A. Baldev, P.B. Kirti, T. Mohapatra, S.R. Bhat, and S. Prakash. 2006. Organization and expression of mitochondrial genome in CMS (*Moricandia*) *Brassica juncea*: Nuclear-mitochondrial incompatibility results in differential expression of mitochondrial *atpA* gene. *Plant Breed.*, 125:623–628.

Ghosh, S.K., S.C. Gulati, and R. Raman. 2002. Combining ability and heterosis for seed yield and its components in Indian mustard (*Brassica juncea*). *Indian J. Genet.*, 62:29–93.

Grelon, M. F. Budar, S. Bonhomme, and G. Pelletier. 1994. Ogura cytoplasmic male sterility (CMS)-associated *orf138* is translated into a mitochondrial membrane polypeptide in male sterile *Brassica* cybrids. *Mol. Gen. Genet.*, 243:540–547.

Harberd, D.J. 1972. A contribution to cytotaxonomy of *Brassica* (Cruciferae) and its allies. *Bot. J. Linn. Soc.*, 65:1–23.

Heyn, F.W. 1976. Transfer of restorer genes from *Raphanus* to cytoplasmic male sterile *Brassica napus. Cruciferae Newslett.*, 1:15–16.

Hinata, K. and N. Konno. 1979. Studies on a male-sterile strain having the *Brassica campestris* nucleus and the *Diplotaxis muralis* cytoplasm. *Japan J. Breed.*, 29:305–311.

Hu, Q., Y. Li, D. Mei, X. Fang, L.N. Hansen, and S.B. Andersen. 2004. Establishment and identification of cytoplasmic male sterility in *Brassica napus* by intergeneric somatic hybridization. *Scientia Agricultura Sinica*, 37:333–338.

Jain, A., S. Bhatia, S.S. Banga, S. Prakash, and M. Lakshmikumaran. 1994. Potential use of random amplified polymorphic DNA (RAPD) to study the genetic diversity in Indian mustard (*Brassica juncea* (L) Czern and Coss) and its relationship with heterosis. *Theor. Appl. Genet.*, 88:116–122.

Janeja, H.S., S.K. Banga, P.B. Bhasker, and S.S. Banga. 2003. Alloplasmic male sterile *Brassica napus* with *Enarthrocarpus lyratus* cytoplasm: introgression and molecular mapping of an *E. lyratus* chromosome segment carrying a fertility restoring gene. *Genome*, 46:792–797.

Jarl, C.I. and C.H. Bornman. 1988. Correction of chlorophyll-defective, male sterile winter oilseed rape (*Brassica napus*) through organelle exchange: phenotypic evaluation of progeny. *Hereditas*, 108:97–102.

Jourdan, P.S., E.D. Earle, and M.A. Mutschler. 1989a. Atrazine-resistant cauliflower obtained by somatic hybridization between *Brassica oleracea* and ATR-*B. napus. Theor. Appl. Genet.*, 78:271–279.

Jourdan, P.S., E.D. Earle, and M.A. Mutschler. 1989b. Synthesis of male sterile, triazine-resistant *Brassica napus* by somatic hybridization between cytoplasmic male sterile *B. oleracea* and atrazine-resistant *B. campestris. Theor. Appl. Genet.*, 78:445–455.

Kameya, T., H. Kanzaki, S. Toki, and T. Abe. 1989. Transfer of radish (*Raphanus sativus* L.) chloroplasts into cabbage (*Brassica oleracea* L.) by protoplast fusion. *Japan. J. Genet.*, 64:27-34.

Kao, H.M., W.A. Keller, S. Gleddie, and G.G. Brown. 1992. Synthesis of *Brassica oleracea/Brassica napus* somatic hybrid plants with novel organelle DNA compositions. *Theor. Appl. Genet.*, 83:313–320.

Kirti, P.B., S.B. Narasimhulu, T. Mohapatra, S. Prakash, and V.L. Chopra. 1993. Correction of chlorophyll deficiency in alloplasmic male sterile *Brassica juncea* through recombination between chloroplast genome. *Genet. Res. Camb.*, 62:11–14.

Kirti, P.B., S.S Banga, S. Prakash, and V.L. Chopra. 1995a. Transfer of Ogu cytoplasmic male sterility to *Brassica juncea* and improvement of male sterile through somatic cell fusion. *Theor. Appl. Genet.*, 91:517–521.

Kirti, P.B., T. Mohapatra, S. Prakash, and V.L. Chopra. 1995b. Development of a stable cytoplasmic male sterile line of *Brassica juncea* from somatic hybrid *Trachystoma ballii* + *Brassica juncea. Plant Breed.*, 114:434–438.

Kirti, P.B., T. Mohapatra, H. Khanna, S. Prakash, and V.L. Chopra. 1995c. *Diplotaxis catholica* + *Brassica juncea* somatic hybrids: molecular and cytogenetic characterization. *Plant Cell Rep.*, 14:593–597.

Kirti, P.B., A. Baldev, K. Gaikwad, S.R Bhat, V. Dineshkumar, S. Prakash, and V.L. Chopra. 1997. Introgression of a gene restoring fertility to CMS (*Trachystoma*) *Brassica juncea* and the genetics of restoration. *Plant Breed.*, 116:259–262.

Kirti, P.B., S. Prakash, K. Gaikwad, S.R. Bhat, V. Dineshkumar, and V.L. Chopra. 1998. Chloroplast substitution overcomes leaf chlorosis in *Moricandia arvensis* based cytoplasmic male sterile *Brassica juncea. Theor. Appl. Genet.*, 97:1179–1182.

Koizuka, N., R. Imai, M. Iwabuchi, T. Sakai, and J. Inamura. 2000. Genetic analysis of fertility restoration and accumulation of ORF125 mitochondrial protein in the Kosena radish (*Raphanus sativus* cv. Kosena) and a *Brassica napus* restorer line. *Theor. Appl. Genet.*, 100:949–955.

Koizuka, N., R. Imai, H. Fujimoto, et al. 2003. Genetic characterization of pentatricopeptide repeat gene, *orf687*, that restorers fertility in the cytoplasmic male sterile Kosena radish. *Plant J.*, 34:407–415.

Krishnasamy, S. and C.A. Makaroff. 1993. Characterization of radish mitochondrial orfB locus: possible relationship with male sterility in Ogura radish. *Curr. Genet.*, 24:156–163.

Krishnasamy, S. and C.A. Makaroff. 1994. Organ-specific reduction in the abundance of a mitochondrial protein accompanies fertility restoration in cytoplasmic male sterile radish. *Plant Mol. Biol.*, 26:935–946.

Leino, M., R. Teixeira, M. Landgren, and K. Glimelius. 2003. *Brassica napus* lines with rearranged *Arabidopsis* mitochondria display CMS and a range of developmental aberrations. *Theor. Appl. Genet.*, 106:1156–1163.

Liu, J.H., M. Landgren, and K. Glimelius. 1996. Transfer of the *Brassica tournefortii* cytoplasm to *B. napus* for the production of cytoplasmic male sterile *B. napus*. *Physiol. Plant.*, 96:123–129.

Mackenzie, S. and L. McIntosh. 1999. Higher plant mitochondria. *Plant Cell*, 11:571–585.

Makaroff, C.A. and J.D. Palmer. 1988. Mitochondrial DNA rearrangements and transcriptional alterations in the male sterile cytoplasm of Ogura radish. *Mol. Cell. Biol.*, 8:1474–1480.

Malik, M., P. Vyas, N.S. Rangaswamy, and K.R. Shivanna. 1999. Development of two new cytoplasmic male-sterile lines of *Brassica juncea* through wide hybridization. *Plant Breed.*, 118:75–78.

Mathias, R. 1985. Transfer of cytoplasmic male sterility from brown mustard (*Brassica juncea* L. Czern.) into rapeseed (*Brassica napus* L.). *Z. Pflanzenzüchtg*, 95:371–374.

Matsuzawa, Y., S. Mekiyanon, Y. Kaneko, S.W. Bang, K. Wakui, and Y. Takahata. 1999. Male sterility in alloplasmic *Brassica rapa* L. carrying *Eruca sativa* cytoplasm. *Plant Breed.*, 118:82-84.

McCollum, G. 1981. Induction of an alloplasmic male sterile *Brassica oleracea* by substituting cytoplasm from "Early Scarlet Glove" (*Raphanus sativus*). *Euphytica*, 30:855–859.

McGrath, J.M. and C.F. Quiros. 1992. Genetic diversity at isozyme and RFLP loci in *Brassica campestris* as related to crop type and geographical origin. *Theor. Appl. Genet.*, 83:783–790.

Mei, D., Y. Li, and Q. Hu. 2003. Study of male sterile line derived from intergeneic hybrids of *Brassica napus* + *Orychophragmus violaceus* and *B. napus* + *Sinapis arvensis*. *Chinese J. Oil Crop Sci.*, 25:72–75.

Meur, G., K. Gaikwad, S.R. Bhat, S. Prakash, and P.B. Kirti. 2006. Homeotic-like modification of stamens to petals is associated with aberrant mitochondrial gene expression in ctytoplasmic male sterile Ogura *Brassica juncea*. *J. Genet.*, 85:133–139.

Narain, A. and S. Prakash. 1972. Investigations on the artificial synthesis of amphidiploids of *Brassica tournefortii* Gouan with other elementary species of *Brassica*. I. Genomic relationships. *Genetica*, 43:90–97.

Ogura, H. 1968. Studies on a new male-sterility in Japanese radish, with special reference to utilization of this sterility towards the practical raising of hybrid seeds. *Mem. Fac. Agr. Kogoshima Univ.*, 6:39–78.

Pathania, A., S.R. Bhat, V. Dinesh Kumar, Asutosh, S. Prakash, and V.L. Chopra. 2003. Cytoplasmic male sterility in alloplasmic *Brassica juncea* carrying *Diplotaxis catholica* cytoplasm: molecular characterization and genetics of fertility restoration. *Theor. Appl. Genet.*, 107:455–461.

Pathania, A., R. Kumar, V. Dinesh Kumar, Ashutosh, K.K. Dwivedi, P.B. Kirti, S. Prakash, V.L. Chopra, and S.R. Bhat. 2007. A duplication of *CoxI* gene is associated with CMS (*Diplotaxis catholica*) *Brassica juncea* derived from somatic hybridization with *Diplotaxis catholica*. *J. Genet.*, 86:93–101.

Paulmann, W. and G. Röbbelen. 1988. Effective transfer of cytoplasmic male fertility from radish (*Raphanus sativus* L.) to rape (*Brassica napus* L.). *Plant Breed.*, 100:299–309.

Pearson, O.H. 1972. Cytoplasmically inherited male sterility characters and flavor components from the species cross *Brassica nigra* (L.) Koch × *B. oleracea* L. *J. Am. Soc. Hort. Sci.*, 97:397–402.

Pellan-Delourme, R. and M. Renard. 1987. Identification of maintainer genes in *Brassica napus* L. for male sterility inducing cytoplasm of *Diplotaxis muralis* L. *Plant Breed.*, 99:89–97.

Pellan-Delourme, R. and M. Renard. 1988. Cytoplasmic male sterility in rapeseed (*Brassica napus* L.): female fertility of restored rapeseed with "Ogura" and cybrid cytoplasms. *Genome*, 30:234–238.

Pelletier, G., C. Primard, F. Vedel, P. Chétrit, R. Rémy, P. Rousselle, and M. Renard. 1983. Intergeneric cytoplasmic hybridization in Cruciferae by protoplast fusion. *Mol. Gen. Genet.*, 191:244–250.

Pelletier, G., M. Ferrault, D. Lancelin, and L. Boulidard. 1989. CMS *Brassica oleracea* cybrids and their potential for hybrid seed production. *12th Eucarpia Congr.* Göttingen. 7:15.

Plieske, J. and D. Struss. 2001. Microsatellite markers for genome analysis in *Brassica*. I. Development in *Brassica napus* and abundance in *Brassicaceae* species. *Theor. Appl. Genet.*, 102:689–694.

Pradhan, A.K., A. Mukhopadhyay, and D. Pental. 1991. Identification of putative cytoplasmic donor of CMS system in *Brassica juncea*. *Plant Breed.*, 106: 204–208.

Pradhan, A.K., S. Prakash, A. Mukhopadhyay, and D. Pental. 1992. Phylogeny of *Brassica* and allied genera based on variation in chloroplast and mitochondrial DNA patterns: molecular and taxonomical classifications are incongruous. *Theor. Appl. Genet.*, 85:331–340.

Pradhan, A.K., Y.S. Sodhi, A. Mukhopadhyay, and D. Pental. 1993. Heterosis breeding in Indian mustard (*Brassica juncea* L. Czern. and Coss): analysis of component characters contributing to heterosis for yield. *Euphytica*, 69:219–229.

Prakash, S. 2001. Utilization of wild germplasm of *Brassica* allies in developing male sterility-fertility restoration systems in Indian mustard-*Brassica juncea*. In: L. Houli and T.D. Fu (Eds.), *Proc. Int. Symp. Rapeseed Science*, Huazhong Agricultural Univ., China, Science Press, New York. p. 73–78.

Prakash, S. and V.L. Chopra. 1988. Synthesis of alloplasmic *Brassica campestris* as a new source of cytoplasmic male sterility. *Plant Breed.*, 101:235–237.

Prakash, S. and V.L. Chopra. 1990. Male sterility caused by cytoplasm of *Brassica oxyrrhina* in *B. campestris* and *B. juncea. Theor. Appl. Genet.*, 79:285–287.

Prakash, S., I. Ahuja, H.C. Uprety, V.D. Kumar, S.R. Bhat, P.B. Kirti, and V.L. Chopra. 2001. Expression of male sterility in alloplasmic *Brassica juncea* with *Erucastrum canariense* cytoplasm and development of fertility restoration system. *Plant Breed.*, 120:178–182.

Prakash, S., P.B. Kirti, S.R. Bhat, K. Gaikwad, V. Dineshkumar, and V.L Chopra. 1998. A *Moricandia arvensis* based cytoplasmic male sterility and fertility restoration system in *Brassica juncea. Theor. Appl. Genet.*, 97:488–492.

Prakash, S, S.R. Bhat, C.F. Quiros, P.B. Kirti, and V.L. Chopra. 2009. *Brassica* and its close allies: cytogenetics and evolution. *Plant Breed. Rev.*, 31:21–187.

Primard, C., J.P. Poupard, R. Horvais, et al. 2005. A new recombined double low restorer line for the Ogu-INRA CMS in rapeseed. *Theor. Appl. Genet.*, 111.736–745.

Rao, G.U. and K.R. Shivanna. 1996. Development of a new alloplasmic CMS *Brassica napus* in the cytoplasmic background of *Diplotaxis siifolia. Cruciferae Newslett.*, 18:68–69.

Rao, G.U., V.S. Batra, S. Prakash, and K.R. Shivanna. 1994. Development of a new cytoplasmic male sterile system in *Brassica juncea* through wide hybridization. *Plant Breed.*, 112:171–174.

Rawat, D.S. and I.J. Anand. 1979. Male sterility in Indian mustard. *Indian J. Genet.*, 39:412–415.

Renard, M., R. Delourme, P. Vallee, and J. Pierre. 1998. Hybrid rapeseed breeding and production. *Acta Hort.*, 459:291–298.

Riaz, A., G. Li, Z. Quresh, M.S. Swati, and C.F. Quiros. 2001. Genetic diversity of oilseed *Brassica napus* inbred lines based on sequence-related amplified polymorphism and its relation to hybrid performance. *Plant Breed.*, 120:411–415.

Rouselle, P. 1982. Premiers résultats d'un programme d'introduction de l'androstérilité 'Ogura' du redis chez le colza. *Agronomie*, 2:859–864.

Rouselle, P. and F. Dosba. 1985. Restauration de la fertilite pour l'androsterilite genocytoplasmique chez le colza (*Brassica napus* L.). Utilization des *Raphano-Brassica. Agronomie*, 5:431–437.

Sakai, T. and J. Inamura. 1990. Intergeneric transfer of cytoplasmic male sterility between *Raphanus sativus* (CMS line) and *Brassica napus* through cytoplast-protoplast fusion. *Theor. Appl. Genet.*, 80:421–427.

Sakai, T., H. J. Liu, M. Iwabuchi, J. Kohno-Murase, and J. Inamura. 1996. Introduction of a gene from fertility restored radish (*Raphanus sativus*) into *Brassica napus* by fusion of x-irradiated protoplasts from a radish restorer line and iodacetoamide-treated protoplasts from a cytoplasmic male-sterile cybrid of *B. napus. Theor. Appl. Genet.*, 93:73–379.

Seyis, F., W. Friedt, and W. Lühs. 2006. Yield of *Brassica napus* L. hybrids developed using resynthesized rapeseed material sown at different locations. *Field Crops Res.*, 96:176–180.

Shen, J., G. Lu, T. Fu, and G. Yang. 2003. Relationship between heterosis and genetic distance based on AFLPs in *Brassica napus*. In: *Proc 11th Int. Rapeseed Congr.* Copenhagen, Denmark. 6–10 July, 2003. p. 1:343–345.

Shiga, T. and S. Baba. 1973. Cytoplasmic male sterility in oilseed rape (*Brassica napus* L.), and its utilization to breeding. *Japan J. Breed.*, 23:187–193.

Sodhi, Y. A. Pradhan, J. Verma, N. Arumugam, A. Mukhopadhyay, and D. Pental. 1994. Identification and inheritance of fertility restorer genes for "*tour*" CMS in rapeseed (*Brassica napus* L.). *Plant Breed.*, 112:223–227.

Stiewe, G. and G. Röbbelen. 1994. Establishing cytoplasmic male sterility in *Brassica napus* by mitochondrial recombination with *B. tournefortii. Plant Breed.*, 113:294–304.

Walters, T., M. Mutschler, and E. Earle. 1992. Protoplast fusion-derived Ogura male sterile cauliflower with cold tolerance. *Plant Cell Rep.*, 10:624–628.

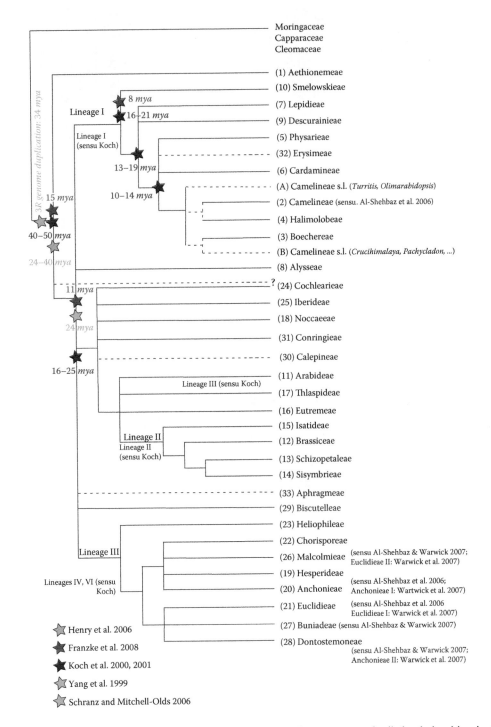

FIGURE 1.1 Synopsis of phylogenetic hypothesis from various sources of tribal relationships in the Brassicaceae family (for details, refer to the text). Lineage I–III are described in Beilstein et al. (2006). Koch et al. (2007) used different numbers given also as "sensu Koch," and we suggest the use of Beilstein's version only to avoid future confusion. Dashed lines indicate uncertain phylogenetic position. However, it should be kept in mind that this synopsis is not derived from one single phylogenetic analysis.

FIGURE 3.1 Morpho-physiological variation in *Diplotaxis*: (A) pre-flowering plants of *D. cretacea*; (B) silique arrangement in *D. cretacea*; (C) flowers of *D. cretacea*; (D) silique structure of *D. cretacea*; (E) a young plant of *D. berthautii*; (F) a young plant of *D. harra*; (G) a young plant of *D. harra* ssp. *Crassifolia*; (H) silique structure of *D. harra* ssp. *Crassifolia*; (I) flowering plants of *D. muralis*; (J) silique arrangement in *D. muralis*; (K) leaf structure of *D. muralis*; (L) silique structure of *D. muralis*; (M) a flowering plant of *D. siettiana*; (N) silique arrangement in *D. siettiana*; (O) silique structure in *D. siettiana*; (P) flowers of *D. siifolia*; (Q) flowering plants of *D. tenuifolia*; (R) leaf structure of *D. tenuifolia*; (S) a flower of *D. tenuifolia*; and (T) silique structure of *D. tenuifolia*.

FIGURE 3.4 Morpho-physiological variation in *Erucastrum*: (A) flowers of *E. abyssinicum;* (B) siliques of *E. abyssinicum;* (C) a young plant of *E. arabicum;* (D) pre-flowering plant of *E. virgatum;* (E) flowers of *E. gallicum;* (F) pod arrangement in *E. gallicum;* (G) leaf structure of *E. gallicum;* (H) flowers of *E. laevigatum;* (I) pre-flowering plant of *E. laevigatum;* (J) leaf structure in *E. laevigatum;* (K) pre-flowering plant of *E. laevigatum* ssp. *Elongatum;* (L) pre-flowering plant of *E. nasturtifolium;* (M) flowers of *E. nasturtifolium;* and (N) a young plant of *E. littoreum*

FIGURE 3.5 Morpho-physiological variation in *Trachystoma, Moricandia, Crambe, Hirschfeldia,* and *Sinapis*: (A) flowers of *T. ballii*; (B) pre-flowering plants of *T. ballii*; (C) leaf structure of *T. ballii*; (D) a flowering plant of *M. arvensis*; (E) plants of *Crambe abyssinica* at reproductive phase; (F) inflorescence *of C. abyssinica*; (G) spherical fruits of *C. abyssinica*; (H) leaf structure of *C. abyssinica*; (I) pre-flowering plants of *H. incana*; (J) flowers of *H. incana*; (K) leaf structure in *H. incana*; (L) pre-flowering plants of *H. incanca* ssp. *cosobrina*; (M) a young plant of *S. alba*; (N) pre-flowering plants of *S. pubescens*; (O) flowers of *S. pubescens*; and (P) leaf structure of *S. pubescens*.

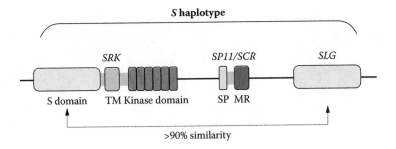

FIGURE 6.1 Linkage of *SRK*, *SLG*, and *SP11/SCR* at the *S* locus. (TM = transmembrane domain; SP = signal peptide; MR = region of mature protein.)

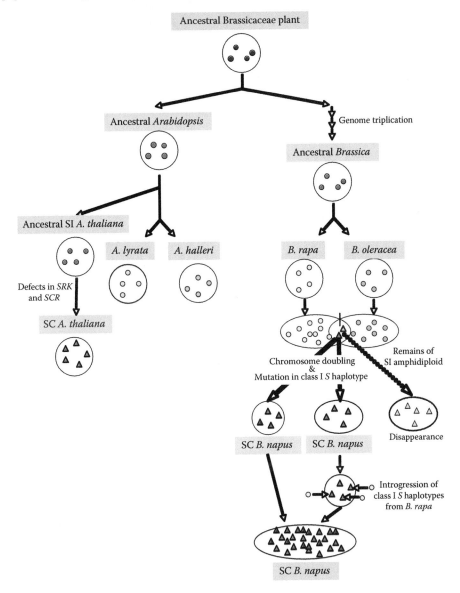

FIGURE 6.2 A model of evolution of *S* haplotypes and establishment of SC species in *Arabidopsis* and *Brassica*. (SC = self-compatibility; SI = self-incompatibility.)

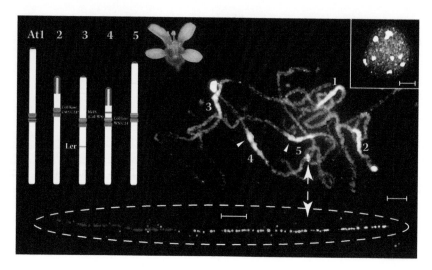

FIGURE 10.1 Idiogram of *Arabidopsis thaliana* (*n* = 5) and fluorescence *in situ* localization (FISH) of several DNA probes to pachytene bivalents isolated from young anthers of *Arabidopsis* accession C24. Left: *Arabidopsis* idiogram showing prominent chromosomal landmarks — NORs containing 45S rDNA on At2 and 4 (red); 5S rDNA loci on At3, 4, and 5 (green; note the variation in number of 5S loci among the accessions); the heterochromatic knob hk4S on the top arm of At4 in accessions Col and WS; the 180-bp pAL repeat (blue) at centromeres embedded within pericentromeric heterochromatin regions (gray), and telomeric repeats at chromosome termini (purple; note the interstitial site at pericentromere of At1). Sizes of chromosome arms and NORs follow Lysak et al. (2006b), localization of 5S rDNA loci follows Fransz et al. (1998) and Sanchez-Moran et al. (2002; labeled by asterisk). Right: *Arabidopsis pachytene* chromosomes after simultaneous *in situ* hybridization of (i) 45S rDNA (NORs of At2 and 4 labeled by red), (ii) 5S rDNA (green signals at pericentromeric heterochromatin of top arms of At4 and 5; arrowheads), (iii) BAC contig spanning ~6.5 Mb of the top arm of At1 (purple), (iv) three BAC clones (72, 95, and 40 kb) of the At5 bottom arm visualized as red, green, and yellow fluorescence, respectively. Bottom: FISH of the same BAC triple to extended DNA fibers. Top right: *Arabidopsis* interphase nucleus with conspicuous chromocenters. Chromosomes and the nucleus counterstained by DAPI. All bars, 5 μm.

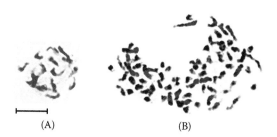

(A)   (B)

FIGURE 10.2 Genomic *in situ* hybridization (GISH) to flower-bud mitotic chromosomes of *Lepidium*. (A) Chromosomes of *L. africanum* (2*n* = 16) probed with fluorescently labeled *L. africanum* genomic DNA, (B) chromosomes of *L. hyssopifolium* (2*n* ≈ 72) probed with fluorescently labeled *L. africanum* genomic DNA. GISH revealed 16 *L. africanum*-derived chromosomes (red) within the *L. hyssopifolium* complement. Chromosomes were counterstained by DAPI and the resulting fluorescence displayed in grey. Bar, 5 μm. (*Source:* Courtesy of T. Dierschke and K. Mummenhoff.)

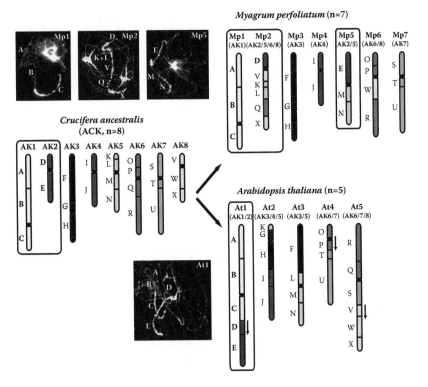

FIGURE 10.3 Ancestral Crucifer Karyotype (ACK, *n* = 8; *Crucifera ancestralis*) and the reconstruction of karyotype evolution on the example of *Myagrum perfoliatum* (*n* = 7; Isatideae, Lineage II) and *Arabidopsis thaliana* (*n* = 5; Camelineae, Lineage I) based on CCP analysis. ACK comprises eight chromosomes (AK1–8) and 24 genomic blocks (A–X; Schranz et al., 2006). Chromosome number was reduced from *n* = 8 in ACK toward *n* = 7 in *M. perfoliatum* (Mp1–7) and *n* = 5 in Arabidopsis (At1–5), respectively. The reductions have been accompanied by gross inversion and translocation events resulting in three and five "fusion" chromosomes and the loss of one and three centromeres in *M. perfoliatum* and *A. thaliana*, respectively. Examples of multicolor chromosome painting analysis of ancestral chromosomes AK1 and AK2 within pachytene chromosome complements of both species. In *M. perfoliatum*, AK1 (blocks A–C) is preserved as chromosome Mp1, whereas AK2 (blocks D and E) participated in the origin of "fusion" chromosomes Mp2 (AK2/5/6/8) and Mp5 (AK2/5). In Arabidopsis, AK1 and AK2 were combined by a reciprocal translocation into At1 (AK1/2). In ACK, genome blocks are considered to be in the upright orientation. Blocks inverted relative to ACK are represented by downward-pointing arrows; NORs and heterochromatic knobs not indicated. Note that pseudocoloring of fluorescently labeled painting probes in microscopic photographs does not correspond to the color coding of AK chromosomes (block Q has not been included in the painting probe in *M. perfoliatum*). Chromosomes counterstained by DAPI. (*Source:* After Lysak et al., 2006; Mandáková and Lysak, unpublished; Schranz et al., 2006.)

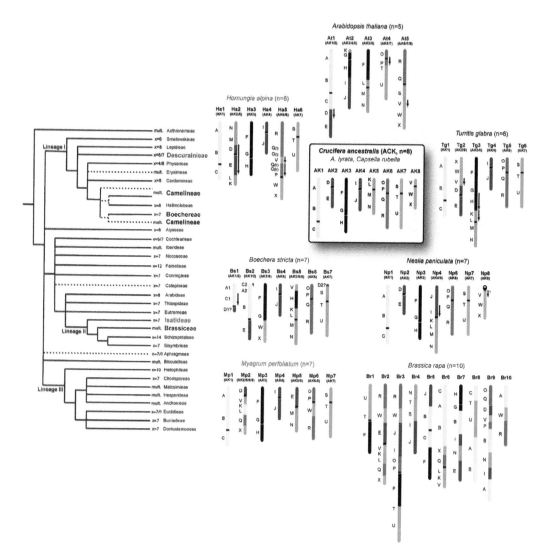

FIGURE 10.4 Phylogenetic relationships among tribes of Brassicaceae (modified from Koch and Al-Shehbaz, Chapter 1 in this book) with the placement of schematic karyotypes currently reconstructed by comparative genetic and chromosome painting analysis. Three major phylogenetic lineages (Lineage I–III) are indicated (sensu Beilstein et al. 2006). Sole or prevailing base chromosome number (x) is given for each tribe (mult. = multiple base chromosome numbers). The color coding of ancestral chromosomes (AK1–8) and 24 conserved genomic blocks (A–X; Schranz et al., 2006) indicate the relationship of reconstructed karyotypes to the Ancestral Crucifer Karyotype (ACK), represented by the *A. lyrata* karyotype. In ACK, genomic blocks are considered to be in the upright orientation; blocks inverted relative to ACK are represented by downward-pointing arrows. NORs and heterochromatic knobs not indicated.

8 Plant–Insect Interaction

D.P. Abrol

CONTENTS

INTRODUCTION

Animal life cannot exist without green plants; and wherever plants grow, insects are also found. Interactions between plants and insects can be both antagonistic and mutualistic (Schoonhoven, 1998), of which the most studied are the antagonistic relationship of insects feeding on plants and the mutualistic relationship of plant pollination by insects. Both interactions have been important in shaping the two interacting groups of organisms throughout evolutionary time. Plants have evolved several different types of defenses against herbivores, and herbivores have evolved different traits to overcome these defenses. The attributes of flowers advertising nectar resources for pollinating insects have also evolved to interplay with insects. When this kind of interdependent evolution leads to properties that are unique for the system, it is defined as coevolution (Allaby, 1998). In the coevolution of plants and herbivores, plants have evolved the ability to produce (sometimes toxic) secondary plant compounds to counteract herbivory (Thompson, 1994; Pilson, 2000). Plants can respond to herbivory either by defense (a decrease in susceptibility to herbivore damage) or tolerance (a decrease in the per-unit effect of herbivory on plant fitness). Plant defense mechanisms include, among others, feeding barriers (e.g., wax layers on leaves) and the production of secondary compounds that act as feeding deterrents or as toxins. It is not uncommon that plants use various strategies simultaneously to minimize herbivory (Nielsen, 1977; Palaniswamy et al., 1998; Pilson, 2000; Shinoda et al., 2002; Agerbirk et al., 2003a). If herbivores feeding on these plants have no alternative host, they need to adapt to these defenses. Their counter-defenses represent natural adaptations that have evolved over a long time — in contrast, for example, to the resistance to pesticides that represents accelerated evolution (Wilson, 2001).

Many phytophagous insects have become specialized in utilizing a rather limited number of plant species. This is thought to reflect the evolution of particular adaptations to shared defensive chemistry across the host plants (Nielsen, 1977; Thompson, 1994; Agerbirk, et al., 2003b). The ability to utilize crucifers (Capparales: *Brassicaceae*) as host plants, for example, has evolved several times within the Chrysomelidae (Coleoptera) (Nielsen, 1988). Many crucifers are crop plants of commercial value. Examples include yellow rocket (*Barbarea vulgaris*), rapeseed (*Brassica napus*), and cabbage and cauliflower (*Brassica oleracea*). Flea beetles feeding on these crucifer species are considered pests. The fact that some (wild) crucifers are unsuitable as host plants has attracted considerable interest. The *B. vulgaris* defense mechanism has proven useful in dead-end trap cropping (Shelton and Nault, 2004), and some of the crucifers are regarded as potential sources of defense genes that might be transferred to the species used as crops to create genetically engineered, herbivory-free cultivars (Palaniswamy and Lamb, 1998; Tattersall et al., 2001). Thus, rather than using insecticides, the spread and numbers of flea beetles can be controlled in a more environmentally friendly way using a manipulation of plant secondary compounds (Gavoski et al., 2000). However, just as insects can evolve resistance to insecticides (*Lucila cuprina* and *Culex pipiens*), they might also become resistant to plant secondary compounds, in a parallel manner to their ability to use novel crucifers. The yellow-striped flea beetle (*Phyllotreta nemorum*) in Denmark serves as a good example to explore these processes in a natural insect-plant system.

Oilseed *Brassica* contribute substantially to the oil economy of the world, including India, in the form of oil yield and their byproducts for industrial use. Rapeseed-mustard is one of the important oleiferous crops and constitutes a major source of edible oil for human consumption and cake for animals. In India, *Brassica* contributes 32% of the total oilseed production, and is the second largest indigenous oilseed crop. Rapeseed (*Brassica campestris* L.) and mustard (*Brassica juncea* L.) are the major *rabi* oilseed crops of India. However, the average productivity of these crops in India is very low (7.7 q/ha) as compared to the worldwide average of 13.4 q/ha [FAO, 1998]. One of the major constraints responsible for the low yield of rapeseed and mustard crops is the attack of insect pests and insufficient population of pollinating insects that are killed due to indiscriminate use of insecticides to control crop pests.

PROBLEM OF INSECT PESTS

Cruciferous crops are subjected to the attack of insect pests, right from the seedling stage through to the pod formation stage, which act as limiting factors in the profitable cultivation of this crop (Sachan and Srivastva, 1972; Yadva and Sachan, 1976; Sachan and Gangwar, 1980; Sharma and Singh, 1990). Of the more than three dozen injurious insect species, painted bug (*Bagrada cruciferarum* (Kirk.)), flea beetle (*Phyllotreta cruciferae* (Goeze)), mustard sawfly (*Athalia proxima lugens* (Klug)), pea leaf-miner (*Chromatomyia horticola* (Goureau)), and mustard aphid (*Lipaphis erysimi* (Kalt.)) are the major and regular ones inflicting severe yield losses to these crops. Some of the sporadic insect pests (e.g., *Agrotis segatum* Denis and Schiffermuller, *Spilosoma* sp., *Plutella xylostella* (Linn.), *Crocidolomia binotalis zeller, Pieris brassicae* (Linn.), and *Myzus persicae* (Sulzer)) are also issues of concern at one or more locations in India (Singh and Singh, 1983; Vora et al., 1985a; Bakhetia and Sekhon, 1989) and abroad. Recently, a pod midge (*Dasineura hisarensis* Sharma and Singh), which was hitherto unknown in Asian countries (Singh et al, 1990) was found to cause up to a 5% loss in yield of these crops in Haryana. Of all these pests, mustard aphid [*Lipaphis erysimi* (Kalt.) (Aphididae: Homoptera)] is a key pest of rapeseed-mustard, causing a 35% to 75% reduction in yield (Bakhetia, 1983; Rohilla et al., 1987; Kumar, 1991) and a 6% reduction in oil content (Singh et al., 1987). Yield losses due to mustard aphid can vary with the variety, agrotechnological practices, and environmental factors (Plates 8.1, 8.2).

PLATE 8.1   (a) Aphid (*Liphaphis erysimi*) infestation on mustard; (b) *Coccinella septempunctata* natural enemy of mustard aphid.

PLATE 8.2   *A. mellifera* working on *Brassica* flowers.

Effective Bioagents of Mustard Aphid

Like other organisms, the mustard aphid has several natural enemies that can be used for its management. They include 11 parasitoids, 52 predators, 2 pathogens, and 1 predatory bird. Manual removal of the mustard aphid infested portions of inflorescence twigs at 15-day intervals initiated after the appearance of the pest was found safer to parasitoids and predators, free from environmental pollution, and more economical than chemical control (Singh et al., 1992).

Chemical Control

A great deal of work has been reported on the chemical control of mustard aphid in India (Kakar and Dogra, 1977). Bakhetia (1984) inferred that the application of granular insecticides at sowing time was not necessary and would be more useful if applied on noticing aphid incidence on the crop. Hazarika and Saharia (1981) concluded that application of granular insecticides kept the aphid population low on mustard crop but was uneconomical on the basis of cost-benefit ratio and yields. Kalra and Gupta (1986) and Singh et al. (1987) opined that oxydemeton methyl (0.025%), phosphamidon (0.03%), and endosulfan (0.035%) can be better utilized in the integrated management of mustard aphid, as these insecticides were found safe to the pollinators. Tripathi et al. (1988) and Khurana and Batra (1989) reported that oxydemeton methyl, dimethoate, and endosulfan were safer to predators and effective in keeping the aphid population under check for longer time periods as compared to malathion and chlorpyriphos. Pandey et al. (1987) reported that the use of 1.5% extracts of *Azadirachta indica* A. juss, *Lantana camera* L., and *Ipomea carnea*. Jacq gave 80%, 66.6%, and 66.6% mortality of *Lipaphis erysimi*, respectively, and also significantly reduced the fecundity. Arora et al. (1982), while studying the juvenile hormone activity of seven terpenoid lactones of *L. erysimi*, observed that a 0.1% quaianolide spray resulted in 98.3% malformed individuals in the first instar and 100% malformed individuals in the second and third instars.

Srivastava and Guleria (2003) evaluated 34 extracts against mustard aphid and *Azadirachta indica* as a check for the treatments. It was found that all treatments showed insecticidal activity against aphid. The extract from *Chrysanthemum, Calotropis procera*, gave results on par with *A. indica*. The other plant extracts (*Zingiber officinale, Ageratum conyzoides, Lantana camera, Pinus roxburghii, Allium sativum, Ricinus communis, Cymbopogon citrates,* and *Hevea brasiliensis*) yielded very good results.

Painted Bug

Painted bug [*Bagrada cruciferarum* (= *Bagrada hilaris* (Kirk.) (Pentatomidae, Hemiptera)] is an occasional pest of cruciferous crops (Vora et al., 1985a). This pest appears at two stages in crop growth — i.e., seedling and maturity/harvesting — and many times the infestation is carried even to the threshing floor. Clean cultivation, balanced doses of nitrogen, and quick threshing of harvested crops or keeping harvested bundles protected with insecticides resulted in low infestation by painted bug. Sarup et al. (1971a, b) reported that fenitrothion proved the best insecticide, followed by mevinphos. Methyl parathion and carbaryl proved less toxic to painted bug. Phorate granules applied at the rate of 1.15 g/m row in furrows, seed treatment with Carbofuran (4 g/100 g), seed and furrow treatment with disulfoton at 1.95 g/m row were 1.83, 1.42, and 1.21 times as effective as lindane (1.95 g/m row length).

Mustard Sawfly

Mustard sawfly [*Athalia lugens proxima* (King) (Tenthredinidae: Hymenoptera)] is a very serious pest of all cruciferous crops in the seedling stage throughout India (Vora et al, 1985a). The affected plants often exhibit quick recovery and a few holes on the leaves are of no economic importance.

Sachan and Sumati (1985) reported that a greater number of *A. proxima* larvae were found on *B. juncea* (Purbiraya) and fewer on *B. napus* and *B. carinata* cultivars. The strains from *B. juncea* and *Sinapis alba* (Linn.) Rabenh showed intermediate reactions. Mustard sawfly can be effectively controlled by several insecticides, for example, carbaryl, quinalphos, sevithion, sevisulf, monocil, and ripcord (sprays) (Sachan et al., 1983b); Carbofuran, endosulfan, and quinalphos (seed treatment) (Sachan and Sharma, 1987); and thiometon, phorate, and aldicarb (granules) (Sarup et al., 1971b; Pareek and Gupta, 1977). It is interesting to note that the use of systemic insecticides as foliar sprays (e.g., phosphamidon and dimethoate) provided protection against sawfly for only 5 days (Teotia and Gupta, 1970). Kumar et al. (1979) found that spraying with 2% bittergourd (*Momordica charantia* Linn.) seed oil resulted in 100% kill of sawfly larvae within 48 hours under field and laboratory conditions. Later, Benerji et al. (1982) reported that a 0.5% alcohol extract of *Derris indica* (Lamk.) seeds; 2.5%, 1.0% and 0.5% petroleum ether extracts of *Acorus calamus* Linn.; and a 2.5% alcohol extract of garlic (*Allium sativum* L.) provided 100% protection of leaves against sawfly under laboratory conditions.

FLEA BEETLE

The flea beetle [*Phyllotreta cruciferae* (Goeze) (Chrysomelidae: Coleoptera)] is a minor pest occurring in all rapeseed-mustard growing regions of India (Rai, 1976). Its damage, coupled with that of painted bug and mustard sawfly, to the seedling and young crop is often very serious. A positive correlation between flea beetle incidence and nitrogen application (0, 20, 40, and 60 kg N/ha) was observed by Rawat et al. (1968) and Makhmoor and Shinde (1986). Singh et al. (1980) observed that application of Carbofuran at sowing time was most effective, followed by phorate and 2-sec-Butylphenyl Methylcarbamate (BPMC).

FLEA LEAF-MINER

Flea leaf-minor [*Chromatomyia horticola* (Goureau) (Agromyzidae: Diptera)] is a polyphagous pest infesting rapeseed-mustard in India (Singh, 1985; Bakhetia and Sekhon, 1989). The pest causes damage to rapeseed-mustard crops, partly due to the numerous punctures made by the females for oviposition, but mainly due to maggots mining into the leaves. The leaf-miner attack has been reported to cause a 15.2% reduction in grain yield (Singh, 1991). Bakhetia and Sandhu (1977) and Singh (1980, 1985) reported that strains of *Brassica tournefortii* Gouan, *B. carinata*, *B. napus*, and *Crambe abyssinica* Fries were resistant, whereas the cultivars from *B. juncea* were susceptible to pea leaf-miner infestation. *Chrysocharis* sp., *Tetrastichus* sp., *Cirrospilus* sp., *Opius lantanae* Bridw, *Opius phaseoli* Fish., and *Sphegigaster* sp. parasitized the larvae of pea leaf-miner. However, Kumar (1984, 1985) reported maggot parasitism ranging from 65.80% to 86.60% by *Chrysonotomyia Formosa* (Westw.), *Diglyphus isaea* (Wlk.), *Opius exiguus* Wesm., and *Sohegigaster* sp.

WILD CRUCIFER–INSECT INTERACTION

Cruciferous crops display enormous diversity and are used as sources of oil, vegetables, mustard condiments, and fodder. The potential of wild crucifer germplasm to provide novel sources of economic traits in breeding programs has increased dramatically in the past 4 or 5 years with the development of biotechnology. Inventories of genetic diversity, such as this product, will facilitate breeding programs, increase the efficiency of locating traits and seed for use in germplasm development in the canola and mustard industries, as well as in the growing areas of diversification and/or alternative uses of crucifer crops for molecular farming, value-added or nutraceutical crops, and phytoremediation of a wide variety of "wild" *Brassica* spp. used for both food and medicinal purposes.

   Shuichi (1994) studied the behavior of insects feeding on ten species of wild crucifer with regard to anti-herbivore defense mechanisms. Most of the crucifer species deterred insect herbivory by disappearing in the summer or by lowering their intrinsic quality as food for insects. Species with these defense mechanisms were exploited by only a few specialized herbivorous insects that seemed to have counter-defenses. The plants without these defense mechanisms were used by many herbivorous insect species. *Rorippa indica* lacked direct defenses but supported a low total density of herbivore individuals. This crucifer has an indirect defense mechanism: ants attracted to floral nectar defended the plant from deleterious herbivores. Crucifers that disappeared seasonally lacked other anti-herbivore defense mechanisms. This suggests that the phenological response is an alternative response to herbivore attack. In a similar study, Shuichi and Naota (1993) examined the plant traits that affect the community structure of herbivorous insects on wild crucifers. They found that within the community of herbivorous insects on the plants with direct defense mechanisms, the number of species and individuals was small and most of the community members were specialists of the plants. On the other hand, within the community on plants without direct defense mechanisms, the number of species was large and the proportion of generalists was high. In addition, the number of individuals was very large on the remaining type B plants, but it was small on type D plants, which were inferred to have indirect defense mechanisms.

Wynne Griffiths et al. (2001) prepared leaf-surface extracts from 18 non-cultivated (wild) plant species derived from Capparidaceae, Cruciferae, Resedaceae, and Tropaeolaceae, ranked them for their ability to stimulate oviposition by the cabbage root fly, and analyzed them for glucosinolates. A total of 28 different glucosinolates were identified. A clear relationship was detected between the indolyl-, benzyl-, and the total glucosinolate compositions on the leaf surface and oviposition preference by cabbage root fly females. However, as the results were not fully explained by differences in leaf-surface glucosinolates, other important oviposition deterrents and stimuli on the leaf surface of these wild crucifers must also be present. Muhamad et al. (2006) fed the larvae of *Plutella xylostella* on five wild crucifers — *Capsella bursa-pastoris, Lepidium virginicum, Cardamine flexuosa, Rorippa indica, R. islandica* — and a crop, cabbage. The developmental period of the immature stages, adult longevity, pre-oviposition period, fecundity, and morphometrical characters of the adults were measured. The flight activity of the adults was also measured by the tethered flight method. All the wild plants except *R. islandica* were less suitable host plants than cabbage, and larvae that were fed on these less-suitable plants emerged as smaller adults with shorter wings. The smaller female adults had lower fecundity but a higher flight activity. Smaller adults measured in terms of their pupal weight among individuals fed on the same host plant had longer wings. These smaller adults with longer wings flew more actively. Abrol (2008) found that of the several insect pests attacking mustard crop, mustard aphid (*Lipaphis erysimi*) was found to be most abundant and attacked the crop in damaging proportions. The screening of wild genotypes of *Brassica* (viz. *B. tournefortii, B. fruticulosa* Sep: *cogonigna, B. oxyrrhina, Diplotaxis muralis, D. harracrassi, D. cretacea, D. teunifolia, D. bartaitlii, D. siettiana, Crambe abyssinica, Erucastrum abyssinicum, E. gallicum, E. virgatum, E. lyratus, Hirschfeldia incana, Sinapis pubescens, S. officinalis,* and *S. orientalis*) showed them to be moderate to highly resistant to aphid attack as compared to the cultivated genotypes (viz. *Brassica napus, B. juncea,* and *B. campestris*). Studies have shown that infestation was lower in the wild species than in the cultivated genotypes. Studies on seasonal abundance, chemical control, and pollinators are in progress.

WILD CRUCIFERS AND FEEDING DETERRENTS

The diamondback moth [*Plutella xylostella* L. (Lepidoptera: Plutellidae)] infests many cruciferous crops and also utilizes many wild crucifers as alternative hosts, especially when the crops are not planted (Talekar and Shelton, 1993). This intimate relationship between crucifers and *P. xylostella* is known to be mediated primarily by glucosinolates and their breakdown products (i.e., isothiocyanates). Glucosinolates act as key stimulants for oviposition to adults (Gupta and Thorsteinson, 1960b; Reed et al., 1989), and as feeding stimulants to larvae (Thorsteinson, 1953). In addition, allyl isothiocyanate stimulates egg production in adults and inhibits dispersal of larvae from host plants (Gupta and Thorsteinson, 1960a). However, not all crucifers are suitable hosts for *P. xylostella*: wintercress [*Barbarea vulgaris* R. Br. (*Brassicaceae*)] is a rare exception. Similarly, *Brassica vulgaris*, is rarely infested even under the high-density population conditions found in the field. *P. xylostella* adults preferably lay eggs on *B. vulgaris*, but hatched larvae do not develop to the second stadium on this plant, suggesting the presence of some feeding deterrents and/or toxins in this plant (Idris and Grafius, 1996).

Serizawa et al. (2001) found two lines of evidence that the leaves of *Barbarea vulgaris* contain potent feeding deterrents against *Plutella xylostella* larvae. Although many glucosinolates act, in general, as feeding stimulants to *P. xylostella* larvae, higher concentrations of particular glucosinolates, such as gluconasturtiin and gluconapin, are reported to be toxic to the larvae (Nayar and Thorsteinson, 1963). Only a few feeding deterrents to crucifer specialists have been identified thus far from the unacceptable crucifers: cucurbitacin E and cucurbitacin I in *Iberis amara,* which deters the flea beetle *Phyllotreta nemorum* (Nielsen et al., 1977); 2-*O*-b-D-glucopyranosyl cucurbitacin E in *Iberis amara,* which deters the cabbage butterfly *Pieris rapae* (Sachdev-Gupta et al., 1993a); and cardenolides in *Erysimum cheiranthoides,* which deter *P. rapae* (Sachdev-Gupta et al., 1993b). However, these substances are not likely feeding deterrents in *B. vulgaris* leaves for the following reasons:

1. The presence of cucurbitacins and cardenolides in the family *Brassicaceae* appears restricted to *Iberis* species and *Erysimum* or *Cheiranthoides* species, respectively, and they have not been reported in *Barbarea* species.
2. *P. xylostella* larvae can develop normally on *E. cheiranthoides* and tolerate the relatively high levels of cardenolides through rapid secretion (Renwick et al., 1991);
3. *Plutella rapae*, which is sensitive to cucurbitacins and cardenolides, can utilize *B. vulgaris* as a common host (Huang et al., 1994).

It is also suggested that *B. vulgaris* contains specific feeding deterrents or toxins to the flea beetle *P. nemorum*, although the compounds have not yet been identified (Nielsen, 1996, 1997a, b; Jonsson, 2005; DeJong and Nielson, 2000; DeJong, et al., 2000). The feeding of *P. nemorum* is also prevented by the cardenolides and cucurbitacins in *Erysimum* and *Iberis* (Nielsen, 1978). Interestingly however, a race of *P. nemorum* that can utilize *B. vulgaris* does not accept *Erysimum* and *Iberis* (Nielsen, 1999), thereby also suggesting the presence of unknown feeding deterrents or toxins in *B. vulgaris*. It will be of interest to see whether the feeding deterrents against *P. xylostella* and *P. nemorum* are identical. *P. xylostella* is a serious worldwide pest of crucifers, and is difficult to control, primarily because of the rapid development of resistance to a wide spectrum of insecticides, including *Bacillus thuringiensis* (Talekar and Shelton, 1993). The development of novel protection methods, alternative to conventional insecticides, is thus in high demand. The feeding deterrent in *B. vulgaris* might be used directly as a repellent to protect susceptible cruciferous crops. Alternatively, resistant cruciferous crops could be developed by introducing the feeding deterrent using various breeding techniques (e.g., traditional crossing, somatic hybridization, or genetic engineering). Because *B. vulgaris* belongs to the family *Brassicaceae*, it may be relatively easy to introduce this character into cruciferous crops. Indeed, intertribal somatic hybrids between *B. vulgaris* and *Brassica napus* have already been generated (Fahleson et al., 1994). The isolation and identification of this compound is now in progress and will be published elsewhere. Holger and Otto (2003) found that the larvae of the pollen beetles (*Meligethes* spp.), the brassica pod midge (*Dasyneura brassicae*), the cabbage seed weevil (*Ceuterrhynchus assimilis*), and the cabbage moth (*Mamestra brassicae*) collected in the field and exposed to entomopathogenic nematodes in the laboratory were easily infected, while infections of pod midge larvae only were observed twice. The number of nematodes produced in cadavers was positively related to the size of the insects.

IMPACT OF CHEMICAL STIMULI ON INSECT PESTS

Renwick (1989) found that Gravid *Pieris rapae* butterflies oviposit on many, but not all, crucifers. Rejection of *Erysimum cheiranthoides* and *Capsella bursa-pastoris* was initially explained by the presence of chemical deterrents in the plants. Analyses and bioassays of plant extracts indicated the absence of oviposition stimulants in *C. bursa-pastoris*, but similar chemical separation of *E. cheiranthoides* extracts revealed the presence of stimulants as well as deterrents. Choice tests illustrate how acceptance or rejection of a plant by an insect may depend on the balance between positive and negative chemical stimuli within the plant. Although some trichome, glucosinolate, or myrosinase QTL co-localize with *Pieris* QTL, none of these traits convincingly explained the resistance QTL, thus indicating that resistance against specialist insect herbivores is influenced by traits other than resistance against generalists.

HOST PLANT RESISTANCE TO INSECT PESTS OF CRUCIFERS

Several insect pests of cruciferous vegetable and oilseed crops are economically important. Chemical control has been the most common method of control of these pests. Although this method of control has been effective against many insects, it has serious drawbacks and a continued reliance on insecticides is not a sustainable pest-control strategy. *Integrated pest management* (IPM) is the most desirable approach for insect pest management, and host-plant resistance is considered a major component of IPM. Developing suitable methodologies, understanding the mechanisms of resistance, and identifying resistance sources and traits are some of the important steps involved in all host-plant resistance programs (Panda and Kush, 1995). In crucifers, all three types of insect resistance modalities (i.e., antixenosis, antibiosis, and tolerance) are known to impart resistance to various insect pests. Cruciferous plants exhibit enormous variation in the level of resistance to insects. Resistant sources and traits have been identified for a number of insects. Genetic engineering and biotechnology offer great potential in the identification and transfer of resistance genes from distant relatives or even unrelated plant species. Studies are in progress to produce transgenic cruciferous crops with genes to produce *Bacillus thuringiensis* toxins and protease inhibitors, and to alter the waxiness and glossiness characteristics to make plants resistant to insect attack. Crop resistance has a number of advantages, one of them being its excellent compatibility with IPM. It helps to promote the stability of the IPM system and to reduce the amount of pesticides used. Crop resistance has some limitations but there are ways to overcome a number of them. Plant breeders and biotechnologists should include insect resistance as a component of their crop improvement programs. Even partial resistance to insects will bring significant benefits, particularly when it is combined with other IPM components. In fact, when durability of resistance is considered, partial resistance is preferable to total resistance. With greater collaboration and commitment from plant breeders, biotechnologists, and entomologists, crop resistance will play a major role in the IPM of insect pests of cruciferous crops.

PROSPECTS FOR IMPROVING INSECT RESISTANCE IN CRUCIFEROUS CROPS

Pieris brassicae and other Pieridae are some of the most serious pests on cruciferous crop plants such as rapeseed, cauliflower, and broccoli (Bonnemaison, 1965). Pfalz et al. (2007) analyzed herbivory by *P. brassicae* larvae with a new recombinant inbred line (RIL) population, obtained from a cross between the parental lines Da (1)-12 and Ei-2. They found no detectable effect of glucosinolates or myrosinase activity on larval herbivory, indicating that the variation in the glucosinolate-myrosinase system present in this RIL population does not contribute to variation in plant damage caused by *Pieris* larvae. Nonetheless, the glucosinolate-myrosinase system does play a role in the interaction between *P. brassicae* and *Alternaria thaliana* or other *Brassicaceae*: Adult *Pieris* females use glucosinolates and their hydrolysis products to locate host plants for oviposition, and hydrolysis products have a stimulating effect on oviposition for *P. brassicae* and other Pieridae (Raybold and

Moyes, 2001; Miles et al., 2005). Likewise, glucosinolate breakdown products serve as a stimulant for larval feeding initiation (Schoonhoven, 1969; Renwick, 2002). This may explain why herbivory by *Pieris rapae*, a close relative of *P. brassicae*, is significantly reduced in *tgg1 tgg2* double mutants, which have very low levels of *Arabidopsis* wild-type myrosinase activity. Hence, a reduction in glucosinolate levels or myrosinase activity in cruciferous crops could potentially reduce plant damage caused by *P. brassicae*. However, a decrease in the effectiveness of the glucosinolate-myrosinase system would very likely render crucifer crops more susceptible to generalist insect herbivores that are sensitive to glucosinolate-based defenses (Lambrix et al., 2001; Kliebenstein et al., 2005). Furthermore, such a manipulation of the glucosinolate-myrosinase system bears the risk that plants could become more attractive to herbivores that usually do not consume crucifers, because these insect species have no effective means to withstand toxic products originating from glucosinolate hydrolysis. Thus, manipulating the glucosinolate-myrosinase system to increase resistance against insect herbivores may be problematic. The detection of QTLs that appear to be independent of the glucosinolate-myrosinase system may provide a way to solve this dilemma. Manipulating the genes that underlie the detected resistance QTL could help increase crop protection against *P. brassicae*, without interfering with a complex defense system that protects crucifers effectively against most herbivorous insects.

RESPONSES OF INSECT PESTS TO VOLATILES PRODUCED BY CRUCIFERS PLANTS

The volatiles produced by oilseed rape (*Brassica napus* L.) influence the behavioral and electrophysiological responses of pest insects and insect natural enemies. The pollen beetle *Meligethes aeneus* is an important pest of *Brassica* oilseed crops. The larvae feed only on the buds and flowers of these plants. The behavioral responses of adult beetles to odors of oilseed rape in bud and flowering stage, as well as to the colors green and yellow, were shown to assist the beetles in finding host plants. Olfactory and visual stimuli were found to interact, and the responses of over-wintered and summer-generation beetles differed in that the summer generation had a higher preference for flower odors and for flower odors in combination with the color yellow (Todd and Baker, 1999).

GENETIC ASPECTS OF INSECT–PLANT INTERACTIONS

The genetic studies have dealt with interactions between the plant *Barbarea vulgaris* and the flea beetle *Phyllotreta nemorum*. The plant occurs in two subspecies in Denmark: ssp. *vulgaris* and ssp. *arcuata*. Furthermore, it is possible to distinguish between two types of ssp. *arcuata*: The G-type has glabrous leaves and is resistant to the most common genotypes of the flea beetle, while the P-type has pubescent leaves and is susceptible to all flea beetle genotypes. Some flea beetle genotypes can live on the G-type, and the ability to survive on this plant is controlled by a few major genes (R-genes). Crossings between resistant and susceptible beetles have been performed (F1, F2, and backcrosses), and the segregation patterns suggest that R-genes are located on autosomes as well as on both sex chromosomes. These genes are abundant in flea beetle populations living on *Barbarea*, but are rare in populations living on other host plants. The R-genes are specific for defenses in *Barbarea* and have no effect on insect fitness of other plant species. The available evidence suggests that the resistance in the plant and the counter-adaptations (R-genes) in the flea beetles are derived characters.

CHEMICAL STIMULANTS AND DETERRENTS REGULATING ACCEPTANCE OR REJECTION OF CRUCIFERS

Current research attempts to unravel the chemical, genetic, and evolutionary aspects of the polymorphism in *Barbarea* and flea beetles. Near-isogenic flea beetle lines that differ in their content of R-genes (autosomal or sex-linked) have been developed by repeated backcrossings to a line without R-genes (6 to 15 generations). Studies on the inheritance of resistance in plants and on the identification of the chemical resistance factor are in progress.

PLANT ATTRIBUTES IN BIOLOGICAL CONTROL

Plants and insect herbivores have long been competing in an evolutionary race in which plants evolve to reduce consumption, while herbivores evolve to increase it (Futuyma and Keese, 1992; Harborne, 1993). Chemical and morphological plant attributes can directly influence the survival, fecundity, and foraging success of natural enemies on hosts or prey. These traits can also have indirect effects by affecting the qualities of an herbivore that, in turn, affect the physiology, behavior, or development of natural enemies. Plant breeding and biological control have mostly been parallel but independent pest management practices in the past (Price, 1986; Van Lenteren et al., 1995; Thomas and Waage, 1996). While plant breeders have almost exclusively focused on selecting varieties with enhanced direct defenses against pests, biological control workers have mainly concentrated on improving natural enemy traits, such as the reproduction and host-finding efficacy, evidently indicating the urgent need for bridging these two pest management practices (Figure 8.1). Yet, a key interaction is still often overlooked in current crop protection strategies: the possibility of manipulating the presence and expression of plant attributes to promote the third trophic level. Therefore, special emphasis should now be placed on breeding crop plants with natural enemy-enhancing traits.

Other plant morphological traits, such as prominent leaf veins or moderate pubescence, can provide sheltered habitats for small natural enemies and promote their abundance (Drowning and Moillet, 1967; Walter and O'Dowd, 1992; Karban et al., 1995; Walter, 1996). In temperate regions, such structures can supply shelter for over-wintering predators and parasitoids, and constitute a key factor in the maintenance of their populations (Hance and Boivin, 1993; Corbett and Rosenheim, 1996; Elkassabany et al., 1996).

A waxy surface and the shape of a leaf are other morphological traits that can affect the prey- or host-finding rate by natural enemies. For example, slipperiness due to a waxy leaf surface caused ladybird beetles to frequently fall off crucifer plants and substantially decreased their consumption rate of aphid prey (Grevstad and Klepetka, 1992). However, leaf shape appeared to counter this effect. Predators did not fall as often from plants that had waxy leaves which had more edges and fewer flat surfaces. In another study (Eigenbrode et al., 1995), predators such as *Chrysoperla carnea* (Stephens), *Orius insidiosus* (Say), and *Hippodamia convergens* Guerin-Meneville were shown to

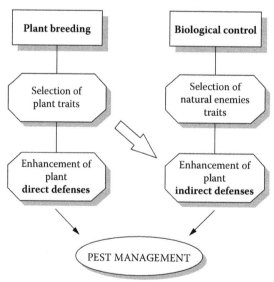

FIGURE 8.1   The necessity of bridging plant breeding and biological control practices to improve crop protection. Crop plants should also be selected to breed for their capacity to enhance natural enemy efficiency, i.e. indirect defenses.

be more effective in reducing populations of *Plutella xylostella* L. on a cabbage (*Brassica oleracea* var. *capitata* L.) variety with glossy surface waxes than on a normal-wax cabbage variety. Increased effectiveness was related to improved mobility of these predators on glossy leaf surfaces. Wax debris that accumulated on tarsae impeded the mobility of predators walking on the normal-wax cabbage variety but not on glossy cabbage (Eigenbrode et al., 1996).

BEE–CRUCIFER INTERACTION

Bees and *Brassica* plants have coevolved during the course of their evolutionary history, each dependent on the other to live — the bees for nectar and pollen and the *Brassica* for pollination. Most oilseed crops are self-incompatible and depend on pollinating insects for seed set. Even those self-fertile species improve qualitatively and quantitatively if pollinated by insects (Table 8.1). The interdependence of bees and *Brassica* manifests from the fact that the pollen of *Brassica* is very sticky and needs insect pollen vectors for its transfer. For *Brassica* plants, bees are marvelously coevolved, pollen-transferring devices.

Most oilseed crops are cross pollinated, and adequate pollination is vital for any significant seed production. The pollinator behavior of different crops is given in Table 8.2. An increase in seed yield as a result of insect pollination has been reported in mustard (Mohammed, 1935; Latif et al., 1960). Meyerhoff (1954) concluded that in 'Lembke' winter rape, honeybees increased the number of pods per plant by 53.2%, pod length by 6.1%, and seeds per pod by 12.6%. Olsson (1952) found that bees excluded 64.7% of the flowers, set seeds, and 1.75 g per podv, but with bees present these values increased to 95.3%, 4.08 seeds, and 2.69 g, respectively, more than doubling total production. Koutensky (1959) also showed that the seed yield of white mustard was increased 66% by honeybee pollination. Olsson (1955) also showed that the presence of bees in cages of white mustard doubled the number of seeds per pod and increased the pod set by 50%.

Latif et al. (1960) showed that rapeseed production in fields with bees was more than double that of fields where bees (*Apis cerana* F.) were absent. They further found that *Apis cerana* colonies kept near var. *sarson* and var. *toria* fields increased the seed yield by 60%. Free and Spencer-Booth (1963) found that seed production was double in *Brassica alba* fields where bees were provided. In *B. juncea*, production increased only 14%, an amount that was not statistically significant in their test but could be of great significance to the grower. Pritsch (1965) also obtained significantly greater yields of white mustard in cages with bees than in cages where bees were excluded.

Deodikar and Suryanaryana (1972) studied the impact of bee pollination on the increase in yield of several oilseed crops (i.e., *Brassica napus, B. campestris* var. *toria, B. campestris* var. *dichotoma, B. juncea,* and *B. alba*) and found that bee pollination significantly increased the seed

TABLE 8.1
Yield in Self-Pollinated and Bee Pollinated Crops

Crop Botanical Name	Reported Range of Seed Increase from Bee Pollinated over Self-Pollinated	
	Percent More	Times More
Brassica napus L. (rape)	12.8–139.3	1.128–2.39
Brassica campestris L. var. toria (Toria)	66.0–220.9	1.66–3.20
Brassica campestris L. var. dichotoma (sarson)	222	3.22
Brassica juncea Czern and Coss. (Rai, Indian mustard)	18.4	1.184
Brassica alba Boiss.(white mustard)	128.1–151.8	2.28–2.51

Source: From Deodikar and Suryanaryana, 1972.

TABLE 8.2
Yield and Oil Potential of *Brassica campestris* var. sarson under Three Conditions of Pollination

Parameter	Open Pollination (X+S.E)	Net Caged (X+S.E)	Muslin Bagged (X+S.E)	Critical Difference	Transformation
% pod setting	95.21+2.46	70.27+4.38	58.81+6.57	(10.09)	Angular
Seeds/pod	12.72+0.82	5.55+1.18	3.95+0.85	(0.73)	√n+1
% healthy seeds	84.78+2.76	41.45+8.25	48.37+12.28	18.41	Angular
Seed weight in mg/1000 seeds	478.50+21.83	459.40+23.41	542.20+11.00	(1.31)	√n
% oil content	39.02+0.29	40.76+0.90	41.64+0.27	1.81	—
Oil yield (mg/pod)	19.03	3.02+	1.95		
Increase in oil yield over muslin bagged	9.76	1.55	1.0		

Source: From Mishra et al., 1988.

yield over self-pollination, and the percentage increase ranged from 12.8 to 222 in different crops (Table 8.1). Langridge and Goodman (1975) found that in var. *Midas*, honeybees increased yields by 2.5% over control plots. They further found that *B. napus* plants kept in the relatively still air of a greenhouse only produced one-third to one-half of normal production. This suggests that either wind or insect pollination was needed to obtain a high seed set.

In case of *Brassica campestris,* pollination benefits in Australia according to Langridge and Goodman (1975) were as follows:

- 775 kg/ha increase over control plot.
- 600 kg/ha increase over the control plot.
- Honeybees increased yield by 61.2% over control plots of var. *Arlo.*
- Honeybees have increased oil production by 10% to 20%.
- Wind helps in pollination.
- Hover flies are also important. For var. *Arlo*, 58% more seeds per plant and 46% greater weight of seed was achieved compared to plots excluded from insects.
- 97% were self-sterile in glasshouse studies. For var. *Golden*, 85% were self-sterile in glasshouse studies.
- The density of honeybees in the field was significantly related to the distance from the apiary (Sweden). Results indicated that higher densities of honeybees increased both seed yield and oil content of the seed (Sweden).

Similar increases in yield have been reported in oilseed rape (Langridge and Goodman, 1975; Kisselhegn, 1977) and *Brassica campestris* var. Jambuck (Kubisova et al., 1980).

Bisht et al. (1980) found that flowers of rapeseed visited by *Apis* species had higher pod set, increased numbers of seeds per pod, and the weight of seed was also higher than those deprived of pollinator visits. Mishra et al. (1988) found that in *Brassica campestris,* percent pod setting, number of pods per plant, and proportion of healthy seeds were significantly higher in open pollinated flowers than in net caged and muslin bagged ones. Similarly, the average weight of seeds and oil content were higher in open pollinated flowers (Table 8.2). *Apis cerana* was the most common pollinating species. The other pollinators observed included *A. mellifera*, syrphid flies, etc. In a similar study, Prasad et al. (1989) found that pollination *of Brassica juncea* by *A. cerana* resulted in increased silique setting, increased length of silique, seed weight increased yield, and had a pronounced effect on oil contents and germination.

Singh et al. (2000) found that silique setting in plots caged with bees resulted in an increase of 239.77% over plots pollinated without insects and 2.65% over open pollinated flowers. Similarly, the number of seeds, seed weight, and seed germination were improved. Moreover, seeds from bee-pollinated plants showed 21.3% higher oil content than plants without bees and 4.1% higher than open pollinated plants. In a similar study, Prasad et al. (1989) found that the pollination of *Brassica juncea* by *Apis cerana* resulted in more silique setting, increased length of silique, seed weight increased yield, and had a pronounced effect on oil content and germination.

Manning and Boland (2000) found that the number of pods per plant decreased as the distance from the apiary increased, with a predicted pod loss of 15.3 pods per plant over a distance of 1000 meters from an apiary. This was equivalent to a 16% loss based on an average of 59 plants/m² and an average pod production of 5666 pods/m². For a 2-t/ha crop, this would be equivalent to about 320 kg/ha, thereby indicating the importance of honeybees in rapeseed pollination.

INSECT POLLINATORS

Rahman (1940) studied the pollinators of *Brassica napus* in India and concluded that the dwarf honey bees of India (*Apis florea* F.), wild bees (*Andrena ilerda* Cam. and *Halictus* sp.), and the fly (*Eristalis tenex* (L.)) were the most important pollinators. Honeybees are the primary pollinators of rape (Belozerova, 1960; Nikitina, 1950; Radchenko, 1964; Vesely, 1962; Ahmad, 1999). The plant is highly attractive to honey bees, providing both nectar and pollen, and the honeybee is of appropriate size for effective transfer of pollen from anthers to stigma. Hammer (1952) reported as many as 20,000 bees per hectare rape in fields 3.5 to 4 km from the apiary. In India, pollinators include *Apis florea*, *A. dorsata*, *A. cerana*, and *Andrena ilerda* (Kapil et al., 1969). Free and Nuttall (1968) studied the activity of honeybees on *B. napus*. They reported that all bees that visited the flowers collected nectar although some also collected pollen. All became covered with pollen but some removed and discarded it. Those that collected the pollen did so primarily during the morning hours. Kapil et al. (1971) recorded 13 species of bees belonging to five families visiting rapeseed mustard flowers in Punjab and Haryana. The bee species identified included *Apis florea*, *A. dorsata*, *A. mellifera*, *Andrena ilerda*, *A. leaena*, *Megachile lanata*, *M. flavipes*, *M. cephalotes*, *Xylocopa fenestrate*, *X. pubescens*, *Nomia* spp., *Halictus* spp., and *Pithitis smargdula*. Of the various bee pollinators, *Apis florea* and *Apis dorsata* were the most important and efficient pollinators of all the crops. Bhalla et al. (1983) recorded seven species of insects visiting rapeseed crop under the mid hill conditions of Himachal Pradesh. Mishra et al. (1988) recorded *Apis cerana* as the most important pollinator of *Brassica campestris* var. Sarson, accounting for 69.47% of the total flower visitors, while *A. mellifera* was the least with 2.95%. The other insect visitors included dipteran flies and syrphids. Prasad et al. (1989) found that among all the insect pollinators, *Apis cerana* was the predominant pollinator of brown mustard. In a similar study, Kumar et al. (1992) observed 18 species of insects foraging on var. *toria*. Among all the foragers, *Apis mellifera* was found to be the dominant one, accounting for 57.45%, but *Andrena ilerda* was the most efficient pollinator of *Brassica campestris* var. *toria*. Mishra et al. (1992) studied the foraging behavior of honeybees on *Brassica campestris* var. *sarson* and found that the average number of *Apis dorsata*, *A. florea*, *A. cerana*, and *A. mellifera* per square meter was 2.5, 9.0, 1.1, and 1.6, respectively.

NUMBER OF COLONIES REQUIRED FOR POLLINATION

Several investigators have attempted to determine the number of colonies of honeybees required for increased yields and their recommendations vary from place to place and crop to crop. For example, Hammer (1963, 1966) recommended three colonies per hectare; Radchenko (1964), two; Downey and Bolton (1961), one; White (1970), two; and Vesely (1962) three to four colonies per hectare.

Although hoverflies appear to play some role in the pollination of rape, we consider the honeybee the more efficient pollinator. White (1970) said that both summer turnip rapes and true rapes depend on bees for maximum production. The data indicate that a heavy bee population on rape would be beneficial, but until more concrete data become available, the one to two strong honeybee colonies per hectare would appear to be logical. The ideal pollinator population and proper distribution of colonies for most efficient pollination of rape must be determined.

POLLINATION RECOMMENDATIONS

Honeybees are the most effective agents involved in the cross pollination of rapeseed mustard crops. In modern agriculture, farm mechanization and high-yielding varieties (HYVs) are very common. In such a condition, to increase the yield, more inputs such as water, fertilizer, and other agrochemicals are in use. Indiscriminate use of pesticides/fungicides often kills a large number of pollinators. In certain cases, a single crop over a vast area is cultivated. This also reduces the number of wild honeybee colonies in those areas. Therefore, the importance of beekeeping in the field is being realized as an important input to increase the production of oilseed crops.

COLONY STRENGTH

Larger and stronger colonies are five times better than the smaller and weaker ones because the former have a higher percentage of older bees as foragers. Thus, good honey-yielding colonies are also better pollinators.

NUMBER AND TIME OF PLACEMENT OF COLONIES

This factor depends on the density of the plant stand, the total number of flowers in inflorescence of each plant, the duration of flowering, the strength of bee colonies, and the number of flowers over an area of 1 hectare of land. Generally, three to five colonies of *Apis mellifera* per hectare of crop in bloom and five strong colonies of *Apis cerana* are recommended for sufficient and efficient pollination.

DISTRIBUTION OF COLONIES IN THE FIELD/ORCHARDS

Honeybees primarily visit the nectar flow source, which is within a 0.3- to 0.5-km radius of the apiary. Beyond the 0.5-km limit, pollination activity diminishes significantly. For efficient pollination, hives should be placed singly, rather than in groups. Bees tend to forage in the area that is closest to their hive, particularly when the weather is not favorable.

TIME AND PLACEMENT OF COLONIES

Generally, colonies should be introduced when 5% to 10% of the crop is in bloom. Earlier placement of bees results in foraging in other weeds and wild plants in the vicinity, and ignoring of the crop bloom. If bees are moved too late, they can only pollinate late and less vigorous plants.

WEATHER CONDITIONS

The failure and success of bee pollination depends on the weather, as it affects equally the crop and the bees. *Apis cerana* can forage at a lower temperature than *A. mellifera*. A wind velocity of more than 15 miles/hour affects the forage behavior; therefore, wind breaks around the orchard or field are recommended. Cool, cloudy weather and storms greatly reduce bee flights.

PROBLEMS ASSOCIATED WITH BEE POLLINATION

Aphid control becomes a problem during pollination when the crop is in bloom and bees are actively foraging in the field (Abrol, 1993, 1997; Abrol and Andotra, 1998, 2003). Insecticides presently registered for use in seed crops are considered highly toxic to bees. Thus, to avoid killing the bees, pollination often must be interrupted. The colonies are either temporarily removed from the field, or applications must be made at night. In either case, several days can pass before the bees resume normal foraging activity. Repeated applications to control aphid populations may cause a significant disruption in pollination as well as reduced seed yield and quality.

Oilseed crops are attacked by aphids, caterpillars, and bugs during the flowering and pod formation stages. This requires the application of insecticides to combat the pest (Sihag, 1986, 1988, 1991; Sihag et al., 1999a, b), which poses serious problems for the foraging activity of honeybees and the developing brood. Evidently, infestation of aphids and other pests, and their management during the flowering period of rapeseed and mustard crops pose a major problem for effective utilization of pollinators for crop production, which also, incidentally, get killed. Aphid control becomes a major problem during pollination when the crop is in bloom and bees are actively foraging in the field.

Furthermore, modern agricultural practices have resulted in the reduction of wild insect pollinators and disturbed the insect-flower relationship by way of disappearing wastelands and uncultivated strips of land, destruction of certain food sources by weed control, and overall changes in the environment. Wild bees are also damaged by pesticides. Poisoning may result from contaminated food as well as from florets, leaves, soil, or other material used by the bees in nesting. The toxicity of a specific insecticide to honeybees and wild bees is not always the same; and even among wild bees, some materials are more toxic to one species than to another. Evidently, the safety of pollinating insects while foraging on crops must be ensured (Abrol, 2007).

CONCLUSION

It is evident from the discussion in this chapter that crucifers depend on two diverse approaches for seed production: one is the attack from pests and the other is the benefit from pollinators. Research has revealed that a number of insect pests cause serious problems during reproductive growth periods in crucifers grown for seed production when plants are actively being pollinated and producing seed. These pests injure the crops primarily by feeding on the developing plant, leaves, flowers, pods, and seeds. Aphids, in particular, feeding in blooms and seed pods pose a serious problem during pollination when the crop is in bloom and bees are actively foraging in the field. Repeated applications to control aphid populations may cause a significant disruption in pollination and reduced seed yield and quality. Therefore, indiscriminate use of insecticides against insect pests of crucifer seed crops during flowering seriously affects the insect-pollinators relationship, often resulting in lower seed yield due to decreased population and activity of these pollinators. In view of the above, the following recommendations can be made:

- Apply pesticides only when needed.
- Use less toxic insecticides.
- The timing of pesticide application is very important; apply in the late afternoon or early evening when bees are not active on the crop.
- Use liquid or granular applications.

To achieve maximum yield, pests must be managed on the one hand and pollinators must be protected on the other. Therefore, an integrated approach could be highly useful. In view of the above, a schematic model (Figure 8.2) has been devised that provides investigators with guidelines for the integrated management of pests and the safety of pollinators.

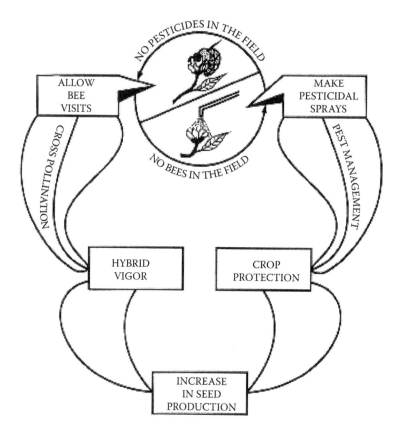

FIGURE 8.2 A schematic model showing the integrated approach involving application of pesticides for the management of crop pests and utilization of bee pollinators for cross pollination and hybrid vigor. The model documents a time lag between the two practices to ensure the safety of bees from pesticides for increased seed production. (*Source:* Redrawn from Abrol, 1997.)

REFERENCES

Abrol, D.P. 1993. Insect pollination and crop production in Jammu and Kashmir. *Curr. Sci.*, 65:265–269.

Abrol, D.P. 1997. *Bees and Beekeeping in India.* Kalyani publishers, Ludhiana.

Abrol, D.P. 2007. Foraging behaviour of *Apis mellifera* and *Apis cerana* as determined by the energetics of nectar production in different cultivars of *Brassica campestris* var. *toria*. *J. Apicultural Sci.,* 51:5–10.

Abrol, D.P. 2008. Evaluation of insecticides against insect pests and beneficial insects of *Brassica* crops. *J. Asia Pacific Entomol.* (in press).

Abrol, D.P. 2007. Honeybee and Rapeseed- A pollinator plant interaction In: *Advances in Botanical Research: Rapeseed Breeding*, Vol. 45. (Ed.) Dr. S.K. Gupta, Elsevier, London, p. 339–367.

Abrol, D.P. and Andotra, R.S. 1998. Impact of pesticides on foraging activity of honeybee, *Apis mellifera* L. on *Brassica campestris* L. var. *toria*. In: *Fourth Int. Asian Apicultural Assoc. Conf.*, March 25–27, Katmandu, Nepal. p. 123.

Abrol, D.P. and Andotra, R.S. 2003. Relative toxicity of some insecticides to *Apis mellifera* L. *J. Asia Pacific Entomol.*, 6:235–237.

Abrol, D.P. and Kapil, R.P. 1996. Insect pollinators of some oilseed corps. *J. Insect., Sci.,* 9:172–174.

Agerbirk, N., Orgaard, M., and Nielsen, J.K. 2003a. Glucosinolates, flea beetle resistance, and leaf pubescence as taxonomic characters in the genus *Barbarea* (*Brassicaceae*). *Phytochemistry*, 63:69–80.

Agerbirk, N., Olsen, C.E., Bibby, B.M., Frandsen, H.O., Brown, L.D. Nielsen, J.K., and Renwick, A.A. 2003b. A saponin correlated with variable resistance of *Barbarea vulgaris* to the diamondback moth *Plutella xylostella*. *J. Chem. Ecol.*, 29:1417–1433.

Ahmad, B. 1999. Insect Pollinator Complex of Rapeseed (*Brassica campestris* var. *toria*). M.Sc. thesis, Assam Agricultural University Jorhat, Assam.

Allaby, M. 1998. *Dictionary of Ecology*. Oxford University Press, Oxford.

Arora, R., Singh, G., Kailey, J.S., and Kalsi, P.S. 1982. Biological activity of terpenoid lactones as juvenile hormone analogues against the mustard aphid *Lipaphis erysimi*. *Phytoparasitica*, 10:57–60.

Bakhetia, D.R.C. 1983. Losses in rapeseed /mustard due to *Liphaphis erysimi* (Kalt.) in India: a literature study. *Proc. Sixth Int. Rapeseed Conf.*, May 1983, Paris, p. 1142–1147.

Bakhetia, D.R.C. 1984. Chemical control of *Lipaphis erysimi* (Kalt.) on rapeseed mustard crops. *J. Res. Punjab Agric. Univ.*, 21:63–75.

Bakhetia, D.R.C. and Sandhu, R.S. 1977. Susceptibility of some *Brassica* species and *Crambe abyssinica* to the leaf miner *Phytomyza horticola* (Meigen). *Crop Improve.*, 4:221–223.

Bakhetia, D.R.C. and Sekhon, B.S. 1989. Insect-pests and their management in rapeseed-mustard. *J. Oilseeds Res.*, 6:269–299.

Belozerova, E.I. 1960. Bees increase seed crop from winter rape. *Pchelovodstvo*, 37:38–40.

Benerji, R., Misra, G., Nigam, S.K., Prasad, N., Pandey, R.S., and Mathur, Y.K. 1982. Indigenous plants as antifeedeedants. *Indian J. Entomol.* 44:71–76.

Bhalla, O.P., Verma, A.K., and Dhaliwal, H.S. 1983. Insect visitors mustard bloom, their foraging behaviour under mid hill conditions. *J. Entomol. Res.*, 1:15–17.

Bisht, D.D., Naim, M., and Mehrotra, K.N. 1980. Studies on the role of honeybees in rapeseed production. In: *Proc. 2nd Int. Conf. Apiculture in Tropical Climate*. Indian Agricultural Research Institute, New Delhi, p. 491–496.

Bonnemaison, L. 1965. Insect pests of crucifers and their control. *Annu. Rev. Entomol.*, 10:233–256

Corbett, A. and Rosenheim, J.A. 1996. Impact of a natural enemy overwintering refuge and its interaction with the surrounding landscape. *Ecol. Entomol.* 21:155–164.

Cortesero, A.M., Stapel, J.O, and Lewis, W.J. 2000. Understanding and manipulating plant attributes to enhance. *Biological Control*, 17, 35–49.

De Jong, P.W. and Nielsen, J.K. 2000. Reduction in fitness of flea beetles which are homozygous for an autosomal gene conferring resistance to defences in *Barbarea vulgaris*. *Heredity*, 84:20–28.

De Jong, P.W., Frandsen, H.O., Rasmussen, L., and Nielsen, J.K. 2000. Genetics of resistance against defences of the host plant *Barbarea vulgaris* in a Danish flea beetle population. *Proc. Royal Society*, London B267:1663–1670.

Deodikar, G.B. and Suryanaryana, M.C. 1972. Crop yields and bee pollination. *Ind. Bee J.*, 34:52–63.

Downey, R.K. and Bolton, J.L. 1961. Production of [Polish and Argentine] rape in Western Canada. *Canada Dept. Agr. Res. Br. Pub. 1021*.

Drowning, R.S. and Moillet, T.K. 1967. Relative densities of predaceous and phytophagous mites on three varieties of apple trees. *Can. Entomol.* 99:738–741.

Eigenbrode, S.D., Castagnola, T., Roux, M.B., and Steljes, L. 1996. Mobility of 3 generalist predators is greater on cabbage with glossy leaf wax than on cabbage with a wax bloom. *Entomol. Exp. Appl.* 81:335–343.

Eigenbrode, S.D., Moodie, S., and Castagnola, T. 1995. Predators mediate host-plant resistance to a phytophagous pest in cabbage with glossy leaf wax. *Entomol. Exp. Appl.* 77:335–342.

Elkassabany, M., Ruberson, J.R., and Kring, T.J. 1996. Seasonal distribution and overwintering of *Orius insidiousus*. *J. Entomol. Sci.* 31:76–88.

Fahleson, A., Eriksson, I., and Glimelius, K. 1994. Intertribal somatic hybrids between *Brassica napus* and *Barbarea vulgaris* — production of *in vitro* plantlets. *Plant Cell Rep.*, 13:411–416.

FAO. 1998. *FAO Production Year Book*. 52:41.

Free, J.B. and Nuttall, P.M. 1968. The pollination of oilseed rape (*Brassica napus*).and the behaviour of bees on the crop. *J. Agric. Sci.Camb.*, 71:91–94.

Free, J.B. and Spencer-Booth, Y. 1963. The pollination of mustard by honeybees. *J. Apic. Res.*, 2:69–70.

Futuyma, D.J. and Keese, M.C. 1992. Evolution and coevolution of plants and phytophagous arthropods. In *Herbivores: Their Interactions with Secondary Plant Metabolites, Vol. 2: Evolutionary and Ecological Processes* Eds. G.A. Rosenthal and M.R. Berenbaum.

Gavoski, J.E., Ekuere, U., Keddie, A., Dosdall, L., Kott, L., and Good, A.G. 2000. Identification and evaluation of flea beetle (*Phyllotreta cruciferae*) resistance within Brassicaceae. *Canadian Journal of Plant Science*, 80:881–887.

Gupta, P.D. and Thorsteinson, A.J. 1960a. Food plant relationship of diamondback moth (*Plutella maculipennis* (Curt.)). I. Gustation and olfaction in relation to botanical specificity of larva. *Ent. Exp. Appl.*, 3:241–250.

Gupta, P.D. and Thorsteinson, A.J. 1960b. Food plant relationship of diamondback moth (*Plutella maculipennis* (Curt.)). II. Sensory relationship of oviposition of the adult female. *Ent. Exp. Appl.*, 3:305–314.

Hammer, O. 1952. Rape growing, bees and seed production. *Dansk Landbr.,* 71:67–69.

Hammer, O. 1963. Summer Rape as a Competitor Affecting the Pollination of Clovers. Dansk froavl no. 14.

Hammer, O. 1966. Some problems of competition between summer rape and clover, in relation to pollination. In *2nd Int. Symp. Pollination,* London, 1964. *Bee World,* 47(Suppl. 1):99–106.

Hance, T. and Boivin, G. 1993. Effec of parasitism by *Anaphes* sp. (Hymenoptera:Mymaridae) on the cold hardiness of Lisronotus oregonensis (Coleoptera:Curculionidae) eggs. *Can. J. Zool.* 71:759–764.

Harborne, J.B. 1993. *Introduction to Ecological Biochemistry. 4th ed.* Academic Press. London.

Hazarika, J. and Saharia, D. 1981. Control of mustard aphid *Liphaphis erysimi* (Kalt by soil and foliar application of certain insecticides. *J. Res. Assam. Univ.,* 2:108–110.

Holger, P. and Otto, N. 2003. Host potential of insects from cruciferous crops to entomopathogenic nematodes and augmentation of nematodes through oil seed rape growing. Paper presented at *10th Meeting of the Int. Org. Biological Control,* Athens, Greece, May 2001; published in *IOBC Bulletin* 26, p. 141–146.

Huang, X., J. Renwick, A.A., and Sachdev-Gupta, K. 1994. Oviposition stimulants in *Barbarea vulgaris* for *Pieris rapae* and *P. napi oleracea:* isolation, identification and differential activity. *J. Chem. Ecol.,* 20:423–438.

Idris, A.B. and Grafius, E. 1996. Effects of wild and cultivated host plants on oviposition, survival, and development of diamondback moth (Lepidoptera: Plutellidae) and its parasitoid *Diadegma insulare* (Hymenoptera: Ichneumonidae). *Environ. Entomol.,* 25:825–833.

Jönsson, M. 2005. Responses to Oilseed Rape and Cotton Volatiles in Insect Herbivores and Parasitoids. Doctor's dissertation. Swedish University of Agricultural Sciences ALNARP 2005. Department of Crop Science, SLU, SE-230 53, Alnarp, Sweden.

Kakar, K.L. and Dogra, G.S. 1977. Comparative efficacy of soil and foliar applications of systemic insecticides against *Phytomyza atricornis* (Meigen), infesting mustard. *Ind. J. Agric. Sci.,* 47:405–407.

Kalra, V.K. and Gupta, D.S. 1986. Chemical control of mustard aphid, *Lipaphis erysimi* (Kalt.) *Ind. J. Entomol.,* 48:148–155.

Kapil, R.P., Grewal, G.S., Kumar, S., and Atwal, A.S. 1971. Insect pollinators of rape seed and mustard. *Ind. J. Entomol.,* 33:61–66.

Kapil, R.P., Grewal, G.S., Kumar, S., and Atwal, A.S. 1969. Insect pollinators of rape seed and mustard. (Abstr.) *56th Indian Sci. Congr. Proc.,* Pt. 3, p. 509.

Karban, R., Englishloeb, G., Walker, M.A., and Thaler, J. 1995. Abundance of phytoseiid mites on *Vitis* species— Effects of leaf hairs, domatia, prey abundance, and plant phylogeny. *Exp. Appl. Acar.* 19:189–197.

Khurana, A.D. and Batra, G.R. 1989. Bioefficacy and persistence of insecticides against *Lipaphis erysimi* (Kalt.) on mustard under late sown conditions. *J. Insect Sci.,* 2:139–145.

Kisselhagen, S. 1977. Seed yield is increased if there are bees on the rape. *Tidsskrift for Biavl,* 111:66–67.

Kliebenstein, D.J., Kroymann, J., Mitchell-Olds, T. 2005. The glucosinolate-myrosinase system in an ecological and evolutionary context. *Curr. Opin. Plant Biol.* 8:264–271.

Koutensky, J. 1959. The pollinating effect of the honey bee (*Apis mellifera* L.). On the increase in rape and white mustard yields per hectare *Ceskoslov. Akad. Zemedel. Ved, sborn. Rostlinna vyroba,* 32:571–582.

Kubisova, S., Nedbalova, V., Plesnik, R. 1980. The pollinating activity of honeybees on rape. *J. Polnohospodarstvo,* 26:744–754.

Kumar, A. 1984. Incidence of pupal parasitism on *Chromatomyia horticola* (Gor.) (Diptera: Agromyzidae) in some parts of Uttar Pradesh. *Ann. Entomol.,* 2:37–44.

Kumar, A. 1985. Estimation of damage caused to *Chromatomyia horticola* population on *Brassica campestris* by hymenopteran parasites. *Ann. Entomol.,* 10:49–53.

Kumar, A., Tewari, G.S., and Pandey, N.D. 1979. Antifeeding and insecticidal properties of bittergourd, *Momordica charantia* Linn. against *Athalia proxima* Klug. *Indian J. Entomol.,* 41:103-106.

Kumar, P.R. 1991. Constraints and available agro-technology for increasing rapeseed-mustard production. In: V. Ranga Rao and M.V.R. Prasad, Eds., *Proc. Natl. Sem. Strategies for Making India Self-reliant in Vegetable Oils.* ICAR, New Delhi, p . 57–63.

Kumar, V., Singh, D., Singh, Hari, and Singh, Harvir. 1990. Combining ability analysis of resistance parameters mustard aphid, *Lipaphis erysimi* Kalt. in Indian mustard, *Brassica juncea* L. Czern and Coss.). *Crop Res.,* 3:204–210.

Kumar, V. 1990. Genetic analysis of tolerance of mustard aphid, *Lipaphis erysimi* Kalt. In Indian Mustard, *Brassica* juncea L. Czern and Coss. M.Sc. thesis, Haryana Agricultural University, Hisar.

Lambrix, V., Reichelt, M., Mitchell-Olds, T., Kliebenstein, D.J., and Gershenzon, J. 2001. The *Arabidopsis* epithiospecifier protein promotes the hydrolysis of glucosinolates to nitriles and influences *Trichoplusia ni* herbivory. *Plant Cell,* 13:2793–2807.

Langridge, D.F. and Goodman, R.D. 1975. A study on pollination of oilseed rape (*Brassica campestris*). *Austral. J. Exptl. Agric. Anim. Husband.*, 15:285–288

Latif, A., Qayyum, A., and Abbas, M. 1960. The role of *Apis indica* in the pollination of [oil seeds] "toria" and "sarson" (*Brassica campestris* var. Toria and dichotoma). *Bee World*, 41:283–286.

Makhmoor, H.D. and Shinde, C.D. 1986. Preliminary study on the effect of different doses of N. P. and K. and their combination in mustard in on the incidence of *Athalia proxima* Klug., *Crocidolomia binotalis* Zell. and *Phytomyza atriconis*. *Res. Devel. Rep.*, 3:66–71.

Manning, W.G. and Boland, W.T. 2000. A preliminary investigation into honey bee (*Apis mellifera*) pollination of canola (*Brassica napus* cv. Karoo) in Western Australia *Austral. J. Exp. Agric.*, 40:439–442.

Meyerhoff, G. 1954. Investigation on the effect of bee visits on rape. *Arch. F. Geflugelzucht und kleintierkunde*, 3:259–306.

Miles, C.I., del Campo, M.L., Renwick, J.A.A. 2005. Behavioral and chemosensory responses to a host recognition cue by larvae of *Pieris rapae*. *J. Comp. Physiol. A – Neuroethol. Sens. Neur. Behav. Physiol.*, 191:147–155.

Mishra, R.C., Kumar, J., and Gupta, J.K. 1988. The effect of mode of pollination on yield and oil potential of *Brassica campestris* L. var. *sarson* with observation on insect pollinators. *J. Apic. Res.*, 27:186–189.

Mishra, R.C. and Kaushik, H.D. 1992. Effect of cross- pollination on yield and oil content of *Brassica campestris* L. var. sarson with pollination efficiency of honeybees, *Apis* spp. *Ann. Entomol.*, 10:33–37.

Mohammad, A. 1935. Pollination studies in toria (*Brassica napus* var. Dichotoma prain), and sarson (*B. campestris* L. var. Sarson prain). *Ind. J. Agric. Sci.*, 5:125–154.

Muhamad, O., Tsukuda, R., Oki Y., Fujisaki, K., and Nakasuji, F. 2006. Influences of wild crucifers on life history traits and flight ability of the diamondback moth, *Plutella xylostella* (Lepidoptera: Yponomeutidae) *Researches on Population Ecology*, 36:53–62.

Nayar, J.K. and Thorsteinson, A.J. 1963. Further investigations into the chemical basis of insect-host plant relationship in an oligophagous insect *Plutella maculipennis* (Curtis) (Lepidoptera: Plutellidae). *Canad. J. Zool.*, 41:923–929.

Nielsen, J.K. 1977. Host plant relationships of *Phyllotreta nemorum* L. (Coleoptera: Chrysomelidae). I. Field studies. *Z. Angew. Entomol.*, 84:396–407.

Nielsen, J.K. 1978. Host plant selection of monophagous and oligophagous flea beetles feeding on crucifers. *Entomol. Exp. Appl.*, 24:562–569.

Nielsen, J.K. 1997a. Genetics of the ability of *Phyllotreta nemorum* larvae to survive in an atypical host plant, *Barbarea vulgaris* ssp. arcuata. *Entomologia Experimentalis et Applicata*, 82:37–44.

Nielsen, J.K. 1997b. Variation in defenses of the plant *Barbarea vulgaris* and in counteradaptations by the flea beetle *Phyllotreta nemorum*. *Entomologia Experimentalis et Applicata*, 82:25–35.

Nielsen, J.K. 1996. Intraspecific variation in adult flea beetle behaviour and larval performance on an atypical host plant. In E. Städler, M. Rowell-Rahier, and R. Baur (Eds.), *Proc. 9th Int. Symp. Insect-Plant Relationships*, p. 160–162.

Nielsen, J.K. 1999. Specificity of a Y-linked gene in the flea beetle *Phyllotreta nemorum* for defenses in *Barbarea vulgaris*. *Entomologia Experimentalis et Applicata*, 91:359–368.

Nielsen, J.K., Larsen, L.M., and Sorensen, H. 1977. Cucurbitacin E and I in *Iberis amara*: feeding inhibitors for *Phyllotreta nemorum.*. *Phytochemistry*, 16:1519–1522.

Nikitina, A.I. 1950. Honeybees raise seed yields of turnips and rutabaga. *Pchelovodstvo*, 27:271–274.

Olsson, G. 1952. Investigations of the degree of cross-pollination in white mustard and rape. *Sverig. Utsadesfpren. Tidskr.*, 62:311–322.

Olsson, G. 1955. Wind pollination of cruciferous oil plants. *Sverig. Utsadesforen. Tidskr.*, 65:418–422.

Palaniswamy, P. and Lamb, R.J. 1998. Feeding preferences of a flea beetle, *Phyllotreta cruciferae* (Coleoptera: Chrysomelidae), among wild crucifers. *Canad. Entomol.*, 130:241–242.

Palaniswamy, P., Lamb, R.J., and Bodnaryk, R.P. 1998. Resistance to the flea beetle *Phyllotreta cruciferae* (Coleoptera: Chrysomellidae) in false flax, *Camelina sativa* (Brassicaceae). *Canad. Entomol.*, 130:235–240.

Panda, N. and Khush, G.S. 1995. *Host Plant Resistance to Insects*. CAB Int., Wallingford, U.K.

Pandey, N.D., Singh, M., and Tewari, G.C. 1977. Antifeeding, repellent and insecticidal properties of some indigenous plant material against mustard sawfly, *Athalia proxima* Klug. *Ind. J. Entomol.*, 30:60–64.

Pareek, L. and Gupta, H.C. 1977. Evaluation of some granular insecticides against *Athali proxima* Klug. infesting mustard. *Ind. J. Entomol.*, 39:392–406.

Pfalz, M., Vogel, H., Mitchell-Olds, T., Kroymann, J., 2007. Mapping of QTL for resistance against the crucifer specialist herbivore *Pieris brassicae* in a new *Arabidopsis* inbred line population. *Da(1)-12 x Ei-2*, PLoS ONE (2007), e578. doi:10.1371/journal.pone.0000578.

Pilson, D. 2000. The evolution of plant response to herbivory: simultaneously considering resistance and tolerance in *Brassica rapa*. *Evolut. Ecol.,* 14:457–489.

Prasad, D., Hameed, S.F., Singh, R., Yazdani, S.S., and Singh, B. 1989. Effect of bee pollination on the quantity and quality of rai crop (*Brassica juncea* Coss. *Indian Bee J.,* 51:447.

Price, P.W. 1986. Ecological aspects of host plant resistance and biological control: interactions among tritrophic levels. In *Interactions of Plant Resistance and Parasitoids and Predators of Insects,* D.J. Boethel and R.D. Eikenbary, (Eds.). Ellis Horwood, Chichester, p. 11–30.

Pritsch, G. 1965. Increasing the yield of oil plants by using honey bees. *Ved. Prace vyzkam. Ustav. Vcelar csazv,* 4:157–163.

Radchenko, T.H. 1964. The influence of pollination on the crop and the quality of seed of winter rape. *Bdzhil'nitstvo,* 1:68–74.

Rahman, K.A. 1940. Insect pollinators of toria (*Brassica napus* Linn., var. Dichotoma prain) and sarson (*B. Campestris* Linn., var. Sarson prain) at Lyallpur. *Ind. J. Agric. Sci.,* 10:422–447.

Rai, B.K. 1976. Pests of Oilseeds Crops in India and Their Control. Indian Council of Agricultural Research, New Delhi.

Rawat, R.R., Mishra, U.S., Thakare, A.V., and Dhamdhere, S.V. 1968. Preliminary study on the effect of different doses of nitrogen on the incidence of major pests of mustard. *Madras Agric. J.,* 55:363-366.

Raybold, A.F. and Moyes, C.L. 2001. The ecological genetics of aliphatic glucosinolates. *Heredity,* 87:383–391.

Reed, D.W., Pinvick, K.A., and Underhill, E.W. 1989. Identification of chemical oviposition stimulants for the diamondback moth, *Plutella xylostella,* present in three species of *Brassicaceae. Ent. Exp. Appl.,* 53:277–286.

Renwick, J.A.A. 1989. Chemical ecology of oviposition in phytophagous insects. Experientia, 45, 223–228.

Renwick, J.A.A. 2002. The chemical world of crucivores: lures, treats and traps. *Entomologia Experimentalis et Applicata* 104:35–42.

Renwick, J.A.A., Radke, C.D., and Sachdev-Gupta, K. 1991. Tolerance of cardenolides in *Erysimum cheiranthoides* by diamondback moth, *Plutella xylostella.* In *Symp. Biol. Hung. 39.* T. Jermy and A. Szentesi (Eds.). Akademiai Kiado, Budapest, p. 527–528.

Rohilla, H.R., Singh, H., Kalra, V.K., and Kharub, S.S. 1987. Losses caused by mustard aphid, *Lipaphis erysimi* (Kalt.) in different *Brassica* genotypes. *Proc. 7th Int. Rapeseed Congr.,* Poland. 5:1077–1083.

Sachan, J.N. and Gangwar, S.K. 1980. Vertical distribution of important insect-pest of cole crops in Meghalaya as influenced by the environment factors. *Ind. J. Entomol.,* 42:414–421.

Sachan, G.C. and Sharma, S. 1987. Effect of some insecticides on germination and seedling vigour of Toria seed. *Ind. J. Pl. Prot.,* 15:65–67.

Sachan, G.C. and Sumati, A. 1985. Incidence of *Athalia proxima* Klug. On species and varieties of *Brassica. Cruciferae Newslett.,* 10:128–129.

Sachan, G.C., Pathak, P.K., and Chibber, R.C. 1983a. Effect of nitrogen levels and date of sowing on the incidence of *Lipaphis erysimi* and yield of mustard. In: *Annual Report Res. Directorate Expt. Stn. G.B. Pant Univ. Agric. Tech.,* Pantnagar.

Sachan, G.C. Pathak, P.K., and Chibber, R.C. 1983b. Screening of rapeseed germplasm and breeding lines against *Athalia proxima* Klug. In: *Annual Report Res. Directorate Expt. Stn. G.B. Pant Univ. Agric. Tech.,* Pantnagar.

Sachan, J.N. and Srivastava, B.P. 1972. Studies on seasonal incidence of insect pest of cabbage. *Ind. J. Entomol.,* 34:123–127.

Sachdev-Gupta, K., Radke, C.D., and Renwick, J.A.A. 1993a. Antifeedant activity of cucurbitacins from *Iberis amara* against larvae of *Pieris rapae. Phytochemistry,* 33:1385–1388.

Sachdev-Gupta, K., Radke, C.D., Renwick, J.A.A., and Dimock, M.B. 1993b. Cardenolides from *Erysimum cheiranthoides*: feeding deterrents to *Pieris rapae* larvae. *J. Chem. Ecol.,* 19:1355–1369.

Sarup, P., Singh, D.S., Sircar, P., Amarpuri, Lal, R., Saxena, V.S., and Srivastva, V.S. 1971a. Relative toxicity of some important pesticides to the adults of *Bagrada cruciferarum* Kirk. (Pentatomidae: Hemiptera). *Ind. J. Entomol.,* 33:452–456.

Sarup, P., Sircar, P., Sharma, D.N., Amarpuri, S., Dewan, R.S., and Lal, R. 1971b. Effect of formulation on the toxicity of pesticidal granules to some important pests of mustard. *Ind. J. Entomol.,* 33:82–89.

Schoonhoven, L.M. 1969. Gustation and foodplant selection in some Lepidopterous larvae. *Entomol. Exp. Appl.,* 12:555–561.

Schoonhoven, L.M., Jermy, T., and Van Loon, J.J.A., 1998. *Insect-Plant Biology*, Chapman & Hall, London.

Serizawa, H., Shinoda, T., and Kawai, A. 2001. Occurrence of a feeding deterrent in *Barbarea vulgaris* (Brassicales: *Brassicaceae*), a crucifer unacceptable to the diamondback moth, *Plutella xylostella* (Lepidoptera: Plutellidae). *Appl. Entomol. Zool.*, 36:465–470.

Sharma, R.M. and Singh, Harvir 1990. A new species of *Dasineura* (Diptera: Cecidomyiidae) injurious to buds of *Brassica* spp. (Cruciferae) in Haryana. *J. Bombay Nat. Hist. Soc.*, 87:429–432.

Shelton, A.M. and Nault, B.A. 2004. Dead-end trap cropping: a technique to improve management of the diamondback moth, *Plutella xylostella* (Lepidoptera: Plutellidae). *Crop Protection*, 23:497–503.

Shinoda, T., Nagao, T., Nakayama, H., Serizawa, H., Koshioka, M., Okabe, H., Kawai, A. 2002. Identification of a triterpenoid saponin from a crucifer, *Barbarea vulgaris*, as a feeding deterrent to the diamondback moth, Plutella xylostella. *J. Chem. Ecol.*, 28:587–599.

Shuichi, Y. 1994. Ecological and evolutionary interactions between wild crucifers and their herbivorous insects. *Plant Species Biol.*, 9:137–143.

Shuichi, Y. and Naota, O. 1993. The phenology and intrinsic quality of wild crucifers that determine the community structure of their herbivorous insects. *Researches on Population Ecology*, 35:151–170.

Sihag, R.C. 1986. Insect pollination increases seed production in cruciferous and umbelliferous crops. *Apic. Res.*, 25:121–126.

Sihag, R.C. 1988. Effect of pesticides and bee pollination on seed yield of some crops in India. *J. Apicult. Res.*, 27:49–54.

Sihag, R.C. 1991. Ecology of European honeybee (*Apis mellifera* L.) in semi arid and subtropical climates. 1. Melliferous flora and over — seasoning of the colonies. *Korean J. Apicult.*, 5:31–43.

Sihag, R.C., Khatkar, S., and Khatkar, S. 1999a. Foraging pattern of three honeybee species on eight cultivars of oilseed crops. 2. Foraging during the entire blooming period of the crops. *Int. J. Tropical Agric.*, 17:253–261.

Sihag, R.C., Khatkar, S., and Khatkar, S. 1999b. Synchronization of foraging activity of honeybees and flowering/anthesis in oilseeds crops. *Ann. Agric. Biolog. Res.*, 4:263–269.

Singh, H. and Singh, Z. 1983. New records of insect-pests of rapeseed-mustard. *Ind. J. Agric. Sci.*, 53:970.

Singh, Hari., Singh, D., Singh, Harvir, and Kumar, V. 1990. Basis of aphid tolerance and combining ability analysis in Indian mustard. In: *Abstract National Seminar on Genetics of Brassicas*. August 8–9,1990. R.A.U. Agricultural Research Station, Durgapura, Jaipur, Rajasthan, p. 15.

Singh, H., Rohilla, H.R., and Yadava, T.P. 1987. Comparative efficacy of aldrin (EC) and BHC (WP and dust) for the control of termite, *Odontotermes obesus* (Rambur) in groundnut (*Arachis hypogea* Linn.)z. *Indian J. Entomol.*, 46:409–411.

Singh, Hari. 1991. A Success Story of Five-fold Increase of Oilseeds Production in Haryana. Haryana Agricultural University, Hisar, India.

Singh, M.P., Singh, K.I., and Devi, C.S. 2000. Role of *Apis cerana himalaya* pollination on yield and quality of rapeseed and sunflower. In: *Asian Bees and Beekeeping* (Eds., Matsuka, M., Verma, L.R., Wongsiri, S., Shrestha, K.K., and Pratap, U.) Oxford and IBH Publishing Co., New Delhi, p. 186–189.

Singh, O.P., Rawat, R.R., and Chaudhary, B.S. 1980. Efficacy of some granular insecticides against flea beetle, *Phyllotreta cruciferae* on mustard. *Ind. J. Plant Protection*, 8:54–56.

Singh, P. 1980. Influence of Host Plant on the Biology of Pea Leaf Miner *Phytomyza horticola* (Goureau). M.Sc. thesis, Punjab Agricultural University, Ludhiana, India.

Singh, P. 1985. Morphological, Anatomical and Biochemical Bases of Resistance in Rapeseed and Mustard to the Leaf Miner *Phytomyza horticola* (Goureau). Ph.D. dissertation, Punjab Agricultural University, Ludhiana, India.

Srivastava, A. and Guleria, S. 2003. Evaluation of botanicals for mustard aphid, Lipaphis erysimi (Kalt.) control in *Brassica Himachal. J. Agric. Res.*, 29(1–2):116–118.

Talekar, N.S. and Shelton, A.M. 1993. Biology, ecology and management of the diamondback moth. *Annu. Rev. Entomol.*, 38:275–301.

Tattersall, D.B., Bak, S., Jones, P.R., Olsen, C.E., Nielsen, J.K., Hansen, M.L., Hoj, P.B., and Moller, B.L. 2001. Resistance to an herbivore through engineered cyanogenic glucoside synthesis. *Science*, 293:1826–1828.

Teotia, T.P.S. and Gupta, G.P. 1972. Effect of host plants on the susceptibility of larvae of *Athalia proxima* Klug. to insecticides. *Ind. J. Entomol.*, 32:140–144.

Thomas, M. and Waage, J.K. 1996. *Integration of Biological Control and Host–Plant Resistance Breeding: A Scientific and Literature Review.* CTA, Wageningen, The Netherlands.

Thompson, J.N. 1994. *The Coevolutionary Process*, Chicago and London: The University of Chicago Press.

Thorsteinson, A.J. 1953. The chemotactic responses that determine host specificity in an oligophagous insect (*Plutella maculipennis* (Curt.): Lepidoptera). *Can. J. Zool.*, 31:52–72.

Todd, J.L. and Baker, T.C. 1999. Function of peripheral olfactory organs. In: *Insect Olfaction*. B.S. Hansson (Ed.). Springer-Verlag Berlin, p. 67–96.

Tripathi, N.L.M., Sachan, G.C., and Verma, S.K. 1988. Relative toxicity and safety of some insecticides to *Coccinella septumpunctata*. *Ind. J. Plant Protection,* 16:57–58.

Van Lenteren, J.C., Li Zhao, H., Kamerman, J.W., and Xu, R. 1995. The parasite–host relationship between *Encarsia formosa* (Hym., Aphelinidae) and *Trialeurodes vaporariorum* (Hom., Aleyrodidae). 26. Leaf hairs reduce the capacity of *Encarsia* to control greenhouse whitefly on cucumber. *J. Appl. Entomol.,* 119:553–559.

Vesely, V. 1962. The economic effectiveness of bee pollination on winter rape (*Brassica napus* l., var. Oleifera metz.). Min. Zemedel. Lesn. A vodniho hospodar. *Ust. Vedtech. Inform. Zemedel. Ekon.,* 8:659–673.

Vora, V.J., Bharodia, R.K., and Kapadia, M.N. 1985a. Pests of oilseed crops and their control on rapeseed and mustard. *Pesticides,* 19:38–40.

Vora, V.J., Bharodia, R.K., and Kapadia, M.N. 1985b. Pests of oilseed crops and their control on Seasamum. *Pesticides,* 19:11–12.

Walter, D.E. 1996. Living on leaves—Mites, tomenta, and leaf domatia. *Annu. Rev. Entomol.* 41:101–114.

Walter, D.E. and O'Dowd, D.J. 1992. Leaf morphology and predators: Effect of leaf domatia on the abundance of predatory mites (Acari: Phytoseiidae). *Environ. Entomol.* 21:478–484.

White, B. 1970. Pollination of commercial rape seed crops. *Australasian Beekeeper,* 72:99–100.

Wilson, T.G. 2001. Resistance of *Drosophila* to toxins. *Annu. Rev. Entomol.,* 46:545–571.

Wynne Griffiths, D., Deighton, N., Nicholas, A., Birch, E., Patrian, B., Baur, R., and Städler, E. 2001. Identification of glucosinolates on the leaf surface of plants from the Cruciferae and other closely related species. *Phytochemistry, 57*:693–700.

Yadva, P.R. and Sachan, J.N. 1976. Residual toxicity of some common insecticides to *Lipaphis erysimi* (Kalt.) (Hemiptera: Aphididae) as serious pest of cauliflower in Rajasthan. *Pesticides,* 2:39–44.

9 Seed Dormancy and Viability

M.S. Naeem, D. Liu, R. Raziuddin,
G.L. Wan, G.X. Tang, and W.J. Zhou

CONTENTS

INTRODUCTION

The *Brassicaceae* family comprises about 3000 species of herbaceous plants within more than 300 genera, the majority of which are found in the Northern Hemisphere. Many common agricultural weeds, such as *Brassica nigra* (L.) Koch, *B. rapa* L., *Cardaria draba* (L.) Desv., *Raphanus raphanistrum* L., and *Sinapis arvensis* L., also belong to this family. The most important crop species from this family are the oilseed *Brassicas* — *Brassica napus* L., *B. rapa* L. (syn. *B. campestris* L.), and *B. juncea* Coss., which are generally referred to as rapeseed, oilseed rape, or canola. Other widely cultivated species in this family include *B. oleracea* L. (cabbage, kale, kohlrabi, Brussels sprouts, cauliflower, and broccoli), *B. chinensis* L. (syn. *B. napus* var. chinensis; Chinese cabbage), *Raphanus sativus* L. (radish), and *Armoracia rusticana* Gaertn. (horseradish). The fruits of these species are usually dehiscent pod-like capsules; if they are longer than their width, they are called a silique; if they are as broad as they are long, they are called a silicula.

Dormancy is a potential problem for most accessions of the Cruciferae. The seeds have a curved embryo and no endosperm. Germination is very species specific; some seeds require light, some require darkness, some need a minimum temperature around 20°C, while others need cool temperatures to germinate. Some require vernalization or exposure to high heat. Seeds possess germination and dormancy characteristics that depend on their genetic nature, and germination and dormancy vary radically among species. Germination occurs under specific environmental conditions with some variability. Seed of crop plants such as *Brassica napus* (Swede rape or Argentine canola), *B. rapa* (Turnip rape or Polish canola), *B. juncea* (brown mustard), and *Sinapis alba* (yellow mustard) germinate within 4 to 5 days in the presence of suitable moisture at temperatures as low as 5°C, and within 24 hours as temperatures approach 20°C. The germination tends to be fairly consistent, and these seeds are often referred to as being non-dormant.

Dormancy, germination, and viability are crucial phenomena in the life cycle of all plants from the point of view of the developmental and regulatory processes involved in the transition from a developing seed through dormancy and into germination and seedling growth. This chapter examines the complexity of the environmental, physiological, molecular, and genetic interactions that occur throughout the life cycle of seeds, along with the concepts and approaches used to analyze seed dormancy and germination behavior. This chapter also identifies the current challenges and remaining questions for future research; it is an effort to gather information about the dormancy and viability of crucifers.

SEED DORMANCY AND CRUCIFERS

Seed dormancy is considered the failure of viable seeds to germinate, even under conditions that favor the normal growth and development of the seedling. There is a distinction between dormant and quiescent seeds. In the latter, no germination events take place, usually due to low moisture contents. They are alive and have metabolism ongoing at a hardly detectable rate, but lack some environmental factors necessary for germination to commence. Viable seeds that are in an environment optimal for germination (including optimal water, temperature, light, and oxygen) and yet fail to complete germination are termed dormant seeds and the phenomenon is called dormancy.

CATEGORIES OF DORMANCY

Seed dormancy is classified into primary and secondary seed dormancy (Harper, 1957; Nikolaeva, 1977; Baskin and Baskin, 1985) (Table 9.1). The classification is based on the time at which seed dormancy occurs. Primary and secondary seed dormancy are further subdivided according to the degree of dormancy.

Primary dormancy has been defined as a state where germination of the progeny is prevented while maturing on the mother plant and for some time after the seed has separated from its parent (Karssen,

---

TABLE 9.1

Categories of Seed Dormancy

Categories	Groups	Caused By	Mechanism
Primary dormancy	Exogenous (outside embryo)	• Maternal tissues and/or endosperm or perisperm	• Inhibition of water uptake (physical dormancy) • Mechanical restraint to embryo expansion and radical protrusion (mechanical dormancy)
	Endogenous (inside embryo)		• Modification of gas exchange • Prevention of leaching of inhibitors from embryo • Supplying inhibitors to the endosperm (chemical dormancy)
		• Underdeveloped embryo (morphological dormancy)	• Embryo in mature seed has to complete development prior to germination
		• Metabolic blocks (physiological dormancy) • Morpho-physiological dormancy	• Physiological mechanisms largely unknown
	Combinational	• Combination of exogenous and endogenous dormancy	
Secondary dormancy		• Metabolic block induced in non-dormant seeds when germination environment is unfavorable	• Physiological mechanisms largely unknown

Source: Adapted from Nikolaeva (1977) and Baskin and Baskin (1998).

1981; Hilhorst and Toorop, 1997). A period of "after-ripening" is generally required to alleviate primary dormancy. After-ripening is the period of time between seed dissemination and the time when maximum germination percentage can be achieved under optimal germination conditions. Several different mechanisms may cause this response (Baskin and Baskin, 1998). Secondary dormancy is usually defined as a reduction in seed germinability that develops at any time after seed dissemination and may, in some instances, be induced prior to the complete alleviation of primary dormancy.

Subdivisions of this classification system are defined by the degree of dormancy, where either primary or secondary dormancy may be conditional or innate (Baskin and Baskin, 1985). Relative to non-dormant seeds, conditionally dormant seeds germinate under a more limited range of conditions, while seeds in innate dormancy fail to germinate under any conditions. In this model, seeds are presumed to pass through conditional dormancy when going from an innate dormant to a non-dormant state, and vice versa.

Seeds may go from non-dormant to a conditionally secondary dormant to an innately secondary dormant state and back to a non-dormant state in the reverse order within the span of 1 year (Baskin and Baskin, 1985). This cycle can repeat itself for several years in any one seed in the seed bank. It serves as a buffer to rapid genetic adaptation as well as genetic bottlenecks that may result from adverse conditions experienced in the short term in annual species. The timing of this cycle is strongly regulated by temperature (Probert, 2000) and is defined by the lifecycle of a species (summer- vs. winter-annual). Non-dormant seeds that fail to germinate as a result of unfavorable external conditions rather than factors within the seed are categorized as quiescent (Baskin and Baskin, 1985).

Primary Dormancy in *Brassica napus*

In winter and spring *Brassica napus*, low germinability exists exclusively during seed maturation and declines with increasing seed maturity (Finkelstein et al., 1985; Schlink, 1994). Primary dormancy levels in winter *B. napus* seeds range between 10 and 20% (Pekrun et al., 1998a) but may be as high as 60% in some genotypes between 8 and 12 weeks after flowering (Schlink, 1994), whereas, by harvest, primary seed dormancy is no longer present.

In contrast, most cruciferous crops, including *Brassica napus*, require a period of after-ripening before high germination percentages are achieved (Tokumasu et al., 1981; Tokumasu and Kato, 1987). Storage of cruciferous seeds within the siliques decreases the maximum germination percentage as well as the rate at which primary dormancy is released (Tokumasu et al., 1981). This may have implications for seed bank persistence if undehissed siliques enter the seed bank. Furthermore, the optimum storage conditions for the release of primary dormancy have been found at relative humidities below 35%, while optimum germination temperatures range between 15 and 35°C (Tokumasu et al., 1981). Research on *B. napus* showed that seed dormancy (presumably primary) was released by low temperatures in conjunction with washing and pricking the seeds (Sugiyama, 1949, cited in Takahashi and Suzuki, 1980) or by treatment with either thiourea or urea (Hori and Sugiyama, 1954, cited in Takahashi and Suzuki, 1980).

Secondary Dormancy in *Brassica napus*

Although primary dormancy is negligible in fully mature *Brassica napus*, seeds of this species may develop secondary seed dormancy (Schlink, 1994; Pekrun, 1994; Pekrun et al., 1997a). In the laboratory, incubation in darkness for up to 4 weeks in conjunction with the use of polyethylene glycol (PEG-8000) solutions with an initial water potential of −1.5 MPa were most successful in inducing secondary dormancy (Pekrun, 1994; Schlink, 1994). Seed dormancy expression tends to be lower at higher initial osmotic potentials of the imbibing solution, given equal exposure intervals (Pekrun et al., 1998b). Low O_2 concentrations (3% O_2:97% N_2), simulating wet soil conditions, in combination with darkness also have been reported to induce secondary dormancy in *Brassica napus*, albeit to a far lesser extent than osmotic stress in combination with darkness (Pekrun, 1994; Pekrun et al., 1997c; Momoh et al., 2002).

The importance of temperature in seed dormancy development in *Brassica napus* is less clear than that of osmotic potential. It has been established that increasing the diurnal temperature variations during dormancy induction tends to decrease secondary seed dormancy development (Pekrun et al. 1997b). This behavior is analogous to dormancy release in other species and has been suggested as a depth and gap sensing mechanism (Thompson and Grime, 1983; Goedert and Roberts, 1986). The effect of static temperatures on dormancy induction is less clear. Investigations have suggested higher mean seed dormancy induction among a group of genotypes induced at 20°C compared to 12°C, although the observed differences were not statistically significant (Momoh et al., 2002). Moreover, the temperature difference between the induction temperature and the subsequent germination test temperature appears to influence secondary dormancy development (Pekrun et al., 1997c; Momoh et al., 2002). When the absolute difference between these temperatures is increased, secondary seed dormancy tends to decrease. These observations suggest a high sensitivity of *B. napus* seed dormancy to temperature, although empirical data confirming this remain still sparse. A genetic component to secondary seed dormancy potential in *B. napus* has been suggested. Among 25 spring and 21 winter *B. napus* cultivars tested in Europe, the mean proportion of seeds induced into secondary dormancy ranged from 0.7% to 76.1% (Pekrun et al., 1997a). Among the two groups of cultivars tested, the average potential for secondary dormancy was similar in spring and winter *B. napus* cultivars. In a different study, however, higher maximum levels of secondary seed dormancy were observed in spring compared to winter *B. napus* genotypes (Momoh et al., 2002).

In *Brassica napus*, secondary seed dormancy is readily reversed by several factors. Schlink (1994) demonstrated that a single exposure to a camera flash with duration of 0.002 seconds was sufficient to increase mean germination in dormant seeds of two *B. napus* cultivars from 13.3% to 63.0%. A final germination percentage of 98.1% was attained when dormant seeds were exposed to continuous light. The observed phenomenon is similar to the well-documented light requirement for germination of lettuce seed (*Lactuca sativa* L.) (Brothwick et al., 1954) and various small seeded weeds (Wesson and Wareing, 1969a, b). Consequently, the involvement of the phytochrome system has been implicated in *B. napus* seed (López-Granados and Lutman, 1998). A seasonal response of germination to light exposure in *B. napus* seed has been observed in winter *B. napus* where the readily germinable proportion of the viable seed bank was lower under light from seed exhumed during summer compared to seed exhumed at any other time of the year (Schlink, 1995). In contrast, Bazanska and Lewak (1986) found that germination of *B. napus* may be inhibited by continuous exposure to white light at lower temperatures in combination with moisture stress. A similar light inhibition of germination also has been shown in mustard (*Sinapis alba* L.) seeds (MacDonald and Hart, 1981).

A stratification treatment at 2 to 4°C for 3 days also has proven effective in releasing secondary seed dormancy in *Brassica napus* (Pekrun et al., 1998b). In addition, exogenous applications of gibberellic acid (0.2 mg/L) have been used to reverse secondary seed dormancy (Pekrun et al., 1998b). The effectiveness of these methods in relieving seed dormancy in other species is well established in the literature (Bewley and Black, 1994; Baskin and Baskin, 1998).

Factors Affecting the Development of Dormancy

Freshly harvested seeds of oilseed rape are thought to exhibit little or no primary dormancy (Lutman, 1993; Schlink, 1994). The development of dormancy is thus brought about by the prolonged absence of environmental cues that favor seed germination. These conditions include: temperature, air, light, water potential, seed age, and storage conditions.

Temperature

The effect of temperature on the induction of seed dormancy is one that cannot be treated as an independent factor, as it is found to interact significantly with other environmental and soil conditions. An increase in temperature increases the rate of after-ripening of air-dry seeds and it also affects the rate of induction of secondary dormancy in imbibed seeds (Karssen, 1982; Baskin et al., 1984). Pekrun et al. (1996, 1997a, b) observed the induction of secondary dormancy at low temperature and other conditions. In contrast, Landbo and Jorgensen (1977) found little or no dormancy at all in some cultivars of *Brassica napus* and *B. rapa* at low temperatures. These inconsistencies were explained in terms of the large cultivar differences in the development of secondary dormancy (Squire, 1999). However, temperature does greatly affect the rate of induction of dormancy. Pons (1991) observed that temperature has a significant effect on the induction of dark dormancy. At low temperatures, the process is slow, which is consistent with the slow rate of dark reversion of phytochrome far red. Temperature fluctuation was observed to lower the percent of surviving seeds (Lutman et al., 1998). Seeds induced into dormancy via incorporate into dry soil would experience diurnal temperature fluctuations if they were near the soil surface, but these variations would be greatly reduced if the seeds were buried at depth.

High temperatures can also be a problem. While many growers have come to recognize the minimum temperature limit in the spring, it remains difficult to understand that soil temperature is equally important in the summer when high temperatures can cause dormancy, and even the death of seeds. When seeds do germinate in high-temperature soil, the seedlings may die from the heat; when they do not, they may either germinate and die, or enter into dormancy. This is visible at low soil water potential (Pekrun, et al., 1997a; Zhou and Kristiansson, 2000; Momoh et al., 2002). The impact of high temperature is not limited exclusively to soil temperature. Seed stored at high temperatures may result in slow, erratic germination or in poor seedling development. Both the onset and breaking of secondary dormancy is strongly influenced by temperature. Dormancy may be lost in seeds under moist conditions at cold — not freezing — temperatures and a brief high temperature (stratification). Duality in the role of temperature in the release of secondary dormancy was reported by Bouwmeester and Karssen (1992). Temperature is the one determining factor in the timing of dormancy cycles over the year by long-term changes in field temperature. Fresh seeds are dormant at maturity. When buried and exposed to natural fluctuations in seasonal temperature, they show an annual dormant and non-dormant cycle. Burial in the summer causes fresh seeds and those buried for 1 year after ripening to require light for germination. During fall and winter, seeds reenter dormancy and during the following summer they become non-dormant again (Baskin and Baskin, 1984).

Zhang and Hampton (1999) reported that high temperature (45°C) and seed moisture content (20%) of the controlled deterioration (CD) test for vigor induced secondary dormancy in six swede (*Brassica napus*) seed lots when they were germinated, post-CD at 20°C without pre-chilling. Pre-chilling and/or germination at 20 to 30°C broke secondary dormancy.

Air

The metabolic processes in plant growth and development depend heavily on oxygen. However, growth can also occur for a limited period under anaerobic or near anaerobic conditions. The requirements for oxygen and other gases in combination for the induction and breaking of dormancy are not clear. In some research, it had been observed that anoxia or hypoxia might reduce the germination of many seeds. Parasher and Singh (1984, 1985) and Symon et al. (1986) noted that some weed seeds, such as those of *Avena fatua* and *Phalaris minor,* did not germinate well in anoxic or hypoxic conditions. The seeds did tolerate anaerobic conditions by entering into secondary dormancy. In the barnyard grass (*Echinochloa crus-galli*), Honek and Martinkova (1992) reported that hypoxia did not induce secondary dormancy at an incubation temperature of 15°C or prevent dormancy termination at 7°C, indicating that oxygen deficiency might increase the proportion of dormant seeds in the soil, and affect the dynamics of the soil seed bank. Benvenuti and Macchia (1997) conducted experiments to investigate the effect of hypoxia on seed germination of *Datura stramonium* in Petri dishes and then buried at various depths in soil. Hypoxia was found to cause a decrease in germination capacity and germination rate of the tested seeds. This inhibition was partially alleviated by the daily exchange of hypoxic gas surrounding the seeds during incubation.

Similarly, seed scarification allowed maintenance of a higher germination capacity under conditions of low oxygen availability, showing that the seed coat was only partially gas permeable. The report also stated that oxygen deficiency led to a decrease in respiratory capacity. However, this was probably compensated for by induction of fermentation metabolism. Thus, daily nitrogen flushing partially eliminated this inhibition, even under conditions of low external oxygen availability. It was therefore postulated that the main depth-derived inhibition was not caused directly by oxygen deficiency but by the increasing difficulty in eliminating toxic fermentation products, which were found proportional to the degree of hypoxia. Finally, incubation for several days under completely anaerobic conditions induced secondary dormancy. Reporting on the physiology of anaerobiosis in *Phalaris minor* Retz. and *Avena fatua* L. seeds, Parasher et al. (1984), observed that anaerobic conditions were not conducive to seed germination in *P. minor* and *A. fatua* at optimum temperature. The magnitude of either delay in germination or loss of seed viability depended on the species, as well as the duration of seed storage under anaerobic conditions. *P. minor* seeds tolerated anaerobic conditions by entering into secondary dormancy and by avoiding anaerobic decomposition. Although *A. fatua* entered into weak secondary dormancy, it was susceptible to anaerobic decomposition. Chemical status of seeds under anaerobic conditions suggested that *P. minor* seeds resisted oxygen stress more than *A. fatua* seeds. This indicates that the response of seeds to anaerobic conditions might also be cultivar/species dependent.

Light

The effects of light vary from genera to genera and sometimes even between species and varieties. Some seeds are stimulated by light, and others are inhibited by light during germination. Knowing the light requirements of a seed to germinate determines seed depth at planting and the types of seed treatments. One of the more striking disclosures in plant physiology involves the effect of light quality on the ability of many species, seeds to complete germination, or be forced into dormancy. In *Arabidopsis thaliana*, no fewer than five phytochromes were found to be present. Investigating the action of these chromophores became intense due to the marked effect that phytochrome B, and possibly others, have on seed germination. When imbibed seeds of lettuce were illuminated with a period of far-red (FR) light, the percentage of seeds that subsequently completed germination in the dark was very low (Khan, 1960; Casal and Sanchez, 1998). If the period of far-red illumination was followed by a period of red light illumination, the seeds subsequently completed germination to almost 100%. If, however, the second red light illumination was followed by another period of far-red illumination, seed germination was again drastically inhibited in the dark. This cycle of germination inhibition and stimulation can continue *ad infinitum* until, at some point much advanced in seed germination, the seeds "escape" from phytochrome control and complete germination in the dark regardless of the illumination they perceived last. Lopez-Granados and Lutman, (1998) and Casal and Sanchez (1998) reinforced the involvement of phytochromes in the photoreaction of oilseed rape seed. They noted the response demonstrated by the seed to FR. Their investigation further disclosed that dormant seed was not produced after imbibition in polyethylene glycol (PEG) under white light. This was in agreement with previous reports by Pekrun (1997a) and Lutman et al. (1998).

Water Potential

Water stress has been confirmed to reduce both the rate and percentage of seed germination. There is, however, a wide range of responses among species based on their sensitivity to resistance or tolerance (Bewley and Black, 1994). All seeds except for those with intact, water-impermeable seed coats imbibe when exposed to water. The rate of imbibition was found to be governed by three categories of factors: (1) seed properties, (2) soil properties, and (3) degree of contact between the seed and the soil. The soil water potential determines the final seed water potential. Several researchers have reported on the effect of low soil water potential on seed germination and the subsequent influence on induction of secondary dormancy. Hilhorst (1998) noted that secondary dormancy is predominantly associated with seed behavior in soil seed banks. Temperature and possibly soil water potential appear to be the predominant factors that determine the annual cycling of dormancy. Prolonged imbibition under conditions of low soil osmotic potential has been shown to cause induction of secondary dormancy in many species (Berrie et al., 1974; Khan and Karssen, 1980; Pons, 1991).

Reporting on the germination behavior of dormant oilseed rape seeds in relation to temperature, Pekrun et al. (1997a) noted that seeds of oilseed rape were induced into secondary dormancy by imbibing them for 2 to 4 weeks in an osmotic solution at −1500 KPa in darkness. In their work, treated seeds were transferred into pots or Petri dishes, where they were given adequate water for germination. These pots and Petri dishes were exposed to various temperature regimes during the following 6 to 14 months. The number of non-germinated seeds remained constant in most treatments but declined when seeds were exposed to pronounced daily temperature alterations. This decline was due to germination and not to mortality from other causes. Seeds exposed to a constant temperature for 1 year appeared to become more dormant with time; it was reported that after prolonged imbibition under water stress conditions, seeds exhibited light sensitivity, with higher percentages of dormant seeds at stress of −15 bars. Lutman et al. (1998), in another experiment, noted that a critical issue relating to the development of secondary dormancy in seeds of oilseed rape is the dryness of the soil. In their work, seeds subjected to high water potential of −2 bars showed very low potential development for secondary dormancy. On the other hand, seeds subjected to a stronger osmotic stress of 20 bars failed to germinate, thus increasing the potential development of secondary dormancy in those seeds.

Seed Age and Storage Conditions

The effect of seed storage on the potential development of secondary dormancy in oilseed rape seeds has not been fully investigated. However, the effect of the length and method of seed storage of other seeds can be used as a basis for further studies on oilseed rape. Kalmbacher et al. (1999) observed the role of seed storage in its final germination performance. They observed that although harvest time of the seed greatly determines its subsequent germination, storage temperature proved to be the determining factor. In another report on the effects of storage conditions on dormancy and vigor of *Picea abies* seeds, Leinonen (1998) observed that storage at 75% relative humidity (RH) and 12°C decreased germination to nearly zero in *Picea abies;* furthermore, he suggested that germination of seeds stored at 75% RH could be stimulated by a short accelerated aging period.

The effect of storage conditions on dormancy release and induction of secondary dormancy in weeds was also investigated by Kim et al. (1996). In two experiments carried out on seeds of nine weed species, they investigated (1) the most effective storage conditions for breaking the seed dormancy of each weed species, and (2) whether germinability decreased by transferring seeds to dry storage conditions. The dormancy of some seed samples was broken under dry conditions, while that in others was broken by soaking in water. Other weed species were released from dormancy by storage under wet conditions. When seeds stored under wet or soaking conditions were air-dried and then restored at room or low temperature, a tendency for the germination rate to decrease was particularly observed for seeds of *Persicaria vulgaris.* The low germination rate of *P. vulgaris* after 3 months of storage did not appear to be caused by drying, because a decrease in germination rate was observed with increasing storage period under all the tested storage conditions (Haferkamp et al., 1994; Campbell et al., 1992).

GENOTYPIC DIFFERENCES IN THE DEVELOPMENT OF DORMANCY

Oilseed rape exhibits a wide variation in the development of secondary dormancy between and within genotypes (Pekrun et al., 1997c). Lutman et al. (1998) reported clear differences in the potential development of secondary dormancy in different cultivars of oilseed rape. These ranged from virtually no dormancy whatsoever to high potential development. This makes it possible to select cultivars on the basis of low or zero dormancy. In Pekrun et al. (1997c), inter- and intravarietal variation in the development of secondary dormancy in oilseed rape seeds was investigated by repeatedly testing 47 cultivars in a standard Petri dish test, using an osmotic solution and darkness to impose dormancy. Cultivars showed a wide range of response, ranging from below 2% dormant seeds (e.g., Falcon, Acrobat, and Industry) to over 50% (e.g., Apex, Nimbus, and Mars). The experiments demonstrate that varietal choice can have a substantial effect on subsequent volunteer rape populations and should be taken into consideration by growers when selecting oilseed cultivars.

The factors affecting secondary seed dormancy expression are still unclear in *Brassica napus.* Although a genetic component has been implicated, its importance relative to other factors such as environment during seed maturation (Gutterman, 1980/81) remains unclear. Furthermore, the influence of seed storage on secondary seed dormancy characteristics has not yet been directly investigated in this species (Pekrun et al., 1997a; Momoh et al., 2002), and the role of constant temperature on seed dormancy induction rates is also unclear (Momoh et al., 2002).

ABSCISIC ACID (ABA) AND SEED DORMANCY

CORRELATION OF ABA WITH SEED DORMANCY AND OTHER MARKERS

Abscisic acid (ABA) has been implicated in the regulation of a number of plant responses, including dormancy. Experiments examining the effects of exogenous applications of ABA and gibberellins on seed have shown antagonism between these two growth regulating substances. High ratios of ABA: gibberellins have been suggested to promote dormancy, while a low ratio of these hormones tends to lead to the completion of germination (Wareing and Saunders, 1971).

Recently, the role of ABA in seed dormancy regulation has been indicated in several species. In *Nicotiana plumbaginifolia* Viv., an ABA-deficient mutant displayed a non-dormant phenotype, and endogenous ABA concentrations decreased during after-ripening (Grappin et al., 2000). Fluridone, an ABA biosynthesis inhibitor (Gamble and Mullet, 1986; Xu and Bewley, 1995; Song et al., 2005, 2006), was equally effective at breaking seed dormancy at exogenous applications of gibberellic acid (GA_3) (Grappin et al., 2000). Fluridone inhibits phytoene desaturase, a key enzyme in the carotenoid synthesis pathway that provides the precursors to ABA. Similarly, ABA-deficient and insensitive mutants of *Arabidopsis thaliana* L. Heynh., a species closely related to *Brassica napus*, have implicated ABA in the regulation of seed dormancy in this species (Schwartz et al., 1997; Finkelstein et al., 2002; Brocard-Clifford et al., 2003). In *B. napus*, endogenous ABA and GA levels were manipulated via exogenous hormone applications (Fu and Lu, 1991). They further noted that endogenous ABA levels correlated negatively with the seed germination rate ($r = -0.9486$), while endogenous gibberellin levels correlated positively with the seed germination rate ($r = 0.9666$). Whether such a correlation exists during secondary seed dormancy induction and release in *B. napus* remains to be investigated. In other species, however, endogenous levels of ABA and gibberellins have not adequately explained seed germination behavior (Wareing and Saunders, 1971; Bewley and Black, 1994; Ramagosa et al., 2001), indicating that a low ratio between these two hormones does not ubiquitously release seed dormancy. Cytokinins and ethylene also may promote germination in some species (Wareing and Saunders, 1971) and may be more effective than gibberellins at antagonizing ABA to promote germination (Wareing and Saunders, 1971).

Most investigations on the role of ABA in seed dormancy have been with respect to primary seed dormancy in crop species (Gosling et al., 1981; Walker-Simmons, 1987; Wang et al., 1995; Benech-Arnold et al., 1999). Nevertheless, ABA also has been implicated in the maintenance of thermo-dormancy, a secondary dormancy induced by exposure to high temperatures during imbibition. When seeds of these species are imbibed at supra-optimal temperatures, reexposure to temperatures that would have resulted in germination prior to the high-temperature treatment result in seed dormancy. This phenomenon is well documented in lettuce seed. In this species, thermo-dormancy is readily released by 30 μm fluridone (Yoshioka et al., 1998), suggesting a role of ABA in thermo-dormancy maintenance. Further investigations have shown a different route for ABA metabolism in thermo-dormant lettuce seeds compared to non-dormant seeds (Chiwocha et al., 2003). In cereals, thermo-dormancy correlates well with the release of ABA from the seed coat (Corbineau and Côme, 2000). In other cases, thermo-dormancy may be regulated by the embryo or a combination of the two (Bewley and Black, 1994). An increase in ABA levels during high-temperature exposure, however, is not ubiquitous among all seeds, as decreases in ABA levels also have been observed during warm as well as cold stratification (Subbaiah and Powell, 1992; Ren et al., 1997; Chien et al., 1998).

Correlating other metabolic indicators to seed dormancy also has been attempted. Indicators that have been correlated to seed dormancy with some success include carbohydrate status (Foley et al., 1992; Nichols et al., 1993) and redox charge (Gallais et al., 1998), as well as changes in membrane properties (Di Nola et al., 1990; Faust et al., 1997; Hilhorst, 1998). Whether these indicators are a cause, consequence, or merely correlate with seed dormancy and its release is less clear.

ABA Synthesis and Metabolism

Recent investigations have revealed that ABA levels alone may not suffice in revealing its role in seed dormancy. ABA content may be similar in dormant and non-dormant seeds (Yoshioka et al., 1998) but the rate and route of ABA inactivation may differ between dormant and non-dormant seeds (Chiwocha et al., 2003). Whether this is of importance in secondary seed dormancy in *Brassica napus* has yet to be investigated. The biochemical pathway(s) of ABA synthesis and metabolism have been difficult to establish due to the low *in vivo* concentration of the molecules involved. It has been shown that the precursors of ABA are pyruvate and glyceraldehyde-3-phosphate (Lichtentahler, 1999). Prior to these findings, it was generally accepted that ABA was primarily synthesized from mevalonic acid (Cutler and Krochko, 1999). The early steps of ABA synthesis up to and including xanthoxin occur in plastids, while the remaining steps occur in the cytosol. Therefore, only the (+)-stereoisomer of ABA may be formed from one of three immediate precursors (Cutler and Krochko, 1999), while ()-ABA is not synthesized in plants (Yamamoto and Oritani, 1996). Alternative biosynthesis pathways of ABA also have been suggested but appear to play only a minor role (Cutler and Krochko, 1999).

ABA can degrade via a number of pathways. The primary route of ABA degradation appears to be via hydroxylation in the 8'-position, followed by rapid conversion to (−)-phaseic acid (PA). A further reduction can occur, forming dihydrophaseic acid (DPA) (Cutler and Krochko, 1999). Other metabolites also have been observed in relatively small amounts, including (+)-7'-hydroxy-ABA (7'OH-ABA) (Zeevaart and Creelman, 1988). The hydroxylated metabolites of ABA may still play an active role in regulation (Walker-Simmons et al., 1997) but PA and DPA appear to be inactive relative to ABA and its hydroxylated catabolites. ABA, PA, DPA, and 7'OH-ABA also may be inactivated via conjugation, thus forming glucose esters, glucosides, or other conjugates. All conjugates are believed to accumulate in the vacuole (Zeevaart and Creelman, 1988) where they exhibit no biological activity. ABA is readily translocated within plants (Sauter et al., 2001), and therefore a cell's ABA content may be a function of *de novo* synthesis as well as import/export. This, in conjunction with the continuous degradation of this hormone, results in ABA content dynamics that are difficult to decipher from single measurements of the hormone and its metabolites (Cutler and Krochko, 1999; Schmitz et al., 2002; Chiwocha et al., 2003).

ABA and Seed Maturation in *Brassica napus*

Changes in endogenous ABA levels during seed maturation in *Brassica napus* are well documented. Similar to many other species (Bewley and Black, 1994), endogenous ABA levels increase during seed maturation, followed by a decrease during desiccation (Juricic et al., 1995). Depending on the genotype, one or two peaks in ABA content may be observed during seed maturation. The first peak is maternally derived, while a second peak, when present, is the result of *de novo* ABA synthesis in the embryo (Karssen et al., 1983). In developing *B. napus* seeds, a spike in endogenous ABA levels also has been observed 1 day after a mild freezing stress, after which ABA levels declined again to control levels (Green et al., 1998). The influence of the freezing stress on seed germinability and sensitivity to ABA were not investigated.

In *Brassica napus*, endogenous ABA levels correlated well with germination rates of excised embryos early during seed maturation (Finkelstein et al., 1985). As seed ABA levels decline during desiccation, however, low seed water content appears to be the primary factor preventing germination (Finkelstein et al., 1985).

Sensitivity to ABA

In addition to endogenous hormone levels, the ability of tissues to respond to plant hormones (tissue sensitivity) must be considered (Trewavas, 1982). Indeed, this is also important in seed dormancy.

Lower sensitivity to ABA has been shown in non-dormant wheat (*Triticum aestivum* L.) seed relative to dormant seed (Walker-Simmons, 1987; Morris et al., 1989), as well as in barley (*Hordeum vulgare* L.) (Wang et al., 1995) and sunflower (*Helianthus annuus* L.) (Bianco et al., 1994). In seeds of some species, sensitivity to ABA may be influenced by temperature, as decreases in the sensitivity to ABA have been reported during cold stratification (Singh and Browning, 1991; Jarvis et al., 1997).

During seed maturation, differences in embryo sensitivity to ABA also have been observed among three genotypes of *Brassica napus* (Juricic et al., 1995). Changes in sensitivity, however, were not correlated to changes in endogenous ABA levels. Interestingly, manipulating GA levels during maturation also altered seed sensitivity to ABA (Juricic et al., 1995).

MOLECULAR APPROACHES

Some promise lies in the application of molecular techniques in elucidating the mechanisms and regulation of seed dormancy (Chen et al., 2002; Foley, 2002; Horvath and Anderson, 2002). The generation of ABA synthesis mutants (under- and over-production relative to the wild type) (Schwartz et al., 1997) and those insensitive to ABA (selected by germination at normally inhibiting concentrations of ABA) are beginning to reveal the complexities of the signaling cascades involving this hormone. At least two of the six *Arabidopsis thaliana* mutants that are insensitive to ABA (ABI1 and ABI2) have been shown to express lower levels of seed dormancy (Finkelstein et al., 2002; Brocard-Clifford et al., 2003). Furthermore, lower levels of seed dormancy have been observed in *Arabidopsis* mutants that produce decreased levels of ABA relative to the wild type (Finkelstein et al., 2002). These findings also strongly suggest a link between ABA synthesis and seed dormancy in this species.

Mutant analysis has indicated a considerable degree of communication between signaling cascades influenced by ABA and other signaling cascades, indicating a high degree of complexity in the regulation of plant responses involving hormones. For example, in *Arabidopsis*, ABA-induced seed dormancy may be alleviated by sucrose, glucose, and only partially by fructose, while seed dormancy expression was unaffected by mannitol and sorbitol (Finkelstein and Lynch, 2000; Brocard-Gifford et al., 2003).

Interactions between ABA and ethylene, as well as other plant hormones, also may influence the expression of seed dormancy (Small and Gutterman, 1992; Ghassemian et al., 2000; Rock, 2000; Chiwocha et al., 2003). These ongoing discoveries reveal complexities in seed dormancy mechanisms and regulation that are difficult, if not impossible, to discern using traditional techniques and may aid in explaining results obtained from hormone studies that investigate seed dormancy. An understanding of the metabolic events that result in secondary seed dormancy in *Brassica napus* may ultimately be used to minimize "weedy" attributes of this crop.

SEED VIABILITY AND LONGEVITY IN CRUCIFERS

BACKGROUND INFORMATION ABOUT VIABILITY

Seeds leak a wide range of compounds when imbibed in water, including carbohydrates, proteins, and inorganic ions (Samad and Pearce, 1978), as well as larger intracellular substances such as starch, grains, and protein bodies (Spaeth, 1987). Differential leakage of specific compounds has been associated with seed viability in some species. The percentage of water-soluble carbohydrates was negatively correlated with germination in carrot (*Daucus carota* L.) (Dadlani and Agrawal, 1983). Fructose, glucose, sucrose, maltose, raffinose, and stachyose were identified in the leachate from non-viable rape (*Brassica napus* L.) seeds (Takayanagi and Murakami, 1969). Viable seeds leaked only trace amounts of fructose and glucose. Artificially aged soybean (*Glycine max* L. Merr.) seeds were observed to leak higher amounts of compounds that absorbed light at 260 nm than non-aged seeds (Schoettle and Leopold, 1984).

Phenolic compounds have been found in seeds. Sinapine, the choline ester of sinapic acid (3,5-dimethoxy-4-hydroxycinnamic acid) (Austin and Wolff, 1968), has been found among various species of *Brassicaceae* (Cruciferae) (Schultz and Gmelin, 1952), and is the main phenolic constituent of rape seed occurring at levels from 1.0% to 2.5% dry matter (Blair and Reichert, 1984). The compound is fluorescent under UV light and becomes yellow at high pH (>10). Hydrolysis products of sinapine are sinapic acid and choline, which are metabolized during germination (Tzagoloff, 1963).

Non-viable or deteriorated seeds have been reported to leak more solutes than viable or vigorous seeds when placed in water (Simon and Harun, 1972; Leopold, 1980; Powell and Matthews, 1981). Aged, damaged, or nonfunctional cellular membranes (Simon and Harun, 1972; McKersie and Stinson, 1980; Murphy and Noland, 1982) and cellular rupture caused by imbibition damage (Powell and Matthews, 1979; Duke and Kakefuda, 1981) have been suggested as major causes.

Among the solutes leaked from seeds of various species are free amino acids (Harman and Granett, 1972), proteins, sugars, and phenolics. Ions and inorganic compounds of potassium, phosphate, and magnesium are also leaked. The two most common methods of measuring seed leakage are electrical conductivity and light absorption at specific wavelengths (Duke et al, 1983; Hepburn et al., 1984).

Viability Determination in Crucifers

Invention Description

The viability of crucifer seeds (Taylor et al., 1990) — for example, cabbage seeds — is determined by detecting leakage of sinapine, which leaks from non-viable seeds but is only exuded in small amounts, if at all, from viable seeds. Crucifer seeds include agronomically significant crop seeds such as cabbage, cauliflower, broccoli, canola, and others.

The method of determination can be practiced in a number of ways. Examples include detecting leaked sinapine by UV3 irradiation, and detecting relative amounts of sinapine in an aqueous soak medium by color change of the medium. The method is not critical so long as it relates sinapine leakage to non-viable seeds as compared to viable seeds.

Preferably, crucifer seeds are coated with an inert or beneficial seed coat that absorbs sinapine, thereby retaining the sinapine and allowing for easy detection of non-viable seed and its separation from viable seed. In a particularly attractive version of this method, crucifer seeds are first soaked in aqueous media for a time sufficient (usually at least several hours) to foster leakage of sinapine from the non-viable seeds, if present. The soaked seeds are then coated with a sinapine-absorbing coating and then dried. Non-viable seeds are then detected by irradiating with UV light to cause fluorescence of sinapine. The presence and relative amount of non-viable seeds present in a mass of seeds can be determined and, if desired, the significantly fluorescent and thus non-viable seeds can be separated from the mass by suitable means — for example, a directing means that moves the seeds, dropping past a sensing means, in a selected direction based on fluorescence/non-fluorescence.

Some viable crucifer seeds leak low levels of sinapine; thus, non-fluorescence in some cases is actually low-level fluorescence. However, the difference in amounts of fluorescence (at least twice as much sinapine and usually very much more from non-viable seeds) makes separating viable from non-viable seeds, or quantifying relative amounts of both, a relatively easy task.

Where soaking or hydration is employed, that process can be the usual hydration step otherwise employed to enhance seed quality and add beneficial materials to the seed, etc., in a routine manner. Otherwise, it can simply be a soak in aqueous media for the purposes of this method.

Where a coating technique is used, the coating technique might be a coating technique such as is regularly used in the seed industry, or one that is further adapted to accomplish the mission; or it might be a coating specifically formulated to accomplish the goal. The coating can be inert to the seed or beneficial to the seed. For the purposes here, the coating or any component therein must not fluoresce at the same wavelengths as sinapine (so as to confuse or prevent detection of sinapine), and the coating must not react with sinapine (so as to cause an inability to detect sinapine, if present).

A number of beneficial seed coats as well as seed coating methods are known. Pesticides, fungicides, etc. (e.g., thiran) can be applied in such a manner. The coating can be merely a polymer or non-volatile liquid or a particulate, powdery material that adheres to the seed, or it can be a mixture of a binder or carrier with active ingredients. For the purposes here, the active ingredient of the coating is a material that absorbs and/or retains sinapine so that it can be detected by fluorescence. This function can be performed by a binder or carrier itself, or by absorbent filler such as minerals, sand, clays, vermiculite, cellulose, etc. The absorbent binder can be cellulose based, methyl cellulose based, etc. Taylor et al. (1990) have provided a comprehensive review of methods of viability determination in crucifers; three examples include:

Example 1 documents and then identifies a fluorescent compound that leaks from non-viable cabbage (*Brassica oleracea* var. *capitata* L.) seeds.

Example 2 focuses on (1) comparing sinapine and electrolyte leakage from heat-killed non-viable and viable cabbage seeds; and then (2) determining if sinapine detection tests could be more accurate than the conductivity method in distinguishing viable from non-viable cabbage seeds.

Example 3 focuses on (1) quantifying the hydration and leakage characteristics of six cultivars from the *Brassicaceae*, then (2) developing a seed coating system to adsorb sinapine that leaked from individual seeds, and (3) comparing seedling growth characteristics by sowing non-coated, coated, and fluorescence-sorted coated seeds in greenhouse studies.

The study was based on five crops representing both horticultural and agronomic *Brassica*s and one flower species. There were a total of six seedlots (cultivars) studied, as shown in Table. 9.2.

Viability in Storage

The majority of seed species retain their viability when dried; in fact, drying is the normal, final phase of maturation for most seeds growing in temperate climates. Hence, it is common for seeds to be stored in a "dry" state or, more correctly, with low moisture content. There are, however, some seed species that must retain a relatively high moisture content during storage in order to maintain maximum viability.

TABLE 9.2
Crop Cultivars and Their Seed Source

Genus Species	Crop	Cultivar	Seed Source
B. oleracea var. capitata	Cabbage	Danish ballhead	Harris-Moran
B. oleracea var. capitata	Cabbage	King Cole	Ferry-Morse
B. oleracea var. capitata	Cauliflower	Snowball	Harris-Moran
B. oleracea var. capitata	Broccoli	Citation	Harris-Moran
B. napus	Canola	Westar	Agro-King
Ervsimum Hieraciifolium	Wallflower	Orange bedder	Harris-Moran

Source: From Taylor et al., 1990.

FACTORS AFFECTING VIABILITY IN STORAGE

Seeds that can be stored in a state of low moisture content are called recalcitrant, and their viability under certain storage conditions conforms to some general rules, as follows:

1. For each 1% decrease in seed moisture content, the storage life of the seed doubles.
2. For each 10°F (5.6°C) decrease in seed storage temperature, the storage life of a seed doubles.
3. The arithmetic sum of the storage temperature (in degrees F) and the percent relative humidity (RH) should not exceed 100, with no more than half the sum contributed by the temperature.

These "rules of thumb" clearly indicate that the temperature and moisture content of the seed are major factors in determining viability in storage. Now consider these factors, along with others that might play a role, as follows:

Moisture Content

Moisture content is defined by International Seed Testing Association (ISTA) Practices as:

$$\% \text{ Moisture content} = \text{Fresh wt. of seed} - (\text{Dry wt. of seed/Fresh wt. of seed}) \times 100$$

When the moisture content is high (>30%), non-dormant seeds may germinate; and from 18% to 30% moisture content, rapid deterioration by microorganisms can occur. Seeds stored at moisture contents greater than 18% to 20% will respire, and in poor ventilation the generated heat will kill them. Below 8 or 9% moisture content, there is little or no insect activity; and below 4 or 5% moisture content, seeds are immune to attack by insects and storage fungi, but they may deteriorate faster than those maintained at a slightly higher moisture content. The activities of seed storage fungi are ultimately more influenced by the RH of the interseed atmosphere than by the moisture content of the seeds themselves. This is because the moisture content of some seeds (e.g., oilseeds) may be different from that of others (e.g., starchy seeds) even though both are in equilibrium with the same atmospheric RH. For example, all cereals and many of the legumes that are high in starch and low in oil have moisture contents of about 11% at 45% RH, whereas oil-containing seeds (e.g., rape) have a moisture content of only 4.4% at this RH (Figure 9.1). The viability of seeds obviously will be affected by the amount of water they contain.

Temperature

Cold storage of seeds at 0 to 5°C is generally desirable, although this may be inadvisable unless they are sealed in moisture-proof containers or stored in a dehumidified atmosphere. Otherwise, the RH of storage could be high, causing the seeds to gain moisture; if they are then brought to a higher temperature (e.g., for transport), they might deteriorate because of their high moisture content. At moisture contents below 14%, no ice crystals form within cells on freezing; thus, storage of dry seeds at subzero temperatures after freezing in a dry atmosphere should improve longevity. Freeze-drying of certain seeds improves their longevity in storage, but others may be killed by this treatment. A method that is increasing in popularity, particularly for small batches of seeds for genebanks, is to keep them immersed in liquid nitrogen. Under such conditions, seeds should survive indefinitely.

Interrelationships

An accurate balance between moisture and temperature is very crucial for seed storage life in a storage house. To predict an accurate interrelationship, viability equations are used. The viability equations

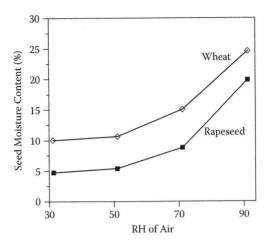

FIGURE 9.1   Changes in equilibrium moisture content of wheat and rapeseed with relative humidity of the ambient air. (*Source:* Based on work of Kreyger, 1972, cited in Thomson, 1979.)

are mathematical models that were developed to predict seed storage life in different environments. The life span of a seedlot, or the time until all the seeds have lost viability, depends on:

- The storage behavior of the species in question — The vast majority of species produce seeds with orthodox behavior that respond in a quantifiable and predictable way to both moisture content and temperature. The viability equations apply only to orthodox or desiccation-tolerant seeds.
- Environmental conditions — In general terms, and within determined limits, the drier and cooler the storage conditions, the longer-lived the seeds.
- The initial seed quality or the proportion of the seeds that are viable at the start of storage.
- The species in question — Although all orthodox seeds respond to relative humidity and storage temperature in a broadly similar way, some species are inherently longer-lived than others.

The relationship between mean storage temperature, moisture content, and mean viability periods is relatively simple but plays very critical role in seed viable storage.

Miscellaneous Factors

In addition to the interrelationships between temperature, moisture content, and time, other factors must be considered when attempting to determine the optimum storage conditions for a particular species:

- *Pre- and post-harvest conditions.* Environmental variation during seed development usually has little effect on the viability of seeds, unless the ripening process is interrupted by premature harvesting. Weathering of maturing seeds in the field, particularly in conditions of excess moisture or freezing temperatures, results in a product with inferior storage potential. Mechanical damage inflicted during harvesting can severely reduce the viability of some seeds. Small seeds tend to escape injury during harvest, and seeds that are spherical tend to suffer less damage than do elongated or irregularly shaped ones. During storage, injured or deeply bruised areas may serve as centers for infection and result in accelerated deterioration. Injuries close to vital parts of the embryonic axis or near the point of attachment of cotyledons to the axis usually bring about the most rapid losses of viability. High temperatures during drying, or drying too quickly or excessively, can dramatically reduce viability.

- *Oxygen pressure during storage.* If seeds are not maintained in hermetic (closed or airtight) storage at low moisture contents, then even under conditions of constant temperature and moisture, the gaseous environment may change as a result of respiratory activity of the seeds and associated microflora. In open storage, seeds maintained in an atmosphere of nitrogen may retain their viability considerably longer than those placed in replenished air or oxygen, although there is probably no advantage over storing seeds hermetically sealed in air. In fact, for storage at relatively low temperatures and moisture content, there is probably little benefit in using controlled atmospheres (i.e., reduced oxygen pressures). There may be advantages for short-term storage in conditions of high temperature or moisture content, but it is probably similar to reducing either the temperature or the moisture content. Hermetic sealing provides a simple and convenient method of controlling seed moisture content after the seeds have been dried adequately.
- *Fluctuating storage conditions.* Fluctuating storage conditions may be harmful to some species but not to others; this must be determined empirically.
- *Cultivar and harvest variability.* Different cultivars and harvests of a particular species may show different viability characteristics under the same storage conditions. These differences are relatively small under good storage conditions; but under adverse conditions (elevated temperature and RH), they can be quite large.

ASSOCIATED MICROFLORA WITH CRUCIFERS AND SEED DETERIORATION

Fungus is one of the most important microflora in seed deterioration. However, in crucifers, seed deterioration differs with period of storage (Figure 9.2), species type, and other storage conditions. Abdel-Mallek (1995) assayed a total of 121 seed samples of cabbage, cauliflower, cress, radish, and turnip for their fungal flora. The highest count of fungi was recorded on cabbage seeds (75%), whereas the lowest count was observed on turnip seeds (33%). Some 35 fungal species and 2 varieties belonging to 16 genera were identified. The broadest number of species (22 species + 1

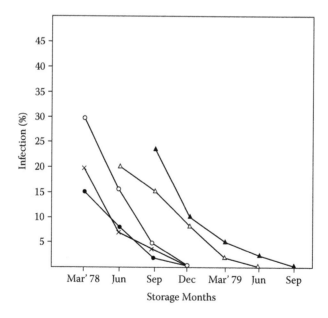

FIGURE 9.2 Infection of oilseeds by storage fungi after different periods of storage. Indian mustard ○, Mustard var. yellow sarson ●, Sesame var. reddish-brown △, Sesame var. black ▲, and linseed ×. (*Source:* From Gopal et al., 1981.)

variety) were isolated from cress seeds. However, only five species, (i.e., *Aspergillus flavus, A. niger, Penicillium chrysogenum, P. funiculosum,* and *Rhizopus stolonifer*) were found to be associated with seeds of the five plants.

Storage fungi, almost exclusively of the genera *Aspergillus* and *Penicillium*, infest seeds only under storage conditions and are never present before, even in seeds of plants left standing in the field after harvesting. Each species of storage fungus has a sharply defined minimum seed moisture content below which it will not grow, although other factors also determine virulence (e.g., ability to penetrate the seed, condition of seed, nutrient availability, and temperature). The major deleterious effects of storage fungi are to (1) decrease viability, (2) cause discoloration, (3) produce mycotoxins, (4) cause heat production, and (5) develop mustiness and caking. Fungi will not grow at seed moisture contents that are in equilibrium with an ambient RH below 68%; hence, they are not responsible for deterioration that occurs at moisture contents below about 13% in starchy seeds and below 7 or 8% in oily seeds.

Deterioration of seeds by insects and mites is a serious problem, particularly in warm and humid climates. Weevils, flour beetles, or borers are rarely active below 8% moisture content and 18-20°C, but are increasingly destructive as the moisture content rises to 15% and the temperature to 30 to 35°C. Mites do not thrive below 60% RH, although they have a temperature tolerance that extends close to freezing.

EFFECT OF AGING AND LOSS OF VIABILITY CAUSES

Viability loss of naturally or artificially aged seeds results mainly from damage to nucleic acids and the deterioration of cellular membranes (Osborne, 1982; Roberts and Ellis, 1982). The damage accumulates progressively, both during natural storage and during accelerated aging, and is affected by temperature and seed moisture content (Villiers, 1974). The extent of damage and its reversibility depend on the efficiency of certain repair mechanisms related to nucleic acid metabolism and on the ability to replace deteriorated membranes (Berjak and Villiers, 1972; Villiers, 1972, 1974; Elder et al., 1987).

Ultra-structural analysis has obvious potential for localizing the aberrant functions within damaged cells. However, it is difficult to evaluate whether ultra-structural changes constitute the cause or consequence of the cellular damage. Several ultra-structural investigations have been carried out on aged seeds (Smith, 1978, 1989; Garcia de Castro and Martinez-Honduvilla, 1984). These studies suggest that the symptoms of cellular aging are evident at the beginning of imbibition and become more obvious as imbibition progresses. The rapid imbibition of viable seeds is accompanied by transient efflux of organic or inorganic compounds into the surrounding medium (Simon and Raja Harun, 1972). This can be interpreted as a result of the discontinuity of cellular membranes that become spontaneously repaired during the later stages of imbibition (Simon, 1978; Elder et al., 1987). This efflux is greater in aged seeds (Parrish and Leopold, 1978; Pesis and Ng, 1983).

Dawidowicz-Grzegorzewska (1991) studied the ultra-structure of non-aged winter rape (*Brassica napus* L.) seeds in comparison with that of artificially aged seeds in which viability was partially or completely impaired as shown in the Figures 9.3 and 9.4. Three ultra-structural symptoms possibly related to age-induced membrane deterioration were observed: (1) a decrease in electron contrast in all cellular membranes excluding plasmalemma; (2) coalescence of small storage lipid bodies to larger units, presumably as a result of the degradation of enclosing half-unit membranes; and (3) the appearance of protoplasmic inclusions inside the storage protein bodies, possibly resulting from rupture of the enclosing unit membranes.

A summary of the variety of metabolic lesions that can occur during storage and that result in a loss of viability appears in Figure 9.5. The reader can appreciate that although any single lesion can result in the loss of viability or reduction of vigor, it is likely that aging elicits a number of changes. Which one occurs first or is the most important is impossible to tell, but it creates a continuing topic for debate.

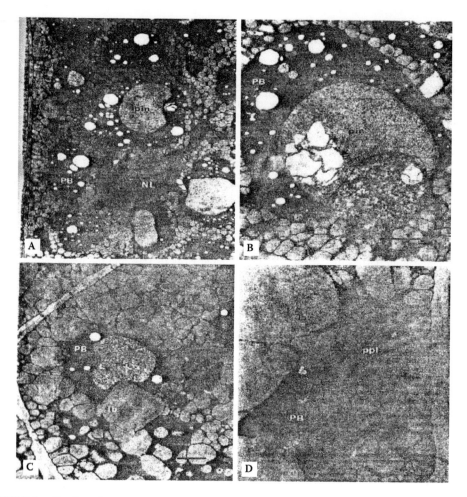

FIGURE 9.3 Portion of ground parenchyma cells in hypocotyl from aged, low-vigor seeds after 5-hour imbibition at 200°C. (A) General view of the cell containing a nucleus with prominent nucleolus (NL) and large fibrillar center (fcr) in the fibrillar component (fc). Protoplasmic inclusions (pin) within protein bodies can be seen in transverse or longitudinal sections (arrow). (B) Cup-shaped protein body containing protoplasmic inclusions and vacuole like regions inside (arrow). (C) Portion of the cell showing several fused lipid bodies (lb), the largest marked with a star. (D) Poorly contrasted proplastids with vesicles, presumably modified plastoglobuli. Lipid bodies are present fused (*) and not fused (lining the plasmalemma). Plasma membrane is well defined (arrow) and tightly appressed to the cell wall. Bars represent 1 μm. (*Source:* From Dawidowicz-Grzegorzewska, 1991.)

Attempts have been made to restore the viability and vigor of aged seeds after storage. One method that has achieved some success is osmopriming, which involves placing dry seeds in solutions of osmotica, thus limiting the rate of imbibition and initially slowing the germination processes. It may be that under conditions of limited water availability, which can be achieved also by placing dry seeds in a water-saturated atmosphere, repair processes are affected and that age-induced damage is sufficiently diminished before the germinative events commence. If the time for, or extent of, repair to cellular damage is inadequate (e.g., when the aged seeds are imbibed in water), then the integrity or metabolism of the cell remains disrupted, and germination not only cannot proceed, but further deterioration occurs following imbibition.

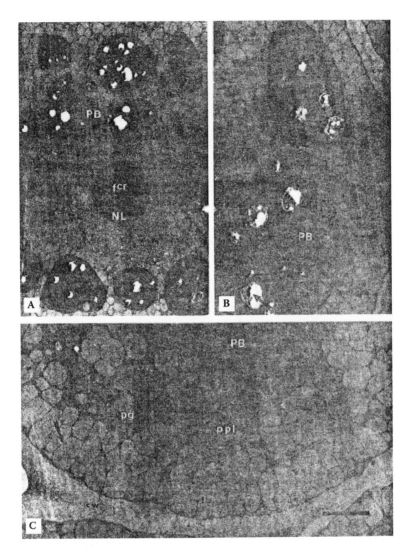

FIGURE 9.4 Portions of ground parenchyma cells in hypocotyl from non-aged, viable winter rapeseeds after 5-hour imbibition at 200°C. (A and B): Abundant protein bodies (PB) containing globoid inclusions (arrow). Lipid bodies (lb) surround protein bodies and line the cytoplasmic side of the plasma-lemma. Lobed nucleus (LN) with prominent chromocenters (chr) close to the nuclear envelope. (C): Poorly contrasted proto-plastids (ppl) with osmiophilic plastoglobules (pg) but lacking thylakoids. Arrow shows half-unit membranes enclosing lipid body. CW, cell wall. Bars represent 1 μm. (*Source:* From Dawidowicz-Grzegorzewska, 1991.)

CONCLUSION

The unique ability of seeds to schedule their dormancy, germination, and viability to coincide with times when environmental conditions are favorable to their survival as seedlings has no doubt contributed significantly to the success of seed-bearing plants. Dormancy, germination, and viability of seeds are crucial phenomena in the seeds of crucifer plants as well.

Primary seed dormancy is generally not observed in *Brassica napus* (Schlink, 1994); *B. napus* may develop secondary seed dormancy (Pekrun, 1994). The factors affecting secondary seed dormancy expression are still unclear in *B. napus*. Although a genetic component has been implicated, its importance relative to other factors such as environment during seed maturation (Gutterman, 1980/81) remains unclear. Furthermore, the influence of seed storage on secondary seed dormancy

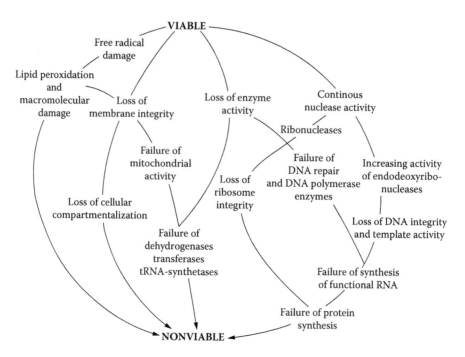

FIGURE 9.5 A diagram that illustrates the variety of causes for the loss of viability in stored seeds (*Source:* From Osborne, 1980.)

characteristics has not yet been directly investigated in this species (Pekrun et al., 1997a; Momoh et al., 2002), and the role of constant temperature on seed dormancy induction rates is also unclear (Momoh et al., 2002). In *B. napus*, secondary seed dormancy is readily reversed by several factors. Schlink (1994) demonstrated that a single exposure to a camera flash with duration of 0.002 seconds was sufficient to increase mean germination in dormant seeds of two *B. napus* cultivars from 13.3% to 63.0%.

Abscisic acid (ABA) has been implicated in the regulation of a number of plant responses, including dormancy. High ratios of ABA:gibberellins have been suggested to promote dormancy, while a low ratio of these hormones tends to lead to the completion of germination (Wareing and Saunders, 1971). The role of ABA in modulating the acquisition of secondary dormancy after seed dissemination is less clear. *De novo* ABA synthesis within the seed is essential for the maintenance of seed dormancy (Yoshioka et al., 1998; Grappin et al., 2000).

The majority of seed species retain their viability when dried; in fact, drying is the normal, final phase of maturation for most seeds growing in temperate climates. Hence, it is common for seeds to be stored in a "dry" state — or more correctly, with low moisture content. There are, however, some seed species that must retain a relatively high moisture content during storage in order to maintain maximum viability. Fungus is one of the most important microflora in seed deterioration. However, in crucifers, seed deterioration is different with regard to the period of storage, species type, and other storage conditions.

Non-viable or deteriorated seeds have been reported to leak more solutes than viable or vigorous seeds when placed in water (Simon and Harun, 1972; Leopold, 1980), which suggested (Taylor et al., 1990) viability determination of crucifer seeds (e.g., cabbage, cauliflower, broccoli, canola, and others) by detecting leakage of sinapine, which leaks from non-viable seeds.

A reasonable overall conclusion can be drawn from the studies discussed in this chapter: that the crucifer seed dormancy and viability phenomenon still require further studies at the molecular level to decide their role in the seed life span.

ACKNOWLEDGMENTS

This work was supported by the National Basic Research Program of China (2006CB101602), the National High Technology Research and Development Program of China (2006AA10A214, 2006AA10Z234), the National Natural Science Foundation of China (20632070, 30600377), the Chinese Academy of Sciences (KGCX3-SYW-203-03), Zhejiang Provincial Natural Science Foundation (R307095), the Science and Technology Department of Zhejiang Province (2008C22078), and the 111 Project from China Ministry of Education and the State Administration of Foreign Experts Affairs (B06014). I (W.J. Zhou) am grateful to Dr. R.N. Lester (University of Birmingham, U.K.), and Prof. H.C. Shen and Prof. Z.Q. Wang (Zhejiang University, China), and my other post-doc fellows and graduate students, past and present, for their contributions to this and related research.

REFERENCES

Abdel-Mallek, A.Y. 1995. Seed-borne fungi of five cruciferous vegetables and relative efficacy of aqueous seed extracts against some associated fungi. *Folia Microbiologica*, 40:493–498.

Austin, F.L. and I.A. Wolff. 1968. Sinapine and related esters in seed meal of *Crambe abyssinica. J. Agr. Food. Chem.*, 16:132–135.

Baskin, J.M. and C.C. Baskin. 1984. Role of temperature in regulating timing of germination in soil seed reserves of *Lamium purpureum* L. *Weed Res.*, 24:341–349.

Baskin J.M. and C.C. Baskin. 1985. The annual dormancy cycle in buried weed seeds: a continuum. *Bio Sci.*, 35:492–498.

Baskin, C.C. and J.M. Baskin. 1998. in *Seeds: Ecology, Biogeography, and Evolution of Dormancy and Germination.* San Diego: Academic Press, p. 5–395.

Bazanska, J. and S. Lewak. 1986. Light inhibits germination of rape seeds at unfavourable temperatures. *Acta Physiol. Plant.*, 8:145–149.

Benech-Arnold, R.L., M.C. Giallorenzi, J. Frank, and V. Rodriguez. 1999. Termination of hull-imposed dormancy in developing barley grains is correlated with changes in embryonic ABA levels and sensitivity. *Seed Sci. Res.*, 9:39–47.

Benvenuti, S. and M. Macchia. 1997. Germination eco-physiology of bur beggar-ticks (*Bidens tripartita*) as affected by light and oxygen. *Weed Sci.*, 45, 696–700.

Berjak, P. and T.A. Villiers. 1972. Aging in plant embryos — Aged-induced damage and its repair during early germination. *New Physiologist*, 71:135–144.

Berrie, A.M.M., J. Paerson, and H.P. West. 1974. Water content and responsivity of lettuce seeds to light. *Physiol. Plant.*, 31:90–96.

Bewley, J.D. and M. Black. 1994. In: *Seeds Physiology of Development and Germination, 2nd ed.* Plenum Press, New York.

Bianco, J., G. Garello, and M.T. Le Page-Degivry. 1994. Release of dormancy in sunflower embryos by dry storage: involvement of gibberellins and abscisic acid. *Seed Sci. Res.*, 4:57–62.

Blair, R. and R.D. Reichert. 1984. Carbohydrate and phenolic constituents in a comprehensive range of rapeseed and canola fractions: nutritional significance for animals. *J. Sci. Food Agric.*, 35:29–35.

Bouwmeester, H.I. and C.M. Karssen. 1992. The dual role of temperature in the regulation of the seasonal changes in dormancy and germination of seeds of *Polygonum persicaria* L. *Oecologia Berlin*, 90:88–94.

Brocard-Clifford, I.M., T.J. Lynch, and R.R. Finkelstein. 2003. Regulatory networks in seeds integrating developmental, abscisic acid, sugar, and light signalling. *Plant Physiol.*, 131:78–92.

Brothwick, H.A., S.B. Hendricks, E.H. Toole, and V.K. Toole. 1954. Action of light on lettuce-seed germination. *Bot. Gaz.*, 115:205–225.

Campbell, P., J. Van Staden, C. Stevens, and M.I. Whitwell. 1992. The effects of locality, season and year of seed collection on the germination of bug-weed (*Solanum auritianum* Scop.) seeds. *South African J. Bot.*, 58:310–316.

Casal, J.J. and R.A. Sanchez. 1998. Phytochromes and seed germination. *Seed Sci. Res.*, 8:317–329.

Chen, T.H.H., G.T. Howe, and H.D. Bradshaw, Jr. 2002. Molecular genetic analysis of dormancy related traits in poplars. *Weed Sci.*, 50:232–240.

Chien, C.T., L.L. Huo-Huang, and T.P. Limm. 1998. Changes in ultrastructure and abscisic acid level, and response to applied gibberellins in Taxus meirei seeds terated with warm and cold stratificaiton. *Ann. Bot.* 81:41–47.

Chiwocha, S.D.S., S.R. Abrams, S.J. Ambrose, A.J. Cutler, M. Loewen, A.R.S. Ross, and A.R. Kermode. 2003. A method for profiling classes of plant hormones and their metabolites using liquid chromatography-electrospray ionization tandem mass spectrometry: an analysis of hormone regulation of thermo-dormancy of lettuce (*Lactuca sativa* L.) seeds. *Plant J.,* 35:402–417.

Corbineau, F. and D. Côme. 2000. Dormancy of cereal seeds as related to embryo sensitivity to ABA and water potential. In: J.D. Viémont and J. Crabbé, (Eds.), *Dormancy in Plants: From Whole Plant Behaviour to Cellular Control,* CABI Publishers, New York, p. 183–194.

Cutler, A.J. and J.E. Krochko. 1999. Formation and breakdown of ABA. *Trends Plant Sci.,* 4:472–478.

Dadlani, V. and V. Agrawal. 1983. Sinapine leakage from non-viable cabbage seeds. *Scientia Hort.,* 19:39–44.

Dawidowicz-Grzegorzewska, A. and A. Podstolski. 1991. Age-related changes in ultra-structure and membrane properties of *Brassica napus* L. seeds. *Ann. Bot.,* 69:39–46.

DiNola, L., C.F. Mishke, and R.B. Taylorson. 1990. Changes in the composition and synthesis of proteins in cellular membranes of *Echinochloa crus-galli* (L.) Beauv. seeds during the transition from dormancy to germination. *Plant Physiol.,* 92:427–433.

Duke, S.H. and G. Kakefuda. 1981. Role of the testa in preventing cellular rupture during imbibition of legume seeds. *Plant Physiol.,* 67:449–456.

Duke, S.H., G. Kakefuda, and H. Tamara. M. 1983. Differential leakage of intracellular substances from imbibing soybean seeds. *Plant Physiol.,* 72:919–924.

Elder, R.H., A. Dell Aquila, M. Mezzina, A. Sarasin, and D.J. Osborne. 1987. DNA ligase in repair and replication in the embryos of rye, *Secale cereale. Mutation Res.,* 181:61–71.

Faust, M., A. Erez, and L.J. Rowland. 1997. Bud dormancy in perennial fruit trees: physiological basis for dormancy induction, maintenance, and release. *Hortsci..* 32:623–629.

Finkelstein, R.R. and T.J. Lynch. 2000. Abscisic acid inhibition of radicle emergence but not seedling growth is suppressed by sugars. *Plant Physiol.,* 122:1179–1186.

Finkelstein, R.R., K.M. Tenbarge, J.E. Shumway, and M.L. Crouch. 1985. Role of ABA in maturation of rapeseed embryos. *Plant Physiol.,* 78:630–636.

Finkelstein, R.R., S.S.L. Gampala, and C.D. Rock. 2002. Abscisic acid signalling in seeds and seedlings. *Plant Cell.,* 14(Suppl.):S15–S45.

Foley, M.E. 2002. Weeds, seeds, and buds — opportunities and systems for dormancy investigations. *Weed Sci.,* 50:267–272.

Foley, M.E., M.O. Bancal, and M.B. Nichols. 1992. Carbohydrate status in dormant and a f t e r - r i p e n e d excised wild oat embryos. *Physiol. Plant.,* 85:461–466.

Fu, S.Z. and Z.J. Lu. 1991. Regulations of ABA and GA on seed germination and dormancy in *Brassica. Plant Physiol. Commun.,* 27:358–360.

Gallais, S., M.A.P. de Crescenzo, and D.L. Laval-Martin. 1998. Pyridine nucleotides and redox charges during germination of non-dormant and dormant caryopses of *Avena fatua* L. *J. Plant Physiol.,* 153:664–669.

Gamble, P.E. and J.E. Mullet. 1986. Inhibition of carotenoid accumulation and abscisic acid biosynthesis in fluridone-treated dark-grown barley. *Eur. J. Biochem.,* 160:117–121.

Garcia de Castro, M.F. and C.J. Martinez-Honduvilla. 1984. Ultra-structural changes in naturally aged *Pinus pinea* seeds. *Physiol. Plant.,* 62:581–588.

Ghassemian, M., E. Nammbara, S. Cutler, H. Kawaide, Y. Kamiya, and P. McCourt. 2000. Regulation of abscisic acid signalling by ethylene response pathway in *Arabidopsis. Plant Cell.,* 12:1117–1126.

Goedert, C.O. and E.H. Roberts. 1986. Characterization of alternating-temperature regimes that remove seed dormancy in seeds of *Brachiaria humidicola* (Rendle) Schweickerdt. *Plant Cell Environ.,* 9:521–525.

Gopal, C.M., Dipak, N., and N. Balen. 1981. Studies on deterioration of some oil seeds in storage. I. Variation in seed moisture, infection and germinability. *Mycologia,* 73(1): 157–166.

Gosling, P.G., R.A. Butler, M. Black, and J. Chapman. 1981. The onset of germination ability in developing wheat. *J. Exp. Bot.,* 32:621–627.

Grappin, P., D. Bouinot, B. Sotta, E. Miginiac, and M. Jullien. 2000. Control of seed dormancy in *Nicotiana plumbaginifolia*: post-imbibition abscisic acid synthesis imposes dormancy maintenance. *Planta,* 210:279–285.

Green, B.R., S. Singh, I. Babic, C. Bladen, and A.M. Johnson-Flanagan. 1998. Relationship of chlorophyll, seed moisture and ABA levels in the maturing *Brassica napus* seed and effect of a mild freezing stress. *Physiol. Plant.,* 104:125–133.

Gutterman, Y. 1980/81. Influences on seed germinability: phenotypic maternal effects during maturation. *Isr. J. Bot.,* 29:105–117.

Haferkamp, M.R., M.G. Karl, and M.D. Macneil. 1994. Influence of storage, temperature, and light on germination of Japanese brome seed. *J. Range Management,* 47:140–144.

Harman, G.E. and A.L. Granett. 1972. Deterioration of stored pea seed: changes in germination, membrane permeability and ultra-structure resulting from infection by *Aspergillus* rubber and from aging. *Physiolog. Plant Path.,* 2:271–278.

Harper, J.L. 1957. The ecological significance of dormancy and its importance in weed control. *Proc. Int. Congr. Crop Protect.,* 4:415–420.

Hepburn, H.A., A.A. Powell, and S. Matthews. 1984. Problems associated with the routine application of electrical conductivity measurements of individual seeds in the germination testing of peas and soyabeans. *Seed Sci. Tech.,* 12:403–413.

Hilhorst, H.W.M. 1998. The regulation of secondary dormancy. The membrane hypothesis revisited. *Seed Sci. Tech.,* 8:77–90.

Hilhorst, H.W.M. and P.E. Toorop. 1997. Review on dormancy, germinability, and germination in crop and weed species. *Adv. Agron.,* 61:111–165.

Honek, A. and Z. Martinkova. 1992. The induction of secondary seed dormancy by oxygen deficiency in a barnyard grass *Echinochloa crus-galli, Experientia,* 48:904–906.

Hori, H. and Sugiyama T. 1954. *J. Jap. Soc. Hort. Sci.,* 22:223–229. [Cited by Takahashi and Suzuki, 1980.]

Horvath, D.P. and J.V. Anderson. 2002. A molecular approach to understanding root bud dormancy in leafy spurge. *Weed Sci.,* 50:227–231.

Jarvis, S.B., M.A. Taylor, J. Bianco, F. Corbineau, and H.V. Davis. 1997. Dormancy breakage in seeds of Douglas fir (*Pseudotsuga menziesii* (Mirb.)). *J. Plant Physiol.,* 151:457–464.

Juricic, S., S. Orlando, and M.T. Lepage-Degivry. 1995. Genetic and ontogenic changes in sensitivity to abscisic acid in *Brassica napus* seeds. *Plant Physiol. Biochem.,* 33:593–598.

Kalmbacher, R.S., West, S.H., and F.G. Martin. 1999. Seed dormancy and aging in *Atra paspalum. Crop Sci.,* 39:1847–1852.

Karssen, C.M. 1980. Environmental conditions and endogenous mechanisms involved in secondary dormancy of seeds. *Isr. J. Bot.,* 29:45–64.

Karssen, C.M. 1982. Seasonal patterns of dormancy in weed seeds. In: Khan A. (Ed,), *The Physiology and Biochemistry of Seed Development, Dormancy and Germination.* Elsevier Biomedical Press, Amsterdam, p. 243–270.

Karssen, C.M., D.L.C. Brinkhorst van der Swan, A.E. Breekland, and M. Koornneef. 1983. Induction of dormancy during seed development by endogenous abscisic acid: studies on abscisic acid deficient genotypes of *Arabidopis thaliana* (L.) Heynh. *Planta,* 157:158–165.

Khan, A. 1960. An analysis of dark-osmotic inhibition of germination of lettuce seeds. *Plant Physiol.,* 35:1–7.

Khan, A.A. and Karssen, C.M. 1980. Induction of secondary dormancy in *Chenopodiumbonus henricus* L. seeds by osmotic and high temperature treatments and its prevention by light and growth inhibitors. *Plant Physiol.,* 66:175–181.

Kim, J.S., I.T. Hwang, and K.Y. Cho. 1996. Effect of storage conditions on the dormancy release and the induction of secondary dormancy in weed seeds. *Korean J. Weed Sci.,* 16:200–209.

Landbo, L. and R.B. Jorgenssen. 1977. Seed germination in weedy *Brassica campestris* and its hybrid with *Brassica napus*: Implication for risk assessment for transgenic oilseed rape. *Euphytica,* 97:209–216.

Leinonen, K. 1998. Effects of storage conditions on dormancy and vigor of *Picea abies* seeds. *New Forests,* 15:231–249.

Leopold, A.C. 1980. Temperature effects on soybean imbibition and leakage. *Plant Physiol.,* 6 65:1096–1098.

Lichtenthaler, H.K. 1999. The 1-Deoxy-D-xylulose-5-phosphate pathway of isoprenoid biosynthesis in plants. *Annu. Rev. Plant Physiol. Plant Mol. Biol.,* 50:47–65.

López-Granados, F. and P.J.W. Lutman. 1998. Effect of environmental conditions on the dormancy and germination of volunteer oilseed rape seed (*Brassica napus*). *Weed Sci.,* 46:419–423.

Lutman, P.J.W. 1993. The occurrence and persistence of volunteer oilseed rape (*Brassica napus*). *Asp. Appl. Biol.,* 35:29–36.

Lutman, P.J., C. Pekrun, and A. Albertini. 1998. *Dormancy and Persistence of Volunteer Oilseed Rape.* Home-Grown Cereals Authority, Caledonia House, London. Research and Development Project Report OS32.

MacDonald, I.R. and Hart J.W. 1981. An inhibitory effect of light on the germination of mustard seed. *Ann. Bot.,* 47:275–277.

McKersie, B.D. and R.H. Stinson. 1980. Effect of dehydration on leakage and membrane structure in *Lotus corniculatus* L. seeds. *Plant Physiol.*, 66:316–320.

Momoh, E.J.J., W.J. Zhou, and B. Kristiansson. 2002. Variation in the development of secondary seed dormancy in oilseed rape genotypes under conditions of stress. *Weed Res.*, 42:446–455.

Morris, C.F., J.M. Moffat, R.G. Sears, and G.M. Paulsen. 1989. Seed dormancy and responses of caryopses, embryos and calli to abscisic acid in wheat. *Plant Physiol.*, 90:643–647.

Murphy, J.B. and T.L. Noland. 1982. Temperature effects on seed imbibition and leakage mediated by viscosity and membranes. *Plant Physiol.*, 69:428–431.

Nichols, M.B., M.O. Bancal, M.E. Foley, and J.J. Volenec. 1993. Nonstructural carbohydrates in dormant and after-ripened wild oat caryopses. *Physiol. Plant.* 88:221–228.

Nikolaeva, M.G. 1977. Factors controlling the seed dormancy pattern. p. 51-74 in A.A. Kahn, (Ed.) *The Physiology and Biochemistry of Seed Dormancy and Germination*. North-Holland New York, NY, USA.

Osborne, D.J. 1980. In: *Senescence in Plants* (K. V. Thimann, Ed.), CRC Press, Boca Raton, FL, p. 13–37 (review of seed aging).

Osborne, D.J. 1982. Deoxyribonucleic acid integrity and its repair in seed germination. *In The Physiology and Biochemistry of Seed Development and Germination*, A.A. Khan, (Ed.) Elsevier, New York, p. 435–463.

Parasher, V. and O.S. Singh. 1984. Physiology of anaerobiosis in *Phalaris minor* Retz. and *Avena fatua* L. seeds. *Seeds Res.*, 12:1–7.

Parasher, V. and O.S. Singh. 1985. Mechanism of anoxia induced secondary dormancy in canary grass (Plalaris *minor* Retz.) and wild oat (*Avena fatua* L.). *Seed Res.*, 13:91–97.

Parrish, D.J. and A.C. Leopold. 1978. On the mechanism of ageing in soybean seeds, *Plant Physiol.*, 61:365–368.

Pekrun, C. 1994. Untersuchungen zur secondären Dormanz bei Raps (*Brassica napus* L.). Ph.D. thesis, University of Göttingen, Germany.

Pekrun, C., J.D.J. Hewitt, and P.J.W. Lutman. 1998b. Cultural control of volunteer oilseed rape (*Brassica napus*). *J. Agric. Sci.*, 130:155–163.

Pekrun, C., P.J.W. Lutman, and F. Lopezgranados. 1996. Population dynamics of volunteer rape and possible means of control. *Proc. 2nd Int. Weed Control Congr.*, Brown, H. et al. (Ed.) 1–4.

Pekrun, C., P.J.W. Lutman, and K. Baeumer. 1998a. Research on volunteer rape: a review. *Pflanzenbauwissenschaften*, 2:84–90.

Pekrun, C., P.J.W. Lutman, and K. Baeumer. 1997b. Germination behaviour of dormant oilseed rape seeds in relation to temperature. *Weed Res.*, 37:419–431.

Pekrun, C., P.J.W. Lutman, and K. Baeumer. 1997c. Induction of secondary dormancy in rape seeds (*Brassica napus* L.) by prolonged imbibition under conditions of water stress or oxygen deficiency in darkness. *Eur. J. Agron.*, 6:245–255.

Pekrun, C., T.C. Potter, and P.J.W. Lutman. 1997a. Genotypic variation in the development of secondary dormancy in oilseed rape and its impact on the persistence of volunteer rape. In *Proc. Brighton Crop Protection Council Conference — Weeds*. Farnham, Surrey, U.K.: British Crop Protection Council, p. 243–248.

Pesis, E. and T.J. Ng. 1983. Viability, vigor and electrolytic leakage of muskmelon seeds subjected to accelerated ageing. *Hort. Sci.*, 18:242–244.

Pons, T.L. 1991. Induction of dark dormancy in seeds: its importance for seed bank in the soil. *Functional Ecol.*, 5:669–625.

Powell, A.A. and S. Matthews. 1979. The influence of testa condition on the imbibition and vigour of pea seeds. *J. Exp. Bot.* 30:193–197.

Powell, A.A. and S. Matthews. 1981. A physical explanation for solute leakage from dry pea embryos during imbibition. *J. Exp. Bot.*, 32:1045–1050.

Probert, R.J. 2000. The role of temperature in the regulation of seed dormancy and germination. In M. Fenner, (Ed.), *Seeds: The Ecology of Regeneration in Plant Communities, 2nd ed.* CABI Publishers, New York.

Ramagosa, I., D. Prada, M.A. Moralejo, A. Sopena, P. Muñoz, A.M. Casas, J.S. Swanton, and J.L. Molina-Cano. 2001. Dormancy, ABA content and sensitivity of a barley mutant to ABA application during seed development and after-ripening. *J. Exp. Bot.*, 52:1499–1506.

Rem, G. F. Chen, H. Liam, J. Zhan, and X. Gam. 1997. Changes in hormone content of *Panix quinquefolium* seeds during stratification. *J. Hort. Sci.*, 71:901–906.

Roberts, E.H. and R. Ellis. 1982. Physiological, ultra-structural and metabolic aspects of seed viability. In *The Physiology and Biochemistry of Seed Development, Dormancy and Germination*, A. A. Khan, (Ed.) Elsevier, New York, p. 465–483.

Rock, C. 2000. Pathways to abscisic acid-regulated gene expression. Tansley Review No. 120. *New Phytol.,* 148:357–396.

Samad, I.M.A. and R.S. Pearce, 1978. Leaching of ions, organic molecules, and enzymes from seeds of peanut (*Arachis hypogea* L.) imbibing without testas or with intact testas. *J. Exp. Bot.,* 29:1471-1478.

Sauter, A., W.J. Davies, and W. Hartung. 2001. The long-distance abscisic acid signal in the droughted plant: the fate of the hormone on its way from root to shoot. *J. Exp. Bot.,* 52:1991–1997.

Schlink, S. 1994. Ökologie der Keimung und Dormanz von Körneraps (*Brassica napus* L.) und ihre Bedeutung für eine Überdauerung der Samen im Boden. Ph.D. thesis, University of Göttingen, Germany.

Schlink, S. 1995. Überdauerungsvermögen und Dormanz von Rapssamen (*Brassica napus* L.) im Boden. In *9th Eur. Weed Research Soc. Symp.,* Budapest. Doorwerth, The Netherlands: European Weed Research Society, p. 65–73.

Schmitz, N., S.R. Abrams, and A.R. Kermode. 2002. Changes in ABA turnover and sensitivity that accompany dormancy termination of yellow cedar (*Chamaecyparis nootkatensis*) seeds. *J. Exp. Bot.,* 53:89–101.

Schoettle, A.W. and A.C. Leopold, 1984. Solute Leakage from Artificially Aged Soybean Seeds after Imbibition. *Crop Sci.,* 24:835–838.

Schultz, O.E., and R. Gmelin. 1952. Papierchromatographie der senfolglucosid-drogen. *Z. Naturforschung,* 7:500–508.

Schwartz, S.H., K.M. Léon-Kloosterziel, M. Koornneef, and J.A.D. Zeevaart. 1997. Biochemical characterization of the aba2 and aba3 mutants in *Arabidopsis thaliana*. *Plant Physiol.,* 114:161–166.

Simon, E.W. 1978. Membranes in dry and imbibed seeds. In *Dry Biological Systems,* J. H. Crowe and J. S. Clegg, (Eds.) Academic Press. New- York, p. 205–224.

Simon, E.W. and R.M. Harun. 1972. Leakage during seed imbibition. *J. Exp. Bot.,* 23:1076–1085.

Singh, Z. and G. Browning. 1991. The role of ABA in the control of apple seed dormancy reappraised by combined gas-chromatography mass-spectrometry. *J. Exp. Bot.,* 42:269–275.

Small, J.G.C. and Y. Gutterman. 1992. Effects of sodium chloride on prevention of thermo-dormancy, ethylene and protein synthesis and respiration in Grand Rapids lettuce seeds. *Physiol. Plant.,* 84:35–40.

Smith, M.T. 1978. Cytological changes in artificially aged seeds during imbibition. *Proc. Electron Microsc. Soc. Southern Africa,* 8:107–108.

Smith, M.T. 1989. The ultra-structure of physiological necrosis in cotyledons of lettuce seeds (*Lactuca sativa* L.). *Seed Sci. Tech.,* 11:453–462.

Song, W.J., W.J. Zhou, Z.L. Jin, D.D. Cao, D.M. Joel, Y. Takeuchi, and K. Yoneyama. 2005. Germination response of *Orobanche* seeds subjected to conditioning temperature, water potential and growth regulator treatments. *Weed Res.,* 45:467–476.

Song, W.J., W.J. Zhou, Z.L. Jin, D. Zhang, K. Yoneyama, Y. Takeuchi, and D.M. Joel. 2006. Growth regulators restore germination of *Orobanche* seeds that are conditioned under water stress and suboptimal temperature. *Aust. J. Agric. Res.,* 57:1195–1201.

Spaeth, S.C. 1987. Pressure-driven extrusion of intracellular substances from bean and pea cotyledons during imbibition. *Plant Physiol.,* 85:217–223.

Squire, G. 1999. Temperature and heterogeneity of emergence time in oilseed rape. *Ann. Applied Biol.,* 135:439–447.

Subbiah, T.K. and L.E. Powell. 1992. Absciric acid relationships in the chill related dormancy mechanism in apple seeds. *Plant Growth Regul.* 11:115–1223.

Sugiyama, T. 1949. *J. Jap. Soc. Hort. Sci.,* 13:1–7. [cited in Takahashi and Suzuki, 1980]

Takahashi, N. and Y. Suzuki. 1980. Dormancy and seed germination. In S. Tsunoda, K. Hinata, and C. Gómez-Campo, (Eds.) *Brassica Crops and Wild Allies: Biology and Breeding.* Tokyo, Japan: Japan Scientific Societies Press, p. 323–337.

Takayanagi and Muralkami. 1969. Determining seed viability. *Proc. International Seed Testing Association.* 34:243–252. Patent EP0364952.

Taylor, A.G., H.J. Hill, X. Huang and T.G. Min. 1990. Determining seed viability. U.S. Patent 4975364.

Thompson, K. and J.P. Grime. 1983. A comparative study of germination responses to diurnally fluctuating temperatures. *J. Appl. Ecol.,* 20:141–156.

Thomson, J.R. 1979. Seed storage methods. In *An Introduction to Seed Technology,* Wiley, New York.

Tokumasu, S. and M. Kato. 1987. The effect of fruits on the prolongation of seed dormancy and its relation to mustard oil content in cruciferous crops. *Acta Hortic.,* 215:131–138.

Tokumasu, S., F. Kakihara, and M. Kato. 1981. The change of dormancy of seeds stored in desiccators and/or harvested fruits in cruciferous crops. *J. Jap. Soc. Hortic. Sci.,* 50:208–214.

Trewavas, A. 1982. Growth substance sensitivity: the limiting factor in plant development. *Plant. Physiol.,* 55:60–72.

Tzagoloff, A. 1963. Metabolism of sinapine in mustard plants. I. Degradation of sinapine into sinapic acid & choline. *Plant Physiol.,* 38:202-206.

Van Staden, J., Gilliland, M.G., and N.A.C. Brown, . 1975a. Ultra-structure of dry viable and non-viable *Protea compacta* embryos. *Z. Pflanzenphysiologie,* 16:28-35.

Villiers, T.A., 1972. Cytological changes in dormancy. II. Pathological ageing changes during prolonged dormancy and recovery upon dormancy release. *New Phytologist,* 71:145–152.

Villiers, T.A. 1974. Seed ageing: chromosome stability and extended viability of seeds stored fully imbibed. *Plant Physiol.,* 53:875–878.

Walker-Simmons, M. 1987. ABA levels and sensitivity in developing wheat embryos of sprouting resistant and susceptible cultivars. *Plant Physiol.,* 84:61–66.

Walker-Simmons, M.K., L.D. Holappa, G.D. Abrams, and S.R. Abrams. 1997. ABA metabolites induce group 3 LEA mRNA and inhibit germination in wheat. *Physiol. Plant.,* 100:474–480.

Wang, M., S. Heimovaara-Dijkstra, and B. Van Duijn.1995. Modulation of germination of embryos isolated from dormant and non-dormant barley grains by manipulation of endogenous abscisic acid levels. *Planta,* 195:586–592.

Wareing, P. F. and P.F. Saunders. 1971. Hormones and dormancy. *Annu. Rev. Plant Physiol.,* 22:261–288.

Wesson, G. and P.F. Wareing. 1969a. The role of light in the germination of naturally occurring populations of buried weed seeds. *J. Exp. Bot.,* 20:402–413.

Wesson, G. and P.F. Wareing. 1969b. The induction of light sensitivity in the seed by burial. *J. Exp. Bot.,* 20:414–425.

Xu, N. and J.D. Bewley. 1995. The role of abscisic acid in germination, storage protein synthesis and dessication tolerance in alfalfa (*Medicago sativa* L.) seeds, as shown by inhibition of its synthesis by fluridone during development. *J. Exp. Bot.,* 46:687–694.

Yamamoto, H. and Oritani T. 1996. Stereo-selectivity in the biosynthetic conversion of xanthoxin into abscisic acid. *Planta,* 200:319–325.

Yoshioka, T., T. Endo, and S. Satoh. 1998. Restoration of seed germination at supraoptimal temperatures by fluridone, an inhibitor of abscisic acid biosynthesis. *Plant Cell Physiol.,* 39:307–312.

Zeevaart, J.A.D. and R.A. Creelman. 1988. Metabolism and physiology of abscisic acid. *Annu. Rev. Plant Physiol. Plant Mol. Biol.,* 39:439–473.

Zhang, T. and J.G. Hampton. 1999. The controlled deterioration test induces dormancy in swede (*Brassica napus* var. naprobrassica) seed. *Seed Sci. Technol.,* 27:1033–1036.

Zhou, W.J. and B. Kristiansson. 2000. Induction of secondary dormancy in rapeseed under conditions of stress in darkness. *J. Zhejiang Univ. (Agric. & Life Sci.),* 26(5):477–478.

10 Comparative Cytogenetics of Wild Crucifers *(Brassicaceae)*

Martin A. Lysak

CONTENTS

INTRODUCTION

Although *Brassicaceae* is not one of the more cytogenetically tractable plant families, the discreet charm of small crucifer chromosomes provoked interest in a few enthusiastic scholars over the past century. However, in contrast to present-day comparative crucifer cytogenetics, which is dominated by *Arabidopsis*-centered studies, the early days of crucifer cytogenetics did not focus on such a plant with an "exceedingly anomalous" chromosome number (Manton, 1932). Nevertheless, the development of *Arabidopsis* genomic resources and cytogenetic tools has ushered in a new era of plant molecular cytogenetics with *Arabidopsis thaliana* as the prime cytogenetic model (Koornneef et al., 2003). Indeed, many of the most crucial findings are inferred from cytogenetic studies in crucifers.

By looking back at early family-wide cytological surveys, it is clear that the work was driven by two aims. First, there was an attempt to describe, compare, and understand the karyological variability represented by chromosome numbers; and second, the strong hope that cytological data would be helpful in resolving phylogenetic relationships and taxonomic problems (Jaretzky, 1928, 1932; Manton, 1932). This was clearly expressed by Manton, who defined the aim of her extensive cytological analysis as "testing the applicability of cytology to the solution of taxonomic problems and of obtaining evidence on the relation between chromosome changes and the natural evolution of species and genera" (Manton, 1932). In three contemporary cytological surveys (Jaretzky, 1928, 1932; Manton, 1932), the potential correspondence between the chromosome number variation observed across *Brassicaceae* and phylogenetic relationships inferred mainly from morphological and anatomical characters (Hayek, 1911) was put to the test. For example, Jaretzky (1928) questioned whether chromosomal data could resolve the classification of *Arabidopsis thaliana* as a member of *Arabis* or as a separate genus *Arabidopsis* (*Stenograma*). However, in only a few genera and lineages (e.g., Hesperideae; Jaretzky, 1932) did the cytological data seem to support the proposed taxonomic treatments (Hayek, 1911). In the majority of taxa studied, patterns of chromosome number variation were often ambiguous, preventing clear-cut conclusions from being reached regarding the systematics and phylogeny of analyzed taxa. This was partly due to the belief at the time that base chromosome numbers could delimit taxa, because the existence of polybasic genera (e.g., *Diplotaxis*, *Iberis*) and tribes (Brassiceae, Sisymbrieae) had only just been realized. Moreover, taxonomic treatments of that time, largely based on morphological characters prone to convergent evolution, did not reflect the true phylogenetic relationships within the family (Al-Shehbaz et al., 2006). Consequently, chromosomal data often did not support or were even in conflict with ill-defined taxonomic concepts (Jaretzky, 1928, 1932; Manton, 1932).

Nevertheless, at a time when even the chromosome number of *Arabidopsis* was still debated, significant steps were being made in the cytogenetics of *Brassica* crop species. Cytogenetic analyses and crossing experiments carried out between 1922 and 1937 by Karpechenko, Morinaga, and U (Prakash and Hinata, 1980 and references therein) elucidated the origin and interspecific relationships of the six *Brassica* species known today as the species of U's triangle (U, 1935). For many decades following this, *Brassica* species were the prime models for crucifer cytogenetics. It is only recently that they have been replaced by the sequenced *Arabidopsis* genome (Koornneef et al., 2003; Lysak and Lexer, 2006).

This chapter focuses primarily on recent advances in comparative cytogenetics and phylogenomics of wild crucifers, starting from the first cytogenetic studies using fluorescence *in situ* hybridization (FISH) at the beginning of the 1990s. There is brief summary of the history and present-day understanding of *Arabidopsis* cytogenetics, which has played an invaluable role in laying the foundations for modern comparative cytogenetics of *Brassicaceae*. This is followed by a section outlining the major achievements of comparative cytogenetics in crucifers, including a critical review of prominent cytogenetic markers, analytical techniques, and the integration of cytogenetics, genomics, and phylogenetics into phylogenomics.

ARABIDOPSIS CYTOGENETICS

FROM DIM BEGINNINGS TO THE CYTOGENETIC MODEL

Laibach (1907) was the first to report the correct chromosome number in *Arabidopsis thaliana* ($2n = 10$). However, because *Arabidopsis*, like most other *Brassicaceae* taxa, is characterized by small chromosomes that were difficult to distinguish using the simple cytogenetic techniques available in the early 1900s, a certain ambiguity existed over this chromosome count for some time. This uncertainty was also fueled by the chromosome count itself because it was unique among chromosome numbers known for other crucifers at that time (e.g., Jaretzky, 1928). Even Manton (1932) considered the chromosome number as unusual almost three decades later.

Analysis of mitotic as well as meiotic chromosomes in euploid, polyploid, as well as trisomic *Arabidopsis* lines confirmed five chromosome pairs in the euploid *Arabidopsis thaliana* complement with one (Steinitz-Sears, 1963) or two nucleolus organizing region (NOR)-bearing chromosomes (Sears and Lee-Chen, 1970), respectively. Later, *Arabidopsis* mitotic chromosomes were analyzed by C-banding using Giemsa staining (Ambros and Schweizer, 1976). However, the tiny and condensed mitotic chromosomes or similarly sized meiotic chromosomes at late prophase, metaphase, and anaphase I were found unsuitable for a detailed cytogenetic analysis. Improvements to cytological analysis were made by Klášterská and Ramel (1980), who showed that it was possible to increase spatial resolution several-fold using extended Giemsa-stained pachytene chromosomes. Surface spreads of *Arabidopsis* prophase I meiocytes were also shown to be a feasible technique for the preparation of synaptonemal complex (SC) spreads for analysis by electron microscopy (Albini, 1994). In the SC karyotype, chromosome lengths could be determined and the two shortest chromosomes were identified as NOR-bearing chromosomes (At2 and At4). However, although SC spreads offer a superior spatial resolution, the technique is laborious and technically demanding, and thus it is not widely used in *Arabidopsis* cytogenetics. Recently, the segregation of homologous chromosomes in several meiotic mutants has been analyzed using SC spreads (Pradillo et al., 2007).

Decisive for a detailed characterization of the *Arabidopsis* karyotype was fluorescent DAPI (4′,6-diamidino-2-phenylindole) staining of chromosomes. DAPI-stained chromosomes along with FISH of repetitive sequences (ribosomal DNA, pAL centromeric repeat) provided a relatively easy way to identify individual *Arabidopsis* chromosomes (Maluszynska and Heslop-Harrison, 1991, 1993). These initial studies ushered in the decade marked by the astounding development of *Arabidopsis* molecular cytogenetics from its rather dim beginnings to *Arabidopsis* becoming the preeminent plant cytogenetic model. The pinnacle of this is demonstrated by the publication of a special issue of *Chromosome Research* dedicated to *Arabidopsis* cytogenetic research ("*Arabidopsis* as a cytogenetics model," *Chromosome Research*, 11:3, 2003). As shown later on, the advances in *Arabidopsis* molecular cytogenetics have been facilitated by several favorable circumstances: (1) wide application of the FISH technology in plant cytogenetics; (2) use of extended meiotic chromosomes; (3) availability of *Arabidopsis* DNA probes, including repetitive elements, chromosome-specific large-insert clones, and single-copy probes; and (4) integration of cytogenetic data with genetic and genomics datasets resulting from the sequencing and annotation of the *Arabidopsis* genome.

As *Arabidopsis* molecular cytogenetics and epigenetics have been recently reviewed several times (Fuchs et al., 2006; Fransz et al., 2003, 2006; Heslop-Harrison et al., 2003; Koornneef et al., 2003; Lysak and Lexer, 2006; Siroky, 2008), only aspects significant for comparative cytogenetics of *Brassicaceae* are highlighted here.

ARABIDOPSIS KARYOTYPE

A detailed description of the *Arabidopsis* karyotype has only been made possible as a result of the introduction of the chromosome spreading technique (Ross et al., 1996). This easy and fast preparation method enables meiotic and mitotic chromosomes to be spread from enzymatically softened individual anthers or whole flower buds on microscopic slides. In particular, DAPI-stained extended pachytene bivalents provide sufficient spatial resolution for *Arabidopsis* cytogenetic studies (Fransz et al., 1998; Lysak et al., 2006b; Ross et al., 1996).

Tandem repetitive elements (e.g., pAL; Martinez-Zapater, 1986) were among the first *Arabidopsis* sequences to be isolated, characterized, and subsequently localized on chromosomes by FISH. Initial FISH localizations of the most abundant *Arabidopsis* tandem repeats (rDNA: 5S and 45S rRNA genes, and pAL, a 180-bp centromeric repeat) were made on mitotic chromosomes and interphase nuclei (Maluszynska and Heslop-Harrison, 1991, 1993; Murata et al., 1997). However, neither the small condensed mitotic chromosomes nor interphase nuclei possessed the required spatial resolution needed for precise FISH mapping. This changed with the introduction of DAPI-stained pachytene chromosomes by Fransz et al. (1998). Such differential DAPI staining revealed conspicuous heterochromatin arrays along extended pachytene bivalents, including centromeric regions, NORs, and a heterochromatic knob that could be used as chromosomal landmarks. This seminal article not only reports for the first time the exact chromosomal localization of the most abundant tandem repeats (5S and 45S rDNA, pAL), but was also the first to use large-insert DNA clones as FISH probes [cosmids, single yeast artificial chromosome (YAC), and bacterial artificial chromosome (BAC)].

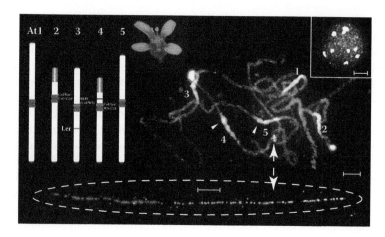

FIGURE 10.1   (See color insert following page 128.) Idiogram of *Arabidopsis thaliana* (*n* = 5) and fluorescence *in situ* localization (FISH) of several DNA probes to pachytene bivalents isolated from young anthers of *Arabidopsis* accession C24. Left: *Arabidopsis* idiogram showing prominent chromosomal landmarks — NORs containing 45S rDNA on At2 and At4 (red); 5S rDNA loci on At3, At4, and At5 (green; note the variation in number of 5S loci among the accessions); the heterochromatic knob hk4S on the top arm of At4 in accessions Col and WS; the 180-bp pAL repeat (blue) at centromeres embedded within pericentromeric heterochromatin regions (gray), and telomeric repeats at chromosome termini (purple; note the interstitial site at pericentromere of At1). Sizes of chromosome arms and NORs follow Lysak et al. (2006b), localization of 5S rDNA loci follows Fransz et al. (1998) and Sanchez-Moran et al. (2002; labeled by asterisk). Right: *Arabidopsis pachytene* chromosomes after simultaneous *in situ* hybridization of (i) 45S rDNA (NORs of At2 and At4 labeled by red), (ii) 5S rDNA (green signals at pericentromeric heterochromatin of top arms of At4 and At5; arrowheads), (iii) BAC contig spanning ~6.5 Mb of the top arm of At1 (purple), (iv) three BAC clones (72, 95, and 40 kb) of the At5 bottom arm visualized as red, green, and yellow fluorescence, respectively. Bottom: FISH of the same BAC triple to extended DNA fibers. Top right: *Arabidopsis* interphase nucleus with conspicuous chromocenters. Chromosomes and the nucleus counterstained by DAPI. All bars, 5 μm.

The detailed cytogenetic data arising from this led to the first high-resolution description of the *Arabidopsis* karyotype (Figure 10.1). The karyotype consists of three (sub)metacentric (At1, At3, and At5) and two acrocentric chromosomes (At2, At4). *Arabidopsis* is characterized by a very small genome (157 Mb; Bennett et al., 2003, Table 10.1) with a small number of repetitive DNA elements (approx. 10%; Koornneef et al., 2003). DNA repeats are mainly located in conspicuous heterochromatin arrays (pericentromeres, NORs) revealed with DAPI staining, whereas chromosome arms are largely depleted of large blocks of repetitive elements. In interphase nuclei, pericentromeric and NOR-related heterochromatin form discrete DAPI-positive loci called chromocenters (Figure 10.1), characteristic for most *Brassicaceae* species. The centromeric regions exhibit a tripartite structure, with the core centromeric region comprising primarily a 180-bp satellite repeat pAL and retrotransposon-derived 106B repeat, with both sides flanked by pericentromeric heterochromatin harboring retrotransposons, transposons, and some genes (Fransz et al., 2003). The two acrocentric chromosomes bear large heterochromatic NORs on their top (short)* arms comprising 45S rDNA repetitive arrays. The 5S rDNA loci are usually located interstitially in the pericentromeric heterochromatin, and their numbers differ among *Arabidopsis* accessions (Figure 10.1). In some *Arabidopsis* accessions (e.g., Columbia), a heterochromatic knob (hk4S) is located on the top arm of At4 (Fransz et al., 1998), and its origin and structure has been the subject of a comprehensive study (Fransz et al., 2000).

---

* The chromosome arm nomenclature follows that of the Arabidopsis Genome Initiative (Table 1 in AGI, 2000). We proposed that the terms "top" and "bottom" be used instead of "short" and "long" (Lysak et al., 2003), which are ambiguous in the case of (sub)metacentric chromosomes (At1, 3, and 5). Nevertheless, no consensus on the terminology has been reached and all four terms can be found in the *Arabidopsis* cytogenetic literature (cf. Lysak et al., 2006a).

Differential DAPI staining and FISH have enabled localized cytogenetic markers, heterochromatin chromosomal profile, and the sequence data (McCombie et al., 2000) to be integrated into a detailed cytogenetic map of the whole top arm of At4, including the pericentromeric region and terminal NOR. This has revealed that the knob originated via a pericentric inversion, relocating part of the pericentromeric heterochromatin into the top arm of At4 (Fransz et al., 2000).

Arabidopsis Repetitive Elements As Cytogenetic Markers

Repetitive DNA elements were among the first cytogenetic markers to be used to identify individual *Arabidopsis* chromosomes. Since the seminal papers of Maluszynska and Heslop-Harrison (1991, 1993) and Fransz et al. (1998, 2000), the number of studies using DNA repeats has steadily grown and only a few key ones are described here. In many *Arabidopsis* studies, DNA repetitive elements serve as chromosome and/or heterochromatin identifiers. Ribosomal 5S and 45S probes, often in combination with the centromeric pAL repeat, are frequently used to identify individual chromosomes and centromeric regions, respectively, within interphase nuclei (Baroux et al., 2004; Berr and Schubert, 2007) or during meiotic chromosome pairing (Pradillo et al., 2007; Sanchez-Moran et al., 2002; Santos et al., 2003). In some FISH experiments, these three repetitive probes were combined with repeat-rich BACs, revealing pericentromeric heterochromatin (e.g., Fransz et al., 2003).

Several studies have analyzed the spatial organization and dynamics of tandem and dispersed (e.g., Ty1-*copia*- and Ty3-*gypsy*-like retrotransposons; Soppe et al., 2002) repetitive elements located within interphase chromocenters. For example, chromocenter reduction and expansion of methylated heterochromatin comprising pericentromeric repeats have been observed during dedifferentiation of mesophyll cells (Tessadori et al., 2007a) or during the switch from vegetative to reproductive growth (Tessadori et al., 2007b). The relationship between DNA methylation and the organization of pericentromeric heterochromatin (chromocenters) was analyzed in wild-type *Arabidopsis* plants as well as in hypomethylated mutants by FISH with several satellites, transposons, and retrotransposons as probes (Soppe et al., 2002).

The *Arabidopsis* telomeric repeat (TTTAGGG) has been shown to occur on all chromosome termini as well as in interstitial locus at the pericentromere of At1 (Figure 10.1). Chromosome ends marked by the telomeric repeat were shown as clustered around the nucleolus prior to homoeologous chromosome pairing at prophase I (Armstrong et al., 2001; Armstrong and Jones, 2003). Recently, it was shown that a synthetic analog of the *Arabidopsis* telomeric repeat, peptide-nucleic acid (PNA), is a more sensitive probe for the visualization of *Arabidopsis* telomeres (Mokros et al., 2006; Vespa et al., 2007).

BAC FISH and Chromosome Painting

Sequencing of the *Arabidopsis* genome has been accomplished due to the use of genomic DNA libraries of high-capacity vectors including cosmids, YACs, BACs, bacteriophage P1-derived artificial chromosomes (PACs, P1 clones), and transformation-competent artificial chromosomes (TACs) (AGI, 2000). The sequencing project led to the assembly of chromosome-arm-specific contigs of large-insert clones that became available to *Arabidopsis* cytogeneticists. Two BAC libraries (IGF library, Mozo et al., 1998; TAMU library, Choi et al., 1995) provided chromosome-specific minimum tiling paths of BAC clones suitable for the *in situ* visualization of corresponding chromosomal regions.

First, fluorescently labeled cosmid, YAC, and BAC clones were applied as chromosomal landmarks to characterize the *Arabidopsis* karyotype (Fransz et al., 1998) and the heterochromatic knob hk4S (Fransz et al., 2000). *Arabidopsis* cosmids and BAC clones were also successfully hybridized to extended DNA fibers (Fransz et al., 1996; Jackson et al., 1998). Extended DNA fibers exhibit a several-fold higher spatial resolution than pachytene chromosomes (Figure 10.1) with discernible DNA probes lying less than 1 kb apart (De Jong et al., 1999). Fiber FISH has been shown to be helpful in the detailed spatial analysis of *in situ* hybridized probes. The DNA condensation ratio (kb/μm) characteristic for extended fibers can be used to estimate the physical size of a probe or its DNA target, as well as to analyze the structure of adjacent, overlapping, or interspersed DNA probes (e.g., Fransz et al., 2000).

These reports ushered in a new era of *Arabidopsis* molecular cytogenetics dominated by multicolor FISH mapping using BAC clones and DNA repeats. The terms "BAC FISH" and "chromosome painting" (CP) were coined to specify the use of BAC clones and BAC contigs (i.e., a continuous series of BACs) as FISH probes. Whereas BAC FISH refers to the application of a single or a few BAC clones (Figure 10.1), CP in the context of *Arabidopsis* and crucifer cytogenetics, describes the use of BAC contigs comprising from several up to more than a hundred clones covering large chromosome regions or whole chromosomes (Figure 10.1; Lysak et al., 2001, 2006b; Lysak and Lexer, 2006; Schubert et al., 2001).

Chromosome-specific BACs often combined with rDNA and pAL repeats were proven to be useful as chromosome identifiers. For example, BAC FISH was applied to identify chromosomes and chromosome arms involved in chromosome fusions in *Arabidopsis* mutants with telomere dysfunction (Mokros et al., 2006; Siroky, 2008; Siroky et al., 2003; Vannier et al., 2006; Vespa et al., 2007). Large-scale CP with BAC contigs covering both arms of chromosome At4 enabled *in situ* visualization of the chromosome during mitosis and meiosis, as well as within interphase nuclei (Lysak et al., 2001). This was historically the first report on the painting of an entire chromosome in a euploid plant. The painting of other *Arabidopsis* chromosomes (Lysak et al., 2003) facilitated the simultaneous visualization of all chromosomes by multicolor CP using differentially labeled BAC contigs covering all ten chromosome arms (Lysak et al., 2006b; Lysak and Lexer, 2006; Pecinka et al., 2004). Two-color and particularly multicolor CP enabled tracing individual chromosomes and specific chromosome regions in all stages of the cell cycle and meiosis. Furthermore, BAC clones provide the flexibility in preparing *ad hoc* painting probes for any chromosome region. The technique has served as a tool to study the spatial organization of interphase chromosomes leading to the concept of the chromocenter-loop model (Fransz et al., 2002) and to elucidate the three-dimensional (3D) arrangement of chromosome territories within interphase nuclei (Pecinka et al., 2004). More recently, CP based on whole-mount FISH (whole-mount CP) was used to study the 3D organization of chromosome territories in interphase nuclei of diverse cell types and tissues in *Arabidopsis* seedlings (Berr and Schubert, 2007). CP has been shown to be useful also for detecting chromosome rearrangements mediated by T-DNA integration. T-DNA insertion into the *Arabidopsis* nuclear genome is frequently linked with inversions and translocations, although the exact mechanism of these integration/rearrangement events remains to be elucidated (Laufs et al., 1999). Such a ~4-Mb paracentric inversion following the integration of two transgenes has been revealed by CP (Pecinka et al., 2005).

The power of comparative chromosome painting (CCP) using *Arabidopsis* BAC contigs to disclose the range of cross-species chromosome synteny will be demonstrated in the following section.

COMPARATIVE CYTOGENETICS IN WILD CRUCIFERS

CHROMOSOME NUMBER AND GENOME SIZE VARIATION

Brassicaceae species are notoriously known for their small chromosomes (Jaretzky, 1928, 1932; Manton, 1932). Most crucifer species possess chromosomes of a few micrometers in size (e.g., 1.5 to 2.8 μm in *Arabidopsis thaliana*, Koornneef et al., 2003; 2 to 5 μm in *Brassica*, Cheng et al., 1995; and 1.2 to 5.8 μm in *Boechera*, Kantama et al., 2007). Larger chromosomes can be found in species with a low chromosome number (*n* = 4–7) and large genome size (e.g., *Bunias*, *Matthiola*, *Physaria*).

Chromosome numbers recently compiled by Warwick and Al-Shehbaz (2006) are known for approximately 70% genera and 40% crucifer species. Chromosome numbers in *Brassicaceae* vary over 32-fold with the lowest chromosome number of $n = 4$ found only in *Physaria* (approximately 8 spp., tribe Physarieae)* and *Stenopetalum* (2 spp., Camelineae); five chromosome pairs ($n = 5$) were observed in *Arabidopsis thaliana* (Camelineae), *Matthiola* (1 sp., Anchonieae), *Physaria* (approximately 21 spp.), and *Stenopetalum* (5 spp.) (Warwick and Al-Shehbaz, 2006). The highest chromosome numbers were reported by Montgomery (1955), Easterly (1963), and Harriman (1965) in North American polyploid *Cardamine* (formerly *Dentaria*) species (*C. angustata*, *C. concatenata*, *C. diphylla*, *C. dissecta*, and *C. maxima*). Among these polyploid species, the highest counts were obtained in *C. concatenata* (*D. laciniata*; $2n = \pm240$, Montgomery, 1955; and $2n = 256$, Easterly, 1963) and *C. diphylla* ($2n \approx 256$; Harriman, 1965). However, the counts must be considered *a priori* as approximate and inaccurate due to the clumping of the extremely high number of very small chromosomes (Harriman, 1965). This is illustrated by the more than threefold variation in chromosome counts ($2n = 74$–256) between different root tips in *C. diphylla* (Harriman, 1965). A future thorough analysis of DAPI-stained meiotic chromosomes, coupled with GISH and/or FISH localization of centromeric satellite repeats, can elucidate chromosome number variation and the origin of the high polyploid *Cardamine* species.

Base chromosome numbers vary from x = 4 to x = 17, with more than a third of the taxa based on x = 8 (Warwick and Al-Shehbaz, 2006), implying that the eight is most likely an ancestral chromosome number of the whole family. Several crucifer genera (e.g., *Brassica*, *Camelina*, *Diplotaxis*, *Erysimum*, *Physaria*) are polybasic, that is, characterized by multiple base chromosome numbers. This is due to ancient whole-genome duplications followed by diploidization via karyotype rearrangements resulting in descending (and secondary perhaps also ascending) dysploidy. Base numbers as low as x = 4 or 7 are most likely secondary numbers reflecting the chromosome number reduction after ancient polyploid events (Lysak et al., 2005, 2007; Mandáková and Lysak, unpublished). Therefore, the use of base chromosome numbers often not reflecting the true genome evolution of crucifer species, appears to be superfluous in some instances, and haploid (n) or diploid ($2n$) chromosome numbers should be given instead.

Chromosome number and nuclear genome size are two sides of the same coin. Both characters provide primary information on the mass and structure of a nuclear genome. Interspecific genome size variation is directly linked to chromosomal and karyotypic variation as well as to the amount and diversity of repetitive sequences. Hence, genome size variation must be taken into account when evolutionary processes reshuffling crucifer genomes are being unveiled. Genome size data also serve more practical purposes such as, for example, to estimate how much DNA to sequence or how many BAC clones should have a large-insert genomic library for complete genome coverage.

Currently available genome size data for *Brassicaceae* species are rather scarce, including approximately 72 taxa, i.e., 1.9% of all crucifers (Lysak et al., 2008). Genome sizes of crucifer species were compiled in the Plant DNA C-values database (Bennett and Leitch, 2005; www.kew.org/genomesize/homepage.html) and by Johnston et al. (2005). Recently, a dataset comprising newly acquired as well as published genome size values for 185 *Brassicaceae* taxa from all but one of the 25 tribes (sensu Al-Shehbaz et al., 2006) was collated by Lysak et al. (2008). Despite the 16-fold variation in holoploid genome size (C-value)† across *Brassicaceae*, the majority of crucifer species possess very small (≤1.4 pg) or small (≤3.5 pg/C; Leitch et al., 1998) genomes with a mean C-value of 0.63 pg (Lysak et al., 2008). Table 10.1 shows genome sizes for the most prominent crucifer model and crop species. The smallest C-values were found in *Arabidopsis thaliana* (0.16 pg/C; Bennett et al., 2003) and some *Sphaerocardamum* species of the Halimolobeae (0.15–0.16 pg/C, $2n = 16$; Bailey, 2001). The monotonous pattern of genome size variation in *Brassicaceae* is disrupted by

---

* The tribal classification used throughout the text follows that of Chapter 1 by Koch and Al-Shehbaz (2008). See Figures 1.1 and 10.4 for all tribes and three major phylogenetic lineages (Lineage I–III) recognized by Beilstein et al. (2006) and Al-Shehbaz et al. (2006, Figure 1).

† The 1C-value represents the DNA amount in an unreplicated gametic nucleus regardless of ploidy level (holoploid genome size). The C_x represents the amount of DNA in one chromosome set of an organism with a base chromosome number x (monoploid genome size). It is calculated by dividing the 2C value by ploidy level.

TABLE 10.1
Genome Sizes (C-values) of Selected Model and Crop
***Brassicaceae* species**

Species	Tribe	2n	1C-value (pg/Mbp)[a]
Arabidopsis thaliana	Camelineae	10	0.16 / 157[b]
A. arenosa (2x, 4x)		16, 32	0.20 / 0.40, 196 / 392
A. halleri		16	0.24 / 235
A. lyrata subsp. *lyrata*		16	0.25 / 245
A. suecica		26	0.35 / 343
Capsella rubella		16	0.22 / 216
Boechera stricta	Boechereae	14	0.24 / 235[c]
Cardamine hirsuta	Cardamineae	16	0.23 / 225
Laevenworthia alabamica		22	0.50 / 490[c]
Arabis alpina	Arabideae	16	0.38 / 372
Thellungiella halophila	Eutremeae	14	0.66 / 647[c]
Noccaea caerulescens	Noccaeeae	14	0.61 / 598[c]
Brassica nigra	Brassiceae	16	0.65 / 637
B. oleracea		18	0.71 / 696
B. rapa		20	0.54 / 529
B. napus		38	1.15 / 1127

Note: Genome size data taken from Lysak et al. (2008) unless otherwise stated.

[a] 1pg = 980 Mbp.

[b] Bennett et al. (2003)

[c] Lysak and Bureš, unpublished data

several relatively high genome size values. The highest C-value was found in *Bunias orientalis* (2.43 pg/C, 2n = 14, Buniadeae), followed by high genome sizes in *Matthiola* (2n = 12, 14) and some *Physaria* (2n = 8, 56) species (Lysak and Lexer, 2006; Lysak et al., 2008).

Large C-values in species having a low number of large chromosomes are puzzling. Such genomes might have undergone one or more whole-genome duplications (i.e., genome size increase), followed by the diploidizing karyotype reshuffling toward low diploid-like chromosome numbers (Lysak and Lexer, 2006). Alternatively, a massive retrotransposon amplification led to genome and chromosome size increase without a change in chromosome number in Anchonieae, Buniadeae, and Physarieae. Another intriguing evolutionary riddle is the overall prevalence of very small genomes across *Brassicaceae*, despite frequent hybridization, polyploidization, and genome reshuffling. It can only be speculated that crucifers (at best exemplified by the *Arabidopsis* genome) possess an effective mechanism of DNA removal (e.g., illegitimate recombination, unequal homologous crossing over) and/or a mechanism impeding the amplification of transposable elements (see Lysak et al., 2008, for discussion). Further work is needed to resolve this issue.

SPECIAL CHROMOSOME FORMS

Special chromosome forms refer to chromosomes deviating from average autosomes by their size, morphology, or structure. B chromosomes are usually heterochromatic and smaller than the regular chromosomes of a complement. Up to now, no survey of crucifer species with B chromosomes has been compiled. Among 1558 species with a known chromosome number listed by Warwick and Al-Shehbaz (2006), 38 species were reported as having at least one B chromosome (*note:* up to 6

Bs were found in two *Cochlearia* species). The scarce data on the frequency of B chromosomes and the lack of any in-depth analysis do not allow drawing conclusions on the origin and evolutionary significance of B chromosomes in crucifer species.

Much attention has been paid to chromosome number variation linked to the occurrence of presumed B chromosomes in North American genus *Boechera* (Dobes et al., 2006, and references therein; Kantama et al., 2007; Sharbel et al., 2004). The best investigated karyologically is the *B. holboellii* complex comprising *B. holboellii*, *B. stricta*, and their presumed hybrid *B. divaricarpa*. All three taxa exist as sexual diploids ($2n = 14$), apomictic triploids ($2n = 21$), higher-level polyploids, and aneuploids. Aneuploid chromosomes revealed in both diploid and triploid *Boechera* plants are smaller and largely heterochromatic compared to the euploid chromosome set, and hence they have been classified as B chromosomes (Dobes et al., 2006). The structure and origin of B chromosomes in the *B. holboellii* complex were analyzed by Kantama et al. (2007). Interestingly, apomicts with $2n = 14$ possessed a chromosome with large blocks of heterochromatin, called *Het*. Apomicts with $2n = 15$ have the *Het* chromosome and one small (deletion) chromosome called *Del*. The *Het* chromosome can be considered a transition form between the B and Y chromosome with a potential role in the genetic control of apomixis as being exclusively present in $2n = 14$ complements of apomictic plants. *Del* was suggested to result from a translocation event between *B. holboellii* (or related taxon) and *B. stricta* chromosomes. Obviously, karyologically diverse *Boechera* species provide an as-yet unexplored playground for in-depth studies of chromosome and karyotype diversification in wild crucifers.

In *Arabidopsis*, a telocentric minichromosome derived from the top arm of At4 has been found in addition to the $2n = 10$ complement (Murata et al., 2005). The mini4S chromosome most likely possessed a functional centromere within the 180-bp pAL repeat array flanked by a telomere and has been transmitted to about 30% of the selfed progeny. A ring form of another minichromosome comprising a centromere and a part of the top At2 arm has been revealed in a T-DNA *Arabidopsis* line (Murata et al., 2007). Despite being ring and dicentric, the minichromosome was stably transmitted to next generations.

Altogether, these findings suggest that aneuploidy occurring in some wild crucifers (e.g., *Cardamine pratensis*; Lökvist, 1956) can be associated with the occurrence of special chromosome forms such as B chromosomes or minichromosomes. Mechanisms underlying the extensive aneuploidy in some crucifers can be now deciphered using new cytogenetic tools such as comparative chromosome painting.

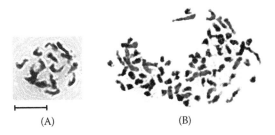

(A) (B)

FIGURE 10.2 (See color insert following page 128.) Genomic *in situ* hybridization (GISH) to flower-bud mitotic chromosomes of *Lepidium*. (A) Chromosomes of *L. africanum* ($2n = 16$) probed with fluorescently labeled *L. africanum* genomic DNA, (B) chromosomes of *L. hyssopifolium* ($2n \approx 72$) probed with fluorescently labeled *L. africanum* genomic DNA. GISH revealed 16 *L. africanum*-derived chromosomes (red) within the *L. hyssopifolium* complement. Chromosomes were counterstained by DAPI and the resulting fluorescence displayed in grey. Bar, 5 μm. (*Source:* Courtesy of T. Dierschke and K. Mummenhoff.)

ALLOPOLYPLOIDY REVEALED BY GENOMIC *IN SITU* HYBRIDIZATION (GISH)

Whole-genome duplication — polyploidy — is accepted as one of the fundamental mechanisms of genome evolution and plant speciation. In *Brassicaceae*, approximately 40% of species are considered polyploid (Warwick and Al-Shehbaz, 2006), whereby both autopolyploidy and allopolyploidy have been documented. Numerous reports on multiple intraspecific cytotypes (for examples, see Warwick and Al-Shehbaz, 2006) indicate that autopolyploidy is not rare in the family. Similarly, there are several relatively recent examples of interspecific hybridization and allopolyploid speciation (reviewed by Marhold and Lihová, 2006), for example, in *Boechera* (Dobes et al., 2006), *Cardamine* (Marhold and Lihová, 2006), *Draba* (Brochmann, 1992; Brochmann et al., 2004), *Lepidium* (Mummenhoff et al., 2004; Figure 10.2), or *Rorripa* (Bleeker, 2007). In many cases, the origin of hybrids and allopolyploids can be unveiled by molecular markers (e.g., AFLP, cpDNA, isozymes) or genomic *in situ* hybridization (GISH).

GISH allows discriminating parental genomes in allopolyploid species and interspecific hybrids by simultaneous or subsequent hybridization of fluorescently labeled genomic DNA (gDNA) of the presumed parental genotypes to chromosomes of the composite genome. The potential cross-hybridization of repetitive sequences shared by both parental genomes can be suppressed by including an excess of the unlabeled gDNA of one of the donor genomes.

GISH-mediated discrimination between parental genomes is feasible due to the sequence divergence of dispersed repetitive elements. Consequently, parental genomes cannot be discerned in hybrids between closely related taxa due to the lack of sufficiently divergent genome-specific repeats. If the allopolyploid speciation event is too ancient, one of the parental species can become extinct or genome-specific repetitive elements of parental and/or the derived allopolyploid species diverged to such an extent that they no longer exhibit the homology sufficient for GISH. Furthermore, in flowering plants with small genomes such as the *Brassicaceae*, GISH is impeded by a low amount of genome-specific dispersed repeats concentrated chiefly in pericentromeric heterochromatin arrays. In such species, GISH is almost exclusively based on differential hybridization of species-specific centromeric satellites included in a respective gDNA probe, whereas whole euchromatic chromosome arms remain virtually unlabeled with no possibility of detecting potential intergenomic translocations (Ali et al., 2004; Lysak and Lexer, 2006). This technical drawback can be partially overcome using large amounts of gDNA probes, longer hybridization times, and/or the addition of (peri)centromeric repeats to decrease hybridization to pericentromeres (Ali et al., 2004). The critical genome size below which GISH labeling of entire chromosomes usually fails was proposed as ~0.6 pg/C_x (Raina and Rani, 2001). It remains unclear whether GISH labeling of whole chromosomes can be more efficient in *Brassicaceae* hybrids and allopolyploids with large genomes (Lysak et al., 2008).

In *Brassicaceae*, most GISH studies concentrated on the natural and synthetic allotetraploid *Brassica* species (*B. carinata*, *B. juncea*, and *B. napus*) and artificial intergeneric hybrids between *B. napus* and other crucifer species mainly from the tribe Brassiceae (Chèvre et al., 2007; reviewed by Lysak and Lexer, 2006, and Snowdon, 2007). GISH data corroborated the phylogenetic relationships among the three "diploid" donor *Brassica* species with the two lineages separating *B. nigra* (BB genome, 2n = 16) from *B. oleracea* (CC, 2n = 18) and *B. rapa* (AA, 2n = 20) (Warwick and Sauder, 2005, and references therein). GISH discriminated between A and C genomes in *B. carinata* (BBCC, 2n = 34), and between A and B genomes in *B. juncea* (AABB, 2n = 36) (Snowdon et al., 1997; Maluszynska and Hasterok, 2005), whereas A and C genomes showing a high level of sequence homology cannot be discerned in *B. napus* (AACC, 2n = 38) (Snowdon et al., 1997). The failure of GISH due to the close genome similarity between some *Brassica* species is not followed in other crucifer genera. Despite a very close phylogenetic relationship between *Arabidopsis thaliana* (TT, 2n = 10) and *A. arenosa* (AA, 2n = 16; AAAA, 2n = 32), the modified GISH protocol (see above) has been successfully applied to identify both parental genomes in the natural and synthetic *A. suecica* (AATT, 2n = 26), as well as in an artificial hybrid between *A. thaliana* and *A. suecica*

(Ali et al., 2004; Lysak and Lexer, 2006). GISH was also used to corroborate the hybridogenous origin of Australian and New Zealand polyploid *Lepidium* species suggested by conflicting chloroplast (cpDNA) and ribosomal internal transcribed spacer (ITS) phylogenetic data. In some species, the hybrid origin was indicated by cpDNA shared with Californian *Lepidium* species, whereas ITS was of the South African type. This pattern has been explained by transoceanic dispersals of *Lepidium* from California and Africa to Australia/New Zealand, followed by allopolyploidization (Mummenhoff et al., 2004). Indeed, GISH confirmed that a South African species *L. africanum* ($2n = 16$) or its close relative is one of the parental genomes in the allopolyploid Australian species *L. hyssopifolium* ($2n = 72$) (T. Dierschke, et al., unpublished). Recently, GISH was used to elucidate the genome composition of sexual ($2n = 14$) and apomictic ($2n = 15$) genotypes from the *Boechera holboellii* complex (Kantama et al., 2007). This species complex, including *B. holboellii, B. stricta,* and their presumed hybrid *B. divaricarpa*, exhibits extensive karyological variation due to recurrent hybridization, introgression, and apomixis. Two-color GISH analysis using *B. holboellii* and *B. stricta* gDNAs revealed that the analyzed apomicts represent interspecific hybrids with different contributions of *B. holboellii*- and *B. stricta*-derived chromosomes. As in most *Brassicaceae* species, GISH in *Boechera* was principally based on genome-specific pericentromeric repeats (Kantama et al., 2007). GISH in *Lepidium* and *Boechera* illustrates the untapped capacity of GISH to reveal the origin of hybridogenous taxa within taxonomically complicated crucifer groups (e.g., *Draba*, Grundt et al., 2006; reviewed by Marhold and Lihová, 2006).

CENTROMERIC AND rDNA TANDEM REPEATS

Among different types of tandem (satellite) repeats characterized in *Brassicaceae* species, centromeric satellites and tandem arrays of rRNA genes (rDNA) are the most prominent ones used as cytogenetic landmarks (reviewed by Lysak and Lexer, 2006).

In crucifers, as in other plants with monocentric chromosomes, centromeric chromosome regions comprise the core centromere region flanked on both ends by adjacent pericentromeric regions. The whole centromeric region is usually heterochromatic, methylated, and exhibiting reduced homologous recombination. The core centromere is characterized by tandem arrays of a ~180-bp satellite repeat and centromeric variant of histone H3 (CENP-A). Heterochromatic pericentromere regions are largely composed of transposons, retrotransposons and pseudogenes, as well as several transcribed genes. Centromeric 180-bp satellites have been isolated and their chromosomal localization analyzed in several crucifer species (Berr et al., 2006; Hall et al., 2005; Kawabe and Nasuda, 2005, 2006; Maluszynska and Heslop-Harrison, 1993). The overall picture emerging from these studies is determined by two partly diverse trends. The rapid diversification of centromeric satellites toward species- and chromosome-specific repeats is opposed by the occurrence of the same repeat type in some congeneric species (*Arabidopsis* spp., *Brassica* spp.). However, the accelerated diversification of centromeric tandem repeats is obviously a prevailing evolutionary pathway observed in different plant lineages (Carroll and Straight, 2006; Ma et al., 2007). The same repeat shared among species is indicative of recent speciation events that were not yet accompanied by sequence divergence of centromeric satellites.

The best-known crucifer centromeric satellite is the 180-bp repeat (pAL) localized in all centromeres of *Arabidopsis thaliana* (Maluszynska and Heslop-Harrison, 1991; Murata et al., 1994). The pAL repeat is species-specific and has not been identified in closely related taxa, including other $n = 8$ *Arabidopsis* species (Heslop-Harrison et al., 2003; Kawabe and Nasuda, 2005; Maluszynska and Heslop-Harrison, 1993). Only in the allotetraploid species *A. suecica* ($2n = 26$) are the pAL and a homologous 180-bp satellite pAa isolated in *A. arenosa* present at 10 and 16 centromeres, respectively (Comai et al., 2003; Kamm et al., 1995). Both centromeric satellites were successfully used as a cytogenetic tool kit for identifying *A. thaliana*- and *A. arenosa*-derived chromosomes in natural and synthetic *A. suecica* lines (Comai et al., 2003; Madlung et al., 2005; Pontes et al., 2004;

Wang et al., 2006), and are major components of genomic DNA probes mediating species-specific chromosome discrimination in *A. suecica* by GISH (Ali et al., 2004).

Various centromeric repeats found in other *Arabidopsis* species are indicative of a more complex evolution at centromeres of these species. pAge1 and pAge2 repeats, together with the pAa repeat, have been revealed in three different subspecies of *A. lyrata* and two subspecies of *A. halleri* (Berr et al., 2006; Kawabe and Nasuda, 2005, 2006; Kawabe and Charlesworth, 2007). Any two of the three repeat types were found to be co-localized in some centromeres of *A. halleri* and *A. lyrata* (Berr et al., 2006; Kawabe and Nasuda, 2005). The dynamic nature of centromeric satellite repeats in *Arabidopsis* species is underlined by the significant intraspecific variation in chromosomal (co) localization of the three satellites in different populations of *A. halleri* (subsp. *gemmifera*) (Kawabe and Nasuda, 2006). Because pAa is found in all three *Arabidopsis* species analyzed, it is most likely the eldest centromeric repeat shared by these species (Berr et al., 2006; Kawabe and Charlesworth, 2007). Furthermore, the occurrence of the three repeat types in *A. halleri* and *A. lyrata* strongly suggests their close phylogenetic relationship, and a more distant relationship of both species to *A. arenosa* (Lysak and Lexer, 2006). Chromosome-specific satellite types were also isolated and localized by FISH in *Olimarabidopsis pumila* ($2n = 32$, Camelineae) and *Sysimbrium irio* ($2n = 28$, Sysimbrieae) by Hall et al. (2005).

In *Brassica*, two 176-bp centromeric satellite repeats sharing approximately 80% sequence homology have been isolated in *B. rapa* (Harrison and Heslop-Harrison, 1995; Koo et al., 2004; Lim et al., 2005). In *B. rapa* (AA genome, $2n = 20$), the two satellites, called CentBr1 and CentBr2, reside at eight and two centromeres, respectively (Lim et al., 2005). Both CentBr1 and CentBr2 were also found in *B. oleracea* (CC genome, $2n = 18$), with the former repeat hybridizing to all centromeres and the latter localized on five chromosomes. Neither of the CentBr satellites have been detected in *B. nigra* (BB genome, $2n = 16$); however, a 329-bp tandem repeat cloned from *B. nigra* showed B-genome specificity also in the AABB genome of *B. juncea* ($2n = 36$) (Schelfhout et al., 2004). The expected number of chromosomes with CentBr1 and CentBr2 repeats corresponding to AA and CC genomes were revealed in the allotetraploid *Brassica* species (Lim et al., 2007). The species-specific differences in the presence of the centromeric repeats is reflecting the phylogenetic split within the genus *Brassica* into the Rapa/Oleracea (*B. rapa* and *B. oleracea*) and the Nigra (*B. nigra*) lineages, respectively (Lysak et al., 2005; Warwick and Black, 1991; Warwick and Sauder, 2005). Similar 176-bp repeats were also identified in other genera of the tribe Brassiceae, including *Diplotaxis*, *Raphanus* and *Sinapis* (Snowdon, 2007, and references therein).

Despite the rapid evolution of centromeric satellite repeats, the same repeat type shared within a species group is pointing to a common ancestry of the group. Hence, centromere-specific repeats can be considered phylogenetically informative cytogenetic markers, corroborating molecular phylogenetic reconstructions or resolving conflicting phylogenetic signals.

Ribosomal DNA (rDNA) is organized in tandem repeated arrays of 5.8S, 18S, and 25S rRNA, and 5S rRNA genes (Neves et al., 2005; Volkov et al., 2007). 5.8S, 18S, and 25S rRNAs are commonly referred to as 45S rDNA in plant cytogenetic literature (a.k.a. 25S or 35S rDNA). Both 5S and 45S rDNA tandem arrays comprising from hundreds to thousands of copies per genome are localized on one or more chromosome pairs. In *Brassicaceae*, rDNA arrays correspond to specific heterochromatic regions revealed upon DAPI staining. 5S rDNA loci can be revealed as heterochromatic knob-like structures (chromomeres) at interstitial or terminal positions on DAPI-stained pachytene chromosomes. Similarly, nucleolar organizing regions (NORs) harboring 45S rDNA arrays can be located at interstitial positions; however, more frequently they are positioned terminally. Terminal NORs, having a conspicuous heterochromatic structure visible on meiotic as well as mitotic chromosomes after conventional or DAPI staining, serve as important chromosomal landmarks. However, interstitial rDNA heterochromatic regions are usually not discernible on condensed mitotic and meiotic (from diplotene onwards) chromosomes (cf. Ali et al., 2005).

rDNA repeats were among the first FISH probes used to characterize DAPI-stained mitotic and meiotic karyotypes and identify individual chromosome types in *Arabidopsis thaliana* and

Brassica species (e.g., Armstrong et al., 1998; Fransz et al., 1998; Hasterok et al., 2001; Howell et al., 2002; Koo et al., 2004; Lim et al., 2005; Maluszynska and Heslop-Harrison, 1993; Snowdon et al., 2002). Later, fluorescently labeled rDNA probes were used to characterize karyotypes in other crucifer species (Ali et al., 2005; Berr et al., 2006; Kantama et al., 2007; Kawabe and Nasuda, 2006; Pontes et al., 2004; Schrader et al., 2000). The use of rDNA probes in comparative cytogenetics of *Brassicaceae* is limited, as shown by several recent studies. The number and localization of rDNA loci (especially 45S rDNA) are prone to intraspecific variation, making cross-species comparisons problematic. Intraspecific variation in the number of rDNA loci has been reported in *A. thaliana* (Fransz et al., 1998; Sanchez-Moran et al., 2002) and *Brassica* species (Armstrong et al., 1998; Hasterok and Maluzsynska, 2000; Kamisugi et al., 1998), but a stable rDNA pattern was revealed among five different populations of *A. halleri* (Kawabe and Nasuda, 2006). In some genera and groups, the uniformity of rDNA patterns does not provide a sufficient level of variation for cross-species comparisons. Furthermore, the number and localization of rDNA arrays are not correlated with molecular phylogenetic relationships within the family and cannot be used to infer the direction of genome and chromosomal evolution (Ali et al., 2005; Berr et al., 2006; Lysak and Lexer, 2006).

This holds true also for the origin of polyploid species, which should theoretically inherit a sum of rDNA loci from the respective parental genomes and the rDNA-bearing parental chromosomes should be identifiable. Nevertheless, this assumption was generally proven to be false as the actual number of rDNA loci in polyploids often differs from the expected one. Hence, the identification of rDNA-bearing parental chromosomes within polyploid genomes can be ambiguous as shown in *Brassica* allopolyploids (e.g., Hasterok et al., 2001; Snowdon et al., 2002). In some crucifer polyploids, the number of rDNA loci is significantly decreased up to diploid-like numbers (Ali et al., 2005). The study by Pontes et al. (2004) helped illuminate the potential causes of the rDNA instability revealed in *Brassicaceae* polyploids. In the natural plants of *Arabidopsis suecica*, they observed one pair of *A. thaliana* NOR-bearing chromosomes and one pair of *A. arenosa*-derived 5S rDNA missing. On the contrary, in the synthetic allotetraploid plants, pairs of *A. thaliana* NORs are gained *de novo*, lost, and/or transposed to *A. arenosa* chromosomes, and an *A. arenosa*-derived 5S rDNA locus is lost.

We conclude that rDNA repeats should be viewed as valuable chromosomal landmarks, useful for chromosome identification within a species or between phylogenetically closely related taxa. In cross-species comparisons, the true chromosome homeology inferred from the presence/absence of rDNA must be corroborated by other cytogenetic markers less prone to homoplasy. The frequent intraspecific variation in number and location of rDNA loci precludes the physical localization of rRNA genes as a reliable characteristic in the phylogenetics of *Brassicaceae*.

OTHER REPEATS AS CYTOGENETIC MARKERS

Repetitive elements other than centromeric satellites and rDNA repeats have been used only rarely as chromosomal markers in comparative cytogenetics of *Brassicaceae*. *Arabidopsis*-type telomeric repeat was localized in all six *Brassica* species of U's triangle. The probe labeled chromosome termini in all species analyzed with no apparent signals at interstitial positions (Hasterok et al., 2005). Recently, a telomere-like repeat was cloned from *B. oleracea* (dos Santos et al., 2007). FISH analysis of the repeat showed its localization at terminal as well as presumably interstitial positions on *B. oleracea* chromosomes, whereas there was almost no hybridization to chromosomes of *B. rapa*. In *B. napus* ($2n = 38$), the repeat marked 18 *B. oleracea*-derived chromosomes and up to 6 *B. rapa*-derived chromosomes when a total of 18 to 24 chromosomes were labeled. However, applying FISH to small and condensed mitotic chromosomes does not provide sufficient detail to evaluate the exact organization of presumed interstitial sites along *B. oleracea* chromosomes. In *Arabidopsis*, chromosomal positions of interstitial telomere and telomere-like repeats (Armstrong et al., 2001; Uchida et al., 2002) do not coincide with the chromosomal breakpoints involved in the origin of the *Arabidopsis* karyotype ($n = 5$) from the Ancestral Crucifer Karyotype ($n = 8$; Lysak et al., 2006a).

It remains to be seen if future cytomolecular studies in other crucifer species will reveal interstitial telomere-like loci as signatures of ancient chromosomal breakpoints (chromosome fusions).

Plant transposable elements (TEs) are generally divided into two major groups according to their transposition intermediate. Retrotransposons are amplified via an RNA intermediate and further classified as non-LTR-retrotransposons (long/short interspersed nuclear elements: LINEs/SINEs) and LTR-retrotransposons, which are further subdivided into Ty1-*copia* and Ty3-*gypsy* elements. DNA transposons transpose via a DNA intermediate. Among plant TEs, LTR-retrotransposons are considered the most abundant component of plant genomes responsible for the extensive genome size variation (Vitte and Bennetzen, 2006). Compared to centromeric satellite repeats undergoing rapid sequence divergence, many LTR-retrotransposons are shared within a genus, tribe, or family due the sequence conservation of their coding domains. Hence, the ubiquity of retroelements can be used as a phylogenetically informative marker reflecting a shared ancestry of a given lineage. On the other hand, this same ubiquity precludes many retrotransposons from being used as chromosome- or species-specific cytogenetic markers, as retroelement probes often label chromosome complements within a group of related species uniformly.

In addition to *Arabidopsis*, major progress in molecular and cytogenetic characterization of TEs has been made in *Brassica*. In *B. rapa*, Lim et al. (2007) identified centromere-specific LTR-retrotransposons (CRB) and pericentromere-specific Ty3-*gypsy*-like retrotransposons (PCRBr). The CRB retrotransposons were found nested within the centromeric CentBr satellite arrays in all six species of the U's triangle. In contrast, the PCRBr localized only in three chromosomes of *B. rapa*, whereas U's was absent in *B. oleracea* and *B. nigra*, and is apparently an A-genome-specific retrotransposon. Interestingly, the PCRBr retroelement has been found to label additional chromosomes in the *Brassica* allotetraploids, including B- and C-genome-derived chromosomes (Lim et al., 2007).

Alix et al. (2005) investigated the abundance and chromosomal distribution of Ty1-*copia*-like and Ty3-*gypsy*-like retrotransposons (including one Athila-related element) of *Brassica oleracea* by FISH in *B. oleracea* and *B. rapa*. The reverse transcriptase probes of the analyzed retroelements provided specific hybridization signals along *B. oleracea* chromosomes. One *copia* probe hybridized along the full length of chromosomes with some interstitial clustering, whereas another *copia* element (interestingly labeling only 16 chromosomes), one *gypsy* probe, and an Athila-like *gypsy* probe hybridized along entire chromosomes with strong signals at (peri)centromeric regions.

The present data on the use of TEs as cytogenetic or phylogenomic markers in *Brassicaceae* are too scant. Comprehensive cross-species analyses of genome and chromosomal distribution of TEs within phylogenetic frameworks are lacking. However, some data indicating that the same repertoire of TEs is shared by all *Brassica* species (Alix et al., 2004) as well as by Camelineae (*Arabidopsis*) and Brassiceae (*Brassica*) (Zhang and Wessler, 2004), challenge the existence of taxon/lineage-specific transposable elements.

BAC FISH IN *BRASSICA* AND WILD CRUCIFERS

As shown in *Arabidopsis*, BAC libraries are the most common source of large-insert clones in today's cytogenetics of crucifers. In addition to *Brassica* species, BAC libraries have been constructed for only a few crucifer species, including *Arabidopsis arenosa*, *Arabis alpina*, *Boechera holboellii*, *B. stricta*, *Capsella rubella*, *Olimarabidopsis pumila,* and *Sisymbrium irio* (Hall et al., 2005; Schranz et al., 2007a). Consequently, exploiting BAC libraries in species other than *Arabidopsis* has been very limited. In the cytogenetics of non-*Arabidopsis* crucifers, BAC libraries were used in two principal ways: (1) as a source of repetitive elements and/or (2) chromosome-specific probes containing coding and low-copy sequences. In centromere research, BAC clones containing centromere-specific satellite repeats and retrotransposons were successfully used to analyze the structure of (peri)centromeric regions in *B. rapa* (Lim et al., 2007), *A. arenosa*, *C. rubella*, *O. pumila*, and *S. irio* (Hall et al., 2005, 2006), including FISH localization on chromosomes.

The extensive use of BAC clones in cytogenetic mapping has been shown for *Brassica* species. Howell et al. (2002, 2005) successfully assigned nine linkage groups in *B. oleracea* ($n = 9$) to the corresponding chromosomes using chromosome-specific BACs. Immense progress has been achieved in cytomolecular mapping in *B. rapa* within the framework of the Korea *Brassica* Genome Project (Lim et al., 2006; Yang et al., 2005). Two *B. rapa* subsp. *pekinensis* BAC libraries serve as a source of chromosome-specific probes for the integration of genetic and cytogenetic maps, and *in silico* comparisons with the *Arabidopsis* genome (Yang et al., 2005). Yang et al. (2006) identified *B. rapa* BACs containing paralogs of *Arabidopsis Flowering Locus C* (*FLC*) genes, which were localized on *B. rapa* chromosomes by BAC FISH. It is envisaged that ongoing genome sequencing initiatives along with decreasing costs of genome analyses will facilitate the generation of BAC libraries and their wider application in crucifer cytogenetics.

COMPARATIVE GENETIC MAPPING

Comparative genetic maps are an invaluable source of information on the extent of interspecific genome and chromosomal colinearity as well as major chromosomal rearrangements differentiating two or more genomes. In crucifer genomics, comparative genetic mapping studies significantly facilitated cross-species cytogenetic analyses. *Arabidopsis* sequence data along with the wealth of genetic markers have been crucial in placing *A. thaliana* as a central reference genome in crucifer comparative genetics. Comparative genetics in *Brassicaceae* has been reviewed on several occasions (e.g., Koch and Kiefer, 2005; Lysak and Lexer, 2006; Schmidt et al., 2001; Snowdon, 2007; Quiros, 1999); here we focus on genetic mapping studies having a direct relevance for comparative cytogenetics of *Brassicaceae*.

Early mapping studies showed extensive genome colinearity among the three "diploid" *Brassica* species (*B. nigra*, $n = 8$; *B. oleracea*, $n = 9$; *B. rapa*, $n = 10$), and between *Arabidopsis thaliana* and *B. nigra* (Lagercrantz and Lydiate, 1996; Lagercrantz, 1998). The latter comparison indicated that the *B. nigra* genome (and genomes of the other two *Brassicas*) comprises three copies of an ancestral genome as a consequence of a presumed ancient hexaploidy event. However, subsequent *Arabidopsis-B. oleracea* comparisons (Babula et al., 2003; Lan et al., 2000; Li et al., 2003; Lukens et al., 2003) were inconclusive as to the nature of the genome duplication found in *Brassica*. In addition to numerous triplicated chromosome segments, single, duplicated, and multiple (up to seven) syntenic regions were also detected in the *B. oleracea* genome, thus challenging the purported hexapolyploid origin of *Brassica* species (Li et al., 2003; Lukens et al., 2003, 2004). Only recently has there been convincing evidence of large-scale triplication within the parental genomes comprising the allotetraploid *B. napus* genome (AACC, $n = 19$) (Parkin et al., 2005). At least 21 *Arabidopsis* genomic blocks that can be duplicated and rearranged to establish the extant *B. napus* genome were revealed by mapping a large set of *B. napus* RFLP markers in *A. thaliana*. The structure and number of duplicated conserved blocks within the *B. napus* genome strongly suggest that "diploid" *Brassica* species have undergone an ancient whole-genome triplication (Parkin et al., 2005; Schranz et al., 2006), discussed here at greater length in connection with comparative painting studies (see "Ancient polyploidy revealed by comparative cytogenetic analysis").

Although the *Arabidopsis-Brassica* genetic comparisons provided much-needed insight into the genome evolution of *Brassicas*, the triplicated structure of *Brassica* genomes did not permit reaching any clear-cut conclusion as to chromosome number, genome structure, and evolution of an ancestral genome common to *Arabidopsis*, *Brassica*, and other crucifers. This has changed with the publication of two comparative genetic maps between *Arabidopsis* and two $n = 8$ Camelineae species: *A. lyrata* subsp. *petraea* (Kuittinen et al., 2004) and *Capsella rubella* (Boivin et al., 2004). Later, both maps were compared to the results of genetic mapping between *Arabidopsis* and *A. lyrata* subsp. *lyrata* (n = 8; Yogeeswaran et al., 2005). Both *A. lyrata* subspecies are reported as diploid ($2n = 16$) and tetraploid ($2n = 32$) cytotypes, whereby subsp. *petraea* is predominantly distributed in Europe and subsp. *lyrata* in North America (Al-Shehbaz and O´Kane, 2002). *C. rubella*

($2n = 16$), an annual self-fertile species occurring mainly in the Mediterranean, has been chosen for genetic mapping as the genus *Capsella* was shown to be phylogenetically very closely related to the genus *Arabidopsis* (Koch et al., 1999, 2001).

In contrast to *Arabidopsis-Brassica* maps (see above), eight linkage groups of *Arabidopsis lyrata* and *Capsella rubella* showed a strikingly high extent of colinearity with the five *Arabidopsis* chromosomes (Boivin et al., 2004; Kuittinen et al., 2004; Yogeeswaran et al., 2005). Despite some discrepancies, particularly in the number of inferred inversion events differentiating all three Camelineae karyotypes caused by the different marker density, all three maps were largely congruent (Koch and Kiefer, 2005; Yogeeswaran et al., 2005; see also Lysak and Lexer, 2006). The three-way (*Arabidopsis–A. lyrata–C. rubella*) comparison concluded that karyotypes of *A. lyrata* and *C. rubella* are, in fact, identical and differ from the *Arabidopsis* karyotype by two reciprocal translocations, three chromosome fusions, and at least three major inversions (Koch and Kiefer, 2005). These rearrangements mediated the reduction in chromosome number in *A. thaliana* from an ancestral $n = 8$ karyotype resembling karyotypes of *A. lyrata* and *C. rubella*. Although the three fusions *Arabidopsis* chromosomes were identified, the mechanism of these fusions remained elusive. The sequence of chromosome rearrangements involved in chromosome fusions was explained only later based on cytogenetic (Lysak et al., 2006a) and genetic data (Kawabe et al., 2006a, b).

The overall similarity of *Capsella* and *Arabidopsis lyrata* karyotypes (Boivin et al., 2004; Koch and Kiefer, 2005; Kuittinen et al., 2004; Yogeeswaran et al., 2005) indicated that they may resemble an ancestral $n = 8$ karyotype shared by Camelineae species, including *Arabidopsis*. Hence, these genetic maps served as a basis for the first comparative chromosome painting studies in *Brassicaceae*.

COMPARATIVE CHROMOSOME PAINTING (CCP)

Comparative chromosome painting (CCP) refers to the FISH-based visualization of large homeologous chromosome regions and/or whole chromosomes shared by two or more species. The extent of shared chromosomal homeology reflects interspecies relatedness and the character of chromosomal rearrangements generating the extant karyotypic variation. In *Brassicaceae*, fluorescently labeled chromosome-specific *Arabidopsis* BAC contigs have been shown as feasible painting probes to unveil homeologous chromosomal regions in other crucifer species (Lysak et al., 2003). Currently, *Brassicaceae* is the only plant family in which CCP is feasible due to the available *Arabidopsis* resources and the specific organization of repetitive DNAs in most crucifer taxa. Only *Arabidopsis* BACs are being used as painting probes for CCP, as chromosome-specific BAC libraries are not available for other crucifer species. This situation will most likely change with the ongoing sequencing projects in *Brassica rapa* (Yang et al., 2005), *Arabidopsis lyrata,* and *Capsella rubella* (Lysak and Lexer, 2006), as well as in other crucifers, providing contigs of anchored BAC clones with a low percentage of dispersed repetitive elements potentially suitable for CCP.

Technically, *Arabidopsis* BAC DNA is labeled by a fluorochrome or DNA hapten and then pooled into a BAC contig, or BAC DNAs are pooled first and labeled in the next step (see Lysak et al., 2006b, for details). Heterologous *in situ* hybridization of *Arabidopsis* BAC contigs to (most frequently) pachytene chromosomes or extended DNA fibers of another crucifer species reveal shared homeologous chromosome regions/chromosomes. *Arabidopsis* BAC contigs *a priori* comprise clones free of dispersed repeats, positioned along chromosome arms. Therefore, painting probes not including BACs from *Arabidopsis* pericentromere regions cannot discern homeologous centromeres in other cruciferous species. As all BACs possess exact chromosome coordinates, the clones may be combined and differently labeled according to a required experimental scheme.

Initial CCP analyses were confined to the use of several BAC clones (Jackson et al., 2000) up to BAC contigs comprising more than a hundred clones (Comai et al., 2003; Lysak et al., 2003). Jackson et al. (2000) localized six *Arabidopsis* BAC clones (431 kb) on mitotic chromosomes as well as DNA fibers of *Brassica rapa*. The BAC contig labeled four to six *B. rapa* chromosomes, indicating a duplication of the *Brassica* genome proven by large-scale CCP some years later (Lysak et al.,

2005; Ziolkowski et al., 2006). Cross-species BAC painting on DNA fibers was followed later by the localization of eight *Arabidopsis thaliana* BACs on DNA fibers in three other *Arabidopsis* species to study the expansion of the *Arabidopsis* FLC region (Sanyal and Jackson, 2005). Both studies showed the untapped potential of CCP on DNA fibers in comparative cytogenetics of *Brassicaceae*.

In the first large-scale CCP analyses, a BAC contig covering both arms of At4 was hybridized to pachytene chromosomes of nine species with $n = 8$, 13 (*A. suecica*) and $n = 16$ assigned to Camelineae and Arabideae (Comai et al., 2003; Lysak et al., 2003). CCP analysis revealed grossly conserved chromosome homeology shared by all the analyzed species as well as species-specific chromosome reshuffling in *Arabis alpina* and *Crucihimalaya walichii*, apparently congruent with the currently established phylogenetic position of Arabideae (vs. Camelineae) and the polyphyletic character of Camelineae, respectively (Bailey et al., 2006; Koch and Al-Shehbaz, 2008).

RECONSTRUCTING KARYOTYPE EVOLUTION IN BRASSICACEAE: ANCESTRAL CRUCIFER KARYOTYPE

Compared to many other crucifer species and groups, the reduced *Arabidopsis* karyotype ($n = 5$) has an evolutionary-derived structure characterized by specific chromosome rearrangements (Koch and Kiefer, 2005; Lysak et al., 2006a). Inevitably, each comparison to the *Arabidopsis* karyotype is complicated by accounting for these species-specific rearrangements. Therefore, a karyotype having more ancestral structure — and thus more suitable as a reference point for cross-species comparisons — has been needed.

The overall similarity between karyotypes of *Arabidopsis lyrata* and *Capsella Rubella*, revealed by comparative genetic mapping and the fact that x = 8 is the most common base number found in Camelineae as well as across *Brassicaceae* (Warwick and Al-Shehbaz, 2006), became a basis for the concept of a hypothetical ancestral karyotype with eight chromosomes ($n = 8$) shared by *Arabidopsis*, Camelineae, and the closest phylogenic clades (Lysak et al., 2006a). The concept of a fictitious extinct species *Crucifera ancestralis* possessing Ancestral Crucifer Karyotype (ACK) has been successfully adopted in the cytogenetic analysis of karyotype evolution in species with presumably reduced chromosome numbers from $n = 8$ toward $n = 7$, 6, and 5, including *Arabidopsis* (Lysak et al., 2006a). The study aimed to (1) corroborate comparative genetic data (i.e., *A. lyrata* and *C. rubella* maps), (2) unveil the extent of chromosomal colinearity shared by species with diverse chromosome numbers, and (3) identify mechanisms of karyotypic changes causing the extant diversity in chromosome numbers ($n = 5$–8). CCP analysis using *A. thaliana* BAC contigs, arranged according to the eight linkage groups (or parts thereof) of *A. lyrata* and *C. rubella*, revealed homeologous chromosome regions in *A. lyrata*. Not only were comparative genetic data confirmed by CCP, but more importantly, identifying centromere (and NOR) positions allowed for the reconstruction of karyotype evolution in *A. thaliana*. The evolution of the *Arabidopsis* karyotype has been marked by a reduction of chromosome number from $n = 8$ toward $n = 5$ through three chromosome fusions, two reciprocal translocations, and at least three inversions (Figure 10.3; Lysak et al., 2006a; 2008). The reconstruction of karyotype evolution in *A. thaliana* posed a question if a tentative ACK has been shared by other crucifer species with varying chromosome numbers. In *Neslia paniculata* ($n = 7$), *Turritis glabra* ($n = 6$, both Camelineae), and *Hornungia alpina* ($n = 6$, Descurainieae), for which no genetic data were available, CCP revealed largely preserved chromosomal colinearity between these species and ACK (Lysak et al., 2006a). Despite some inversion events, six homeologous chromosomes in *Neslia* and four in *Hornungia* and *Turritis* resembled the structure of ancestral chromosomes. Although some ancestral chromosomes (e.g., AK5) participated in chromosome fusions more often than others (e.g., AK1, AK7), the chromosome fusion events were unique in all three species and *Arabidopsis* (Figure 10.4). Altogether, these data follow the scenario proposed by Jaretzky (1928) some 80 years ago — that chromosome reductions from an ancestral karyotype with $n = 8$ toward evolutionary derived karyotypes with $n = 7$, 6, and 5 were independent and recurrent events.

As no signs for centromere retention were found either by sequence analysis in *Arabidopsis* (AGI 2000) or cytogenetically in *Arabidopsis* and the other taxa (Lysak et al., 2006a), a loss of one to three centromeres has been implied. The centromere elimination has been explained by a reciprocal translocation between a (sub)metacentric and acrocentric (telocentric) chromosome with translocation breakpoints located close to the chromosome ends. The translocation generates a "fusion" chromosome and a minichromosome comprising mainly centromere of the acrocentric and two telomeres. The latter translocation product is supposed to be meiotically unstable and eliminated (Lysak et al., 2006a; Schranz et al., 2006; Schubert, 2007). About 85% breakpoints mediating chromosome rearrangements in the species analyzed by Lysak et al. (2006a) included repeat-rich centromeric and terminal chromosome regions. Ectopic (non-allelic) and intrachromosomal recombinations within the repeat arrays are the most likely molecular mechanisms of the described large-scale rearrangements. In particular, megabase-long tracts of rRNA gene repeats at terminal NORs are suspected as potential sites of translocations (Lysak et al., 2006a; Schubert, 2007). This is supported by the significantly higher association frequency of NOR-bearing chromosomes compared to other chromosomes in interphase nuclei of *Arabidopsis thaliana* (Pecinka et al., 2004) and *A. lyrata* (Berr et al., 2006).

ABCs: The Conserved Blocks of Crucifer Genomes

The ACK concept was further expanded by defining apparently conserved chromosomal blocks that make up individual ancestral chromosomes (Schranz et al., 2006). Conserved ancestral blocks were revealed through interspecific genetic mapping between *Arabidopsis* and *Brassica napus*. Parkin et al. (2005) identified a minimum of 21 conserved chromosomal segments within the *Arabidopsis* genome, which can be duplicated and rearranged to build up the allopolyploid genome of *B. napus*. These chromosomal blocks were largely identical with the colinear chromosomal regions revealed by genetic mapping (e.g., Koch and Kiefer, 2005) and comparative painting (Lysak et al., 2006a) as shared between *Arabidopsis thaliana*, *A. lyrata*, *Capsella rubella*, and other Camelineae and Descurainieae species. Hence, these data have been integrated with Parkin et al.'s block system into a set of 24 (A–X) conserved genomic blocks within the ACK (Schranz et al., 2006; Figures 10.3 and 10.4).

Since the definition of conserved genomic blocks, the importance of ACK as a reference point in comparative phylogenomic studies across *Brassicaceae* has further increased. The reshuffling of crucifer building blocks in the genome of *Boechera stricta* ($n = 7$) was analyzed by constructing a comparative genetic map between *Arabidopsis* and *B. stricta* (Schranz et al., 2007b). In *Boechera*, three chromosomes possess the ancestral structure, whereas the remaining five AK chromosomes were rearranged to form four *Boechera* chromosomes. Interestingly, two or three ancestral blocks (A, C, and D?) were apparently split between two different *Boechera* chromosomes (Figure 10.4). These findings, along with the author's unpublished data, suggest that the proposed set of 24 conserved blocks should not be viewed as a rigid system. On the contrary, the blocks can be split and reshuffled by independent rearrangements using different breakpoints. Further cross-species comparisons will shed light on the fate of ancestral chromosomal blocks in different crucifer taxa.

Ancient Polyploidy Revealed by Comparative Cytogenetic Analysis

With the *Arabidopsis* genome sequenced, large intra- and interchromosomal duplications have been revealed and interpreted as relics of an ancient whole-genome duplication (AGI, 2000; Henry et al., 2006). Although somewhat controversial, the available data suggest that the ancestor of *A. thaliana*, and thus the whole family, underwent three rounds of whole-genome duplication referred to as R1 (γ), R2 (β), and R3 (α) events (Bovers et al., 2003; De Bodt et al., 2005; Henry et al., 2006; Maere et al., 2005). The youngest R3 duplication event occurred approx. 24 to 40 mya (million years ago; Blanc et al., 2003) or 25 to 26.7 mya (Blanc and Wolfe, 2004), after the divergence of *Brassicales* and *Malvales* (represented by cotton) but before the *Arabidopsis-Brassica* split (Adams and Wendel, 2005; Blanc et al., 2003; De Bodt et al., 2005).

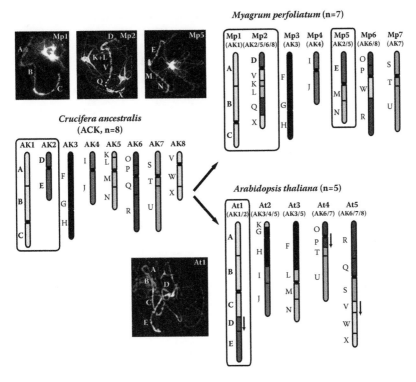

FIGURE 10.3 (See color insert following page 128.) Ancestral Crucifer Karyotype (ACK, *n* = 8; *Crucifera ancestralis*) and the reconstruction of karyotype evolution on the example of *Myagrum perfoliatum* (*n* = 7; Isatideae, Lineage II) and *Arabidopsis thaliana* (*n* = 5; Camelineae, Lineage I) based on CCP analysis. ACK comprises eight chromosomes (AK1–8) and 24 genomic blocks (A–X; Schranz et al., 2006). Chromosome number was reduced from *n* = 8 in ACK toward *n* = 7 in *M. perfoliatum* (Mp1–7) and *n* = 5 in *Arabidopsis* (At1–5), respectively. The reductions have been accompanied by gross inversion and translocation events resulting in three and five "fusion" chromosomes and the loss of one and three centromeres in *M. perfoliatum* and *A. thaliana*, respectively. Examples of multicolor chromosome painting analysis of ancestral chromosomes AK1 and AK2 within pachytene chromosome complements of both species. In *M. perfoliatum*, AK1 (blocks A–C) is preserved as chromosome Mp1, whereas AK2 (blocks D and E) participated in the origin of "fusion" chromosomes Mp2 (AK2/5/6/8) and Mp5 (AK2/5). In Arabidopsis, AK1 and AK2 were combined by a reciprocal translocation into At1 (AK1/2). In ACK, genome blocks are considered to be in the upright orientation. Blocks inverted relative to ACK are represented by downward-pointing arrows; NORs and heterochromatic knobs not indicated. Note that pseudocoloring of fluorescently labeled painting probes in microscopic photographs does not correspond to the color coding of AK chromosomes (block Q has not been included in the painting probe in *M. perfoliatum*). Chromosomes were counterstained by DAPI. (*Source:* Lysak et al., 2006a; Mandáková and Lysak, 2008; Schranz et al., 2006.)

The two ancient events were dated roughly to 100 to 168 mya and 66 to 110 mya, respectively (De Bodt et al., 2005). The most recent polyploidization presumably shared by all crucifers has been probably paralleled by an independent whole-genome duplication event in the ancestry of Cleomaceae, a sister family to *Brassicaceae*. This paleopolyploid event was most likely a whole-genome triplication, younger (ca. 20 mya) than the R3 duplication in *Brassicaceae* (Schranz and Mitchell-Olds, 2006). Current genomic analyses suggest that paleopolyploid events were frequent, occurred independently in different phylogenetic lineages of flowering plants (Cui et al., 2006), and were followed by genome diploidization in repeating cycles (Adams and Wendel, 2005). The extremely compact *Arabidopsis* genome provides an excellent example as to the extent to which paleopolyploid genomes can be eroded by the diploidization process. Genome diploidization acting over a million years has masqueraded the paleopolyploid nature

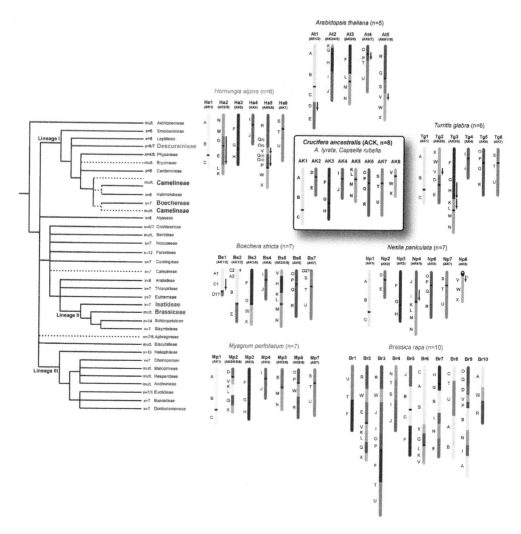

FIGURE 10.4   (See color insert following page 128.) Phylogenetic relationships among tribes of Brassicaceae (modified from Koch and Al-Shehbaz, Chapter 1 in this book) with the placement of schematic karyotypes currently reconstructed by comparative genetic and chromosome painting analysis. Three major phylogenetic lineages (Lineage I–III) are indicated (sensu Beilstein et al. 2006). Sole or prevailing base chromosome number (x) is given for each tribe (mult. = multiple base chromosome numbers). The color coding of ancestral chromosomes (AK1–8) and 24 conserved genomic blocks (A–X; Schranz et al., 2006) indicate the relationship of reconstructed karyotypes to the Ancestral Crucifer Karyotype (ACK), represented by the *A. lyrata* karyotype. In ACK, genomic blocks are considered to be in the upright orientation; blocks inverted relative to ACK are represented by downward-pointing arrows. NORs and heterochromatic knobs not indicated.

of *Brassicaceae* species. Hence, the assessed percentage of polyploid crucifer taxa (ca. 40%), primarily inferred from chromosome numbers (polyploidy defined as $n \geq 14$; Warwick and Al-Shehbaz, 2006), is grossly underestimated.

The most recent paleopolyploid event (R3) revealed in *Arabidopsis* as large-scale intra- and inter-chromosomal segmental duplications cannot be detected cytogenetically by chromosome painting due to the significant divergence of the duplicated regions (Lysak et al., 2001). However, younger mesopolyploid events can be detected by CCP analysis, as shown for the tribe Brassiceae (Lysak et al., 2005, 2007; Ziolkowski et al., 2006). As discussed previously, comparative genetic mapping between *Arabidopsis* and *Brassica* species suggested that *Brassica* genomes were triplicated via an ancient hexaploidization event (e.g., Lagercrantz, 1998; Parkin et al., 2005). CCP analysis of the chromosomal block U of the ACK revealed triplication of the block in a number of Brassiceae species, including *bona fide* diploid species with low chromosome numbers ($2n$ = 14, 16, 18; Lysak et al., 2005). The hexaploidization event has been dated as occurring 8 to 15 mya (Lysak et al., 2005) or 13 to 17 mya (Yang et al., 2006), after the *Brassica-Arabidopsis* split (24 mya, Lysak et al., 2005; 17 to 18 mya, Yang et al., 2007). In *B. oleracea* and ten species traditionally treated as members of Brassiceae, the ancient triplication was identified by CCP as three copies (or three copies fused as two chromosomes) of other ancestral genomic blocks, respectively (Lysak et al., 2007; Ziolkowski et al., 2006). A whole-genome duplication, but not the Brassiceae-specific triplication, has been identified in the genome of *Orychophragmus violaceus* ($2n$ = 24) (Lysak et al., 2007), repeatedly assigned to Brassiceae.

A similar whole-genome duplication was recently revealed by CCP analysis in some endemic Australian Camelineae species with diploid-like chromosome numbers (n = 4–6) (Mandáková and Lysak, unpublished). Altogether, cytogenetic data suggest that such whole-genome duplication events might be more common than previously thought, even in *bona fide* diploid species. Future CCP analyses will probably unearth more independent whole-genome duplications in other crucifer lineages and clades.

As demonstrated above, three levels of polyploidy can be recognized in the evolution of *Brassicaceae*. The ancient whole-genome duplications (*paleopolyploidy*) that occurred 25 mya or more are detected only by sequence comparisons but not by cytogenetic methods. Duplications detectable cytogenetically (*mesopolyploidy*) are younger than the paleopolyploid events (ca. 8 to 17 mya for the Brassiceae-specific triplication). The most recent polyploid events (*neopolyploidy*), including both autopolyploidy and allopolyploidy, are easily recognized by increased chromosome numbers; and in allopolyploids, the parental genomes can be revealed by molecular markers and by GISH. The origin of several crucifer neopolyploids is usually dated to the Pleistocene (Marhold and Lihová, 2006), 1.8 million to 11,500 years before present. For example, the origin of the allotetraploid *Arabidopsis suecica* has been dated to between 50,000 to 10,000 years ago (Jakobsson et al., 2007).

KARYOTYPE EVOLUTION IN BRASSICACEAE WITHIN THE PHYLOGENETIC CONTEXT

Considering that the *Brassicaceae* family comprises some 3700 species, we have information on the karyotype structure of only a handful of species with a bias toward the tribe Camelineae. Despite this limitation, some preliminary conclusions on karyotype evolution could be drawn from comparisons of cytogenetic data with current phylogenetic hypotheses (Bailey et al., 2006; Beilstein et al., 2006; Koch et al., 2007; Koch and Al-Shehbaz, 2008 [Chapter 1]; Warwick et al., 2007). The ACK has been shown as a probable ancestral karyotype of Camelineae with species-specific scenarios of chromosome number reduction (Lysak et al., 2006a), perhaps corresponding to the polyphyletic character of Camelineae (Bailey et al., 2006; Koch and Al-Shehbaz, 2008 [Chapter 1]). The ACK was also proven as an ancestral genome for Boechereae (Schranz et al., 2007) and Descurainieae (Lysak et al., 2006a). These findings suggest that all clades (tribes) within Lineage I descended from the ACK (Figure 10.4).

Although conserved genomic blocks shared between Lineage II (Beilstein et al., 2006) represented by *Brassica napus* (Brassiceae; Parkin et al., 2005) and Lineage I (Camelineae) were clearly identified, the number of ancestral Brassiceae chromosomes remains elusive due to the complex mesopolyploid history of this tribe (Lysak et al., 2005, 2007; Parkin et al., 2005; Yang et al., 2006). Nonetheless, currently acquired CCP data suggest the ACK as an ancestral genome of Sisymbrieae and Isatideae, and thus most likely of the whole Lineage II (Figure 10.4), as well as closely related tribes Calepineae, Conringieae, Eutremeae, and Noccaeeae (Mandáková and Lysak, 2008).

Currently, no genetic or cytogenetic data are available for the tribes comprising Lineage III. Also, it remains to be tested if CCP using *Arabidopsis* BAC contigs is applicable to Aethionemeae, which is a sister group to the remaining *Brassicaceae* tribes (Figure 10.4), and whether the ACK can be identified as an ancestral genome shared by the whole family.

Phylogenetic Significance of Cytogenetic Signatures

Current genomic research on wild crucifers represents a disciplinary crossroads between molecular cytogenetics, genomics, and phylogenetics termed cytogenomics or phylogenomics (Dobigny and Yang, 2008; Lysak and Lexer, 2006). Sequence-based phylogenetic trees play an invaluable role in the navigation of cytogenetic analyses, and cytogenomic data (cytogenetic signatures) overlaid onto phylogenetic trees may provide novel markers corroborating the existing phylogenies or resolving conflicting phylogenetic signals. Chromosome rearrangements show only a low tendency toward convergent evolution and are generally considered rare genomic changes (RGCs; Rokas and Holland, 2000), although rare reversals and convergence due to the reuse of the same breakpoints (Dobigny and Yang, 2008) has not been investigated in plants and cannot be ruled out. The parsimony assumption implies that associations of conserved blocks can be disrupted by rearrangements occurring independently in different lineages, whereas the same combination of chromosomal blocks (syntenic association) in two independent lineages is unlikely (Faraut, 2008).

In *Brassicaceae*, even the scant data on karyotype evolution provided some cytogenetic signatures supporting and redefining the proposed phylogenetic topologies and intrafamiliar taxa. For example, the whole-genome triplication revealed by genetic and CCP analysis in *Brassica* and other Brassiceae species (Lysak et al., 2005, 2007; Parkin et al., 2005; Ziolkowski et al., 2006) is an RGC characterizing the whole tribe. Cytogenetic data were congruent with phylogenetic analysis using a cpDNA marker indicating the monophyly of Brassiceae. This clade-specific signature has not been identified in closely related tribes such as Calepineae, Conringieae, and Sisymbrieae (Lysak et al., 2005), as well as in the genus *Orychophragmus* (Lysak et al., 2007), suggesting its exclusion from Brassiceae. Within Brassiceae, a reciprocal translocation transposing two of the three homeologous copies of the analyzed block onto one chromosome is a rearrangement apparently specific for the subtribe Zillineae (Lysak et al., 2007). In tribes Brassiceae, Isatideae, and Sisymbrieae of Lineage II, and closely related tribes Calepineae, Conringieae, Eutremeae, and Noccaeae, comparative painting analysis revealed the syntenic association of blocks V, K, L, Q, and X from ancestral chromosomes AK5, AK6, and AK8 (Mandáková and Lysak, 2008). This block association, exemplified here by chromosome Mp2 in *M. perfoliatum* and chromosomes Br2 and Br6 (with a secondary rearrangement in Br6) in *Brassica rapa* (Figure 10.4), may be a unique signature shared by Lineage II and several tribes within the currently unresolved polytomy suggesting their common ancestry.

Future comparative cytomolecular studies in not yet analyzed crucifer species and groups should uncover other phylogenetically informative cytogenetic signatures.

ACKNOWLEDGMENTS

The author thanks T. Mandáková, P. Mokroš, T. Dierschke, B. Krizek, K. Mummenhoff, and M. Koch for their help with the preparation of figures and sharing unpublished data. I am indebted to P. Bureš for his help with genome size estimation (Table 10.7) and K. Matůšová for technical assistance. The work was supported by grants from the Czech Ministry of Education (MSM0021622415) and the Grant Agency of the Czech Academy of Science (KJB601630606).

REFERENCES

Adams KL, Wendel JF (2005) Polyploidy and genome evolution in plants. *Current Opinion in Plant Biology,* 8:135–141.

AGI, The Arabidopsis Genome Initiative (2000) Analysis of the genome sequence of the flowering plant *Arabidopsis thaliana. Nature* 408:796–815.

Albini SM (1994) A karyotype of the *Arabidopsis thaliana* genome derived from synaptonemal complex analysis at prophase I of meiosis. *Plant J.* 5:665–672.

Al-Shehbaz IA, O'Kane SL Jr. (2002) Taxonomy and phylogeny of *Arabidopsis* (*Brassicaceae*). In: Somerville CR, Meyerowitz EM (Eds.), *The Arabidopsis Book,* American Society of Plant Biologists, Rockville, MD, pp. doi/10.1199/tab.0001, http://www.aspb.org/publications/arabidopsis/.

Al-Shehbaz IA, Beilstein MA, Kellogg EA (2006) Systematics and phylogeny of the *Brassicaceae* (*Cruciferae*): an overview. *Plant Syst Evol* 259:89–120.

Ali HBM, Lysak MA, Schubert I (2004) Genomic *in situ* hybridization in plants with small genomes is feasible and elucidates the chromosomal parentage in interspecific *Arabidopsis* hybrids. *Genome* 475:954–960.

Ali HBM, Lysak MA, Schubert I (2005) Chromosomal localization of rDNA in the *Brassicaceae*. *Genome* 48:341–346.

Alix K, Heslop-Harrison JS (2004) The diversity of retroelements in diploid and allotetraploid *Brassica* species. *Plant Mol Biol* 54:895–909.

Alix K, Ryder C, Moore J, King GJ, Heslop-Harrison JS (2005) The genomic organization of retrotransposons in *Brassica oleracea*. *Plant Mol Biol* 59:839–851.

Ambros P, Schweizer D (1976) The Giemsa C-banded karyotype of *Arabidopsis thaliana* (L.) Heynh. *Arabidopsis Inf. Serv.* 13:167–171.

Armstrong SJ, Fransz P, Marshall DF, Jones GH (1998) Physical mapping of DNA repetitive sequences to mitotic and meiotic chromosomes of *Brassica oleracea* var. *alboglabra* by fluorescence *in situ* hybridisation. *Heredity* 81:666–673.

Armstrong SJ, Franklin FCH, Jones GH (2001) Nucleolus associated telomere clustering and pairing precede meiotic chromosome synapsis in *Arabidopsis thaliana*. *J Cell Sci* 114:4207–4217.

Armstrong SJ, Jones GH (2003) Meiotic cytology and chromosome behaviour in wild-type *Arabidopsis thaliana*. *J Exp Bot* 54:1–10.

Babula D, Kaczmarek M, Barakat A, Delseny M, Quiros CF, Sadowski J (2003) Chromosomal mapping of *Brassica oleracea* based on ESTs from *Arabidopsis thaliana*: complexity of the comparative map. *Mol Genet. Genomics* 268:656–665.

Bailey CD (2001) Systematics of *Sphaerocardamum* (*Brassicaceae*) and Related Genera. Ph.D. thesis, Cornell University, Ithaca, NY.

Bailey CD, Koch MA, Mayer M, Mummenhoff K, O'Kane SL Jr, Warwick SI, Windham MD, Al-Shehbaz IA (2006) Toward a global phylogeny of the *Brassicaceae*. *Mol Biol Evol* 23:2142–2160.

Baroux C, Fransz P, Grossniklaus U (2004) Nuclear fusions contribute to polyploidization of the gigantic nuclei in the chalazal endosperm of *Arabidopsis, Planta* 220:38–46.

Beilstein MA, Al-Shehbaz IA, Kellogg EA (2006) *Brassicaceae* phylogeny and trichome evolution. *Am J Bot* 93:607–619.

Bennett MD, Leitch IJ, Price HJ, Johnston JS (2003) Comparisons with *Caenorhabditis* (~100 Mb) and *Drosophila* (~175 Mb) using flow cytometry show genome size in arabidopsis to be ~157 Mb and thus ~25% larger than the Arabidopsis Genome Initiative of ~125 Mb. *Ann Bot* 91:1–11.

Bennett MD, Leitch IJ (2005) Plant DNA C-values database (release 4.0, Oct. 2005). http://www.kew.org/genomesize/homepage.

Berr A, Pecinka A, Meister A, Kreth G, Fuchs J, Blattner FR, Lysak MA, Schubert I (2006) Chromosome arrangement and nuclear architecture but not centromeric sequences are conserved between *Arabidopsis thaliana* and *A. lyrata*. *Plant J* 48:771–783.

Berr A, Schubert I (2007) Interphase chromosome arrangement in *Arabidopsis thaliana* is similar in differentiated and meristematic tissues and shows a transient mirror symmetry after nuclear division. *Genetics* 176:853–863.

Blanc G, Wolfe KH (2004) Widespread paleopolyploidy in model plant species inferred from age distributions of duplicate genes. *Plant Cell* 16:1667–1678.

Blanc G, Hokamp K, Wolfe KH (2003) A recent polyploidy superimposed on older large-scale duplications in the *Arabidopsis* genome. *Genome Res* 13:137–144.

Bleeker W (2007) Interspecific hybridization in *Rorippa* (*Brassicaceae*): patterns and processes. *Systematics Biodiversity* 5:311–319.

Boivin K, Acarkan A, Mbulu R-S, Clarenz O, Schmidt R (2004) The *Arabidopsis* genome sequence as a tool for genome analysis in *Brassicaceae*. A comparison of the *Arabidopsis* and *Capsella rubella* genomes. *Plant Physiol* 135:735–744.

Bowers JE, Chapman BA, Rong J, Paterson AH (2003) Unravelling angiosperm genome evolution by phylogenetic analysis of chromosomal duplication events. *Nature* 422:433–438.

Brochmann C, Soltis PS, Soltis DE (1992) Multiple origins of the octoploid Scandinavian endemic *Draba cacuminum* — electrophoretic and morphological evidence. *Nordic J Bot* 12:257–272.

Brochmann C, Brysting AK, Alsos IG, Borgen L, Grundt HH, Scheen A-C, Elven R (2004) Polyploidy in arctic plants. *Biol J Linn Soc* 82:521–536.

Carroll CW, Straight AF (2006) Centromere formation: from epigenetics to self-assembly. *Trends Cell Biol* 16:70–78.

Cheng BF, Heneen WK, Chen BY (1995) Mitotic karyotypes of *Brassica campestris* and *Brassica alboglabra* and identification of the *B. alboglabra* chromosome in an addition line. *Genome* 38:313–319.

Chèvre AM, Adamczyk K, Eber F, Huteau V, Coriton O, Letanneur JC, Laredo C, Jenczewski E, Monod H (2007) Modelling gene flow between oilseed rape and wild radish. I. Evolution of chromosome structure. *Theor Appl Genet* 114:209–221.

Choi S, Creelman RA, Mullet JE, Wing R (1995) Construction and characterization of a bacterial artificial chromosome library of *Arabidopsis thaliana*. *Plant Mol Biol Rep* 13:124–128.

Comai L, Tyagi AP, Lysak MA (2003) FISH analysis of meiosis in *Arabidopsis* allopolyploids. *Chromosome Res* 11:217–226.

Cui LY, Wall PK, Leebens-Mack JH, Lindsay BG, Soltis DE, Doyle JJ, Soltis PS, Carlson JE, Arumuganathan K, Barakat A, Albert VA, Ma H, dePamphilis CW (2006) Widespread genome duplications throughout the history of flowering plants. *Genome Res* 16:738–749.

De Bodt S, Maere S, Van de Peer Y (2005) Genome duplication and the origin of angiosperms. *Trends Ecol Evol.* 20:591–597.

De Jong JH, Fransz P, Zabel P (1999) High resolution FISH in plants — techniques and applications. *Trends Plant Sci* 4:258–263.

Dobes C, Koch M, Sharbel TF (2006) Embryology, karyology, and modes of reproduction in the North American genus *Boechera* (*Brassicaceae*): a compilation of seven decades of research. *Ann Missouri Botanical Garden* 93:517–534.

Dobigny G, Yang F (2008) Comparative cytogenetics in the genomics era: cytogenomics comes of age. *Chromosome Res* 16:1–4.

dos Santos KGB, Becker HC, Ecke W, Bellin U (2007) Molecular characterisation and chromosomal localisation of a telomere-like repetitive DNA sequence highly enriched in the C genome of *Brassica*. *Cytogenet Genome Res* 119:147–153.

Easterly NW (1963) Chromosome numbers of some northwestern Ohio *Cruciferae*. *Castanea* 28:39–42.

Faraut T (2008) Addressing chromosome evolution in the whole-genome sequence era. *Chromosome Res* 16:5–16.

Fransz PF, Alonso-Blanco C, Liharska TB, Peeters AJM, Zabel P, de Jong JH (1996) High-resolution physical mapping in *Arabidopsis thaliana* and tomato by fluorescence *in situ* hybridization to extended DNA fibres. *Plant J* 9:421–430.

Fransz P, Armstrong S, Alonso-Blanco C, Fischer TC, Torres-Ruiz RA, Jones G (1998) Cytogenetics for the model system *Arabidopsis thaliana*. *Plant J* 13:867–876.

Fransz P, Armstrong S, de JongJ H, Parnell LD, van Drunen G, Dean C, Zabel P, Bisseling T, Jones GH (2000) Integrated cytogenetic map of chromosome arm 4S of *A. thaliana*: structural organization of heterochromatic knob and centromere region. *Cell* 100:367–376.

Fransz P, de Jong JH, Lysak M, Castiglione MR, Schubert I (2002) Interphase chromosomes in *Arabidopsis* are organized as well defined chromocenters from which euchromatin loops emanate. *Proc Natl Acad Sci USA* 99:14584–14589.

Fransz P, Soppe W, Schubert I (2003) Heterochromatin in interphase nuclei of *Arabidopsis thaliana*. *Chromosome Res* 11:227–240.

Fransz P, ten Hoopen R, Tessadori F (2006) Composition and formation of heterochromatin in *Arabidopsis thaliana*. *Chromosome Res* 14:71–82.

Fuchs J, Demidov D, Houben A, Schubert I (2006) Chromosomal histone modification patterns — from conservation to diversity. *Trends Plant Sci* 11:199–208.

Grundt HH, Kjølner S, Borgen L, Rieseberg LH, Brochmann C (2006) High biological species diversity in the arctic flora. *Proc Natl Acad Sci USA* 103:972–975.

Hall SE, Luo S, Hall AE, Preuss D (2005) Differential rates of local and global homogenization in centromere satellites from *Arabidopsis* relatives. *Genetics* 170:1913–1927.

Hall AE, Kettler GC, Preuss D (2006) Dynamic evolution at pericentromeres. *Genome Res* 16:355–364.

Harriman NA (1965) The genus *Dentaria* L. (*Cruciferae*) in eastern North America. Ph.D. thesis, Vanderbilt University.

Harrison GE, Heslop-Harrison JS (1995) Centromeric repetitive DNA sequences in the genus *Brassica*. *Theor Appl Genet* 90:157–165.

Hasterok R, Maluszynska J (2000) Cytogenetic analysis of diploid *Brassica* species. *Acta Biol Cracoviensia, ser. Botanica* 42:145–153.

Hasterok R, Jenkins G, Langdon T, Jones RN, Maluszynska J (2001) Ribosomal DNA is an effective marker of *Brassica* chromosomes. *Theor Appl Genet* 103:486–490.

Hasterok R, Ksiazczyk T, Wolny E, Maluszynska J (2005) FISH and GISH analysis of *Brassica* genomes. *Acta Biol Cracoviensia, ser Botanica* 47:185-192.

Hayek A (1911) *Entwurf eines Cruciferensystems auf phylogenetischer Grundlage. Beih Bot Centralbl* 27:127–335.

Henry Y, Bedhomme M, Blanc G (2006) History, protohistory and prehistory of the *Arabidopsis thaliana* chromosome complement. *Trends Plant Sci* 11:267–273

Heslop-Harrison JS, Brandes A, Schwarzacher T (2003) Tandemly repeated DNA sequences and centromeric chromosomal regions of *Arabidopsis* species. *Chromosome Res* 11:241–253.

Howell EC, Barker GC, Jones GH, Kearsey M J, King GJ, Kop EP, Ryder CD, Teakle GR, Vicente JG, Armstrong SJ (2002) Integration of the cytogenetic and genetic linkage maps of *Brassica oleracea*. *Genetics* 161:1225–1234.

Howell EC, Armstrong SJ, Barker GC, Jones GH, King GJ, Ryder CD, Kearsey MJ (2005) Physical organization of the major duplication on *Brassica oleracea* chromosome O6 revealed through fluorescence *in situ* hybridization with *Arabidopsis* and *Brassica* BAC probes. *Genome* 48:1093–1103.

Jackson SA, Ming L, Goodman HM, Jiang J, Ming LW, Jiang JM (1998) Application of fiber-FISH in physical mapping of *Arabidopsis thaliana*. *Genome* 41:566–572.

Jackson SA, Cheng Z, Wang ML, Goodman HM, Jiang J (2000) Comparative fluorescence *in situ* hybridization mapping of a 431-kb *Arabidopsis thaliana* bacterial artificial chromosome contig reveals the role of chromosomal duplications in the expansion of the *Brassica rapa* genome. *Genetics* 156:833–838.

Jakobsson M, Säll T, Lind-Halldén, Halldén C (2007) The evolutionary history of the common chloroplast genome of *Arabidopsis thaliana* and *A. suecica*. *J Evol Biol* 20:104–121.

Jaretzky R (1928) *Untersuchungen über Chromosomen und Phylogenie bei einigen Cruciferen. Jahrb f wiss Botanik* 68:1–45.

Jaretzky R (1932) *Beziehungen zwischen Chromosomenzahl und Systematik bei den Cruciferen. Jahrb wiss Bot* 76:485–527.

Johnston JS, Pepper AE, Hall AE, Chen ZJ, Hodnett G, Drabek J, Lopez R, Price HJ (2005) Evolution of genome size in *Brassicaceae*. *Ann Bot* 95:229–235.

Kamisugi Y, Nakayama S, O'Neil CM, Mathias RJ, Trick M, Fukui K (1998) Visualization of the *Brassica* self-incompatibility S-locus on identified oilseed rape chromosomes. *Pl Molec Biol* 38:1081–1087.

Kamm A, Galasso I, Schmidt T, Heslop-Harrison JS (1995) Analysis of a repetitive DNA family from *Arabidopsis arenosa* and relationship between *Arabidopsis* species. *Pl Molec Biol* 27:853–862.

Kantama L, Sharbel TF, Schranz ME, Mitchell-Olds T, de Vries S, de Jong H (2007) Diploid apomicts of the *Boechera holboellii* complex display large-scale chromosome substitutions and aberrant chromosomes. *Proc Natl Acad Sci USA* 104:14026–14031.

Kawabe A, Charlesworth D (2007) Patterns of DNA variation among three centromere satellite families in *Arabidopsis halleri* and *A. lyrata*. *J Mol Evol* 64:237–247.

Kawabe A, Nasuda S (2005) Structure and genomic organization of centromeric repeats in *Arabidopsis* species. *Mol Gen Genomics* 272:593–602.

Kawabe A, Nasuda S (2006) Polymorphic chromosomal specificity of centromere satellite families in *Arabidopsis halleri* ssp. *gemmifera*. *Genetica* 126:335–342.

Kawabe A, Hansson B, Forrest A, Hagenblad J, Charlesworth D (2006a) Comparative gene mapping in *Arabidopsis lyrata* chromosomes 6 and 7 and *A. thaliana* chromosome. IV. Evolutionary history, rearrangements and local recombination rates. *Genet Res* 88:45–56.

Kawabe A, Hansson B, Hagenblad J, Forrest A, Charlesworth D (2006b) Centromere locations and associated chromosome rearrangements in *Arabidopsis lyrata* and *A. thaliana*. *Genetics* 173:1613–1619.

Klášterská I, Ramel C (1980) Meiosis in PMCs of *Arabidopsis thaliana*. *Arabidopsis Inf Serv* 17:1–10.

Koch MA, Kiefer M (2005) Genome evolution among cruciferous plants: a lecture from the comparison of the genetic maps of three diploid species — *Capsella rubella*, *Arabidopsis lyrata* subsp. *petraea*, and *A. thaliana*. *Am J Bot* 92:761–767.

Koch M, Bishop J, Mitchell-Olds T (1999) Molecular systematics and evolution of *Arabidopsis* and *Arabis*. *Plant Biol* 1:529–539.

Koch M, Haubold B, Mitchell-Olds T (2001) Molecular systematics of the *Brassicaceae*: evidence from coding plastidic *matK* and nuclear *Chs* sequences. *Am J Bot* 88:534–544.

Koch MA, Dobes C, Keifer C, Schmickl R, Klimes L, Lysak MA (2007) Supernetwork identifies multiple events of plastid *trn*F(GAA) pseudogene evolution in the *Brassicaceae*. *Molec Biol Evol* 24:63–73.

Koo D-H, Plaha P, Lim YP, Hur Y, Bang J-W (2004) A high-resolution karyotype of *Brassica rapa* ssp. *pekinensis* revealed by pachytene analysis and multicolor fluorescence *in situ* hybridization. *Theor Appl Genet* 109:1346–1352.

Koornneef M, Fransz P, de Jong H (2003) Cytogenetic tools for *Arabidopsis thaliana*. *Chromosome Res* 11:183–194.

Kuittinen H, de Haan AA, Vogl C, Oikarinen S, Leppala J, Koch M, Mitchell-Olds T, Langley CH, Savolainen O (2004) Comparing the linkage maps of the close relatives *Arabidopsis lyrata* and *A. thaliana*. *Genetics* 168:1575–1584.

Lagercrantz U (1998) Comparative mapping between *Arabidopsis thaliana* and *Brassica nigra* indicates that *Brassica* genomes have evolved through extensive genome replication accompanied by chromosome fusions and frequent rearrangements. *Genetics* 150:1217–1228.

Lagercrantz U, Lydiate D (1996) Comparative genome mapping in *Brassica*. *Genetics* 144:1903–1910.

Laibach F (1907) *Zur Frage nach der Individualität der Chromosomen im Pflanzenreich. Beih Botan Zentralbl* 22:191–210.

Lan T-H, DelMonte TA, Reischmann KP, Hyman J, Kowalski SP, McFerson J, Kresovich S, Paterson AH (2000) An EST-enriched comparative map of *Brassica oleracea* and *Arabidopsis thaliana*. *Genome Res* 10:776–788.

Laufs P, Autran D, Traas J (1999). A chromosomal paracentric inversion associated with T-DNA integration in *Arabidopsis*. *Plant J* 18:131–139.

Leitch IJ, Chase MW, Bennett MD (1998) Phylogenetic analysis of DNA C-values provides evidence for a small ancestral genome size in flowering plants. *Ann Bot* 82(Suppl. A):85–94.

Li G, Gao M, Yang B, Quiros CF (2003) Gene for gene alignment between the *Brassica* and *Arabidopsis* genomes by direct transcriptome mapping. *Theor Appl Genet* 107:168–180.

Lim K-B, de Jong H, Yang T-J, Park J-Y, Kwon S-J, Kim JS, Lim M-H, Kim JA, Jin M, Jin Y-M, Kim SH, Lim YP, Bang J-W, Kim H-I, Park B-S (2005) Characterization of rDNAs and tandem repeats in the heterochromatin of *Brassica rapa*. *Mol Cells* 19:436–444.

Lim YP, Plaha P, Choi SR, Uhm T, Hong CP, Bang JW, Hur YK (2006) Toward unraveling the structure of *Brassica rapa* genome. *Physiol Plant* 126:585–591.

Lim KB, Yang TJ, Hwang YJ, Kim JS, Park JY, Kwon SJ, Kim JA, Choi BS, Lim MH, Jin M, Kim HI, de Jong H, Bancroft I, Lim Y, Park BS (2007) Characterization of the centromere and peri-centromere retrotransposons in *Brassica rapa* and their distribution in related *Brassica* species. *Plant J* 49:173–183.

Lövkvist B (1956) The *Cardamine pratensis* complex. Outline of its cytogenetics and taxonomy. *Symbolae Botanicae Upsalienses* 14:1–131.

Lukens L, Zou F, Lydiate D, Parkin I, Osborn T (2003) Comparison of a *Brassica oleracea* genetic map with the genome of *Arabidopsis thaliana*. *Genetics* 164:359–372.

Lukens LN, Quijada PA, Udall J, Pires JC, Schranz ME, Osborn TC (2004) Genome redundancy and plasticity within ancient and recent *Brassica* crop species. *Biol J Linn Soc* 82:665–674.

Lysak MA, Lexer C (2006) Towards the era of comparative evolutionary genomics in *Brassicaceae*. *Pl Syst Evol* 259:175–198.

Lysak MA, Fransz PF, Ali HBM, Schubert I (2001) Chromosome painting in *Arabidopsis thaliana*. *Plant J* 28:689–697.

Lysak MA, Pecinka A, Schubert I (2003) Recent progress in chromosome painting of *Arabidopsis* and related species. *Chromosome Res* 11:195–204.

Lysak MA, Koch MA, Pecinka A, Schubert I (2005) Chromosome triplication found across the tribe *Brassiceae*. *Genome Res* 15:516–525.

Lysak MA, Berr A, Pecinka A, Schmidt R, McBreen K, Schubert I (2006a) Mechanisms of chromosome number reduction in *Arabidopsis thaliana* and related *Brassicaceae* species. *Proc Natl Acad Sci USA* 103:5224–5229.

Lysak M, Fransz P, Schubert I (2006b) Cytogenetic analyses of *Arabidopsis*. In: Salinas J and Sanchez-Serrano JJ (Eds.), *Methods in Molecular Biology*, Vol. 323: Arabidopsis Protocols, second edition, p. 173–186, Humana Press Inc., Totowa, NJ.

Lysak MA, Cheung K, Kitschke M, Bures P (2007) Ancestral chromosomal blocks are triplicated in *Brassiceae* species with varying chromosome number and genome size. *Plant Physiol* 145:402–410.

Lysak MA, Koch MA, Beaulieu JM, Meister A, Leitch IJ (2008) The dynamic ups and downs of genome size evolution in *Brassicaceae*. *Mol Biol Evol* (in press).

Ma J, Wing RA., Bennetzen JL, Jackson SA (2007) Plant centromere organization: a dynamic structure with conserved functions. *Trends Genet* 23:134–139.

Madlung A, Tyagi AP, Watson B, Jiang HM, Kagochi T, Doerge RW, Martienssen R, Comai L (2005) Genomic changes in synthetic *Arabidopsis* polyploids. *Plant J* 41:221–230.

Maere S, De Bodt S, Raes J, Casneuf T, Van Montagu M, Kuiper M, Van de Peer Y (2005) Modeling gene and genome duplications in eukaryotes. *Proc Natl Acad Sci* 102:5454–5459.

Maluszynska J, Heslop-Harrison JS (1991) Localization of tandemly repeated DNA sequences in *Arabidopsis thaliana*. *Plant J* 1:159–166.

Maluszynska J, Heslop-Harrison JS (1993) Molecular cytogenetics of the genus *Arabidopsis*: *in situ* localization of rDNA sites, chromosome numbers and diversity in centromeric heterochromatin. *Ann Bot* 71:47–484.

Maluszynska J, Hasterok R (2005) Identification of individual chromosomes and parental genomes in *Brassica juncea* using GISH and FISH. *Cytogenet Genome Res* 109:310–314.

Manton I (1932) Introduction to the general cytology of the *Cruciferae Ann Bot* 46:509–556.

Marhold K, Lihová J (2006) Polyploidy, hybridization and reticulate evolution: lessons from the *Brassicaceae*. *Plant Syst Evol* 259:143–174.

Martinez-Zapater JM, Estelle MA, Somerville CR (1986) A highly repeated DNA sequence in *Arabidopsis thaliana*. *Mol Gen Genet* 204:417–423.

McCombie WR, de la Bastide M, Habermann K, Parnell L, Dedhia N, Gnoj L, Schutz K, Huang E, Spiegel L, Yordan C, Sehkon M, Murray J, Sheet P, Cordes M, Threideh J, Stoneking T, Kalicki J, Graves T, Harmon G, Edwards J, Latreille P, Courtney L, Cloud J, Abbott A, Scott K, Johnson D, Minx P, Bentley D, Fulton B, Miller N, Greco T, Kemp K, Kramer J, Fulton L, Mardis E, Dante M, Pepin K, Hillier L, Nelson J, Spieth J, Simorowski J, May B, Ma P, Preston R, Vil D, See LH, Shekher M, Matero A, Shah R, Swaby I, O'Shaughnessy A, Rodriguez M, Hoffman J, Till S, Granat S, Shohdy N, Hasegawa A, Hameed A, Lodhi M, Johnson A, Chen E, Marra M, Wilson RK, Martienssen R (2000) The complete sequence of a heterochromatic island from a higher eukaryote. *Cell* 100:377–386.

Mokros P, Vrbsky J, Siroky J (2006) Identification of chromosomal fusion sites in *Arabidopsis* mutants using sequential bicolour BAC-FISH. *Genome* 49:1036–1042.

Montgomery FH (1955) Preliminary studies in the genus *Dentaria* in eastern North America. *Rhodora* 57:161–173.

Mozo T, Fischer S, Meier-Ewert S, Lehrach H, Altmann T (1998) Use of the IGF BAC library for physical mapping of the *Arabidopsis thaliana* genome. *Plant J* 16:377–384.

Mummenhoff K, Linder P, Friesen N, Bowman JL, Lee J-Y, Franzke A (2004) Molecular evidence for bicontinental hybridogenous genomic constitution in *Lepidium* sensu stricto (*Brassicaceae*) species from Australia and New Zealand. *Am J Bot* 91:254–261.

Murata M, Ogura Y, Motoyoshi F (1994) Centromeric repetitive sequence in *Arabidopsis thaliana*. *Japan J Genet* 69:361–370.

Murata M, Heslop-Harrison JS, Motoyoshi F (1997) Physical mapping of the 5S ribosomal RNA genes in *Arabidopsis thaliana* by multi-color fluorescence *in situ* hybridization with cosmid clones. *Plant J* 12:31–37.

Murata M, Shibata F, Yokota E (2005) The origin, meiotic behavior, and transmission of a novel minichromosome in *Arabidopsis thaliana*. *Chromosoma* 115:311–319.

Murata M, Yokota E, Shibata F, Kashihara K (2007) A ring minichromosome generated by T-DNA insertion in *Arabidopsis thaliana*. *Chromosome Res* 15(Suppl. 2):71–72.

Neves N, Delgado M, Silva M, Caperta A, Morais-Cecilio L, Viegas W (2005) Ribosomal DNA heterochromatin in plants. *Cytogenet Genome Res* 109:104–111.

Parkin IA, Sharpe AG, Lydiate DJ (2003) Patterns of genome duplication within the *Brassica napus* genome. *Genome* 46:291–303.

Pecinka A, Schubert V, Meister A, Kreth G, Klatte M, Lysak MA, Fuchs J, Schubert I (2004) Chromosome territory arrangement and homologous pairing in nuclei of *Arabidopsis thaliana* are predominantly random except for NOR-bearing chromosomes. *Chromosoma* 113:258–269.

Pecinka A, Kato N, Meister A, Probst AV, Schubert I, Lam E (2005) Tandem repetitive transgenes and fluorescent chromatin tags alter local interphase chromosome arrangement in *Arabidopsis thaliana*. *J Cell Sci* 118:3751–3758.

Pontes O, Neves N, Silva M, Lewis MS, Madlung A, Comai L, Viegas W, Pikaard CS (2004) Chromosomal locus rearrangements are a rapid response to formation of the allotetraploid *Arabidopsis suecica* genome. *Proc Natl Acad Sci USA* 101:18240–18245.

Pradillo M, Lopez E, Romero C, Sanchez-Moran E, Cunado N, Santos JL (2007) An analysis of univalent segregation in meiotic mutants of *Arabidopsis thaliana*: a possible role for synaptonemal complex. *Genetics* 175:505–511.

Prakash S, Hinata K (1980) Taxonomy, cytogenetics and origin of crop Brassicas, a review. *Opera Bot* 55:1–57.

Quiros CF (1999) Genome structure and mapping. In: Gomez-Campo C (Ed.) *Biology of Brassica Coenospecies.* p. 217–245, Elsevier, Amsterdam.

Raina SN, Rani V (2001) GISH technology in plant genome research. *Meth Cell Sci* 23:83–104.

Rokas A, Holland WH (2000) Rare genomic changes as a tool for phylogenetics. *Trends Ecol Evol* 15:454–459.

Ross K, Fransz PF, Jones GH (1996) A light microscopic atlas of meiosis in *Arabidopsis thaliana*. *Chromosome Res* 4:507–516.

Sanchez-Moran E, Armstrong SJ, Santos JL, Franklin FCH, Jones GH (2002) Variation in chiasma frequency among eight accessions of *Arabidopsis thaliana*. *Genetics* 162:1415–1422.

Santos JL, Alfaro D, Sanchez-Moran E, Armstrong SJ, Franklin FCH, Jones GH (2003) Partial diploidization of meiosis in autotetraploid *Arabidopsis thaliana*. *Genetics* 165:1533–1540.

Sanyal A, Jackson SA (2005) Comparative genomics reveals expansion of the FLC region in the genus *Arabidopsis*. *Mol Gen Genomics* 275:26–34.

Schelfhout CJ, Snowdon RJ, Cowling WA, Wroth JM (2004) A PCR based B-genome specific marker in *Brassica* species. *Theor Appl Genet* 109:917–921.

Schmidt R, Acarkan A, Boivin K (2001) Comparative structural genomics in the *Brassicaceae* family. *Plant Physiol Biochem* 39:253–262.

Schrader O, Budahn H, Ahne R (2000) Detection of 5S and 25S rRNA genes in *Sinapis alba*, *Raphanus sativus* and *Brassica napus* by double fluorescence *in situ* hybridization. *Theor Appl Genet* 100:665–669.

Schranz ME, Mitchell-Olds T (2006) Independent ancient polyploidy events in the sister families *Brassicaceae* and *Cleomaceae*. *Plant Cell* 18:1152–1165.

Schranz ME, Lysak MA, Mitchell-Olds T (2006) The ABC's of comparative genomics in the *Brassicaceae*: building blocks of crucifer genomes. *Trends Plant Sci* 11:535–542.

Schranz ME, Song BH, Windsor AJ, Mitchell-Olds T (2007a) Comparative genomics in the *Brassicaceae*: a family-wide perspective. *Curr Opin Plant Biol* 10:168–175.

Schranz ME, Windsor AJ, Song B, Lawton-Rauh A, Mitchell-Olds T (2007b) Comparative genetic mapping in *Boechera stricta*, a close relative of *Arabidopsis*. *Plant Physiol* 144:286–298.

Schubert I, Fransz PF, Fuchs J, de Jong JH (2001) Chromosome painting in plants. *Meth Cell Sci* 23:57–69.

Schubert I (2007) Chromosome evolution. *Curr Opin Plant Biol* 10:1–7.

Sears LMS, Lee-Chen S (1970) Cytogenetic studies in *Arabidopsis thaliana*. *Can J Genet Cytol* 12:217–223.

Sharbel TF, Voigt ML, Mitchell-Olds T, Kantama L, de Jong H (2004) Is the aneuploid chromosome in an apomictic *Boechera holboellii* a genuine B chromosome? *Cytogen Genome Res* 106:173–183.

Siroky J (2008) Chromosome landmarks as tools to study the genome of *Arabidopsis thaliana Cytogenet Genome Res* 120 (DOI:10.1159/000121068).

Siroky J, Zluvova J, Riha K, Shippen DE, Vyskot B (2003) Rearrangements of ribosomal DNA clusters in late generation telomerase-deficient Arabidopsis. *Chromosoma* 112:116–123.

Snowdon RJ (2007) Cytogenetics and genome analysis in *Brassica* crops. *Chromosome Res* 15:85–95.

Snowdon RJ, Köhler W, Friedt W, Köhler A (1997) Genomic *in situ* hybridization in *Brassica* amphidiploids and interspecific hybrids. *Theor Appl Genet* 95 :1320–1324.

Snowdon RJ, Friedrich T, Friedt W, Köhler W (2002) Identifying the chromosomes of the A- and C-genome diploid *Brassica* species *B. rapa* (syn. *campestris*) and *B. oleracea* in their amphidiploid *B. napus*. *Theor Appl Genet* 104:533–538.

Soppe WJ, Jasencakova Z, Houben A, Kakutani T, Meister A, Huang MS, Jacobsen SE, Schubert I, Fransz PF (2002) DNA methylation controls histone H3 lysine 9 methylation and heterochromatin assembly in *Arabidopsis*. *EMBO J* 21:6549–6559.

Steinitz-Sears LM (1963) Chromosome studies in *Arabidopsis thaliana*. *Genetics* 48:483–490.

Tessadori F, Chupeau M-C, Chupean Y, Knip M, Germann S, van Driel R, Fransz P, Gaudin V (2007a) Large-scale dissociation and sequential reassembly of pericentric heterochromatin dedifferentiated *Arabidopsis* cells. *J Cell Sci* 120:1200–1208.

Tessadori F, Schulkes RK, van Driel R, Fransz P (2007b) Light-regulated large-scale reorganization of chromatin during the floral transition in Arabidopsis. *Plant J* 50:848–857.

U N (1935) Genomic analysis of *Brassica* with special reference to the experimental formation of *B. napus* and peculiar mode of fertilization. *Jpn J Bot* 7:389–452.

Uchida W, Matsunaga S, Sugiyama R, Kawano S (2002) Telomere-like repeats in the *Arabidopsis thaliana* genome. *Genes Genet Syst* 77:63–67.

Vannier JB, Depeiges A, White C, Gallego ME (2006) Two roles for Rad50 in telomere maintenance. *EMBO J* 25:4577–4585.

Vespa L, Warrington RT, Mokros P, Siroky J, Shippen DE (2007) ATM regulates the length of individual telomere tracts in *Arabidopsis*. *Proc Natl Acad Sci USA* 104:18145–18150.

Vitte C, Bennetzen JL (2006) Analysis of retrotransposon structural diversity uncovers properties and propensities in angiosperm genome evolution. *Proc Natl Acad Sci USA* 103:17638–17643.

Volkov RA, Komarova NY, Hemleben V (2007) Ribosomal DNA in plant hybrids: inheritance, rearrangement, expression. *Systematics Biodiversity* 5:261–276.

Wang JL, Tian L, Lee HS, Wei NE, Jiang HM, Watson B, Madlung A, Osborn TC, Doerge RW, Comai L, Chen ZJ (2006) Genomewide nonadditive gene regulation in *Arabidopsis* allotetraploids. *Genetics* 172:507–517.

Warwick SI, Al-Shehbaz IA (2006) *Brassicaceae*: Chromosome number index and database on CD-ROM. *Pl Syst Evol* 259:237-248.

Warwick SI, Black LD (1991) Molecular systematics of *Brassica* and allied genera (subtribe Brassicinae, Brassiceae) — chloroplast genome and cytodeme congruence. *Theor Appl Genet* 82:81–92.

Warwick SI, Sauder C (2005) Phylogeny of tribe Brassiceae (*Brassicaceae*) based on chloroplast restriction site polymorphisms and nuclear ribosomal internal transcribed spacer (ITS) and chloroplast trnL intron sequences. *Can J Bot* 83:467–483.

Warwick SI, Sauder CA, Al-Shehbaz, Jacquemoud F (2007) Phylogenetic relationships in the tribes Anchonieae, Chorisporeae, Euclidieae, and Hesperideae (*Brassicaceae*) based on nuclear ribosomal its DNA sequences. *Ann Missouri Bot Gard* 94:56–78.

Yang T-J, Kim J-S, Lim K-B, Kwon S-J, Kim J-A, Jin M, Young JP, Lim M-H, Kim H-I, Hyung Kim S, Lim YP, Park B-S (2005) The Korea *Brassica* Genome Project: a glimpse of the *Brassica* genome based on comparative genome analysis with *Arabidopsis*. *Comp Funct Genom* 6:138–146.

Yang T-J, Kim JS, Kwon S-J, Lim K-B, Choi B-S, Kim J-A, Jin M, Park JY, Lim M-H, Kim H-I, Lim YP, Kang JJ, Hong JH, Kim CB, Bhak J, Bancroft I, Park BS (2006) Sequence-level analysis of the diploidization process in the triplicated FLOWERING LOCUS C region of *Brassica rapa*. *Plant Cell* 18:1339–1347.

Yogeeswaran K, Frary A, York TL, Amenta A, Lesser AH, Nasrallah JB, Tanksley SD, Nasrallah ME (2005) Comparative genome analyses of *Arabidopsis* spp.: inferring chromosomal rearrangement events in the evolutionary history of *A. thaliana*. *Genome Res* 15:505–515.

Zhang X, Wessler SR (2004) Genome-wide comparative analysis of the transposable elements in the related species *Arabidopsis thaliana* and *Brassica oleracea*. *Proc Natl Acad Sci USA* 101:5589–5594.

Ziolkowski PA, Kaczmarek M, Babula D, Sadowski J (2006) Genome evolution in *Arabidopsis*/*Brassica*: conservation and divergence of ancient rearranged segments and their breakpoints. *Plant J* 47:63–74.

11 Distant Hybridization

Yukio Kaneko, Sang Woo Bang, and Yasuo Matsuzawa

CONTENTS

INTRODUCTION

The *Brassicaceae* (Cruciferae) comprise ca. 330 genera, including 3700 species, a few dozen of which are domesticated as edible oil, vegetable, spice, ornamental flower, and forage crops. Some of the wild relatives, on the other hand, have been evaluated for their genetic resources to develop more potential varieties with biotic and abiotic tolerance in agricultural practice (Warwick, 1993).

Interspecific and intergeneric hybridization has been carried out among cultivated species to determine mainly the cytogenetics and speciation of these members, and then the wide cross between cultivated species and wild relatives was extensively performed. Cross incompatibility as an isolation barrier in plants, however, has hampered the development of filial hybrids in the wide cross. Based on anatomical studies of the growth of hybrid embryos, interspecific hybrids and their amphidiploidal lines were successfully produced by Nishi et al. (1970) and Inomata (1977) via *in vitro* embryo and ovary culture techniques, respectively. Currently, extensive work in wide cross focuses on gene flow through microspores in the case of transgenic rapeseed (Jodi and Philip, 1994; Warwick et al., 2003; FitzJohn et al., 2007); gene and genome analysis in comparison with the model plant *Arabidopsis* (Yamagishi, 2003); and genetic resources for breeding strategies.

Matsuzawa et al. (1996) suggested a system for the development of five hybrid progenies by wide cross (Figure 11.1). These hybrid lines would be valuable genetic resources not only to breed more productive cultivars with novel agronomic traits, but also to analyze each chromosome and gene concerned. In these research programs, it is prerequisite to grow as many as possible true F_1 hybrids and their progenies. This chapter outlines the wide cross between the *Brassica* crops and their wild relatives, and also the development, preservation, and improvement of the hybrid line.

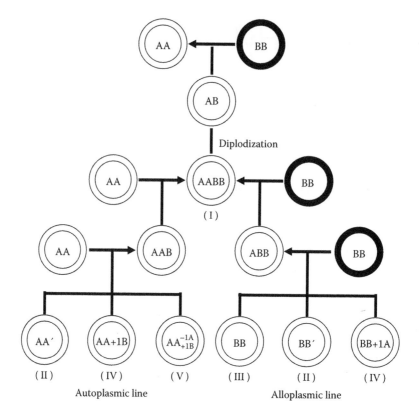

FIGURE 11.1   Schematic diagram of the distant hybridization breeding system between AA and BB genome species concerned. (I), synthetic amphidiploid line (SADL); (II), alien gene(s) introgression line (AGIL); (III), alloplasmic line (ALPL); (IV), monosomic alien chromosome addition line (MAAL); (V), monosomic alien chromosome substitution line (MASL); * AB, AABB, and AAB (ABB) show genomes for amphihaploid, amphidiploid, and sesquidiploid, respectively; * A' (B') means some genetic modification via recomination between each complement of A and B genome; * AA+1(BB+1) means A(B) genome species aded single chromosome of B(A) genome species; *○ and ● show hte difference in cytoplasmic background for A and B genome species, respectively; * A ← B means the hybridization in which A and B genome species are pistilate and pollen parents, respectively; * In (V), a chromosome of A-genome is subsituted by the one of B-genome. (*Source:* From Matsuzawa et al. 1996, revised.)

CROSS INCOMPATIBILITY BETWEEN CULTIVATED SPECIES AND WILD RELATIVES

It is generally understood that the barriers in wide cross operate in both the pre-zygotic and post-zygotic phases of sexual reproduction. Stebbins (1958) suggested that the pre-zygotic and post-zygotic barriers might be due to the failures of pollen germination, pollen tube growth, and pollen tube penetration to ovule, and to the degeneration of hybrid embryo, male and female sterility in hybrid plant, and lethality in hybrid progenies, respectively. Khush and Brar (1992) surveyed the cross incompatibility in distant hybridization and then offered some successful procedures to overcome the above barriers. From comprehensive studies on the interspecific and intergeneric hybridization in *Brassica* crops, Matsuzawa (1983) evaluated the pre-zygotic barrier in terms of the pollen germination index (P.G.I. $\approx$ b + 2c + 3d + 4e/a + b + c + d + e, $0 \leqq$ P.G.I.$\leqq$ 4), where a, b, c, d, and e represent the number of pistils in which there was no pollen grain on stigma, pollen grain not germinating on stigma, pollen tube growing into stigma, pollen tube growing into style, and pollen tube penetrating into ovule, respectively. To evaluate the pre-zygotic incompatibility in distant hybridization within *Brassicaceae* plants, Kerlan et al. (1992) also proposed an equation for the index of pollination compatibility (I) = x + 2y + 3z , where x (0, 1, 2, or 3), y (0, 1, and 2), and z (0, 1, and 2) were the corresponding scores from the values of pollen grains germinated / pollen grains on the stigma, number of pollen tubes growing in style, and number of pollen tubes penetrating into ovules, respectively.

Post-zygotic barrier or incompatibility, on the other hand, could be evaluated by an anatomical survey for the development and growth of embryos made in Brussels sprouts (*Brassica oleracea*) by Wilmar and Hellendoorn (1968), and effective *in vitro* culture techniques with pollinated flower, ovary, ovule, and embryo were successfully applied. Bang (1996) and Bang et al. (1996a, b, 1997, 1998, 2000, 2002, 2003, 2007) and Jeung et al. (2003 to 2008) have investigated the cross incompatibility within the intergeneric hybridization between cultivated species and wild relatives, and then produced many potential hybrid progenies through *in vitro* procedures for breeding and scientific goals.

DEVELOPMENT OF NOVEL HYBRID PLANTS FOLLOWING EMBRYO RESCUE

Production of novel hybrids by reciprocal crosses between *Brassica* crops and wild relatives of *Brassica, Sinapis, Diplotaxis, Moricandia, Eruca,* and *Orichophragmus* is shown in Tables 11.1a through 11.1e.

About 45 species of wild relatives were surveyed from the reported literature. It is observed that wild relatives (viz., *Brassica, Diplotaxis,* and *Sinapis*) were frequently used as parents for hybridization studies. In *B. napus, B. rapa,* and *B. juncea,* hybrids could be produced even in conventional cross when they were used as the pistillate parent. However, reciprocal cross would be seriously difficult. On the other hand, various embryo rescue techniques have been attempted in an effort to overcome hybrid inviability and breakdown in early stages. The results of the cross combinations are summarized below.

BRASSICA NAPUS × WILD RELATIVES

A large number of hybrids were successfully obtained in the wide crosses of *Brassica napus* × *Brassica* or *Sinapis* when *B. napus* was used as the female parent. Hybrid plants have been produced by various workers even by the conventional crossing technique. Many hybrids were produced between *B. napus* and *Rapistrum rugosum* (Heyn, 1977), *Moricandia nitens* (Takahata et al., 1993), or *Mattiola incana* (Luo et al., 2003). Using 14 species of *Diplotaxis* as the female parent, Ringdahl et al. (1987) have produced hybrids using intergeneric hybridization, and these were obtained only in two cross combinations of *D. erucoides* and *D. muralis.* The ovary culture technique for embryo rescue was confirmed as more successful in producing viable hybrid plants.

BRASSICA RAPA × WILD RELATIVES

In *Brassica rapa,* many hybrids were obtained in reciprocal crosses with wild *Brassica,* particularly when *Sinapis* was used as the male parent, and *Diplotaxis* and *Eruca* as the female parents. Chen et al. (2007) obtained the novel hybrids of *B. rapa* × *Capsella bursa-pastoris* by conventional crossing. Hybridization between *B. rapa* and wild relatives has been performed by ovary and/or sequential culture.

BRASSICA OLERACEA × WILD RELATIVES

Hybrid production was somewhat limited in such cross combinations, especially when *Brassica oleracea* was used as the female parent. Hybrid plants, however, could be produced in about half of the cross

TABLE 11.1A
Production of F1 Hybrid using Different Embryo Rescue Methods in Reciprocal Crosses between *B. napus* and Wild Relatives

Cross Combination		Embryo Rescue Methods *						References
Female	Male	Conventional Cross	Embryo	Ovule	Ovary	Ovary → Ovule	Ovary → Embryo	
B. napus (n = 19)	*Hirschfeldia incana* (n = 7)	O,O	—	—	O,O	—	—	Kerlan et al. (1992), Lefol et al. (1996)
ditto	*B. fruticulosa* (n = 8)	O	—	—	—	—	—	Heyn (1977)
ditto	*B. bourgeaui* (n = 9)	—	—	—	O	—	—	Inomata (1993)
ditto	*B. montana* (n = 9)	—	—	—	O	—	—	Inomata (1993)
ditto	*B. cretica* (n = 9)	—	—	—	O	—	—	Inomata (1993)
ditto	*B. tournefortii* (n = 10)	O,O	—	—	—	—	—	Prakash and Narain (1971) Heyn (1977); Nagpal et al. (1996)
ditto	*B. gravinae* (n = 20)	×	—	—	—	O	—	Nanda Kumar et al. (1989)
ditto	*R. raphanistrum* (n = 9)	×,O	—	—	O	—	—	Kerlan et al. (1992); Lefol et al. (1997)
ditto	*E. sativa* (n = 11)	Δ,O	—	—	—	—	—	Heyn (1977); Bijral and Sharma (1996)
ditto	*E. gallicum* (n = 15)	O	—	—	—	—	—	Lefol et al. (1997)
ditto	*D. catholica* (n = 9)	Δ	—	—	—	—	—	Bijral and Sharma (1998)
ditto	*D. siifolia* (n = 10)	×	—	—	×	—	—	Batra et al. (1990)
ditto	*D. tenuifolia* (n = 11)	O	—	—	—	—	—	Heyn (1977)
ditto	*D. muralis* (n = 21)	Δ	—	—	—	—	—	Bijral and Sharma (1996a)
ditto	*S. arvensis* (n = 9)	×, O, ×	—	O	O	—	O	Mizushima (1950); Inomata (1988); Kerlan et al. (1992); Bing et al. (1991, 1995)
ditto	*S. pubescens* (n = 9)	—	—	—	O	—	—	Inomata (1994)
ditto	*S. alba* (n = 12)	×,O	O,×	—	—	—	—	Heyn (1977); Lelivelt et al. (1993)
ditto	*M. arvensis* (n = 14)	×	—	—	O,×	—	—	Delourme et al. (1989); Takahata et al. (1993b)
ditto	*Matthiola incana* (n = 7)	—	—	—	O	—	—	Luo et al. (2003)
ditto	*Rapistrum rugosum* (n = 8)	O	—	—	—	—	—	Heyn (1977)

Species	Cross						References
ditto	E. lyrates (n = 10)	—	—	x	—	—	Gundimeda et al. (1992)
ditto	Orchophragmus violaceus (n = 12)	—	—	—	○	○, ○, ○	Li and Luo (1993); Luo et al. (1994); Li et al. (1995); Hua and Li (2006)
ditto	Capsella bursa-pastoris (n = 16)	—	—	—	—	○	Chen et al. (2007)
Hirschfeldia incana (n = 7)	B. napus (n = 19)	—	x, ○	—	—	x, ○	Kerlan et al. (1992); Lefol et al. (1996)
B. maurourum (= 8)	ditto	—	○	○	—	○	Chrungu et al. (1999)
B. gravinae (n = 20)	ditto	x	x	—	—	x	Nanda Kumar et al. (1989)
R. raphanistrum (n = 9)	ditto	—	○	—	—	x, x	Kerlan et al. (1992); Lefol et al. (1997)
E. gallicum (n = 15)	ditto	○	—	—	—	x, x	Batra et al. (1989); Lefol et al. (1997)
D. erucoides (n = 7)	ditto	○	○, x	—	—	○	Ringdahl et al. (1987);Delourme et al. (1989);Vyas et al. (1995)
D. siettiana (n = 8)	ditto	—	—	—	—	x	Ringdahl et al. (1987)
D. ibicensis (n = 8)	ditto	—	—	—	—	x	Ringdahl et al. (1987)
D. virgata (n = 9)	ditto	—	—	—	—	x	Ringdahl et al. (1987)
D. assurgens (n = 9)	ditto	—	—	—	—	x	Ringdahl et al. (1987)
D. tenuisiliqua (n = 9)	ditto	—	—	—	—	x	Ringdahl et al. (1987)
D. siifolia (n = 10)	ditto	—	○	—	—	x, x	Ringdahl et al. (1987); Batra et al. (1990)
D. viminea (n =10)	ditto	—	—	—	—	x	Ringdahl et al. (1987)
D. tenuifolia (n = 11)	ditto	—	—	—	—	x	Ringdahl et al. (1987)
D. pitardiana (n = 11)	ditto	—	—	—	—	x	Ringdahl et al. (1987)
D. cretacea (n = 11)	ditto	—	—	—	—	x	Ringdahl et al. (1987)

—continued

TABLE 11.1A (Continued)
Production of F1 Hybrid using Different Embryo Rescue Methods in Reciprocal Crosses between *B. napus* and Wild Relatives

Cross Combination		Embryo Rescue Methods*						
		Conventional				Ovary → Ovule	Ovary → Embryo	
Female	Male	Cross	Embryo	Ovule	Ovary			References
D. crassifolia (n = 13)	ditto	×	—	—	—	—	—	Ringdahl et al. (1987)
D. harra (n = 13)	ditto	×	—	—	—	—	—	Ringdahl et al. (1987)
D. muralis (n = 21)	ditto	○	—	—	—	—	—	Ringdahl et al. (1987)
S. arvensis (n = 9)	ditto	×, ×	—	×	○, ×	—	—	Kerlan et al. (1992); Bing et al. (1991, 1995)
S. pubescens (n = 9)	ditto	—	—	—	×	—	—	Inomata (1994)
S. alba (n = 12)	ditto	×, ×	○, ○	—	—	×	○	Lelivelt et al. (1993); Brown et al. (1997); Momotaz et al. (1998)
M. arvensis (n = 14)	ditto	×	—	—	○	—	—	Takahata et al. (1993)
M. nitens (n = 14)	ditto	—	—	○	—	—	—	Rawsthorne et al. (1998)
E. lyratus (n = 10)	ditto	—	—	—	—	○	—	Gundimeda et al. (1992)
O. violaceus (n = 12)	ditto	×	—	—	—	—	—	Li et al. (1995)

*○: Combination obtained hybrid more than a plant, △: combination obtained only embryo development, ×: no hybrid or embryo development.; —: not observed.

TABLE 11.1B
Production of F1 Hybrid using Different Embryo Rescue Methods in Reciprocal Crosses between *B. rapa* and Wild Relative

Cross Combination		Embryo Rescue Methods *						
Female	Male	Conventional Cross	Embryo	Ovule	Ovary	Ovary → Ovule	Ovary → Embryo	References
B. rapa (n = 10)	*Hirschfeldia incana* (n = 7)	O	—	—	—	—	—	Mizushima (1968)
ditto	*B. fruticulosa* (n = 8)	O	—	—	—	—	—	Mizushima (1968)
ditto	*B. bourgeaui* (n = 9)	—	—	—	O	—	—	Inomata (1986, 1993)
ditto	*B. montana* (n = 9)	—	—	—	O	—	—	Inomata (1993)
ditto	*B. cretica* (n = 9)	—	—	—	O	—	—	Inomata (1985, 1993)
ditto	*B. tournefortii* (n = 10)	O, O, ×	—	—	O	—	—	Prakash and Narain (1971); Narain and Prakash (1972); Choudhary et al. (2001)
ditto	*R. raphanistrum* (n = 9)	×	—	—	—	—	—	Lefol et al. (1997)
ditto	*E. sativa* (n = 11)	O	—	—	—	—	—	Mizushima (1950)
ditto	*Erucastrum canariense* (n = 9)	×	—	—	—	—	—	Bhaskar et al. (2002)
ditto	*E. gallicum* (n = 15)	O	—	—	—	—	—	Lefol et al. (1997)
ditto	*D. siettiana* (n = 8)	×	—	—	×	—	—	Nanda Kumar and Shivanna (1993)
ditto	*D. siifolia* (n = 10)	×	—	—	×	—	—	Batra et al. (1990)
ditto	*D. tenuifolia* (n = 11)	—	—	—	—	—	O	Jeung et al. (2007)
ditto	*S. arvensis* (n = 9)	O, ×, ×	—	×	—	×	—	Mizushima (1950); Bing et al. (1991, 1996); Momotaz et al. (1998)
ditto	*S. turgida* (n = 9)	—	—	×	O	O	—	Momotaz et al. (1998)
ditto	*S. alba* (*B. hirta*, n = 12)	—	—	—	O	—	—	Jandnrova and Dolezel (1995)
ditto	*M. arvensis* (n = 14)	—	—	—	O	—	—	Takahata and Takeda (1990)
ditto	*E. lyratus* (n = 10)	—	—	×	—	—	—	Gundimeda et al. (1992)

—continued

TABLE 11.1B (Continued)

Production of F1 Hybrid using Different Embryo Rescue Methods in Reciprocal Crosses between *B. rapa* and Wild Relative

Cross Combination		Embryo Rescue Methods *						
Female	Male	Conventional Cross	Embryo	Ovule	Ovary	Ovary → Ovule	Ovary → Embryo	References
ditto	*O. violaceus* (n = 12)	O	O	—	—	—	—	Li and Heneen (1999)
ditto	*Capsella bursa-pastoris* (n = 16)	O	—	—	—	—	—	Chen et al. (2007)
B. fruticulosa (n = 8)	*B. rapa* (n = 10)	O, Δ, O	—	O	—	O	—	Mizushima (1968; Takahata and Hinata (1983); Nanda Kumar et al. (1988); Chandra et al. (2004b)
B. maurorum (n = 8)	ditto	Δ	—	×	×	O	O	Chrungu et al. (1999); Jeung et al. (2003)
B. maurorum (n = 8)	ditto	Δ	—	×	—	O	—	Garg et al. (2007)
B. oxyrrhina (n = 9)	ditto	—	?	—	—	—	—	Prakash and Chopra (1990)
B. tournefortii (n = 10)	ditto	O, O	—	—	?	—	—	Mizushima (1968), Nagpal et al. (1996), Choudhary and Joshi (2001)
B. barrelieri (n = 10)	ditto	O	—	—	—	—	—	Takahata and Hinata (1983)
B. cossoniana (n = 16)	ditto	—	—	—	—	—	O	Jeung et al (2003)
R. raphanistrum (n = 9)	ditto	×	—	—	—	—	—	Lefol et al. (1997)
E. sativa (n = 11)	ditto	—	—	Δ	O	—	O	Matsuzawa and Sarashima (1986); Agnihotri et al. (1988); Jeung et al (2003)
E. vesicaria (n = 11)	ditto	—	—	—	—	—	O	Jeung et al. (2004)
Erucastrum canariense (n = 9)	ditto	×	—	O	O	O	—	Bhaskar et al. (2002)
Erucastrum cardaminoides (n = 9)	ditto	×	—	—	—	O	—	Chandra et al. (2004a)
E. gallicum (n = 15)	ditto	×	—	×	—	—	—	Batra et al. (1989), Lefol et al. (1997)

Species							References
D. erucoides (n = 7)	ditto	△	—	O	O	O	Vyas et al. (1995), Bhat et al. (2006), Jeung et al. (2006b)
D. erucoides (n = 7)	ditto	—	—	—	O	O	Garg et al. (2007)
D. siettiana (n = 8)	ditto	×	—	O	—	—	Nanda Kumar and Shivanna (1993)
D. virgata (n = 9)	ditto	O	—	—	—	—	Takahata and Hinata (1983)
D. catholica (n = 9)	ditto	—	—	—	O	O	Banga et al. (2003); Jeung et al. (2003)
D. siifolia (n = 10)	ditto	×	—	×	O	—	Batra et al. (1990); Ahuja et al. (2003)
D. tenuifolia (n = 11)	ditto	—	—	—	—	O	Jeung et al. (2006b)
D. muralis (n = 21)	ditto	—	—	—	—	O	Jeung et al. (2006b)
S. arvensis (n = 9)	ditto	×	×	×	×	—	Bing et al. (1991); Momotaz et al. (1998); Jeung et al. (2003)
S. alba (n = 12)	ditto	—	—	—	×	O	Momotaz et al. (1998); Jeung et al. (2003)
M. arvensis (n = 14)	ditto	—	—	O, ×	—	—	Takahata and Takeda (1990); Jeung et al. (2004)
Enarthrocarpus lyratus (n = 10)	ditto	—	—	—	O	·	Gundimeda et al. (1992)
O. violaceus (n = 12)	ditto	×	—	—	—	—	Li and Heneen (1999)

*○: Combination obtained hybrid, more than a plant, △: combination obtained only embryo development, ×: no hybrid or embryo development; —: not observed.

TABLE 11.1C
Production of F1 Hybrid using Different Embryo Rescue Methods in Reciprocal Crosses between *B. oleracea* and Wild Relatives

Cross Combination		Embryo Rescue Methods *						References
Female	Male	Conventional Cross	Embryo	Ovule	Ovary	Ovary → Ovule	Ovary → Embryo	
B. oleracea (n = 9) ×	*Hirschfeldia incana* (n = 7)	—	—	○	—	—	—	Quiros et al. (1988)
ditto	*S. arvensis* (n = 9)	○	—	×	—	—	—	Mizushima (1950); Momotaz et al. (1998)
ditto	*S. turgida* (n = 9)	—	—	○	—	—	—	Momotaz et al. (1998)
ditto	*S. alba* (n = 12)	—	—	○	—	—	—	Momotaz et al. (1998)
ditto	*M. arvensis* (n = 14)	—	—	○ (Placenta → Emb)	—	—	—	Bang et al. (2007)
ditto	*E. lyratus* (n = 10)	—	—	×	—	—	—	Gundimeda et al. (1992)
ditto	*O. violaceus* (n = 12)	○	—	—	—	—	—	Li and Heneen (1999)
B. maurorum (n = 8)	*B. oleracea* (n = 9)	○	—	—	○	○	○	Chrungu et al. (1999); Jeung et al. (2003)
B. tournefortii (n = 10)	ditto	○	—	—	△ (Ova → Ovu → Emb)	—	—	Mizushima (1968); Nagpal et al. (1996)
B. cossoniana (n = 16)	ditto	—	—	—	—	—	○	Jeung et al. (2003)
E. sativa (n = 11)	ditto	—	—	—	—	—	○	Matsuzawa and Sarashima (1986); Jeung et al. (2003)
E. vesicaria (n = 11)	ditto	—	—	—	—	—	○	Jeung et al. (2003)
Erucastrum abyssinicum (n = 16)	ditto	△	—	△	—	○	○	Rao et al. (1996); Sarmah and Sarla (1998)
D. erucoides (n = 7)	ditto	—	—	×	×	○	×, ○	Gundimeda et al. (1992); Vyas et al. (1995); Jeung et al. (2006b)
D. tenuifolia (n = 11)	ditto	—	—	—	×	—	—	Jeung et al. (2006b)

								References
S. arvensis (n = 9)	ditto	—	—	—	—	—	O	Momotaz et al. (1998), Jeung et al. (2003)
S. turgida (n = 9)	ditto	—	—	—	—	—	O	Jeung et al. (2003)
S. alba (n = 12)	ditto	—	×	×	O	—	—	Momotaz et al. (1998); Jeung et al. (2003)
M. arvensis (n = 14)	ditto	—	—	×, O	—	—	O	Jeung et al. (2003, 2004); Bang et al. (2007)
Enarthrocarpus lyratus (n = 10)	ditto	—	O	—	O	—	—	Gundimeda et al. (1992)
O. violaceus (n = 12)	ditto	O	—	—	—	—	—	Li and Heneen (1999)

*○: Combination obtained hybrid more than a plant, △: combination obtained only embryo development, ×: no hybrid or embryo development; —: not observed.

TABLE 11.1D
Production of F1 Hybrid using Different Embryo Rescue Methods in Reciprocal Crosses between *B. juncea* and Wild Relatives

| Cross Combination | | Embryo Rescue Methods * | | | | | | |
Female	Male	Conventional Cross	Embryo	Ovule	Ovary	Ovary → Ovule	Ovary → Embryo	References
B. juncea (n=18)	*B. oxyrrhina* (n = 9)	○	—	—	—	—	—	Bijral and Sharma (1999b)
ditto	*B. tournefortii* (n = 10)	—	—	—	○	—	—	Yadav et al. (1991)
ditto	*B. gravinae* (n = 20)	○	—	—	—	○	—	Nanda Kumar et al. (1989)
ditto	*R. raphanistrum* (n = 9)	×	—	—	—	—	—	Lefol et al. (1997)
ditto	*E. sativa* (n = 11)	○	—	—	—	—	—	Bijral and Sharma (1999a)
ditto	*E. gallicum* (n = 15)	×,×	—	—	×	—	—	Batra et al. (1989); Lefol et al. (1997)
ditto	*D. erucoides* (n = 7)	○	—	—	—	—	—	Inomata (1998)
ditto	*D. virgata* (n = 9)	—	—	—	○	—	—	Inomata (2003)
ditto	*D. siifolia* (n = 10)	×	—	—	×	—	—	Batra et al. (1990)
ditto	*S. arvensis* (n = 9)	○	—	—	—	—	—	Bing et al. (1991)
ditto	*S. pubescens* (n = 9)	—	—	—	○	—	—	Inomata (1988)
ditto	*S. alba* (n = 12)	—	—	○	△	—	—	Bajaj (1990)
ditto	*M. arvensis* (n = 14)	×	—	—	×	—	—	Takahata et al. (1993)
ditto	*Enarthrocarpus lyratus* (n = 10)	—	—	×	×	○	—	Gundimeda et al. (1992)
ditto	*O. violaceus* (n = 12)	○	—	—	—	—	—	Li et al. (1998)
ditto	*Crambe abyssinica* (n = 45)	—	—	—	○	—	—	Wang and Wuo (1998)
B. maurorum (n = 8)	*B. juncea* (n = 18)	—	—	×	×	○	—	Chrungu et al. (1999)
B. gravinae (n = 20)	ditto	×	—	×	×	×	—	Nanda Kumar et al. (1989)
R. raphanistrum (n = 9)	ditto	×	—	—	○	—	—	Lefol et al. (1997)
E. gallicum (n = 15)	ditto	×	—	—	○	—	—	Lefol et al. (1997)
Erucastrum abyssinicum (n = 16)	ditto	—	—	—	△	—	—	Rao et al. (1996)
D. erucoides (n = 7)	ditto	—	—	—	×	○	—	Vyas et al. (1995)

							Reference
D. catholica (n = 9)	ditto	—	—	—	—	O	Banga et al. (2003)
D. siifolia (n = 10)	ditto	×	—	—	O	O	Batra et al. (1990); Ahuja et al. (2003)
S. arvensis (n = 9)	ditto	×	—	—	—	—	Bing et al. (1991)
M. arvensis (n = 14)	ditto	×	—	—	△	—	Takahata et al. (1993)
M. moricandioides (n = 14)	ditto	—	—	—	O	—	Takahata et al. (1993a)
E. lyratus (n = 10)	ditto	—	—	×	—	×	Gundimeda et al. (1992)
O. violaceus (n = 12)	ditto	×	—	—	—	—	Li et al. (1998)

*O: Combination obtained hybrid more than a plant, △: combination obtained only embryo development, ×: no hybrid or embryo development; —: not observed.

TABLE 11.1E
Production of F1 Hybrid using Different Embryo Rescue Methods in Reciprocal Crosses between *R. sativus* and Wild Relatives

Cross Combination		Embryo Rescue Methods *						
Female	Male	Conventional Cross	Embryo	Ovule	Ovary	Ovary → Ovule	Ovary → Embryo	References
R. sativus (n = 9) ×	*B. maurorum* (n = 8)	—	×	—	—	—	—	Bang (1996); Bang et al. (1997)
ditto	*B. oxyrrhina* (n = 9)	—	×	—	—	—	—	Bang (1996); Bang et al. (1997)
(*R. caudatus*) (n = 9) ditto	*B. tournefortii* (n = 10)	×	—	—	—	—	—	Choudhary et al. (2000)
ditto	*B. cossoniana* (n = 16)	—	×,○	—	—	—	—	Bang (1996); Bang et al. (1997)
ditto	*E. sativa* (n = 11)	—	○	—	—	—	—	Bang (1996)
ditto	*E. vesicaria* (n = 11)	—	○	—	—	—	—	Bang (1996)
ditto	*D. erucoides* (n = 7)	—	×	—	—	—	—	Bang (1996)
ditto	*D. catholica* (n = 9)	—	×	—	—	—	—	Bang (1996)
ditto	*D. tenuisiliqua* (n = 9)	—	×	—	—	—	—	Bang (1996)
ditto	*D. tenuifolia* (n = 11)	×	×	—	—	—	—	Bang (1996); Bang et al. (2003)
ditto	*D. muralis* (n = 21)	—	×	—	—	—	—	Bang (1996)
ditto	*S. arvensis* (n = 9)	—	○	—	—	—	—	Bang (1996); Bang et al. (1996b)
ditto	*S. pubescens* (n = 9)	—	×	—	×	—	○	Bang (1996); Bang et al. (1996b)
ditto	*S. turgida* (n = 9)	—	○	—	—	—	—	Bang (1996); Bang et al. (1996b)
ditto	*S. alba* (n = 12)	—	○	—	—	—	—	Bang (1996); Bang et al. (1996b)
ditto	*M. arvensis* (n = 14)	—	×	—	—	—	—	Bang (1996)
B. maurorum (n = 8) ×	*R. sativus* (n = 9)	—	—	—	—	—	○	Bang (1996); Bang et al. (1998)
B. oxyrrhina (n = 9)	ditto	○	—	—	○	—	○	Bang (1996); Bang et al. (1997); Matsuzawa et al. (1997)
B. tournefortii (n = 10)	*R. caudatus* (n = 9)	○,○	—	—	—	—	—	Mizushima (1968); Choudhary et al. (2000)
B. cossoniana (n = 16)	*R. sativus* (n = 9)	—	○	—	—	—	○	Bang et al. (1997)

							Reference
E. sativa (n = 11)	ditto	—	—	—	○	○	Matsuzawa and Sarashima (1986); Bang (1996)
E. vesicaria (n = 11)	ditto	—	—	—	—	○	Bang (1996)
D. erucoides (n = 7)	ditto	—	—	—	—	×	Bang (1996)
D. catholica (n = 9)	ditto	—	—	—	—	○	Bang (1996)
D. tenuisiliqua (n = 9)	ditto	—	—	—	—	×	Bang (1996)
D. tenuifolia (n = 11)	ditto	—	—	—	—	×,○	Bang (1996); Bang et al. (2003)
D. muralis (n = 21)	ditto	—	—	—	—	○	Bang (1996)
S. arvensis (n = 9)	ditto	○	—	—	○	○	Mizushima (1950); Bang (1996); Bang et al. (1996b)
S. pubescens (n = 9)	ditto	—	—	—	—	×	Bang (1996); Bang et al. (1996b)
S. turgida (n = 9)	ditto	○,○	—	—	—	○	Bang (1996); Bang et al. (1996b)
S. alba (n = 12)	ditto	×	—	—	—	○	Bang (1996); Bang et al. (1996b)
M. arvensis (n = 14)	ditto	—	—	—	—	○	Bang et al. (1996a)

*○: Combination obtained hybrid more than a plant, △: combination obtained only embryo development, ×: no hybrid or embryo development; —: not observed.

combinations when wild relatives were used as the female parent. Jeung et al. (2003, 2004, 2006a, b, 2007) have been successful in producing many hybrids using the above cross combinations.

The hybrid plants were recovered through ovule culture when the wild species were used as the female parent. On the other hand, ovary or sequential culture was confirmed to be effective in reciprocal crosses.

BRASSICA JUNCEA × WILD RELATIVES

A number of attempts have been made to produce viable hybrids using *Brassica juncea* as the female parent. The hybrids from *Moricandia moricandioides* (Takahata et al., 1993) and *Crambe abyssinica* (Wang and Wuo, 1998) were obtained when these were crossed with *B. juncea*. Hybrid embryos were rescued using ovary and/or ovary-ovule sequential culture in the above cross combinations.

RAPHANUS SATIVUS × WILD RELATIVES

Many studies on intergeneric hybridization using *Raphanus sativus* and wild relatives have been reported in Japan. When two and four species of *Eruca* and *Sinapis*, respectively, were tried, hybrids were obtained from all combinations except *S. pubescens*. The efficiency could be enhanced in the reciprocal crosses when the wild relatives were used as the female parents. In reciprocal crosses, the ovary-embryo sequential culture was more successful. This culture technique might be an effective method for intergeneric hybridization between *R. sativus* and wild relatives.

CELL FUSION SYSTEM BETWEEN THE BRASSICA AND WILD RELATIVES

Alternatively, a somatic cell fusion technique has been applied to bypass the sexual barriers in wide hybridization. The parasexual hybrid plants developed were confirmed as distinct from the sexual ones with regard to various genetic features, especially in cytoplasmic organella, offering some useful resources for breeding strategies (see Table 11.1f).

INDUCTION OF AMPHIPLOIDAL LINES AND THEIR RELIABLE PRESERVATION

Amphidiploidal lines were developed by doubling the chromosome number of the F_1 hybrid plants with 0.2 % aqueous colchicine. Amphidiploidal plants were also developed sporadically in F_2 generations of amphihaploidal F_1 hybrids. However, reliable preservation or maintenance of these potential amphidiploidal lines was somewhat difficult due to hybrid sterility or hybrid incongruity. As shown in Table 11.2, success in the production and maintenance of each hybrid progeny was extremely low in comparison with parental species.

Hybrid progenies were bred between *Brassica napus, B. juncea,* and *B. rapa* having the basic genome "A" and their wild relatives by selfing and backcrossing with *Brassica* crops and/or open pollination. On the other hand, those lines between *B. oleracea* and *Raphanus sativus* that carried the corresponding "C" and "R" genome and wild relatives were successfully grown through embryo rescue techniques by overcoming the hybrid sterility encountered in interspecific and intergeneric hybridization.

Breeding attempts were made to produce potential hybrid progenies from F_1 hybrids. Many amphiploidal lines were successfully developed from monogenomic species such as *Brassica rapa, B. oleracea,* and *Raphanus sativus* in comparison with digenomic species (*B. napus* and *B. juncea*). In *B. napus, B. rapa,* and *B. juncea,* a large number of hybrid lines were backcrossed for substitution of cytoplasm or nucleus using the wild relatives as female parents.

In *Brassica napus,* F_1 hybrids between *B. napus* and *B. cretica* or *Sinapis arvensis* produced hybrid progenies in all self- or sib-, open-, and backcross-pollinations, including the amphiploidal hybrid progenies. When wild relatives were used as the female parent, progenies were

TABLE 11.1F
Production of F1 Hybrid by Cell Fusion System between *Brassica* Crops and Wild Relatives

Combination for Cell Fusion		F_1 Hybrid[*]	Ref.
B. juncea ($n = 18$)	*M. arvensis* ($n = 14$)	○	Kirti et al. (1992b)
B. juncea ($n = 18$)	*D. catholica* ($n = 9$)	○, ○	Kirti et al. (1995); Mohapatra et al. (1998)
B. juncea ($n = 18$)	*Trachystoma ballii* ($n = 8$)	·	Kirti et al. (1992a)
M. nitens ($n = 14$)	*B. oleracea* (n = 9)	○	Yan et al. (1999)
B. napus ($n = 19$)	*B. tournefortii* (n = 10) irradiated with x-ray	○	Stiewe and Röbbelen (1994)
B. napus ($n = 19$)	*M. arvensis* ($n = 14$)	○	O'Neill et al. (1996)
B. napus ($n = 19$)	*O. violaceus* ($n = 12$)	○	Hu et al. (2002a)
B. napus ($n = 19$)	*A. thaliana* ($n = 5$)	○, ○, ○, ○	Forsberg et al. (1994); Bohman et al. (2002); Yamagishi et al. (2002); Leino et al. (2003)
B. napus ($n = 19$)	*A. thaliana* irradiated or unirradiated with UV	○	Forsberg et al. (1994)
B. napus (n = 19)	*Crambe abyssinica* (n = 45)	○	Wang et al. (2003)
B. napus (n = 19)	*S. arvensis* (n = 9)	○	Hu et al. (2002b)
B. napus (n = 19)	*E. sativa* (n = 11)	○	Fahlesan et al. (1988); Sundberg and Glimelius (1991)
B. napus (n = 19)	*S. alba* (n = 12)	○	Lelivelt et al. (1993)
B. oleracea (n = 9)	*M. arvensis* (n = 14)	○	Toriyama et al. (1978)
M. arvensis (n = 14)	*B. oleracea* (n = 9)	○	Ishikawa et al. (2003)
B. oleracea (n = 9)	*S. turgida* (n = 9)	○	Toriyama et al. (1994)
S. alba (n = 12)	*B. oleracea* (n = 9)	○	Nothnagel et al. (1997)

[*]○:Combination obtained hybrid, more than a plant.

obtained by backcrossing with *B. napus*. F_1 hybrids between *B. rapa* and wild relatives have also been produced.

In the wide cross of *Brassica oleracea*, hybrid progenies have been produced only in the case of the F_1 hybrid between *B. oleracea* and *Moricandia arvensis*. On the other hand, amphiploids and their hybrid progenies could be obtained from some combinations using *B. oleracea* as the male parent. It would be assumed that it might be difficult to maintain the progenies when *B. oleracea* is used as the female parent. These facts have also been recognized in F_1 hybrids, including cultivated crops such as *Raphanus sativus* and *B. nigra* (Sarashima and Matsuzawa, 1989).

In the reciprocal cross between *Brassica juncea* and wild relatives, many F_1 hybrids were obtained and backcrossed with *B. juncea*. On the other hand, production of hybrid progenies from F_1 hybrids between *B. juncea* and wild relatives was limited and fewer amphiploids were developed. In the distant cross of *Raphanus sativus*, the amphidiploids and sesquidiploids were occasionally produced through self- or open-pollination systems. Bang (1996) and Bang et al. (1998, 2000, 2002, 2003) have been able to produce more potential amphidiploids.

TABLE 11.2
Production and Maintenance of Progeny from F₁ Hybrid through Several Crosses

Cross Combination		Production of Progeny from F₁ Hybrid [1]			Production of Amphiploid [2]	Ref.
Female	Male	Self and/or Sib	Open	Backcross		
B. napus (n=19) ×	*B. bourgeaui* (n = 9)	—	—	○	—	Inomata (2002)
ditto	*B. montana* (n = 9)	○	○	○	—	Inomata (2002)
ditto	*B. cretica* (n = 9)	○	○	○	○	Inomata (2002)
ditto	*B. gravinae* (n = 20)	○	○	—	○	Nanda Kumar et al. (1989)
ditto	*R. raphanistrum* (n = 9)	—	—	○	—	Lefol et al. (1997)
ditto	*E. sativa* (n = 11)	△	—	○	○	Heyn (1977)
ditto	*E. gallicum* (n = 15)	○	—	○	—	Chen et al. (2007)
ditto	*D. tenuifolia* (n = 11)	—	—	○	—	Heyn (1977)
ditto	*D. muralis* (n = 21)	○	—	—	—	Bijral and Sharma (1996a)
ditto	*S. arvensis* (n = 9)	○	○	○	○, ○	Mizushima (1950); Inomata (1988); Bing et al. (1991, 1995)
ditto	*S. pubescens* (n = 9)	—	○	—	○	Inomata (1994)
ditto	*S. alba* (n = 12)	—	—	○	—	Heyn (1977)
ditto	*Matthiola incana* (n = 7)	○	—	—	○	Luo et al. (2003)
ditto	*Rapistrum rugosum* (n = 8)	—	—	○	—	Heyn (1977)
ditto	*O. violaceus* (n = 12)	○, ○, ○	—	—	—	Li and Luo (1993); Li et al. (1995); Hua and Li (2006); Wu et al. (1997)
ditto	*Capsella bursa-pastoris* (n = 16)	—	—	○	—	Lefol et al. (1997)
Hirschfeldia incana (n = 7)	*B. napus* (n = 19)	—	—	○	—	Lefol et al. (1996)
B. maurorum (n = 8)	ditto	—	—	○	○	Chrungu et al. (1999)

Cross (♀)	(♂)					References
B. tournefortii (n = 10) ×	ditto	○	—	○,○	—	Nagpal et al. (1996); Pahwa et al. (2004)
D. erucoides (n = 7) ×	ditto	—	—	○,○	—	Ringdahl et al. (1987); Delourme et al. (1989); Vyas et al. (1995)
D. muralis (n = 21) ×	ditto	—	—	○,○	—	Fan et al. (1985); Ringdahl et al. (1987)
S. alba (n = 12) ×	ditto	—	—	○	—	Lelivelt et al. (1993); Brown et al. (1997); Momotaz et al. (1998)
M. arvensis (n = 14) ×	ditto	—	○	○	—	Takahata et al. (1993b)
E. lyratus (n = 10) ×	ditto	—	—	○	—	Gundimeda et al. (1992)
O. violaceus (n = 12) ×	ditto	○	—	—	—	Li et al. (1995)
B. rapa (n = 10) ×	B. fruticulosa (n = 8)	—	—	—	—	Mizushima (1968)
ditto	B. montana (n = 9)	×	○	○	△	Inomata (1993)
ditto	B. bourgeaui (n = 9)	×	○	○	△	Inomata (1993)
ditto	B. cretica (n = 9)	×	○	○	△	Inomata (1993)
ditto	B. tournefortii (n = 10)	?	○	○	—	Prakash et al. (1971); Choudhary and Joshi (2001),
ditto	E. gallicum (n = 15)	—	—	○	—	Lefol et al. (1997)
ditto	D. tenuifolia (n = 11)	○	—	○	○	Jeung et al. (2007)
ditto	S. arvensis (n = 9)	?	—	—	—	Mizushima (1950); Bing et al. (1996); Momotaz et al. (1998)
ditto	S. alba (n = 12)	○	—	—	△	Jandurova and Dolezel (1995)

—continued

TABLE 11.2 (Continued)
Production and Maintenance of Progeny from F$_1$ Hybrid through Several Crosses

Cross Combination		Production of Progeny from F$_1$ Hybrid [1]			Production of Amphiploid [2]	Ref.
Female	Male	Self and/or Sib	Open	Backcross		
ditto	O. violaceus (n = 12)	O	—	—	O	Wu et al. (1996); Li and Heneen (1999),
ditto	Capsella bursa-pastoris (n = 16)	×	—	O	—	Chen et al. (2007)
B. fruticulosa (n = 8) ×	B. rapa (n = 10)	—	O	—	O	Mizushima (1968); Takahata and Hinata (1983); Nanda Kumar et al. (1988)
B. maurorum (n = 8)	ditto	×	—	O	O,O	Takahata and Hinata (1983); Chrungu et al. (1999); Jeung et al. (2003)
B. maurorum (n = 8)	ditto	O	O	—	—	Garg et al. (2007)
B. oxyrrhina (n = 9)	ditto	—	—	O	O	Prakash and Chopra (1990)
B. tournefortii (n = 10)	ditto	?	—	O	O	Mizushima (1968); Choudhary and Joshi (2001); Nagpal et al. (1996)
B. cossoniana (n = 16)	ditto	—	—	O	O	Jeung et al (2003)
E. sativa (n = 11)	ditto	?	—	—	O	Matsuzawa and Sarashima (1986)
Erucastrum canariense (n =9)	ditto	—	—	O	O	Bhaskar et al. (2002)
Erucastrum cardaminoides (n =9)	ditto	?	—	—	O	Chandra et al. (2004a)
D. erucoides (n = 7)	ditto	O	×,O	×	—	Vyas et al. (1995); Garg et al. (2007)
D. erucoides (n = 7)	ditto	—	—	O	O	Jeung et al. (2006b)

Female parent	Male parent					Reference
D. erucoides (n = 7)	ditto	—	—	○	—	Bhat et al. (2006)
D. siettiana (n = 8)	ditto	—	—	○	○	Nanda Kumar and Shivanna (1993)
D. catholica (n = 9)	ditto	○,○,○	—	○,○	○,○	Banga et al. (2003); Jeung et al. (2003)
D. siifolia (n = 10)	ditto	?	—	—	—	Batra et al. (1990); Ahuja et al. (2003)
M. arvensis (n = 14)	ditto	—	—	○	○	Jeung et al. (2006a)
Enarthrocarpus lyratus (n = 10)	ditto	—	—	○	—	Gundimeda et al. (1992); Banga et al. (2003)
B. oleracea (n = 9) ×	M. arvensis (n = 14)	—	—	○	—	Bang et al. (2007)
B. maurorum (n = 8) ×	B. oleracea (n = 9)	○	—	○	○	Jeong et al. (2003)
B. tournefortii (n = 10)	ditto	?	—	○	○	Mizushima(1968); Nagpal et al. (1996)
E. sativa (n = 11)	ditto	?	—	—	—	Matsuzawa and Sarashima (1987)
Erucastrum abyssinicum (n = 16)	ditto	○	—	○	○	Rao et al. (1996); Sarmah and Sarla (1998)
D. erucoides (n = 7)	ditto	—	×	×	○	Vyas et al. (1995); Gundimeda et al. (1992); Jeung et al. (2006b)
S. arvensis (n = 9)	ditto	—	—	×	—	Jeung et al. (2003)
S. turgida (n = 9)	ditto	?	—	×	○	Momotaz et al. (1998), Jeung et al. (2003)
S. alba (n = 12)	ditto	○	—	○	○	Wei et al. (2007)
M. arvensis (n = 14)	ditto	—	○	○	○	Jeung et al. (2006a); Bang et al. (2007)
E. lyratus (n = 10)	ditto	—	○	○	○	Gundimeda et al. (1992)

—continued

TABLE 11.2 (Continued)
Production and Maintenance of Progeny from F₁ Hybrid through Several Crosses

| Cross Combination | | Production of Progeny from F₁ Hybrid[1] | | | Production of Amphiploid[2] | Ref. |
Female	Male	Self and/or Sib	Open	Backcross		
O. violaceus (n = 12)	ditto	×	—		—	Li and Heneen (1999)
B. juncea (n = 18)	B. oxyrrhina (n = 9)	?	—	○	—	Bijral and Sharma (1999b)
ditto	B. tournefortii (n = 10)	?	—	—	—	Vyas et al. (1995)
ditto	B. gravinae (n = 20)	○	○	—	○	Nanda Kumar et al. (1989)
ditto	E. sativa (n = 11)	○	—	○	—	Bijral and Sharma (1999a)
ditto	D. erucoides (n = 7)	—	×	×	—	Inomata (1998)
ditto	D. virgata (n = 9)	—	○	○	○	Inomata (2003)
ditto	S. arvensis (n = 9)	×	×	○	—	Mizushima (1950)
ditto	S. pubescens (n = 9)	×	○	○	—	Inomata (1988)
ditto	O. violaceus (n = 12)	○	○	×	—	Li et al. (1998)
ditto	Crambe abyssinica (n = 45)	?	○	—	—	Wang and Wuo (1998)
B. maurorum (n = 8) ×	B. juncea (n = 18)	—	—	○	—	Chrungu et al. (1999)
E. gallicum (n = 15)	ditto	—	—	○	—	Batra et al. (1989)
Erucastrum abyssinicum (n = 16)	ditto	—	○	○	○	Rao et al. (1996)
D. erucoides (n = 7)	ditto	—	—	○	—	Vyas et al. (1995)
D. siifolia (n = 10)	ditto	○	—	○	—	Batra et al. (1990)
M. arvensis (n = 14)	ditto	—	○	○	○	Takahata et al. (1993b)
E. lyratus (n = 10)	ditto	—	—	○	—	Gundimeda et al. (1992)
R. sativus (n = 9) ×	E. sativa (n = 11)	—	○	○	○	Bang (1996)
ditto	S. arvensis (n = 9)	—	○	—	—	Bang (1996)

	R. sativus (n = 9)					Reference
B. maurorum (n = 8) ×	R. sativus (n = 9)	○	—	—	○	Bang (1996); Bang et al. (1998)
B. oxyrrhina (n = 9)	ditto	s	s	s	s	Matsuzawa et al. (1997)
B. tournefortii (n = 10)	(R. caudatus) (n = 9)	○	—	—	○	Choudhary et al. (2000)
B. cossoniana (n = 16)	ditto	○	—	—	○	Bang et al. (2000)
E. sativa (n = 11)	ditto	?	○	—	—	Matsuzawa and Sarashima (1986); Bang (1996)
E. vesicaria (n = 11)	ditto	—	○	—	—	Bang (1996)
D. tenuifolia (n = 11)	ditto	—	—	○	○	Bang et al. (2003)
S. arvensis (n = 9)	ditto	—	○	—	—	Bang (1996)
S. turgida (n = 9)	ditto	○	○	—	○	Bang (1996); Bang et al. (1996b)
M. arvensis (n = 14)	ditto	—	—	x, ○	○	Takahata et al. (1993b); Bang (1996)

1) ○: Production of progenesis, △ or ?: unidentified production of progenesis, and ×: no progeny production.

2) ○: Production of amphiploid, △ or ?: unidentified production of one, and ×: no one production; —: not observed.

TABLE 11.3
Production and Maintenance of Progeny from F$_1$ Hybrid Derived from Cell Fusion by Several Crosses

Cell Fusion		Production of Progeny from F1 Hybrids [1]			Production of Amphiploid [2]	Ref.
		Self	Open	Backcross		
B. napus (n = 19)	B. tournefortii (n = 10)	O	—	O	—	Stiewe and Röbbelen (1994)
B. napus (n = 19)	E. sativa (n = 11)	O	—	O	—	Sundberg and Glimelius (1991)
B. napus (n = 19)	O. violaceus (n = 12)	—	—	O	O	Hu et al. (2002a)
B. napus (n = 19)	A. thaliana (n = 5)	O	—	O	O	Forsberg et al. (1994, 1998)
B. napus (n = 19)	A. thaliana (n = 5)	O	—	O, O	O	Bohman et al. (2002); Leino et al. (2003, 2004)
B. napus (n = 19)	S. arvensis (n = 9)	O	—	O	—	Hu et al. (2002b)
B. napus (n = 19)	Crambe abyssinica (n = 45)	O	—	O	—	Wang et al. (2003, 2004)
B. juncea (n = 18)	M. arvensis (n = 14)	—	—	O	—	Kirti et al. (1992b)
B. juncea (n = 18)	D. catholica (n = 9)	—	O	O	—	Kirti et al. (1995); Pathania et al. (2003)
B. juncea (n = 18)	Trachystoma ballii (n = 8)	×	—	O	O	Kirti et al. (1992a)
M. nitens (n = 14)	B. oleracea (n = 9)	×	—	O	O	Yan et al. (1999)
M. arvensis (n = 14)	B. oleracea (n = 9)	—	—	O	O	Ishikawa et al. (2003)
S. alba (n = 12)	B. oleracea (n = 9)	—	—	O	O	Nothnagel et al. (1997)
S. alba (n = 12)	B. napus (n = 19)	—	—	O	O	Lelivelt et al. (1993)

[1] O: Production of progenesis, ×: no progeny production; —: not observed.
[2] O: Production of amphiploid.

Amphidiploidal lines developed by somatic cell fusion could offer valuable genetic resources not only for the traits under nuclear genes, but also for those under the cytoplasmic ones contributing to breeding strategies (Table 11.3). Genetic modification in both nuclear and cytoplasmic backgrounds was studied with special reference toward developing the hybrid seed production system in the backcrossed progenies.

PRODUCTION AND MAINTENANCE OF SYNTHETIC AMPHIDIPLOID LINES

Synthetic fertile amphidiploid lines derived from hybridization between cultivated crops and wild relatives have been used as genetic stocks in breeding programs, especially as bridging material for the transfer of desirable traits from wild species to cultivated species. The fertile amphidiploid lines obtained from doubling the chromosome number of the sterile hybrids and somatic hybridization between cultivated crops and wild allied genera were described as novel *Brassica* crops such as "Hakuran" (*Brassica campestris* × *B. oleracea*), "Radicole" (*Raphanus sativus* × *B. oleracea*), and "Raparadish" (*B. campestris* × *R. sativus*) (Namai, 1987). Recently, researchers have produced a number of amphidiploid lines between crop cultivars and wild species, including the amphidiploid lines of *B. fruticulosa* × *R. sativus* (Bang et al., 1997, 2000), *B. maurorum* × *R. sativus* (Bang et al., 1997, 1998), *B. oxyrrhina* × *R. sativus* (Bang et al., 1997; Matsuzawa et al., 1997), *B. oxyrrhina* × *B. rapa* (unpublished results), *B. oxyrrhina* × *B. oleracea* (unpublished results), *B. rapa* × *Diplotaxis tenuifolia* (Jeung et al., 2009, in press), and *B. oleracea* × *D. tenuifolia* (unpublished results). These amphidiploid lines show low pollen and seed fertility due to irregular chromosome behavior in the pollen mother cells (PMCs) during initial generation. And, with the advancement of generations, they attained high pollen and seed fertility resulting from complete meiotic stabilization (Table 11.4). Sarashima and Matsuzawa (1989) proposed a comprehensive nomenclature for the artificial amphidiploid lines between mono-genomic *Brassica* crops, and have maintained these as novel true-breeding lines under the new nomenclature. The amphidiploid line of *B. fruticulosa* × *R. sativus* may be called × *Raphanobrassica* (*Rb*) *fruticulosa-sativus, R. sativus* × *B. rapa* as ×*Rb. sativus-rapa*, and the amphidiploid line of *B. rapa* × *D. tenuifolia* may be called × *Brassicodiplotaxis rapa-tenuifolia*.

In recent studies, these species were used as potential materials for investigating the genetic mechanism of the C_3-C_4 intermediate characteristics, and the breeding of new leafy salad vegetables. Both *Moricandia arvensis* and *Diplotaxis tenuifolia* are unique species among the *Brassicaceae*, in which C_4 species have not been found until now because of intermediate C_3-C_4 photosynthetic activity. The C_3-C_4 intermediate species show a low CO_2 compensation point and Kranz-like leaf anatomy because of the bundle sheath cells (BSCs), which may include numerous chloroplasts and mitochondria. It has been suggested that the Kranz-like leaf anatomy and biochemical components of C_3-C_4 intermediate characteristics are controlled by different genetic mechanisms (Rylott et al., 1998). The four subunits of glycine decarboxylase (GDC), including the P-protein, are encoded by the nuclear genome (Douce and Heldt, 2000), and the development of chloroplast and mitochondria is also under the control of the nuclear genome (Leon et al., 1998). With respect to the expression of C_3-C_4 intermediate characteristics, Apel et al. (1984) first reported that the intergeneric F_1 and BC_1 hybrids between *Moricandia arvensis* and *Brassica alboglabra* (C_3) showed intermediate gas exchange traits, which may further indicate the intermediate expression of C_3 and C_4 characteristics. In recent studies with intergeneric hybrids between *D. tenuifolia* (C_3-C_4) and *Raphanus sativus* (C_3) and the reciprocal intergeneric hybrids between *M. arvensis* (C_3-C_4) and *B. oleracea* (C_3), Ueno et al. (2003; 2007) demonstrated that the C_3-C_4 intermediate characteristics (such as the BSC-dominant expression of the GDC P-protein) reduced the CO_2 compensation point and were controlled by the nuclear genome according to the ratio of genome constitution in their hybrids.

Diplotaxis tenuifolia also has considerable potential as healthy leafy salad vegetables because of bioactive phytochemicals, and is mainly used for human consumption under the common name of rocket together with some *Eruca* species (Martínez-Sánchez et al., 2007). The amphidiploid line

TABLE 11.4
Cytogenetical Stability and Fertility of Intergeneric and Interspecific Amphidiploid Lines Synthesized between Cultivated Species and Wild Ones

Cross Combination (2n)	Generation	2n	IV	III	II	I	Pollen Fertility (%)	Selfing	Backcross To Wild Species	Backcross To Crop Species	Ref.
Brassica fruticulosa (32) × *Raphanus sativus* (18)	F₁	50	0.56 (0–2)	0.12 (0–2)	22.48 (19–25)	2.44 (0–8)	55.1	8 P			Bang et al. (1997, 2000)
	F₂	50			24.6	0.7	50.3 (18.9–84.4)	34.7 (0–88.5)	251.7 (0.5–540)	0	Bang et al. (2000)
	F₃	50								4*	Bang et al. (2000)
B. fruticulosa (16) × *B. campestris* (20)	F₁	36			(14–18)	(0–6)	50.0			**	Nanda Kumar et al. (1988)
× *B. rapa* (20)	F₁	36		(0–1)	18		Normal				Chandra et al. (2004)
B. maurorum (16) × *R. sativus* (18)	F₁	34			16.67 (14–17)	0.66 (0–6)	78.0 (75.5–87.6)	53.3 (47.2–59.3)			Bang et al. (1997, 1998)
	F₂	34			16.8 (14–17)	0.4 (0–6)	90.9 (56.1–98.6)	99.5 (34.4–192.3)			Bang et al. (1998)
	F₃	34			16.9 (16–17)	0.2 (0–2)	87.9 (64.4–95.2)	37.8 (2.9–87.0)	204.5 (5–544.2)	0	Bang et al. (1998)
	F₄	34			16.8 (16–17)	0.3 (0–2)	90.2 (78.8–97.2)	38.8 (0–66.7)	217.0 (0–368.8)	0	Bang et al. (1998)
× *B. nigra* (16)	F₁	32	>0.28 (0–1)		14.7 (12–16)	0.28 (0–2)	64.0			**	Chrungu et al. (1999)
× *B. rapa* (20)	F₁	36	0.15 (0–1)	0.11 (0–1)	17.3 (14–18)	0.47 (0–4)	75.0	**		**	Garg et al. (2007)
× *B. napus* (38)	F₁	54			26.3 (25–27)	1.3 (0–4)	Sterile			**	Chrungu et al. (1999)
× *B. carinata* (34)	F₁	50			24.5 (24–25)	1.0 (0–2)	Sterile			**	Chrungu et al. (1999)
B. napus (38) × *Sinapis srvensis* (18)	F₁	56	0.05 (0–1)		25.7 (23–28)	4.0 (0–10)	66–77				Kerlan et al. (1993)
+ *S. alba* (24)	S₁	62						100		370	Wang et al. (2005)
B. oxyrrhina (18) × *R. sativus* (18)	F₁	36			17.59 (15–18)	0.83 (0–6)	81.6 (70–82)	277 OP		1.3	Bang et al. (1997)

Parentage	Gen.	2n								Reference
B. tournefortii (20) × B. campestris (20)	F₂	36			17.38 (18–16)	1.23 (2–4)	87.5 (80–98)	386.7 (330–420) 330		Nagpal et al. (1996)
	F₃	36						**		
× B. nigra (16)	F₁	40			19.1	1.8 (19–20)	76.0 (0–2)	**		Nagpal et al. (1996)
× B. oleracea (18)	F₁	36			17.2	1.6 (16–18)	93.0 (0–2)	**		Nagpal et al. (1996)
× R. caudatus (18)	F₁	38	0.2 (0–3)	0.01 (0–1)	18.1 / 18.3 (13–19)	1.7 (18–19) / 0.2 (0–2)	10.0 (0–2) / 58.3	High		Choudhary et al. (2000)
		38	0.2 (0–3)		18.6 (15–19)		77.3	420		
Diplotaxis catholica (18) × B. rapa (20)	F₁	38	0.06 (0–1)	0.47 (1–2)	18.03 (16–19)	0.28 (0–1)	73.0			Banga et al. (2003)
D. erucoides (14) × B. campestris (20)	F₁	34		17	16.49			**		Vyas et al. (1995)
× B. rapa (20)	F₁		0.05	0.24		1	75.0	**	0	Garg et al. (2007)
D. siifolia (20) × B. juncea (36)	F₁	56		20	0	60.0	**	**		Batra et al. (1990)
D. tenuifolia (22) × R. sativus (18)	F₁	40			0	75.0	0	0*	6*	Bang et al. (2003)
D. virgata (18) × B. juncea (36)	BC₁	54		24.0	6.1	39.1 (21–26)	74.6 (2–12)	(0–83.2)	538.8	Inomata (2003)
Enarthrocarpus lyratus (20) × B. oleracea (18)	F₁	38	0.03 (0–6)	15.13 (0–1)	(7–19)	82.0	**op			Gundimeda et al. (1992)
Eruca sativa (22) × R. sativus (18)	F₁	40		18.76 (14–20)	2.48 (0–12)	4.0 (0–12)	0		11.5	Bang (1996) (0, 22.9)
Erucastrum Canariense (18) × B. rapa (20)	F₁	38	(0–1)	(18–19)	(0–3)	(65–80)	**	**		Bhaskar et al. (2002)
Lesquerella fendleri + B. napus (38) (12)	S₁	50				5		18		Skarzhinskaya et al. (1996)
M. arvensis (28) × R. sativus (18)	F₁	46			22.72 (22–23)	0.56 (0–2)	66.6	0	0, 6*	Bang et al. (1996)
+ B. oleracea (18)	S₁				(20–23)	(0–6)	50	0	45	Ishikawa et al. (2003) 112 (39–322)
× B. oleracea (18)	F₂						3.7			Bang et al. (2007)
+ B. juncea (36)	S₁		(0–3)		(?–19)	(10–22)	0	0	12*	Kirti et al. (1992) [b]

—continued

TABLE 11.4 (Continued)

Cytogenetical Stability and Fertility of intergeneric and Interspecific Amphidiploid Lines Synthesized between Cultivated Species and Wild Ones

Cross Combination (2n)	Generation	2n	Chromosome Pairing at MI of Meiosis				Pollen Fertility (%)	Seed Setting (%)			Ref.
								Selfing	Backcross To		
			IV	III	II	I			Wild Species	Crop Species	
Sinapis arvensis (18) × *R. sativus* (18)	F_2	36			9.23 (6–11)	17.54 (14–24)	30.8	29[op]		6.5	Bang (1996)
S. turgida (18) × *R. sativus* (18)	F_1	36		-	-	50	50[op]			6.5	Bang et al. (1996)
	F_1	36		0.01 (0–4)	8.72 (3–13)	18.3 (8–30)	32.1 (28.6–35.5)	1.04, 109[op] (0–3.1)	5.2 (2.3–7.1)		
Trachystoma balli (16) + *B. juncea* (36)	S_1	52		(0–2)	26	(0–2)	0 (Very low)	0		**	Kirti et al. (1992)[a]
R. sativus (18) × *S. arvensis* (18)	F_1	36			17.9 (17–18)	0.2 (0–2)	58				Bang et al. (1996)
R. sativus (18) × *S. pubescens* (18)	F_1	36			17.74 (16–18)	0.52 (0–4)	69.9 (68.7–71)	0	0		Bang et al. (1996)

p Number of F_2 plants obtained by selfing and/or sib crossing.

op Number of seeds obtained by open pollination.

* Number of BC_1 hybrids obtained by embryo rescue.

** The F_2 seed was obtained.

of *Brassica rapa* × *D. tenuifolia* has desirable traits, such as morphological characters, vigorous growth, and fragrance specific to *D. tenuifolia*, but its seed fertility was not sufficient to use it as a commercial crop. This line, however, could produce abundant seeds that grew to be sesquidiploid plants, inheriting the traits when backcrossed to the released cultivars of *B. rapa* (Jeung et al., 2009, in press). Therefore, the amphidiploid lines of both *B. rapa* × *D. tenuifolia* and *B. oleracea* × *D. tenuifolia* could be useful genetic stocks for breeding new leafy salad vegetables.

The sesquidiploids with various genome constitutions produced through reciprocal backcrosses of the amphidiploid with parental species have been used as bridging plants to obtain alien introgression lines, alloplasmic lines, alien chromosome addition lines, and alien chromosome substitution lines.

Alien Gene(s) Introgression Lines (AGILs)

Warwick (1993) described the wild genera of the tribe Brassiceae as sources of agronomic traits. A large collection of wild allies as sources of resistance to blackleg or stem canker (*Leptosphaeria maculans*), *Alternaria* leaf spot (*Alternaria* spp.), and clubroot (*Plasmodiophora brassicae*) have been screened. Interspecific and intergeneric hybridization between wild allies and *Brassica* crop species and their progenies would certainly provide valuable breeding materials for introgression of some agronomic traits. The possibilities of using interspecific and intergeneric hybridization between *Brassica napus* and some wild species in the tribe Brassiceae to introduce resistance to the important fungal pathogens were reviewed by Siemens (2002). As indicated by Namai (1987), the introgression of desirable gene(s) from one species to another species of Brassiceae has been accomplished via meiotic allosynthesis in *Brassica* crops. Bang (1996) described the homoeologous paring between the *Raphanus sativus* chromosome and *Sinapis arvensis* chromosome in the *S. arvensis* monosomic addition line of *R. sativus* (MAL-d type). The translation of pollen fertility restorer gene(s) from cytoplasmic donor species to alloplasmic *Brassica* oil crops has been obtained through homoeologous pairing and between the AB and M chromosome in *Moricandia arvensis* MAL of *B. juncea* (Prakash et al., 1998), and between the A and E chromosome in *Enarthrocarpus lyratus* MAL of *B. rapa* (Bang et al., 2003). Recently, intergeneric transfer of resistance to the beet cyst nematode (*Heterodera schachii*) and clubroot (*Plasmodiophora brassicae*) from *R. sativus* to *B. napus* was successfully achieved using the *R. sativus* MALs of *B. napus* (Budahn et al., 2006; Akaba et al., 2008). The analysis of the gene and genome, and efficient induction and identification of recombination for introgression, might be prerequisite to the development of an alien gene introgression line. The genetic linkage maps recently constructed for some *Brassica* crops may offer useful tools for a genetic approach for the introgression.

Alloplasmic Lines (ALPLs)

F_1 hybrid seed production in *Brassica* crops is presently based on self-incompatibility and cytoplasmic male sterility (CMS). The self-incompatibility controlled by *S*-alleles is not sufficiently stable to environmental factors. CMS, which is a maternally inherited trait, has been ascertained as more successful because of the stable expression of pollen sterility without any obvious changes in vegetative growth and female fertility, as well as its suitability for the development of hybrid parental lines. The ALPLs developed by exchanging the cytoplasm through interspecific and intergeneric hybridization and successive backcrossing to nuclear genome donor species acquired the advantages of inducing CMS and generating some agronomic traits, such as herbicide resistance, modification in flavor content, morphogenetic potentials, etc. (Namai, 1987). Sarashima et al. (1990) produced many kinds of alloplasmic lines of mono-genomic crops (*Brassica campestris*, *B. oleracea,* and *Raphanus sativus*), but most of them did not exhibit any agronomically useful trait, except for the CMS of *B. oleracea* carrying *R. sativus* cytoplasm. The CMS in alloplasmic lines can be induced by unsuitable genetic affinity between the mitochondrial genome and nuclear genome. A number of

TABLE 11.5
Characteristics of Various Alloplasmic Lines of Cultivated Species in Wild Species Cytoplasm Background

Cytoplasm Species	Nuclear Species	Origin	Characteristic	Ref.
Arabidopsis thaliana	*B. napus*	*A. thaliana* + *B. napus*	Rearrangements of mt-genome, *B. napus* plastid genome, various types of aberrant growth and flower development	Leino et al. (2003)
D. berthautii	*B. juncea*	*D. berthautii* × *B. campestris*	Male sterility, smaller nectaries, showing four floral types	Malik et al. (1999)
D. catholica	*B. juncea*	cms-*B. juncea* × progenies of *D. catholica*+ *B. juncea*	Fertility restorer line, good pollen fertility, flower normal to apetalous, bilocular	Pathania et al. (2003)
D. erucoides	*B. juncea*	*D. erucoides* × *B. campestris*	Male sterility, developed nectarines, small and pale anthers	Malik et al. (1999)
Enarthrocarpus lyratus	*B. juncea*	(*E. lyratus*) *B. rapa* × *B. nigra*	Complete male sterility, slight leaf yellowing and delayed maturity, introgression *Rf* -gene(s) through *E. lyratus* MAL of *B. juncea*	Banga et al. (2003)
Eruca sativa	*B. campestris*	*E. sativa* × *B. campestris*	Six CMS types; petaloid, antherless, brown anther, partial petaloid, partial antherless, petaloid antherless. seed fertility, normal green leaf and nectary gland development	Matsuzawa et al. (1999)
E. vesicaria	*R. sativus*	*E. vesicaria* × *R. sativus*	Severe chlorosis, flowers with non-dehiscent slender anthers, both male sterile line and male fertile line in B_2F_2, female fertility	Bang (1996)
M. arvensis	*R. sativus*	*M. arvensis* × *R. sativus*	Complete male sterility having pollen grains without fertilizability, light green leaf	Bang et al. (2002)

wild allies, cytoplasm have been induced aggressively as CMS sources in several *Brassica* crops of *B. rapa* (*B. campestris*), *B. juncea,* and *B. napus* through interspecific and intergeneric hybridization and successive backcrossing. There included very diverse wild species as reviewed by Prakash et al. (1999), such as *Diplotaxis muralis, D. catholica, D. siifolia, D. erucoides, B. oxyrrhina, B. tournefortii, Moricandia arvensis,* and *Trachystoma ballii.* Recently, some new CMS sources have been induced in several *Brassica* crops in addition to *R. sativus; Arabidopsis thaliana, D. berthautii, Enarthrocarpus lyratus, Eruca sativa, Eruca vesicaria,* and *M. arvensis* (Table 11.5).

Alloplasmic *Brassica napus* carrying *Arabidopsis thaliana* cytoplasm, which was obtained from six backcrosses of the somatic hybrid to *B. napus,* showed the morphological and functional aberrations of only stamens and petals, while all plants had normal morphology in the sepals and pistils, and with fully female fertility (Leino et al., 2003). Two alloplasmic *B. juncea* were produced using the amphidiploid lines of *Diplotaxis erucoides* × *B. campestris* and *D. berthautii* × *B. campestris* as the cytoplasmic donor plant (Malik et al., 1999). Alloplasmic *B. campestris* with *Eruca sativa* cytoplasm was produced via a sesquidiploid plant obtained from the amphihaploid F_1 hybrid by open pollination (Matsuzawa et al., 1996), and alloplasmic *R. sativus* with *Eruca vesicaria* cytoplasm also was produced via the same method (Bang, 1996). Alloplasmic *R. sativus* with *M. arvensis* cytoplasm produced a sesquidiploid plant that was generated from the backcrossing of amphidiploid F_1 hybrid to *R. sativus* using the embryo rescue technique (Bang et al., 1996, 2002). As indicated by Namai (1976), the sesquidiploid plant seems to be a bridge plant to generate an alloplasmic line with no barrier(s) in the various interspecific and intergeneric hybridizations. Occasionally, the male and female sterility of the sesquidiploid plant and the following pre- and post-fertilization barriers also appear in successive backcrossings, as a result of which the alloplasmic line could not be generated. The pre- and post-fertilization barriers often could be overcome using bud pollination and embryo rescue. However, the female sterility of a sesquidiploid plant may be considered the principal cause of hybrid breakdown. In the sesquidiploid plants between *Brassica* wild allies and *R. sativus,* many progenies involving alloplasmic lines could be generated through repeat pollinations without embryo rescue, although no progeny was obtained by bud pollination and embryo rescue techniques (personal data). Female sterility in the case of sesquidiploid plants caused the delayed maturity of egg cells. As discussed above, both using the synthetic allopolyploid among different combinations as the bridge plant to transfer wild allies cytoplasm, and ascertaining the cause of sexual incompatibility barriers of the hybrid plants as accurately as possible, may be practical improvements for producing more alloplasmic lines.

Monosomic Alien Chromosome Addition Lines (MALs)

Chromosome addition lines such as monosomic alien chromosome addition lines (MALs) were examined for the analysis of agronomic traits and gene(s) that were assumed to be located on the chromosome added from the other species. The advantages of using MALs include the possibility of assigning species-specific gene(s) and/or characteristics to particular chromosomes, and the potential to transfer desirable agronomic traits between species (Namai, 1987; McGrath and Quiros, 1990; Matsuzawa et al., 1996; Prakash et al., 1996). The MAL types of *Raphanus sativus-Brassica oleracea* and *B. rapa-B. oleracea* were produced by Kaneko et al. (1987) and Quiros et al. (1987), respectively; and several others have been successfully bred through interspecific and intergeneric hybridization between crop species and wild allies, including *B. campestris-B. alboglabra* (Chen et al., 1992), *R. sativus-E. sativa* (Bang, 1996), *R. sativus-S. arvensis* (Bang, 1996), *B. campestris-B. oxyrrhina* (Srinivasan et al., 1998), *R. sativus-M. arvensis* (Bang et al., 2002), and *B. napus-S. alba* (Wang et al., 2005).

To produce the MAL types, the existence of cross incompatibility in the interspecific and intergeneric hybridization must be overcome through successive backcrosses. Therefore, it is essential to use specific markers for the successful classification of the MAL types carrying individual chromosome donor species. The cross incompatibility often appearing in successive backcrosses of the amphidiploid (F_1) to background species appears to depend on genetic affinity between background species and chromosome donor species. The kind of barrier(s) appearing in the first backcross of amphidiploid (F_1) to background species may be reflected in the male and female sterility of the amphidiploid and the following pre- and post-fertilization barriers. In the intergeneric hybridization between *Moricandia arvensis* and *Raphanus sativus* (Bang et al., 1996a, b), the amphidiploid plant produced by colchicine treatment showed more stable chromosome association at MI in pollen mother cells (PMCs) and higher pollen fertility than the amphihaploid plant. In the backcross of the amphidiploid to *R. sativus*, it appears that there was no pre-fertilization barrier due to the pollen tubes of *R. sativus* penetrating near or into the ovules of the amphidiploid plant, resulting in effective fertilization, but the post-fertilization barrier appeared as hybrid embryo abortion. The post-fertilization barrier(s) existing in the backcrossing of the amphidiploid to *R. sativus* could be overcome through the application of the embryo rescue technique. Occasionally, the male and female sterility of the BC_1 hybrid plant (sesquidiploid) and the following pre- and post-fertilization barriers also appear in successive backcrossings (BC_2). On the other hand, the barrier(s) appearing in the first and second backcrosses were not observed in several cross combinations, such as *R. sativus-S. arvensis* (Bang, 1996), *R. sativus-E. sativa* (Bang, 1996), and *B. campestris-B. oxyrrhina* (Srinivasan et al., 1998).

Bang (1996) produced nine MAL types of *Raphanus sativus-Sinapis arvensis* that seem to be complete MAL types corresponding to the *S. arvensis* genomic chromosome number ($n = 9$), but two MAL types were obtained from double monosomic addition lines, and one MAL type showed pollen sterility and lower seed setting. In the cross combination of *R. sativus-E. sativa* (Bang, 1996), 13 MAL plants were obtained in the BC_2 generation but two MAL plants showed pollen and seed sterility. The complete MAL types in two cross combinations have not been obtained until more recently. Srinivasan et al. (1998) obtained 24 MAL plants in four successive backcrosses of the synthetic amphidiploid *B. campestris-B. oxyrrhina* to *B. campestris*, and were grouped into seven different synteny groups in addition to three unmarked MAL plants based on morphological similarity and RAPD patterns. Recently, three MAL types carrying an *S. alba* chromosome were obtained in the second backcross of the somatic amphidiploid *B. napus-S. alba* to *B. napus* (Wang et al., 2005). Bang et al. (2002) were able to classify 12 MAL types of *R. sativus-M. arvensis* based on their morphological, physiological and cytogenetic characteristics, and RAPD-specific markers. However, they were also unable to obtain 14 types of MALs corresponding to the *M. arvensis* genome ($n = 14$). The complete MAL types could not be developed from any cross combinations in *Brassicaceae* (Kaneko et al., 1987, 2001; Quiros et al., 1987, 1988; Jahier et al., 1989; Chen et al., 1992). It seems that a genetic system controlling the cross incompatibility in interspecific and intergeneric hybridization occurs between background genome and a few chromosomes of the donor genome.

Among the *Brassicaceae* species, *Moricandia arvensis* ($2n = 28$) and *Diplotaxis tenuifolia* ($2n = 22$), the C_3-C_4 intermediate species, belong to the same "Rapa/Oleracea" lineage at the level of molecular phylogeny using chloroplast DNA restriction site variation (Warwick et al., 1992; Warwick and Black, 1994). This suggests the existence of a common phylogenetic ancestor and monophyletic evolution of the C_3-C_4 intermediate characteristics in *Brassicaceae* (Apel et al., 1997). The C_3-C_4 intermediate species monosomic and disomic addition lines (MALs and DALs) of C_3 species could provide more valuable information in understanding the evolution and the genetic system of the C_3-C_4 intermediate characteristics under the control of individual chromosomes. Bang et al. (2008) produced five *M. arvensis* DAL types of *Raphanus sativus* (autoplasmic DAL types) and seven *M. arvensis* DAL types of *R. sativus* carrying *M. arvensis* cytoplasm (alloplasmic DAL types) through selfing and sib crossing of these MALs and DALs. They also maintained five autoplasmic and twelve alloplasmic MAL types of *R. sativus*, and examined the genetic mechanisms of the inner leaf structure, the intercellular pattern of glycine decarboxylase expression, and the gas exchange characteristics of C_3-C_4 intermediate photosynthesis at individual chromosome levels. Under high light in this study, the CO_2 compensation point of almost all the auto- and alloplasmic MAL and DAL plants were mostly not significantly different from that of *R. sativus* (C_3 plant). However, the CO_2 compensation point of the auto- and alloplasmic DAL plants carrying one pair of *M. arvensis* chromosomes ('b'-DAL type) were significantly different from that of *R. sativus* (C_3), although the auto- and alloplasmic MAL plants with one 'b' *M. arvensis* chromosome were not significantly different from *R. sativus* (C_3). The 'b' chromosome of the *M. arvensis* genome controls the Kranz-like leaf anatomy and biochemical components of C_3-C_4 intermediate photosynthesis, resulting in a lower CO_2 compensation point than for C_3 plants. Based on the results from this (Bang et al., 2008) and previous studies (Ueno et al., 2003, 2006, 2007), the C_3-C_4 intermediate characteristics such as the gas-exchange characteristics might be expressed in the hybrid plants on the basis of the constitution ratio of the parent genomes and each individual chromosome. Therefore, the DAL plant would be potential material for investigating the genetic mechanisms of the C_3-C_4 intermediate characteristics. However, Bang et al. (2009) were able to produce only seven alloplasmic DAL types (a, b, c, d, e, f, and l types, $2n = 20$) because the other five alloplasmic MAL types showed lower pollen fertility or complete male sterility. If the DAL types carrying the remaining seven chromosomes of *M. arvensis* and the DAL series of another C_3-C_4 intermediate species of *D. tenuifolia* could be obtained through another cross combination, the comparative studies of these DAL series may provide additional important information to understand the genetic system and the evolution of the C_3-C_4 intermediate species in *Brassicaceae* at the individual chromosome levels.

Monosomic Alien Chromosome Substitution Lines (MASLs)

Numerous disomic alien chromosome substitution lines (DASLs) have been bred in wheat cultivars, in which one pair of chromosomes was replaced by the homoeologous chromosome of *Secale, Hordium,* and *Aegilops* (Khush, 1973). Such DASLs can be used for studying the genetic effects of individual chromosomes and for estimating the number of genes controlling each trait and their linkage relationships. These lines can be generally developed in the progenies following self-fertilization of the monosomic alien chromosome substitution line (MASL). In the breeding system, MASLs could be generated only when the corresponding two chromosome are homoeologous and complement each other, especially within diploid species. In the tribe Brassiceae, as far as is known, DASLs and MASLs that were generated using the technique described above are not found. In the backcrossing of *Sinapis arvensis* monosomic addition lines (MALs) of *Raphanus sativus* to *R. ativus*, a monosomic *R. sativus* ($2n = 17$) was spontaneously generated and homoeologous paring between *R. sativus* and *S. arvensis* was observed in the *S. arvensis* MAL of *R. sativus* (MAL-d type) (Bang, 1996; Table 11.6). If monosomic *R. sativus* is crossed with the *S. arvensis* DAL of *R. sativus*, the MASL of *R. sativus* could be produced in the following generation, and then the DASL could be generated in the progenies following self-fertilization of the MASL. The chromosome substitution line is also known to occur spontaneously during successive backcrossings following interspecific hybridization. By analyzing the meiotic configuration, Banga (1988) reported the development of C genome chromosome substitution lines in *Brassica juncea,* which arose spontaneously in an interspecific hybridization between *B. juncea* and *B. napus*. Although phylogenetic relationships have been suggested in *Brassica* (Röbbelen, 1960; Armstrong and Keller, 1981, 1982) as well as between *Brassica* and allied genera (Quiros et al., 1988), we could not ever certify the generation of MASLs among segregates of any amphiploid lines, as the techniques are not available for the critical analysis and identification of each chromosome. In this regard, *in situ* hybridization procedures, somatic chromosome maps by imaging methods, and genomic species-specific markers could provide strong clues for the development and analysis of MASLs.

TABLE 11.6

Characteristics of Various Monosomic and Disomic Chromosome Addition Lines of Cultivated Species Carrying Wild Species Chromosomes

Background Nuclear Species	Chromosome Addition Species	Characteristic	MAL Types or MAL Plants													Ref.
			a	b	c	d	e	f	g	h	i	j	k	l	m	
B. campestris	B. oxyrrhina	Trivalent ratio (%)	44.2	0	6.1	0	8.9	11.5	0	0	12.5	13.6				Srinivasan et al. (1998)
		Pollen fertility (%)	0	0	0	0	0	12-16	0	0	0	0				
		Seed setting (%)	94.9	71.3	90.2	74.3	56.9	84.3	89.6	56.6	92.5	50.5				
		Transmission rate (%)	9.7	11.1	10.4	8.1	8.6	11.8	7.1	13.6	15.8	9.5				
B. oleracea	S. alba	Pollen fertility (%)	77.4	68.5												Wei et al. (2007)
B. napus	S. alba	Pollen fertility (%)	82	~	90											Wang et al. (2005) (chromosome Rese.)
		Seed setting (%)	500	~	700											
R. sativus (n=9)	E. sativa (n=11)	9II + 11 pairing (%)	100	100	100	100	100	100	100	100	100	100	100	100	100	Bang (1996)
		Pollen fertility (%)	79.0	89.8	80.4	59.8	92.9	61.5	90.4	91.4	0	84.9	0	81.3	60.2	
		Seed setting (%)	21.1	73.7	98.4	45.0	56.0	264.0	250.0	89.2	0	260.3	0	250.5	161.8	
R. sativus (n=9)	M. arvensis (n=14)	9II + 11 pairing (%)	100	100	100	100	100	100	100	100	100	100	100	100		Bang et al. (2002)
		Pollen fertility (%)	60.2	53.9	85.6	22.9	71.0	53.6	0	0	0	0	12.0	3.4		
		Seed setting (%)	105	232	166	103	96	288	259	120	146	51	163	208		
		Transmission rate (%)	18.0	17.0	11.5	24.0	10.0	32.5	8.0	5.5	8.5	32.5	15.0	20.0		
R. sativus (n=9)	S. arvensis (n=9)	9II + 11 pairing (%)	100	100	100	100	100	100	100	100	100					Bang (1996)
		Pollen fertility (%)	80.5	70.2	90.3	88.3	0	94.5	85.9	58.5	74.8					
		Seed setting (%)	156.2	161.9	70.1	263.2	28.8	211.5	98.7	108.4	84.2					
		Transmission rate (%)	40.0	30.0	20.0*	60.0	60.0	20.0*	10.0	30.0	36.4					

* transmission rate (%) from double monosomic addition line (2n = 20).

CONCLUSION

As is discussed in this chapter, sexual hybrids and their progenies recovered by the aid of embryo rescue have contributed to the genetic diversity for the distant hybridization in the Brassiceae. Such breeding practices could provide useful materials in the following genetic and breeding efforts. The simultaneous or incorporative utilization of both sexually and asexually modified genotypes would certainly contribute to studies on the genetic effects of individual chromosomes on plant traits, inspection of genes controlling a trait, their linkage relationships, and genetic improvement in the Brassiceae.

REFERENCES

Agnihotri A, V Gupta, SM Lakshmikumaran, SA Ranade, KR Shivanna, S Prakash, V Jagannathan (1988): Production of *Eruca Brassica* hybrids by embryo rescue and DNA analysis of the hybrids. *Cruciferae Newslett.,* 13:84–85.

Ahuja I, PB Bhaskar, SK Banga, SS Banga (2003): Synthesis and cytogenetic characterization of intergeneric hybrids of *Diplotaxis siifolia* with *Brassica rapa* and *B. juncea. Plant Breed.,* 122:447–449.

Akaba M, Y Kaneko, K Hatakeyama, M Ishida, SW Bang, Y Matsuzawa (2008): Evaluation of clubroot resistance of radish chromosome using *Brassica napus-Raphanus sativus* monosomic addition line. *Breed. Res.,* 10(Suppl. 1):(2005). (In Japanese)

Apel P, H Bauwe, H Ohle (1984): Hybrids between *Brassica alboglabra* and *Moricandia arvensis* and their photosynthetic properties. *Biochem. Physiol. Pflanzen,* 179:793–797.

Apel P, C Horstmann, M Pfeffer (1997): The *Moricandia* syndrome in species of the *Brassicaceae* — evolutionary aspects. *Photosynthetica,* 33:205–215.

Armstrong KC, WA Keller (1981): Chromosome pairing in haploids of *Brassica campestris. Theor. Appl. Genet.,* 59:49–52.

Armstrong KC, WA Keller (1982): Chromosome pairing in haploids of *Brassica oleracea. Can. J. Genet. Cytol.,* 24:735–739.

Bajaj YPS (1990): Wide hybridization in legumes and oilseed crops through embryo, ovule, and ovary culture. In *Biotechnology in Agriculture and Forestry,* Vol. 10 (Ed. YPS Bajaj), p. 3–37.

Bang SW (1996): Studies on the intergeneric crossability and production of hybrid progenies between radish (*Raphanus sativus* L.) and allied genera. Doctoral thesis. Univ. of Tokyo Agr. and Techni., Tokyo, p. 1–276.

Bang SW, Y Kaneko, Y Matsuzawa (1996a): Production of intergeneric hybrids between *Raphanus* and *Moricandia. Plant Breed.,* 115:385–390.

Bang SW, Y Kaneko, Y Matsuzawa (1996b): Production of intergeneric hybrids between *Raphanus* and *Sinapis* and the cytogenetics of their progenies. *Breeding Sci.,* 46:45–51.

Bang SW, D Iida, Y Kaneko, Y Matsuzawa (1997): Production of new intergeneric hybrids between *Raphanus sativus* and *Brassica* wild species. *Breeding Sci.,* 47:223–228.

Bang SW, Y Kaneko, Y Matsuzawa (1998): Cytogenetical stability and fertility of an intergeneric amphidiploid line synthesized from *Brassica maurorum* Durieu. and *Raphanus sativus* L. *Bull. of Coll. Agri., Utsunomiya Univ.,* 17:23–29.

Bang SW, Y Kaneko, Y Matsuzawa (2000): Ctyogenetical stability and fertility in an intergeneric amphidiploid line synthesized from *Brassica fruticulosa* Cyr. ssp. *Mauritanica* (Coss.) Maire. and *Raphanus sativus* L. *Bull. Coll. Agric., Utsunomiya Univ.,* 17:67–73.

Bang SW, Y Kaneko, Y Matsuzawa, KS Bang (2002): Breeding of *Moricandia arvensis* monosomic chromosome addition lines (2n = 19) of alloplasmic (*M. arvensis*) *Raphanus sativus. Breeding Sci.,* 52:193–199.

Bang SW, Y Mizuno, Y Kaneko, Y Matsuzawa, KS Bang (2003): Production of intergeneric hybrids between the C3-C4 intermediate species *Diplotaxis tenuifolia* (L.) DC. and *Raphanus sativus* L. *Breeding Sci.,* 53:231–236.

Bang SW, K Sugihara, BH Jeung, R Kaneko, E Satake, Y Kaneko, Y Matsuzawa (2007): Production and characterization of intergeneric hybrids between *Brassica oleracea* and a wild relative *Moricandia arvensis. Plant Breed.,* 126:101–103.

Bang SW, O Ueno, Y Wada, SK Hong, Y Kaneko, Y Matsuzawa (2009): Production of *Raphanus sativus* (C$_3$)–*Moricandia arvensis* (C$_3$-C$_4$ intermediate) monosomic and disomic addition lines with each parental cytoplasmic background and their photorespiratory chacteristics. *Plant Prod. Sci* 12:70–79.

Banga SS (1988): C-genome chromosome substitution lines in *Brassica juncea* (L.) Coss. *Genetica,* 77:81–84.

Banga SS, PB Bhaskar, L Ahuja (2003a): Synthesis of intergeneric hybrids and establishment of genomic affinity between *Diplotaxis catholica* and crop *Brassica* species. *Theor. Appl. Genet.*, 106:1244–1247.

Banga SS, JS Deol, SK Banga (2003b): Alloplasmic male-sterile *Brassica juncea* with *Enarthrocarpus lyratus* cytoplasm and the introgression of gene(s) for fertility restoration from cytoplasm donor species. *Theor. Appl. Genet.*, 106:1390–1395.

Batra V, KR Shivanna, S Prakash (1989): Hybrids of wild species *Erucastrum gallicum* and crop *Brassicas*. *Proc. 6th Int. Congr. of SABRAO*, p. 443–446.

Batra V, S Prakash, KR Shivanna (1990): Intergeneric hybridization between *Diplotaxis siifolia*, a wild species and crop brassicas. *Theor. Appl. Genet.*, 80:537–541.

Bhaskar PB, I Ahuja, HS Janeja, SS Banga (2002): Intergeneric hybridization between *Erucastrum canariense* and *Brassica rapa*. Genetic relatedness between Ec and A genomes. *Theor. Appl. Genet.*, 105:754–758.

Bhat SR, P Vijayan, K Ashutosh, K Dwivedi, S Prakash (2006): *Diplotaxis erucoides*-induced cytoplasmic male sterility in *Brassica juncea* is rescued by the *Moricandia arvensis* restorer: genetic and molecular analyses. *Plant Breed.*, 125:150–155.

Bijral JS, TR Sharma (1996a): Intergeneric hybridization between *Brassica napus* and *Diplotaxis muralis*. *Cruciferae Newslett.*, 18:10–11.

Bijral JS, TR Sharma (1996b): Cytogenetics of intergeneric hybrids between *Brassica napus* L. and *Eruca sativa* Lam. *Cruciferae Newslett.*, 18:12–13.

Bijral JS, TR Sharma (1998): Production and cytology of intergeneric hybrids between *Brassica napus* and *Diplotaxis catholica*. *Cruciferae Newslett.*, 20:15.

Bijral JS, TR Sharma (1999a): *Brassica juncea-Eruca sativa* sexual hybrids. *Cruciferae Newslett.*, 21:33–34.

Bijral JS, TR Sharma (1999b): Morpho-cytogenetics of *Brassica juncea* × *Brassica oxyrrhina* hybrids. *Cruciferae Newslett.*, 21:35–36.

Bing DJ, RK Downey, GFW Rakow (1991): Potential of gene transfer among oilseed *Brassica* and their weedy relatives. *GCIRC 1991 Congress*, p. 1022–1027.

Bing DJ, RK Downey, GFW Rakow (1995): An evaluation of the potential of intergeneric gene transfer between *Brassica napus* and *Sinapis arvensis*. *Plant Breed.*, 114:481484.

Bing DJ, RK Downey, GFW Rakow (1996): Assessment of transgene escape from *Brassica rapa* (*B. campestris*) into *B. nigra* or *Sinapis arvensis*. *Plant Breed.*, 115:1–4.

Bohman S, M Wang, C Dixelius (2002): *Arabidopsis thaliana*-derived resistance against *Leptosphaeria maculans* in a *Brassica napus* genomic background. *Theor. Appl. Genet.*, 105:498–504.

Brown J, AP Brown, JB Davis, D Erickson (1997): Intergeneric hybridization between *Sinapis alba* and *Brassica napus*. *Euphytica*, 93:163–168.

Budahn H, O Schrader, H Peterka (2008): Development of a complete set of disomic rape-radish chromosome-addition lines. *Euphytica*, 162:117–128.

Chandra A, ML Gupta, I Ahuja, G Kaur, SS Banga (2004a): Intergeneric hybridization between *Erucastrum cardaminoides* and two diploid crop *Brassica* species. *Theor. Appl. Genet.*, 108:1620–1626.

Chandra A, ML Gupta, SS Banga, SK Banga (2004b): Production of an interspecific hybrid between *Brassica fruticulosa* and *B. rapa*. *Plant Breed.*, 123:497–498.

Chen BY, V Simonsen, C Lanner-Herrera, WK Heneen (1992): A *Brassica campestris- alboglabra* addition line and its use for gene mapping, intergenomic gene transfer and generation of trisomics. *Theor. Appl. Genet.*, 84:592–599.

Chen HF, H Wang, ZY Li (2007): Production and genetic analysis of partial hybrids in intertribal crosses between *Brassica* species (*B. rapa, B. napus*) and *Capsella bursa-pastoris*. *Plant Cell Rep.*, 26:1791–1800.

Choudhary BR, P Joshi, K Singh (2000): Synthesis, morphology and cytogenetics of *Raphanofortii* (TTRR, 2n=38): a new amphidiploid of hybrid *Brassica tournefortii* (TT, 2n=20) *Raphanus caudatus* (RR, 2n=18). *Theor. Appl. Genet.*, 101:990–999.

Choudhary BR, P Joshi (2001): Crossability of *Brassica tournefortii* and *B. rapa*, and morphology and cytology of their F_1 hybrids. *Theor. Appl. Genet.*, 102:1123–1128.

Chrungu B, N Verma, A Mohanty, A Pradhan, KR Shivanna (1999): Production and characterization of interspecific hybrids between *Brassica maurorum* and crop brassicas. *Theor. Appl. Genet.*, 98:608–613.

Delourme R, F Eber, AM Chevre (1989): Intergeneric hybridization of *Diplotaxis erucoides* with *Brassica napus*. I. Cytogenetic analysis of F_1 and BC_1 progeny. *Euphytica*, 41:123–128.

Douce R, HW Heldt (2000): Photorespiration. In Leegood RC, TD Shakey, S. von Caemmerer (Eds.), *Photosynthesis. Physiology and Metabolism*. Kluwer Academic Publishers, Dordrecht, p. 115–136.

Fahlesan J, L Rahlen, K Glimelius (1988): Analysis of plants regenerated from protoplast fusion between *Brassica napus* and *Eruca sativa*. *Theor. Appl. Genet.*, 76:507–512.

Fan Z, W Tai, BR Stefansson (1985): Male sterility in *Brassica napus* L. associated with an extra chromosome. *Can. J. Genet. Cytol.,* 27:467–471.

FitzJohn RG, TT, Armstrong, LE Newstrom-Lloyd, AD Wilton, M Cochrane (2007): Hybridisation within *Brassica* and allied genera: evaluation of potential for transgene escape. *Euphytica,* 158:209–230.

Forsberg J, M Landgren, K Glimelius (1994): Fertile somatic hybrids between *Brassica napus* and *Arabidopsis thaliana. Plant Sci.,* 95:213–223.

Forsberg J, C Dixelius, U Lagercrantz, K Glimelius (1998): UV dose-dependent DNA elimination in asymmetric somatic hybrids between *Brassica napus* and *Arabidopsis thaliana. Plant Sci.,* 131:65–76.

Garg H, S Banga, P Bansal, C Atri, SS Banga (2007): Hybridizing *Brassica rapa* with wild crucifers *Diplotaxis erucoides* and *Brassica maurorum. Euphytica,* 156:417–424.

Gundimeda HR, S Prakash, KR Shivanna (1992): Intergeneric hybrids between *Enarthrocarpus lyratus,* a wild species, and crop brassicas. *Theor. Appl. Genet.,* 83:655–662.

Heyn FW (1977): Analysis of unreduced gametes in the Brassiceae by crosses between species and ploidy levels. *Z. Pflanzenzuchtg.,* 78:13–70.

Hu Q, LN Hansen, J Laursen, C Dixelius, SB Andersen (2002a): Intergeneric hybrids between *Brassica napus* and *Orychophragmus violaceus* containing traits of agronomic importance for oilseed rape breeding. *Theor. Appl. Genet.,* 105:834–840.

Hu Q, SB Andersen, C Dixelius, LN Hansen (2002b): Production of fertile intergeneric somatic hybrids between *Brassica napus* and *Sinapis arvensis* for the enrichment of the rapeseed gene pool. *Plant Cell Rep.,* 21:147–152.

Hua YW, ZY Li (2006): Genomic *in situ* hybridization analysis of *Brassica napus - Orychophragmus violaceus* hybrids and production of *B. napus* aneuploids. *Plant Breed.,* 125:144–149.

Inomata N (1977): Production of interspecific hybrids between *Brassica campestris* and *Brassica oleracea* by culture *in vitro* of excised ovaries. I. Effects of yeast extract and casein hydrolysate on the development of excised ovaries. *Japan J. Breed.,* 27:295–304.

Inomata N (1985): Interspecific hybrids between *Brassica campestris* and *B. cretica* by ovary culture *in vitro. Cruciferae Newslett.,* 10:92–93.

Inomata N (1986): Interspecific hybrids between *Brassica campestris* and *B. bourgeaui* by ovary culture *in vitro. Cruciferae Newslett.,* 11:14–15.

Inomata N (1987): Interspecific hybrids between *Brassica campestris* and *B. montana* by ovary culture *in vitro. Cruciferae Newslett.,* 12:8–9.

Inomata N (1988): Intergeneric hybridization between *Brassica napus* and *Sinapis arvensis* and their crossability. *Cruciferae Newslett.,* 13:22–23.

Inomata N (1991): Intergeneric hybridization in *Brassica juncea* × *Sinapis pubescens* and *B. napus* × *S. pubescens,* and their cytological studies. *Cruciferae Newslett.,* 14/15:10–11.

Inomata N (1993a): Embryo rescue techniques for wide hybridization. In Labana KS, SS Banga, SK Banga (Eds.), *Breeding Oilseed Brassicas.* Springer-Verlag, Berlin, p. 94–107.

Inomata N (1993b): Crossability and cytology of hybrid progenies in the cross between *Brassica campestris* and three wild relatives of *B. oleracea, B. bourgeaui, B. cretica* and *B. montana. Euphytica,* 69:7–17.

Inomata N (1994): Intergeneric hybridization between *Brassica napus* and *Sinapis pubescens,* and the cytology and crossability of their progenies. *Theor. Appl. Genet.,* 89:540–544.

Inomata N (1998): Production of the hybrids and progenies in the intergeneric cross between *Brassica juncea* and *Diplotaxis erucoides. Cruciferae Newslett.,* 20:17–18.

Inomata N (2002): A cytogenetic study of the progenies of hybrids between *Brassica napus* and *Brassica oleracea, Brassica bourgeaui, Brassica cretica* and *Brassica montana. Plant Breed.,* 121:174–176.

Inomata N (2003): Production of intergeneric hybrids between *Brassica juncea* and *Diplotaxis virgata* through ovary culture, and the cytology and crossability of their progenies. *Euphytica,* 133:57–64.

Ishikawa S, SW Bang, Y. Kaneko, Y. Matsuzawa (2003): Production and characterization of intergeneric somatic hybrids between *Moricandia arvensis* and *Brassica oleracea. Plant Breed.,* 122:233–238.

Jahier J, AM Chevre, AM Tanguy, F Eber (1989): Extraction of disomic addition lines of *Brassica napus-B. nigra. Genome,* 32:408–413.

Jandurova OM, J Dolezel (1995): Cytological study of interspecific hybrid between *Brassica campestris* × *B. hirta (Sinapis alba). Sex Plant Reproduction,* 8:37–43.

Jeung BH, SW Bang, S Niikura, Y Kaneko, Y Matsuzawa (2003): Wide cross-compatibility and production of hybrid progenies between *Brassica* crops and wild relatives. *Res. Breed.,* 5(Suppl. 2):282. (In Japanese)

Jeung BH, SW Bang, S Niikura, Y Kaneko, Y Matsuzawa (2004): Intergeneric cross-compatibility and production of hybrid progenies between *Moricandia arvensis* and the inbred lines of *Brassica* crops. *Res. Breed.,* 6(Suppl. 2):194. (In Japanese)

Jeung BH, SW Bang, S Niikura, Y Kaneko, Y Matsuzawa (2006a): Progenies of intergeneric hybrid between *Moricandia arvensis* and *Brassica* crops. *Res. Breed.,* 8(Suppl. 1):208. (In Japanese)

Jeung BH, SW Bang, S Niikura, Y Kaneko, Y Matsuzawa (2006b): Intergeneric cross-compatibility and production of hybrid progenies between *Diplotaxis* ssp. and the inbred lines of *Brassica* crops. *Res. Breed.,* 8(Suppl. 2):136. (In Japanese)

Jeung BH, T Saga, K Okayasu, G Hattori, SW Bang, Y Kaneko, Y Matsuzawa (2009): Production and characterization of an amphidiploid line between *Brassica rapa* L. and a wild relative *Diplotaxis tenuifolia* (L.) DC. *Plant Breed.* (in press).

Jodi AS, JD Philip (1994): Opportunities for gene transfer from transgenic oilseed rape (*Brassica napus*) to related species. *Transgenic Res.,* 3:263–278.

Kaneko Y, Y Matsuzawa, M Sarashima (1987): Breeding of the chromosome addition lines of radish with single kale chromosome. *Jpn. J. Breed.,* 37:438–452.

Kaneko Y, H Yano, SW Bang, Y Matsuzawa (2001): Production and characterization of *Raphanus sativus-Brassica rapa* monosomic chromosome addition lines. *Plant Breed.,* 120:163–168.

Kerlan MC, AM Chevre, F Eber, A Baranger, M Renard (1992): Risk assessment of outcrossing of transgenic rapeseed to related species. I. Interspecific hybrid production under optimal conditions with emphasis on pollination and fertilization. *Euphytica,* 62:145–153.

Kerlan MC, AM Chevre, F Eber (1993): Interspecific hybrids between a transgenic rapeseed (*Brassica napus*) and related species: cytogenetical characterization and detection of the transgene. *Genome,* 36:1099–1106.

Khush GS (1973): *Cytogenetics of Aneuploids.* Academic Press, New York and London.

Khush GS, DS Brar (1992): Overcoming the barriers in hybridization. In Kalloo G, JB Chowdhury (Eds.), *Distant Hybridization of Crop Plants.* Springer-Verlag. Berlin, p. 47–61.

Kirti PB, SB Narasimhulu, S Prakash, VL Chopra (1992a): Production and characterization of intergeneric somatic hybrids of *Trachystoma ballii* and *Brassica juncea. Plant Cell Rep.,* 11:90–92.

Kirti PB, SB Narasimhulu, S Prakash, VL Chopra (1992b): Somatic hybridization between *Brassica juncea* and *Moricandia arvensis* by protoplast fusion. *Plant Cell Rep.,* 11:318–321.

Kirti PB, T Mohapatra, H Khanna, S Prakash, VL Chopra (1995): *Diplotaxis catholica* + *Brassica juncea* somatic hybrids: molecular and cytogenetic characterization. *Plant Cell Rep.,* 14:593–597.

Lefol E, A Fleury, H Darmency (1996): Gene dispersal from transgenic crops. II. Hybridization between oilseed rape and the wild hoary mustard. *Sex Plant Reproduction,* 9:189–196.

Lefol E, G Seguin-Swartz, RK Downey (1997): Sexual hybridization in crosses of cultivated *Brassica* species with the crucifers *Erucastrum gallicum* and *Raphanus raphanistrum*: potential for gene introgression. *Euphytica,* 95:127–139.

Leino M, R Teixeira, M Landgren, K Glimelius (2003): *Brassica napus* lines with rearranged *Arabidopsis* mitochondria display CMS and a range of developmental aberrations. *Theor. Appl. Genet.,* 106:1156–1163.

Leino M, S Thyselius, M Landgren, K Glimelius (2004): *Arabidopsis thaliana* chromosome III restores fertility in a cytoplasmic mail-sterile *Brassica napus* line with *A. thaliana* mitochondrial DNA. *Theor. Appl. Genet.,* 109:272–279.

Lelivelt CLC, EHM Leunissen, HJ Frederiks, JPFG Helsper, FA Krens (1993): Transfer of resistance to the beet cyst nematode (*Heterodera schachtii* Schm.) from *Sinapis alba* L. (white mustard) to the *Brassica napus* L. gene pool by means of sexual and somatic hybridization. *Theor. Appl. Genet.,* 85:688–696.

Leon P, A Arroyo, S Mackenzie (1998): Nuclear control of plastid and mitochondrial development in higher plants. *Annu. Rev. Plant Physiol., Plant Mol. Biol.,* 49:453–480.

Li ZY, P Luo (1993): First intergeneric hybrids of *Brassica napus* × *Orychophragmus violaceus. Oil Crops Newslett.,* 10:27–29.

Li Z, HL Liu, P Luo (1995): Production and cytogenetics of intergeneric hybrids between *Brassica napus* and *Orychophragmus violaceus. Theor. Appl. Genet.,* 91:131–136.

Li Z, JG Wu, Y Liu, HL Liu, WK Heneen (1998): Production and cytogenetics of the intergeneric hybrids *Brassica juncea* × *Orychophragmus violaceus* and *B. carinata* × *O. violaceus. Theor. Appl. Genet.,* 96:251–265.

Li Z, WK Heneen (1999): Production and cytogenetics of intergeneric hybrids between the three cultivated *Brassica* diploids and *Orychophragmus violaceus. Theor. Appl. Genet.,* 99:694–704.

Luo P, ZY Li, YY Wu (1994): Intergeneric hybridization between *Brassica napus* and *Orychophragmus violaceus. Ch. J. Bot.,* 6:86–88.

Luo P, HL Fu, ZQ Lan, SD Zhou, HF Zhou, Q Luo (2003): Phytogenetic studies on intergeneric hybridization between *Brassica napus* and *Matthiola incana*. *Acta Bot. Sin.*, 45:432–436.

Malik M, P Vyas, NS Rangaswamy, KR Shivanna (1999): Development of two new cytoplasmic male-sterile lines in *Brassica juncea* through wide hybridization. *Plant Breed.*, 118:75–78.

Martínez-Sánchez A, L Rafael, IG Maria , F Federico (2007): Identification of new flavonoid glycosides and flavonoid profiles to characterize rocket leafy salads *(Eruca vesicaria* and *Diplotaxis tenuifolia)*. *J. Agric. Food Chem.*, 55:1356–1363.

Matsuzawa Y (1983): Studies on the interspecific and intergeneric crossability in *Brassica* and *Raphanus*. *Special Bull. Coll. Agr. Utsunomiya Univ.*, 39:1–86. (In Japanese)

Matsuzawa Y, M Sarashima (1986): Intergeneric hybridization of *Eruca, Brassica* and *Raphanus*. *Cruciferae Newslett.*, 11:17–18.

Matsuzawa Y, Y Kaneko, SW Bang (1996): Prospects of the wide cross for genetics and plant breeding in Brassicaceae. *Bull. Coll. Agr. Utsunomiya Univ.*, 16:5–10.

Matsuzawa Y, T Minami, SW Bang, Y Kaneko (1997): A new *Brassicoraphanus* (2n=36); the true-breeding amphidiploid line of *Brassica oxyrrhina* Coss. (2n=18) × *Raphanus sativus* L. (2n=18). *Bull. of Coll. Agric., Utsunomiya Univ.*, 16:1–7.

Matsuzawa Y, S Mekiganon,Y Kaneko, SW Bang, K Wakill, Y Takahata. (1999): Male sterility in alloplasmic *Brassica vapal* carrying *Eruca sativa* cytoplasm. *Plant Breed*, 118:82–84.

McGrath JM, CF Quiros (1990): Generation of alien chromosome addition lines from synthetic *Brassica napus*: morphology, cytology, fertility, and chromosome transmission. *Genome*, 33:374–383.

Mizushima U (1950): Karyogenetic studies of species and genus hybrids in the tribe Brassiceae of Cruciferae. *Tohoku J. Agric. Res.*, 1:1–14.

Mizushima U (1968): Phylogenetic studies on some wild *Brassica* species. *Tohoku J. Agric. Res.*, 19:83–99.

Mohapatra T, PB Kirti, V Dinesh Kumar, S Prakash, VL Chopra (1998): Random chloroplast segregation and mitochondrial genome recombination in somatic hybrid plant of *Diplotaxis catholica + Brassica juncea*. *Plant Cell Rep.*, 17:814–818.

Momotaz A, M Kato, F Kakihara (1998): Production of intergeneric hybrids between *Brassica* and *Sinapis* species by means of embryo rescue techniques. *Euphytica,* 103:123–130.

Nagpal R, SN Raina, YS Sodhi, A Mukhopadhyay, N Arumugam, AK Pradhan, D Pental (1996): Transfer of *Brassica tournefortii* (TT) genes to allotetraploid oilseed *Brassica* species (*B. juncea* AABB, *B. napus* AACC, *B. carinata* BBCC): homoeologous pairing is more pronounced in the three-genome hybrids (TACC, TBAA, TCAA, TCBB) as compared to allodiploids (TA, TB, TC). *Theor. Appl. Genet.,* 92:566–571.

Namai H (1976): Cytogenetic and breeding studies on transfer of economic characters by means of interspecific and intergeneric crossing in the tribe Brassiceae of Cruciferae. *Memories Fac. Agr., Tokyo Univ. of Education*, 22:101–171. (In Japanese)

Namai H (1987): Inducing cytogenetical alterations by means of interspecific and intergeneric hybridization in *Brassica* crops. *Gamma Field Symp.,* 26:41–89.

Nanda Kumar PBA, KR Shivanna, S Prakash (1988): Wide hybridization in *Brassica*. Crossability barriers and studies on the F$_1$ hybrid and synthetic amphidiploids of *B. fruticulosa* × *B. campestris*. *Sex Plant Reprod.,* 1:234–239.

Nanda Kumar PBA, S Prakash, KR Shivanna (1989): Wide hybridization in *Brassica*: studies on interspecific hybrids between cultivated species (*B. napus, B. juncea*) and a wild species (*B. gravinae*). *Proc. 6th Int. Congr. SABRAO*, p. 435–438.

Nanda Kumar PBA, KR Shivanna (1993): Intergeneric hybridization between *Diplotaxis siettiana* and crop brassicas for the production of alloplasmic lines. *Theor. Appl. Genet.*, 85:770–776.

Narain A, S Prakash (1972): Investigations on the artificial synthesis of amphidiploids of *Brassica tournefortii* Gouan with the other elementary species of *Brassica*. I. Genomic relationships. *Genetica*, 43:90–97.

Nishi S, M Toda, T Toyoda (1970): Studies on the embryos culture in vegetable crops. III. On the conditions affecting embryo culture of interspecific hybrids between cabbage and Chinese cabbage. *Bull. Hort. Res. Sta., Japan,* Ser. A9:73–88. (In Japanese)

Nothnagel T, H Budahn, P Straka, O Schrader (1997): Successful backcrosses of somatic hybrids between *Sinapis alba* and *Brassica oleracea* with the *Brassica oleracea* parent. *Plant Breed.*, 116:89–97.

O'Neill CM, T Murata, CL Morgan, RJ Mathias (1996): Expression of the C3-C4 intermediate character in somatic hybrids between *Brassica napus* and the C3-C4 species *Moricandia arvensis*. *Theor. Appl. Genet.*, 93:1234–1241.

Pahwa RS, SK Banga, KPS Gogna, SS Banga (2004): *Tournefortii* male sterility system in *Brassica napus*. Identification, expression and genetic characterization of male fertility restorers. *Plant Breed.*, 123:444–448.

Pathania A, SR Bhat, D Dinesh Kumar, Ashutosh, PB Kirti, S Prakash, VL Chopra (2003): Cytoplasmic male sterility in alloplasmic *Brassica juncea* carrying *Diplotaxis catholica* cytoplasm: molecular characterization and genetics of fertility restoration. *Theor. Appl. Genet.*, 107:455–461.

Prakash S, A Narain (1971): Genomic status of *Brassica tournefortii* Gouan. *Theor. Appl. Genet.*, 41:203–204.

Prakash S, VL Chopra (1990): Male sterility caused by cytoplasm of *Brassica oxyrrhina* in *B. campestris* and *B. juncea*. *Theor. Appl. Genet.*, 79:285–287.

Prakash SH, PB Kirti, SH Bhat, K Gaikwad, VD Kumar, VL Chopra (1998): A *Moricandia arvensis* based cytoplasmic male sterility and fertility restoration system in *Brassica juncea*. *Theor. Appl. Genet.*, 97:488–492.

Prakash S, Y Takahata, PB Kirti, VL Chopra (1999): Cytogenetics. In Gomez-Campo C (Ed.), *Biology of Brassica Coenospecies*. Elsevier Science B.V., p.59–106.

Quiros CF, O Ochoa, SF Kianian, D Douches (1987): Analysis of the *Brassica oleracea* genome by the generation of *B. campestris-oleracea* chromosome addition lines: characterization by isozymes and rDNA genes. *Theor. Appl. Genet.*, 74:758–766.

Quiros CF, O Ochoa, DS Douches (1988): Exploring the role of x=7 species in *Brassica* evolution: hybridization with *B. nigra* and *B. oleracea*. *J. Hered.*, 79:351–358.

Rao GU, M Lakshmikumaran, KR Shivanna (1996): Production of hybrids, amphiploids and backcross progenies between a cold-tolerant wild species, *Erucastrum abyssinicum* and crop brassicas. *Theor. Appl. Genet.*, 92:786–790.

Rawsthorne S, CL Morgan, CM O'Neill, CM Hylton, DA Jones, ML Frean (1998): Cellular expression pattern of the glycine decarboxylase P protein in leaves of an intergeneric hybrid between the C3-C4 intermediate species *Moricandia nitens* and the C3 species *Brassica napus*. *Theor. Appl. Genet.*, 96:922–927.

Ringdahl EA, PBE McVetty, JL Sernyk (1987): Intergeneric hybridization of *Diplotaxis* spp. with *Brassica napus*: a source of new CMS systems?. *Can. J. Plant Sci.*, 67:239–243.

Röbbelen G (1960): Beiträge zur analyse des *Brassica*-genome. *Chromosome*, 11:205–228.

Rylott EL, K Metzlaff, S Rawsthorne (1998): Developmental and environmental effects on the expression of the C_3-C_4 intermediate phenotype in *Moricandia arvensis*. *Plant Physiol.*, 118:1277–1284.

Sarashima M, Y Matsuzawa (1989): Cross-compatibility and synthesis of new amphidiploid species in interspecific and intergeneric hybridization of Cruciferae crops. *Bull. Coll. Agr. Utsunomiya Univ.*, 14:85–92. (In Japanese)

Sarashima M, Y Matsuzawa, S Suto (1990): Studies on breeding of the alloplasmic lines in Cruciferae crops. *Bull. Coll. Agr. Utsunomiya Univ.*, 14:1–34. (In Japanese)

Sarmah BK, N Sarla (1998): *Erucastrum abyssinicum* × *Brassica oleracea* hybrids obtained by ovary and ovule culture. *Euphytica*, 102:37–45.

Siemens J (2002): Interspecific hybridization between wild relatives and *Brassica napus* to introduce new resistance traits into the oilseed rape gene pool. *Czech J. Genet. Plant Breed.*, 38:155–157.

Srinivasan K, VG Malathi, PB Kirti, S Prakash, VL Chopra (1998): Generation and characteristics of monosomic chromosome addition lines of *Brassica campestris-B. oxyrrhina*. *Theor. Appl. Genet.*, 97:976–981.

Stebbins GL (1958): The inviability, weakness and sterility of interspecific hybrids. *Adv. Genet.*, 9:147–215.

Stiewe G, G Röbbelen (1994): Establishing cytoplasmic male sterility in *Brassica napus* by mitochondrial recombination with *B. tournefortii*. *Plant Breed.*, 113:294–304.

Sundberg E, K Glimelius (1991): Effects of parental ploidy level and genetic divergence on chromosome elimination and chloroplast segregation in somatic hybrids within *Brassicaceae*. *Theor. Appl. Genet.*, 83:81–88.

Takahata Y, K Hinata (1983): Studies on cytodemes in subtribe Brassicinae (Cruciferae). *Tohoku J. Agric. Res.*, 33:111–124.

Takahata Y, T Takeda (1990): Intergeneric (intersubtribe) hybridization between *Moricandia arvensis* and *Brassica* A and B genome species. *Theor. Appl. Genet.*, 80:38–42.

Takahata Y, T Takeda, N Kaizuma (1993a): Wide hybridization between *Moricandia* and crop species of *Brassica* and *Raphanus*. *XV Int. Botanical Congr.*, p. 516.

Takahata Y, T Takeda, N Kaizuma (1993b): Wide hybridization between *Moricandia arvensis* and *Brassica* amphidiploid species (*B. napus* and *B. juncea*). *Euphytica*, 69:155–160.

Toriyama K, K Hinata, T Kameya (1978): Production of somatic hybrid plants, 'Brassicomoricandia', through protoplast fusion between Moricandia arvensis and Brassica oleracea. Plant Sci., 48:123–128.

Toriyama K, T Kameya, K Hinata (1994): IV.4 Somatic hybridization between Brassica and Sinapis. Biotechnol. Agric. Forestry, 27:334–341.

Ueno O, SW Bang, Y Wada, A Kondo, K Ishihara, Y Kaneko, Y Matsuzawa (2003): Structural and biochemical dissection of photorespiration in hybrids differing in genome constitution between Diplotaxis tenuifolia (C_3-C_4) and radish (C_3). Plant Physiol., 132:1550–1559.

Ueno O, Y Wada, M Wakai, SW Bang (2006): Evidence from photosynthetic characteristics for the hybrid origin of Diplotaxis muralis from a C3-C4 intermediate and C3 species. Plant Biol., 8:253–259.

Ueno O, SW Bang, Y Wada, N Kobayashi, R Kaneko, Y Kaneko, Y Matsuzawa (2007): Inheritance of C_3-C_4 intermediate photosynthesis in reciprocal hybrids between Moricandia arvensis (C_3-C_4) and Brassica oleracea (C_3) that differ in their genome constitution. Plant Prod. Sci., 10:68–79.

Vyas P, S Prakash, KR Shivanna (1995): Production of wide hybrids and backcross progenies between Diplotaxis erucoides and crop brassicas. Theor. Appl. Genet., 90:549–553.

Wang Y, P Luo, X Li, Z Lan (1997): Access and identification of intergeneric hybrid between Brassica juncea × Crambe abyssinica. Acta Bot. Sin., 39:296–301. (In Chinese)

Wang Y, P Wuo (1998): Intergeneric hybridization between Brassica species and Crambe abyssinica. Euphytica, 101:1–7.

Wang YP, K Sonntag, E Rudloff (2003): Development of rapeseed with high erucic acid content by asymmetric somatic hybridization between Brassica napus and Crambe abyssinica. Theor. Appl. Genet., 106:1147–1155.

Wang YP, RJ Snowdon, E Rudloff, P Wehling, W Friedt, K Sonntag (2004): Cytogeneric characterization and fae1 gene variation in progenies from asymmetric somatic hybrids between Brassica napus and Crambe abyssinica. Genome, 47:724–731.

Wang YP, XX Zhao, K Sonntag, P Wehling, RJ Snowdon (2005): Behaviour of Sinapis alba chromosome in a Brassica napus background revealed by genomic in-situ hybridization. Chromosome Res., 13:819–826.

Warwick SI (1993): Guide to the wild germplasm of Brassica and allied crops. Part IV. Wild species in the tribe Brassiceae (Cruciferae) as sources of agronomic traits. Centre for Land and Biological Resources Research, Research Branch, Agriculture Canada, p. 1–19.

Warwick SI, LD Black, I Aguinagalde (1992): Molecular systematics of Brassica and allied genera (subtribe Brassicinae, Brassicaceae) — chloroplast DNA variation in the genus Diplotaxis. Theor. Appl. Genet., 83:839–850.

Warwick SI, LD Black (1994): Evaluation of the subtribes Moricandiinae, Savignyinae, Vellinae, and Zillinae (Brassicaceae, tribe Brassiceae) using chloroplast DNA restriction site variation. Can. J. Bot., 72:1692–1701.

Warwick SI, MJ Simard, A Legere, HJ Beckie, L Braun, B Zhu, P Mason, G Seguin-Swartz, CN Stewart (2003): Hybridization between transgenic Brassica napus L. and its wild relatives: Brassica rapa L., Raphanus raphanistrum L., Sinapis arvensis L., and Erucastrum gallicum (Willd.) O.E. Schulz. Theor. Appl. Genet., 107:528–539.

Wei WH, SF Zhang, LJ Wang, J Li, B Chen, Z Wang, LX Luo, XP Fang (2007): Cytogenetic analysis of F_1, F_2, and BC_1 plants from intergeneric sexual hybridization between Sinapis alba and Brassica oleracea by genomic in situ hybridization. Plant Breed., 126:392–398.

Wilmar JC, M Hellendoorn (1968): Embryo culture of Brussels sprouts for breeding. Euphytica, 17:28–37.

Wu JG, Z Li, Y Liu, HL Liu, TD Fu (1997): Cytogenetics and morphology of the pentaploid hybrid between Brassica napus and Orychophragmus violaceus and its progeny. Plant Breed., 116:251–257.

Yadav RC, PK Sareen, JB Chowdhury (1991): Interspecific hybridization in Brassica juncea × Brassica tournefortii using ovary culture. Cruciferae Newslett., 14/15:84.

Yamagishi H (2003): Organellar genomes of somatic hybrids having Arabidopsis thaliana as a parent. Res. Breed., 5(Suppl 2):38–39. (In Japanese)

Yamagishi H, M Landgren, J Forsberg, K Glimelius (2002): Production of asymmetric hybrids between Arabidopsis thaliana and Brassica napus utilizing an efficient protoplast culture system. Theor. Appl. Genet., p. 106.Yan Z, Z Tian, B Huang, R Huang, J Meng (1999): Production of somatic hybrids between Brassica oleracea and the C3-C4 intermediate species Moricandia nitens. Theor. Appl. Genet., 99:1281–1286.

Zhang C, G Xu, R Huang, C Chen, J Meng (2004): A dominant gdcP-specific marker derived from Moricandia nitens used for introducing the C3-C4 character from M. nitens into Brassica crops. Plant Breed., 123:438–443.

12 Phytoalexins

M. Soledade C. Pedras and Qingan Zheng

CONTENTS

DETECTION, ISOLATION, AND STRUCTURE ELUCIDATION

Plants respond to biotic or abiotic stress with a wide range of mechanisms that involve the production of multiple natural products, also known as secondary metabolites, having diverse roles and biological activities. Phytoalexins are secondary metabolites produced *de novo* by plants in response to stress, including microbial attack, intense heat, or UV radiation (Bailey and Mansfield, 1982). That is, phytoalexins are not present in naturally healthy plant tissues but their primary precursors are recruited into the phytoalexin biosynthetic pathway upon elicitation by an exogenous stimulus. The ecological significance and defensive roles of phytoalexins have been widely demonstrated in various plant families and are not the subject of this review, as several general reviews dealing with these aspects have appeared over the past 20 years. In contrast to phytoalexins, phytoanticipins are constitutive antimicrobial metabolites whose concentration may increase when the plant is under stress (VanEtten et al., 1994).

Because the production of phytoalexins is restricted to the stressed tissues, the detection of phytoalexins can be challenging, even if highly sensitive analytical instruments such as a high-performance liquid chromatograph (HPLC) with diode array detector (DAD) or mass spectrometer detector (MSD) is used. In crucifers, the detection of phytoalexins in tissue extracts has been carried out by both thin layer chromatography (TLC), with biodetection utilizing spores of *Cladosporium* or *Bipolaris* species, and HPLC-DAD (Pedras et al., 2007c). The TLC method is sensitive but may not work in all situations because other antifungal compounds present in extracts can mask the bioactivity of phytoalexins. Similarly, comparison of HPLC chromatograms of extracts of elicited tissues with those of control samples can also be misleading because complete separation of all metabolites may not be possible. Due to the complexity of organic solvent extracts obtained from crucifers, overlap of peaks of constitutive metabolites with phytoalexin peaks can occur. An example of an HPLC-DAD chromatogram of extracts of leaves of *Brassica rapa* elicited with copper chloride and chromatograms of extracts of control leaves is shown in Figure 12.1. Ideally, to ensure that potential phytoalexins are detected, both TLC with biodetection

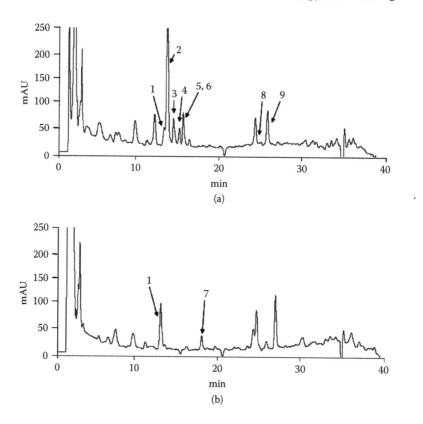

FIGURE 12.1   HPLC-DAD (220 nm) chromatograms of dichloromethane extracts of leaves of *Brassica rapa* cv. *Reward*. a) leaves sprayed with copper chloride and extracted after 2 days; b) control leaves sprayed with water and extracted after 2 days. Peak 1: indole-3-acetonitrile; peak 2: spirobrassinin; peak 3: brassilexin; peak 4: arvelexin; peak 5: rutalexin; peak 6: rapalexin B; peak 7: caulilexin C; peak 8: rapalexin A; peak 9: cyclobrassinin.

and HPLC-DAD-MS methods are employed. An additional analytical method that can provide detection and essential structural information on phytoalexins, without carrying out sample isolation, involves coupling an HPLC to a nuclear magnetic resonance (NMR) spectrometer. This is a very powerful but rather expensive method that requires analysis of NMR spectral data and thus expert knowledge in this area.

Phytoalexin isolation requires extraction of stressed plant tissues (leaves, stems, roots) with organic solvents such as hexane, dichloromethane, ethyl acetate, acetonitrile, or methanol, of up to several kilograms of elicited tissues. Control tissues should be used as well to ensure that the metabolites are in fact produced *de novo*. Upon fractionation of an extract using column chromatography, the fractions are analyzed by HPLC-DAD-MS and bioassayed to confirm the presence of the desired metabolites. Multiple chromatographic separations are usually required to purify the compounds and obtain sufficient quantities of samples for chemical characterization, structure determination, and bioassays. Clearly, depending on the plant species and availability of detection methods, the isolation and purification of crucifer phytoalexins can be rather time consuming and expensive. To obtain substantial amounts of crucifer phytoalexins for biological studies, chemical synthesis is the most reliable method. All crucifer phytoalexins currently known have been synthesized, and their syntheses were reviewed recently (Pedras et al., 2007c); however, some of the synthetic methods are technically demanding and can be achieved in reasonable yields only by well-trained researchers.

Spectroscopic Characterization

In general, the structure determination of organic molecules requires sophisticated and expensive analytical instruments usually common in chemistry laboratories. NMR spectroscopy and high-resolution mass spectrometry (HR-MS) are usually essential for unambiguous structural assignments. However, if the structures of the compounds are known, an HPLC-DAD chromatogram and a low-resolution mass measurement (LR-MS) may be sufficient to establish the identity of the molecule. As shown in Table 12.1, HPLC-DAD-MS data are currently available for the characterization of about half of the currently known crucifer phytoalexins (Pedras et al., 2006a). Additional information on currently known crucifer phytoalexins and related metabolites is expected to become available in the near future to facilitate their identification.

Analytical Quantification

To determine the range of phytoalexins produced by a given plant species, it is essential to test a variety of elicitors, because different elicitors may induce biosynthesis of different phytoalexins. In addition, because some biotic elicitors such as fungi are able to metabolize phytoalexins during the infection process, results of phytoalexin quantitation can be misleading. For example, it may appear that very small amounts of phytoalexins and fewer structural types are produced by an infected plant, but this profile could be due to rapid phytoalexin metabolism by the pathogen. Metabolic processes carried out by cruciferous pathogens are being intensely investigated and have been reviewed recently (Pedras and Ahiahonu, 2005). Therefore, it follows from the above discussion that the quantitative analyses of phytoalexins can pose considerable challenges due to the large number and higher quantities of constitutive metabolites produced by crucifers. Accurate measurements utilizing HPLC (DAD or MS detectors) require multiple sampling and calibration curves of pure phytoalexins.

CHEMICAL STRUCTURES AND BIOLOGICAL ACTIVITY

Structures

The phytoalexins of crucifers isolated thus far contain either an indole or an oxindole moiety with additional substituents. Brassinin, together with cyclobrassinin and 1-methoxybrassinin were reported for the first time in 1986 (Takasugi et al., 1986). The dithiocarbamate group present in brassinin, 1-methoxybrassinin, and 4-methoxybrassinin is the toxophore responsible for their antifungal activity. These phytoalexins appear to be the only naturally occurring compounds containing a dithiocarbamate toxophore, although synthetic dithiocarbamates and thiocarbamates have been commercially available and used as fungicides and pesticides for decades (Pedras et al., 2007c). It is noteworthy that only seven of the forty currently reported crucifer phytoalexins do not contain sulfur (Figure 12.2). Interestingly, three different indolyl-3-acetonitriles (indolyl-3-acetonitrile, caulilexin C, and arvelexin) are found to be phytoalexins in some species and phytoanticipins in other species, as they are constitutive metabolites whose concentration in the stressed plant tissue increases. The high sulfur requirement of crucifer crops is well known and is likely due, at least in part, to the sulfur content of their metabolites. Sulfur itself appeared to be a component of extracts of the plant wasabi, which was shown also to produce a large number of isothiocyanates (Pedras and Sorensen, 1998).

Compared to the number of species in the family *Brassicaceae*, very few species have been screened for phytoalexin production (Pedras et al., 2007c). Yet, among these species, only an even smaller number are wild species (Table 12.2). Recently, it was shown that the phytoalexin rapalexin A is produced in both *Arabidopsis thaliana* and *Thellungiella halophila*, which produces mainly wasalexins A and B and methoxybrassenin B, while mainly camalexin is produced in *A. thaliana* (Pedras and Adio, 2008). These differences are very clear upon examination of the HPLC profiles of

TABLE 12.1

Characterization of Crucifer Phytoalexins and Related Compounds by HPLC-DAD-MS Analyses[a]

Compound Name (t_R min) Molecular Formula	[M]	Positive Mode MS	Negative Mode MS	UV λ_{max} nm
Arvelexin (12.3 min) $C_{11}H_{10}N_2O$	186	209 [M+Na]+ (14), 187 (100), 160 (33), 147 (65), 132 (5)	ND	220, 266, 280, 290
Brassicanal A (9.3 min) $C_{10}H_9NOS$	191	214 [M+Na]+ (6), 192 (100), 164 (14), 117 (24)	190 (100), 175 (9)	218, 258, 269, 325
Brassicanal C (8.8 min) $C_{10}H_9NO_3S$	223	246 [M+Na]+ (28), 192 (100), 174 (71), 164 (64), 148 (73), 146 (70)	222 (100), 192 (15), 193 (13)	215, 247, 310
Brassicanate A (14.4 min) $C_{11}H_{11}N O_2S$	221	244 [M+Na]+ (28), 222 (8), 190 (100)	220 (100), 205 (7)	220, 238, 268, 300
Brassilexin (10.6 min) $C_9H_6N_2S$	174	175 (100), 148 (5)	173 (100)	220, 245, 264
Brassinin (18.3 min) $C_{11}H_{12}N_2S_2$	236	259 [M+ Na]+ (100), 237 (2), 176 (2), 130 (51)	236 (100), 190 (9), 172 (7)	220, 269
Brassitin (11.3 min) $C_{11}H_{12}N_2OS$	220	243 [M+Na]+ (28), 130 (100)	ND	220, 270, 278
Brussalexin (16.3 min) $C_{13}H_{14}N_2OS$	246	269 [M+Na]+ (12), 259 (100), 247 (1), 130 (29)	ND	228, 279
Camalexin (9.1 min) $C_{11}H_8N_2S$	200	201 (100)	199 (100), 190 (6), 172 (7), 142 (3)	215, 278, 318
Caulilexin A (15.3 min) $C_{10}H_9N O_3S$	223	246 [M+Na]+ 224 (31), 176 (100)	222 (5), 176 (100)	213, 252, 318
Caulilexin C (16.3 min) $C_{11}H_{10}N_2O$	186	187 (49), 160 (100), 130 (20)	ND	220, 272
Cyclobrassinin (23.1 min) $C_{11}H_{10}N_2S_2$	234	235 (37), 187 (15), 162 (100)	233 (8), 190 (37), 172 (9), 161 (100)	205, 229, 285, 294
Cyclobrassinin sulfoxide (11.0 min) $C_{11}H_{10}N_2OS_2$	250	273 [M+Na]+ (15), 251 (3), 187 (100)	249 (3), 201 (4), 160 (100)	213, 226, 278, 330
Cyclobrassinone (6.8 min) $C_{11}H_8N_2O_2S$	232	255 [M+Na]+ (6), 233 (100), 176 (3)	231 (100), 174 (13)	218, 278, 312
Dioxibrassinin (6.5 min) $C_{11}H_{12}N_2O_2S_2$	268	291 [M+Na]+ (80), 269 (48), 251 (15), 221 (59), 203 (100), 161 (39)	267 (12), 219 (11), 159 (100), 160 (19)	210, 255
Erucalexin (20.4 min) $C_{12}H_{12}N_2O_2S_2$	280	303 [M+Na]+ (2), 281 (57), 250 (100), 249 (66), 203 (35)	ND	234, 262, 368
Indolyl-3-acetonitrile (10.6 min) $C_{10}H_8N_2$	156	179 [M+Na]+ (2), 157 (5), 130 (100)	ND	220, 272, 285
Indole-3-carboxaldehyde (5.4 min) C_9H_7NO	145	168 [M+Na]+ (5), 146 (70), 118 (100)	144 (100)	210, 245, 260, 300
Isalexin (2.8 min) $C_9H_7NO_3$	177	200 [M+Na]+ (100), 178 (93), 132 (6), 105 (7)	ND	199, 234, 335

—continued

TABLE 12.1 (Continued)

Characterization of Crucifer Phytoalexins and Related Compounds by HPLC-DAD-MS Analyses[a]

Compound Name (t_R min) Molecular Formula	[M]	Positive Mode MS	Negative Mode MS	UV$_{max}$ nm
1-Methoxybrassenin B (23.7 min) $C_{13}H_{14}N_2O_2S_2$	296	317 [M+Na]$^+$ (4), 295 (47), 174 (100), 159 (5)	ND	204, 315
1-Methoxybrassinin (23.6 min) $C_{12}H_{14}N_2OS_2$	266	289 [M+Na]$^+$ (25), 267 (49), 235 (36), 219 (10),160 (100), 146 (68), 128 (30), 117 (87)	265 (100), 220 (47), 190 (48), 144 (68)	218, 268
1-Methoxybrassitin (16.9 min) $C_{12}H_{14}N_2O_2S$	250	251 (4), 160 (100)	ND	219, 247, 273, 287
6-Methoxycamalexin (9.2 min) $C_{12}H_{10}N_2OS$	230	253[M+Na]$^+$ (2), 231 (100)	229 (100), 215 (10)	218, 296, 324
1-Methoxyspirobrassinin (16.1 min) $C_{12}H_{14}N_2OS_2$	280	303 [M+Na]$^+$ (3), 281 (100), 250 (21)	ND	218, 260, 295
1-Methylcamalexin (14.1 min) $C_{12}H_{10}N_2S$	214	215 (100)	ND	214, 274, 318
Rapalexin A (22.9 min) $C_{10}H_8N_2OS$	204	205 (15), 147 (18)	203 (100), 188 (38)	223, 292
Rapalexin B (13.9 min) $C_{10}H_8N_2O_2S$	220	221 (20), 206 (22), 189 (21)	219 (100), 204 (50)	203, 214, 264, 288
Rutalexin (12.9 min) $C_{11}H_8N_2O_2S$	232	255 [M+Na]$^+$ (5), 233 (100), 192 (17), 148 (17)	231 (100), 174 (4)	213, 242, 275
Sinalbin B (33.1 min) $C_{12}H_{12}N_2OS_2$	266	265 (40), 233 (100), 217(8), 192 (60), 160 (26)	ND	231, 275
Sinalexin (19.3 min) $C_{10}H_8N_2OS$	204	205 (39), 174 (100)	ND	225, 247, 262
Spirobrassinin (10.9 min) $C_{11}H_{12}N_2S_2$	250	273 [M+Na]$^+$ (8), 251 (100), 203 (21), 178 (8)	249 (89), 217 (3), 201 (100)	220, 258, 296
Wasalexin A (22.8 min) $C_{13}H_{14}N_2O_2S_2$	294	317 [M+Na]$^+$ (6), 295 (100), 263 (6), 247 (21)	ND	205, 248, 285, 368
Wasalexin B (20.6 min) $C_{13}H_{14}N_2O_2S_2$	294	317 [M+ Na]$^+$ (2), 295 (100), 263 (3), 247 (15)	ND	205, 248, 285, 368

[a] Data were obtained using an HPLC equipped with autosampler, binary pump, degasser, and a diode array detector connected directly to a mass detector (ion trap mass spectrometer) with an electrospray source. Chromatographic separation was carried out at room temperature using an Eclipse XSB C-18 column (5-μm particle size silica, 150 mm × 4.6 mm). The mobile phase consisted of a gradient of 0.2% formic acid in water (A) and 0.2% formic acid in acetonitrile (B) (75% A to 75% B in 35 min, to 100% B in 5 min) and a flow rate of 1.0 mL/min. The ion mode was set as positive and negative. The interface and MSD detector parameters were as follows: nebulizer pressure, 70.0 psi (N_2); dry gas, N_2 (12.0 L/min); dry gas temperature, 350°C; spray capillary voltage, 3500 V; skimmer voltage, 40.0 V; ion transfer capillary exit, 100 V; scan range, m/z 100–500. Ultrahigh pure He was used as the collision gas. Data were acquired in positive and negative modes in a single LC run, using the continuous-polarity switching ability of the mass spectrometer. Samples dissolved in acetonitrile (ca. 0.1–0.3 mg/mL) were injected (10 μL) using an autosampler (Pedras and Adio, 2006).

[b] ND = not detected.

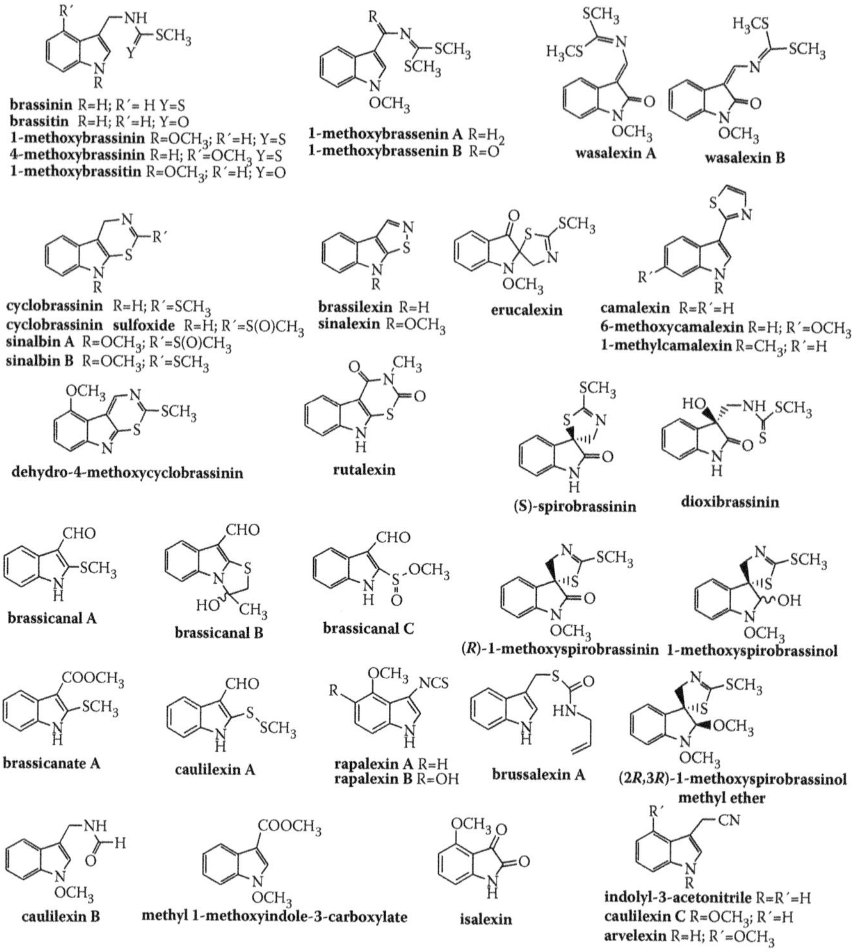

brassinin R=H; R′= H Y=S
brassitin R=H; R′=H; Y=O
1-methoxybrassinin R=OCH₃; R′=H; Y=S
4-methoxybrassinin R=H; R′=OCH₃ Y=S
1-methoxybrassitin R=OCH₃; R′=H; Y=O

1-methoxybrassenin A R=H₂
1-methoxybrassenin B R=O

wasalexin A

wasalexin B

cyclobrassinin R=H; R′=SCH₃
cyclobrassinin sulfoxide R=H; R′=S(O)CH₃
sinalbin A R=OCH₃; R′=S(O)CH₃
sinalbin B R=OCH₃; R′=SCH₃

brassilexin R=H
sinalexin R=OCH₃

erucalexin

camalexin R=R′=H
6-methoxycamalexin R=H; R′=OCH₃
1-methylcamalexin R=CH₃; R′=H

dehydro-4-methoxycyclobrassinin

rutalexin

(S)-spirobrassinin

dioxibrassinin

brassicanal A

brassicanal B

brassicanal C

(R)-1-methoxyspirobrassinin

1-methoxyspirobrassinol

brassicanate A

caulilexin A

rapalexin A R=H
rapalexin B R=OH

brussalexin A

(2R,3R)-1-methoxyspirobrassinol
methyl ether

caulilexin B

methyl 1-methoxyindole-3-carboxylate

isalexin

indolyl-3-acetonitrile R=R′=H
caulilexin C R=OCH₃; R′=H
arvelexin R=H; R′=OCH₃

FIGURE 12.2   Chemical structures of currently known crucifer phytoalexins.

leaf extracts of each species, although the genes of *T. halophila* and *A. thaliana* have high sequence identity (90% to 95% at *c*DNA level; Taji et al., 2004). It is noteworthy that two other species (*Wasabi japonica* syn. *Eutrema wasabi* and *Thlaspi arvense*) produce wasalexins A and B as well (Pedras et al., 1999, 2003), whereas methoxybrassenin B is produced in *Brassica oleracea* (Monde et al., 1991) and rapalexin A in *B. rapa* (Pedras et al., 2007b).

BIOACTIVITY

By definition, phytoalexins are active against microbial plant pathogens; however, considering the broad range of pathogens hosted by crucifer crops, the number of plant pathogens tested against phytoalexins is rather small. Earlier studies determined the biological activity of phytoalexins using microbial species (e.g., *Bipolaris leersiae*) unrelated with the pathogens of the producing plant species (Pedras et al., 2007c). The variety of assays used to test the bioactivity of phytoalexins provides a reasonable method to compare the bioactivity of various structures. In general, phytoalexins are tested at concentrations around 0.1 to 0.5 mM in agar or liquid medium using mycelial cultures of fungi or bacterial cultures. These assays have shown that phytoalexins are not equally toxic to each pathogenic species, and selective antifungal activities have been reported. This is not surprising because the chemical structures of crucifer phytoalexins are quite diverse, despite sharing a

TABLE 12.2
Phytoalexins from Wild Plant Species of the Family Brassicaceae and Their Elicitors

Species (Common Name, Biotic and/or Abiotic Elicitors)	Phytoalexins (Ref.)
Arabidopsis thaliana (thale cress, AgNO$_3$, *Pseudomonas syringae*, CuCl$_2$)	Camalexin (Tsuji et al., 1992), rapalexin A (Pedras and Adio, 2008)
Arabis lyrata (lyrata rock cress, *Cochliobolus carbonum*, *P. syringae*)	Camalexin (Zook et al., 1998)
Brassica atlantica (CuCl$_2$, *Leptosphaeria maculans*)	Brassilexin, cyclobrassinin, 1-methoxybrassinin (Rouxel et al., 1991)
B. montana (CuCl$_2$, *L. maculans*)	Brassilexin, cyclobrassinin, 1-methoxybrassinin (Rouxel et al., 1991)
B. tournefortii (Asian mustard, CuCl$_2$)	No indolyl phytoalexins detected (Sarwar and Pedras, 2006)
B. adpressa (CuCl$_2$, *L. maculans*)	Brassilexin, cyclobrassinin, cyclobrassinin sulfoxide, 1-methoxybrassinin (Rouxel et al., 1991)
Camelina sativa (false flax, *Arabidopsis brassicae*)	Camalexin, 6-methoxycamalexin (Browne et al., 1991)
Capsella bursa-pastoris (shepherd's purse, *Alternaria brassicae*)	Camalexin, 6-methoxycamalexin, 1-methylcamalexin (Jimenez et al., 1997)
Crambe abyssinica (Abyssinian mustard, CuCl$_2$)	Rapalexin B (Pedras et al., 2007b; Sarwar and Pedras, 2006)
Diplotaxis tenuifolia (sand rocket, CuCl$_2$)	Arvelexin (Sarwar and Pedras, 2006)
D. muralis (wallrocket, CuCl$_2$)	Rapalexin A, 1,4-dimethoxyindolyl-3-acetonitrile (Pedras et al., 2007b; Sarwar and Pedras, 2006)
Erucastrum gallicum (dog mustard, *Sclerotinia sclerotiorum*)	Arvelexin, erucalexin, indole-3-acetonitrile, 1-methoxyspirobrassinin (Pedras and Ahiahonu, 2004; Pedras et al., 2006b)
Sinapis arvensis (white mustard, *L. maculans*)	Brassilexin, cyclobrassinin sulfoxide (Stork and Sacristan, 1995)
Sisymbrium officinale (hedge mustard, CuCl$_2$)	Methyl 1-methoxyindole-3-carboxylate (Sarwar and Pedras, 2006)
Thellungiella halophila (salt cress, CuCl$_2$)	Wasalexins A and B and methoxybrassenin B (Pedras and Adio, 2008)
Thlaspi arvense (stinkweed, pennycress, CuCl$_2$, *L. maculans*)	Arvelexin, wasalexin A (Pedras et al., 2003)

common biosynthetic precursor, tryptophan. On the other hand, camalexin has been tested against a much wider variety of microbes, which also is not surprising considering that it was for some time the only phytoalexin reported from the intensively studied model species, *Arabidopsis thaliana* (Pedras et al., 2007c). Phytoalexins are also toxic to plant tissues; thus, their production is restricted to the tissues subject to stress. Damage caused by insects and nematodes has not been reported to induce crucifer phytoalexins.

It was shown that several crucifer phytoalexins (i.e., brassinin, spirobrassinin, brassilexin, camalexin, 1-methoxyspirobrassinin, 1-methoxyspirobrassinol, and methoxyspirobrassinol methyl ether) displayed significant antiproliferative activity against various cancer cells (Pilatova et al., 2005; Sabol et al., 2000). This activity appeared to be associated with the modulation of transcription factors regulating cell cycle, differentiation, and apoptosis. 1-Methoxybrassinin displayed the most potent antiproliferative activity, whereas cyclobrassinin, spirobrassinin, and brassinin displayed cancer chemopreventive activity in models of mammary and skin carcinogenesis. However, because phytoalexins are produced in very low quantities as production is restricted to the areas where elicitation occurs, it is unlikely that these compounds are responsible for the protective role of crucifers against cancer.

BIOSYNTHESIS

A complete understanding of any given metabolic pathway must include not only knowledge of the genes and corresponding enzymes, but also the metabolic products of these enzymes. Thus, complementary biosynthetic investigations can start from either the metabolite end using isotopically labeled intermediates, or from the gene cloning end using a variety of molecular genetics methodologies. As discussed below, to date very few genes have been cloned and no biosynthetic enzymes have been isolated. This is clearly an area of crucifer research where great developments are expected to occur in the next decade.

BIOSYNTHETIC RELATIONSHIP WITH OTHER METABOLITES OF CRUCIFERS

Crucifer phytoalexins are alkaloids biosynthetically derived from the amino acid (*S*)-tryptophan. The structures of brassinins (Figure 12.3) are related to those of indolyl glucosinolates and indolyl acetonitriles with respect to the methoxy substituent location in the indole nucleus and the primary biosynthetic precursor, tryptophan. Glucosinolates are secondary metabolites also characteristic of species within the *Brassicaceae*, containing a carbon attached to an iminosulfate, to a α-D-thioglucose unit, and to another carbon that provides the link to a wide variety of substituents. The iminosulfate group of glucosinolates accounts for their ionic character and polarity, in contrast to brassinins and indolyl acetonitriles that are much less polar and thus relatively soluble in organic solvents. The important biological functions and significant biological activities of glucosinolates have stimulated an enormous amount of research for more than three decades, particularly evident in the past decade (Grubb and Abel, 2006; Hansen and Halkier, 2005; Mikkelsen et al., 2003; Nafisi et al., 2006). Great advances have been made in recent years to determine the biosynthetic genes of glucosinolates, specifically due to the availability of numerous *Arabidopsis thaliana* ecotypes and mutants. Indolyl acetonitriles can result from indolyl glucosinolates, but can also result from other sources, including indolyl acetaldoxymes (Grubb and Abel, 2006).

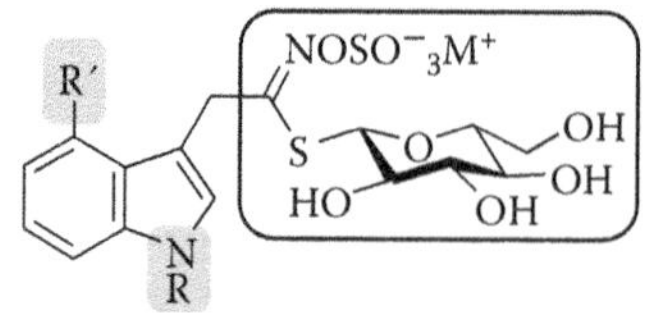

glucobrassicin R=R´=H
neoglucobrassicin R=OCH₃; R´=H
4-methoxyglucobrassicin R=H; R´=OCH₃

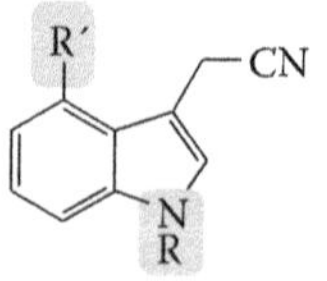

indolyl-3-acetonitrile R=R´=H
caulilexin C R=OCH₃; R´=H
arvelexin R=H; R´=OCH₃

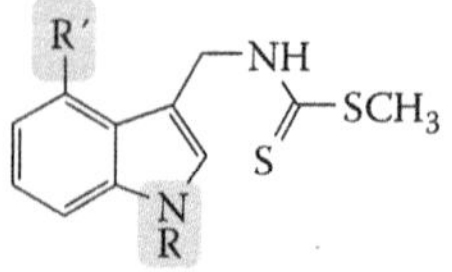

brassinin R=R´= H
1-methoxybrassinin R=OCH₃; R´=H
4-methoxybrassinin R=H; R´=OCH₃

FIGURE 12.3 Chemical structures of indolyl glucosinolates glucobrassicins, indolyl acetonitriles, and phytoalexins brassinins.

BIOSYNTHETIC PATHWAYS

The biosynthetic pathway of crucifer phytoalexins and that of indolyl glucosinolates starts with the conversion of tryptophan to indolyl-3-acetaldoxime and appears to diverge into various branches, including (1) indolyl glucosinolates, (2) brassinins, and (3) camalexin (Pedras and Okinyo, 2008). Currently, camalexin is the only crucifer phytoalexin having some of its encoding genes cloned, as the availability of mutants of *Arabidopsis thaliana* has greatly facilitated this approach (Glawischnig, 2007). CYP79B2 and CYP79B3 are enzymes that catalyze the conversion of (*S*)-tryptophan to indolyl-3-acetaldoxime, whereas CYP71A13 catalyzes the conversion of indolyl-3-acetaldoxime to indole-3-acetonitrile (Nafisi et al., 2007), and CYP71B15 (PAD3) (Glawischnig et al., 2004) catalyzes the oxidative decarboxylation of dihydrocamalexic acid (Figure 12.4). However, considering that the wasalexin and methoxybrassenin pathways are present in *Thellungiella halophila*, whose genome is currently being sequenced (http://www.jgi.doe.gov/sequencing/why/CSP2007/thellungiella.html), and that these pathways are likely to involve a number of genes also present in the economically important genus *Brassica*, these phytoalexin biosynthetic genes are likely to be cloned earlier than others.

The biosynthetic relationships among the various crucifer phytoalexins have been established using isotopically labeled precursors containing deuterium, carbon-13 (^{13}C), carbon-14 (^{14}C), sulfur-34 (^{34}S), and sulfur-35 (^{35}S). Based on these data, a biosynthetic scheme is provided in Figure 12.4. Oxidation of indolyl-3-acetaldoxime provides a putative intermediate that could be converted through a C-S lyase to the corresponding thiohydroxamic acid, possibly the last intermediate common to both pathways. Indole-3-methylisothiocyanate could then be thiomethylated to yield brassinin. A similar pathway could be envisioned for 1-methoxybrassinin or 4-methoxybrassinin via the corresponding 1-methoxy or 4-methoxy aldoximes (Pedras et al., 2007c). Rutalexin is derived from cyclobrassinin, and brassicanate A appears to derive from brassinin; however, the biosynthetic precursors of isalexin, rapalexins A and B, and brussalexin A are not known. Erucalexin, a unique structure among these metabolites in that it has the carbon substituent at C-2 rather than at C-3 position, and 1-methoxyspirobrassinin, are both derived from 1-methoxybrassinin.

APPLICATION AND PROSPECTS

Considering the relatively low number of crucifer species screened for production of phytoalexins, new structures and biological activities are likely to be discovered in the next few years. In fact, it is

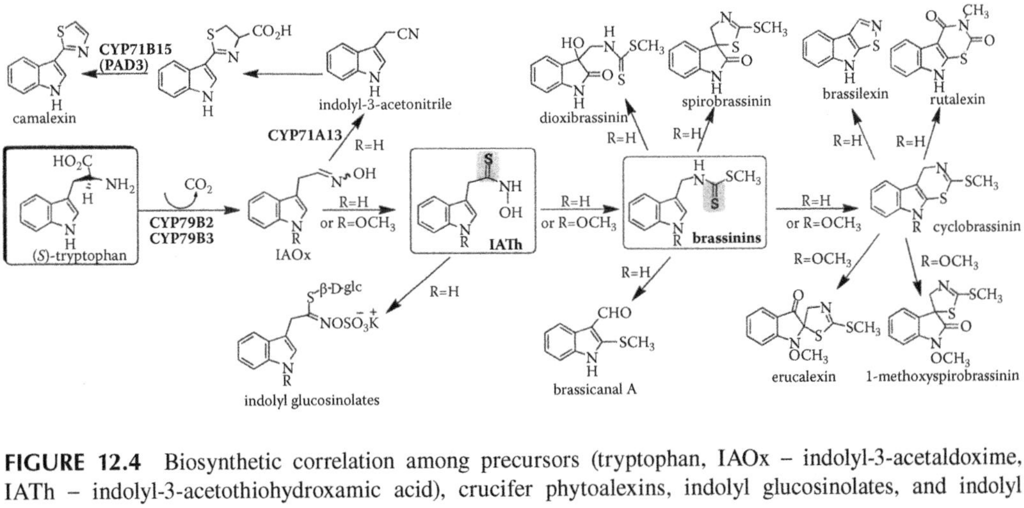

FIGURE 12.4 Biosynthetic correlation among precursors (tryptophan, IAOx – indolyl-3-acetaldoxime, IATh – indolyl-3-acetothiohydroxamic acid), crucifer phytoalexins, indolyl glucosinolates, and indolyl acetonitriles.

important to analyze additional wild crucifer species as they represent potential sources of disease resistance. Despite the protective role and antifungal activity of phytoalexins, some pathogenic fungi of crucifers can detoxify them (Pedras and Ahiahonu, 2005). These detoxification reactions appear to be catalyzed by enzymes that are both substrate and pathogen specific. For example, a fungal enzyme involved in the detoxification of brassinin, brassinin oxidase, could not transform a number of brassinin analogs (Pedras and Jha, 2006; Pedras et al., 2007a). In this context, it is foreseen that further work on mechanisms of phytoalexin detoxification occurring in plant pathogens can provide new targets to control these pathogenic fungi. Toward this end, a term coined "PALDOXIN" has been proposed to designate potential PhytoALexin DetOXification INhibitors (Pedras and Ahiahonu, 2005). By design, paldoxins are nontoxic selective inhibitors to be used as a mixture against specific fungal plant pathogens, potentially synergistic with natural chemical defenses, including phytoalexins.

In addition, it is essential to develop new synthetic strategies to provide more economic and accessible routes to the currently known phytoalexins, particularly to the 1-methoxy-substituted compounds that are essential for biosynthetic and bioactivity studies. As well, biosynthetic studies to establish the intermediates between indolyl-3-acetaldoxime and brassinin, and intermediates common to 1-methoxy- and 4-methoxy-substituted phytoalexins are necessary. Likewise, it is crucial to clone the genes/enzymes of the phytoalexin pathway in *Brassica* species. Genetic manipulation of these complex and unique crucifer defense pathways requires a complete metabolic understanding rather far from our current knowledge.

REFERENCES

Bailey, J.A. and J.W. Mansfield. 1982. *Phytoalexins.* Blackie and Son, Glasgow, U.K., p. 334.

Browne, L.M., K.L. Conn, W.A. Ayer, and J.P. Tewari. 1991. The camalexins: new phytoalexins produced in the leaves of *Camelina sativa* (Cruciferae). *Tetrahedron,* 47:3909–3914.

Glawischnig, E. 2007. Camalexin. *Phytochemistry,* 68:401–406.

Glawischnig E., B.G. Hansen, C.E. Olsen, and B.A. Halkier. 2004. Camalexin is synthesized from indole-3-acetaldoxime, a key branching point between primary and secondary metabolism in *Arabidopsis. Proc. Natl. Acad. Sci. (U. S. A.),* 101:8245–8250.

Grubb, C.D. and S. Abel. 2006. Glucosinolate metabolism and its control. *Trends Plant Sci.,* 11:89–100.

Hansen, B.G. and B.A. Halkier. 2005. New insight into the biosynthesis and regulation of indole compounds in *Arabidopsis thaliana. Planta,* 221:603–606.

Jimenez, L.D., W.A. Ayer, and J.P. Tewari. 1997. Phytoalexins produced in the leaves of *Capsella bursa-pastoris* (shepherd's purse). *Phytoprotection,* 78:99.

Mikkelsen, M.D., B.L. Petersen, E. Glawischnig, A.B. Jensen, E. Andreasson, and B.A. Halkier. 2003. Modulation of CYP79 genes and glucosinolate profiles in *Arabidopsis* by defense signaling pathways. *Plant Physiol.,* 131:298–308.

Monde, K., K. Sasaki, A. Shirata, and M. Takasugi. 1991. Methoxybrassenins A and B, sulphur-containing stress metabolites from *Brassica oleracea* var. capitata. *Phytochemistry,* 30:3921–3922.

Nafisi, M., I.E. Sønderby, B.G. Hansen, et al. 2006. Cytochromes P450 in the biosynthesis of glucosinolates and indole alkaloids. *Phytochem. Rev.,* 5:331–346.

Nafisi, M., S. Goregaoker, C.J. Botanga, et al. 2007. Arabidopsis cytochrome P450 monooxygenase 71A13 catalyzes the conversion of indole-3-acetaldoxime in camalexin synthesis. *Plant Cell,* 19:2039–2052.

Pedras, M.S.C., A.M. Adio, M. Suchy, et al. 2006a. Detection, characterization and identification of crucifer phytoalexins using high-performance liquid chromatography with diode array detection and electrospray ionization mass spectrometry. *J. Chromatogr. A,* 1133:172–183.

Pedras, M.S.C. and A.M. Adio. 2008. Phytoalexins and phytoanticipins from the wild crucifers *Thellungiella halophila* and *Arabidopsis thaliana:* rapalexin A, wasalexins and camalexin. *Phytochemistry,* 69:889–893.

Pedras, M.S.C. and P.W.K. Ahiahonu. 2004. Phytotoxin production and phytoalexin elicitation by the phytopathogenic fungus *Sclerotinia sclerotiorum. J. Chem. Ecol.,* 30:2163–2179.

Pedras, M.S.C. and P.W.K. Ahiahonu. 2005. Metabolism and detoxification of phytoalexins and analogs by phytopathogenic fungi. *Phytochemistry,* 66:391–411.

Pedras, M.S.C. and M. Jha. 2006. Toward the control of *Leptosphaeria maculans*: design, syntheses, biological activity and metabolism of potential detoxification inhibitors of the crucifer phytoalexin brassinin. *Bioorg. Medicinal Chem.*, 14:4958–4979.

Pedras, M.S.C., M. Jha, Z. Minic, and O.G. Okeola. 2007a. Isosteric probes provide structural requirements essential for detoxification of the phytoalexin brassinin by the fungal pathogen *Leptosphaeria maculans*. *Bioorg. Medicinal Chem.*, 15:6054–6061.

Pedras, M.S.C. and D.P. O. Okinyo. 2008. Remarkable incorporation of the first sulfur containing indole derivative: another piece in the puzzle of crucifer phytoalexins. *Org. Biomolec. Chem.*, 6:51–54.

Pedras, M.S.C. and J.L. Sorensen. 1998. Phytoalexin accumulation and production of antifungal compounds by the crucifer wasabi. *Phytochemistry*, 49:1959–1965.

Pedras, M.S.C., J.L. Sorensen, F.I. Okanga, and I.L. Zaharia. 1999. Wasalexins A and B, new phytoalexins from Wasabi: isolation, synthesis, and antifungal activity. *Bioorg. Medicinal Chem. Lett.*, 9:3015–3020.

Pedras, M.S.C., P.B. Chumala, and M. Suchy. 2003. Phytoalexins from *Thlaspi arvense*, a wild crucifer resistant to virulent *Leptosphaeria maculans*: structures, syntheses and antifungal activity. *Phytochemistry*, 64:949–956.

Pedras, M.S.C., M. Suchy and P.W.K. Ahiahonu. 2006b. Unprecedented chemical structure and biomimetic synthesis of erucalexin, a phytoalexin from the wild crucifer *Erucastrum gallicum*. *Org. Biomolec. Chem.*, 4:691–697.

Pedras, M.S.C., Q.A. Zheng, and G.S. Ravi. 2007b. The first naturally occurring aromatic isothiocyanates, rapalexins A and B, are cruciferous phytoalexins. *J. Chem. Soc., Chem. Commun.*, 368–370.

Pedras, M.S.C., Q.A. Zheng, and V.K. Sarma-Mamillapalle. 2007c. The phytoalexins from *Brassicaceae*: structure, biological activity, synthesis and biosynthesis. *Natural Prod. Commun.*, 2:319–330.

Pilatova, M., M. Sarissky, P. Kutschy, et al. 2005. Cruciferous phytoalexins: antiproliferative effects in T-Jurkat leukemic cells. *Leukemia Res.*, 29:415–421.

Rouxel, T., A. Kollmann, L. Boulidard, and R. Mithen. 1991. Abiotic elicitation of indole phytoalexins and resistance to *Leptosphaeria maculans* within Brassiceae. *Planta*, 184:271–278.

Sabol, M., P. Kutschy, L. Siegfried, et al. 2000. Cytotoxic effect of cruciferous phytoalexins against murine L1210 leukemia and B16 melanoma. *Biologia (Bratislava)*, 55:701–707.

Sarwar M.G. and M.S.C. Pedras. 2006. New phytoalexins from crucifers. *Can. J. Plant Pathol.*, 28:336.

Storck, M. and M.D. Sacristan. 1995. The role of phytoalexins in the seedling resistance to *Leptosphaeria maculans* in some crucifers. *Z. Naturforsch.*, 50c:15–20.

Taji, M.S., M. Satou, T. Sakurai, et al. 2004. Comparative genomics in salt tolerance between *Arabidopsis* and *Arabidopsis*-related halophyte salt cress using Arabidopsis microarray, *Plant Physiol.*, **135**:1697–1709.

Takasugi, M., N. Katsui, and A. Shirata. 1986. Isolation of three novel sulphur-containing phytoalexins from the Chinese cabbage *Brassica campestris* L. ssp. *pekinensis* (Cruciferae). *J. Chem. Soc., Chem. Commun.*, 1077–1078.

Tsuji, J., M. Zook, E.P. Jacobson, D.A. Gage, R. Hammerschmidt, and S.C. Somerville. 1992. Phytoalexin accumulation in *Arabidopsis thaliana* during the hypersensitive reaction to *Pseudomonas syringae* pv *syringae*. *Plant Physiol.*, 98:1304–1309.

VanEtten, H.D., J.W. Mansfield, J.A. Bailey, and E.E. Farmer. 1994. Two classes of plant antibiotics: phytoalexins versus "phytoanticipins". *Plant Cell*, 6:1191–1192.

Zook, M., L. Leege, D. Jacobson, and R. Hammerschmidt. 1998. Camalexin accumulation in *Arabis lyrata*. *Phytochemistry*, 49:1393–1404.

13 Introgression of Genes from Wild Crucifers

Hu Qiong, Li Yunchang, and Mei Desheng

CONTENTS

INTRODUCTION

Genus *Brassica* belongs to the tribe Brassiceae in the Cruciferae family. *Brassica* crops are among the most agronomically important crops cultivated throughout the world as vegetables and oil crops, both for human consumption and industrial uses as well as condiments and fodder crops. Among the six cultivated species, three diploids (i.e., *B. nigra, B. campestris,* and *B. oleracea*) are primary species and the other three (i.e., *B. napus, B. juncea,* and *B. carinata*) are amphidiploids resulting from crosses between corresponding pairs of the primary species (Morinaga, 1934). Four species of *Brassica* have been widely cultivated as oilseed crops: *B. carinata, B. campestris, B. juncea,* and *B. napus.* Among these species, *B. napus* is dominantly planted in major production regions due to good adaptability and productiveness. Increasing interest in production — both for edible purposes and raw materials for biodiesel — has led to intensive efforts on breeding. In the oilseed *Brassica* species, varietal development programs for quality with high yield and wide adaptability have stimulated breeders all over the world to exploit potential resources, especially the wild types of the *Brassica* coenospecies.

According to Prakash et al. (2008), *Brassica* coenospecies cover at least 14 genera from Brassicinae, Raphaninae, and Moricandinae subtribes, including *Brassica, Coincya, Diplotaxis, Eruca, Erucastrum, Hirschfeldia, Sinapis, Sinapidendron, Trachystoma, Enarthrocarpus, Raphanus, Moricandia, Pseuderucaria,* and *Rytidocarpus,* with most of them being wild and weedy. Useful genes that confer desirable traits for crop production can usually be found in wild forms. In the long history of variety development of *Brassica* oil crops, genetic integration of traits/genes from wild aliens is a major approach for the introduction of valuable traits, as in many other crops.

Such genetic integration can be achieved both by conventional cross and through biotechnology, depending on the relatedness and crossibility of the donor species with *Brassica* crops. Generally, species with close relatedness can be readily crossed by traditional means, including those within the *Brassica* genus (e.g., *B. napus* × *B. nigra*) for the transfer of blackleg resistance (Chevre et al., 1996, 1997; Snowdon et al., 2000). For species that are not readily crossed, special techniques can be used to aid the hybridization, for example, grafting, mixed pollination, bud pollination, or *in vitro* fertilization. Biotechnological means can be used for the hybridization of remotely related species that are recalcitrant to crossing. Wide hybridization in *Brassica* can be traced back to the early nineteenth century when Sageret (1826) produced inter-subtribal hybrids of *Raphanus sativus* × *B. oleracea*. Since then, with the aid of *in vitro* techniques, a large number of sexual and somatic hybrids have been successfully obtained.

SEXUAL CROSS

Cross barriers were always encountered for the hybridization of different species. Either pre-fertilization barriers due to the inability of pollen tubes to reach the style or post-fertilization barriers with embryo abortion due to the genetic incompatibility between the developing embryo and the endosperm can cause the failure of hybrid production. Although a very large number of pollinations were attempted, production of hybrids was unsuccessful for many combinations. Measures to overcome pre-fertilization barriers include grafting, mixed pollination, bud pollination, stump pollination, and *in vitro* fertilization (Hosoda et al., 1963; Namai, 1971; Kameya and Hinata, 1970). Embryo rescue techniques have been used effectively for overcoming post-fertilization barriers. In fact, sequential culture (culture of ovaries, ovules, and seeds/embryos) is more effective than simple ovary or ovule culture (Shivanna, 1996). With advancements in sexual cross techniques, hybridization between wild and oilseed *Brassica* species for the genetic improvement of oilseed *Brassicas* has been widely used and, as a result, a large number of sexual hybrids comprising interspecific, intergeneric, and intertribal combinations have been obtained (Table 13.1 and Table 13.2). Using *B. rapa* as the paternal parent, as least 20 wild species including five from *Brassica* have been reported to be successfully crossed. The number of species used to cross with *B. juncea*, *B. napus*, and *B. carinata* are 10, 13, and 2, respectively (Table 13.1). With the *Brassica* oilseed crops as the maternal parent, 15, 11, 18, and 10 wild species have been successfully crossed with *B. rapa*, *B. napus*, *B. juncea,* and *B. carinata*, respectively (Table 13.2).

Crossibility as well as chromosome pairing of sexual hybrids generally depends on the relatedness and the ploidy level of the parental species. Hybrids from crosses with parental species which are phylogenetically closely related and/or with higher ploidy levels are easier to produce and to form multivalents of during meiosis. Introgression of desirable traits controlled by nuclear genes from the wild species to cultivated crops can never be possible without multivalent formation for the exchange of genetic material from chromosomes of different origin. A positive relationship between bivalent frequency and chromosome number was observed in many studies. For example, in the cross between *Diplotaxis catholica* ($n = 9$) and *Brassica rapa* ($n = 10$), in addition to univalents, two bivalents or one trivalent, and three bivalents were observed in the hybrids with 19 chromosomes, whereas the hybrid of *D. catholica* ($n = 9$) and *B. juncea* ($n = 18$) with 27 chromosomes had the configuration of 13 I + 7 II–2 III + 7 II + 7 I at meiosis (Mohanty, 1996).

Although a wide range of wild species has been used for the production of sexual hybrids for the introgression of useful traits in oilseed *Brassicas* with wild species as the maternal parent, the production of hybrids with crop species as the maternal parent is generally easier. A typical example is the hybridization between *Brassica* species and *Orychophragmus violaceus* ($2n = 24$). Hybrids of *O. violaceus* with *Brassica* species — with the *Brassica* species as maternal parents — were all readily produced by the aid of ovary culture (Li et al., 1995; Li and Heneen, 1998, 1999; Hua et al., 2006). But when *O. violaceus* was used as the maternal parent, no hybrids were obtained despite extensive efforts.

TABLE 13.1

Sexual Hybridization with *Brassica* Oilseed Species as the Paternal Parent

Wild Species	*Brassica* Species	Ref.
B. barrelieri ($n = 10$)	*B. rapa*	Takahata and Hinata (1983)
B. fruticulosa ($n = 8$)	*B. rapa*	Takahata and Hinata (1983); Nanda Kumar et al. (1988); Chandra et al. (2004)
B. maurorum ($n = 8$)	*B. rapa*	Takahata and Hinata (1983)
B. oxyrrhina ($n = 9$)	*B. rapa*	Prakash and Chopra (1990)
B. tournefortii ($n = 10$)	*B. rapa*	Choudhary and Joshi (2001)
Diplotaxis catholica ($n = 9$)	*B. rapa*	Mohanty (1996)
Diplotaxis erucoides ($n = 7$)	*B. rapa*	Vyas et al. (1995)
Diplotaxis muralis ($n = 21$)	*B. rapa*	Harberd and McArthur (1980)
Diplotaxis siettiana ($n = 8$)	*B. rapa*	Nanda Kumar and Shivanna (1993)
Diplotaxis siifolia ($n = 10$)	*B. rapa*	Ahuja et al. (2003)
Diplotaxis tennuifolia ($n = 11$)	*B. rapa*	Salisbury (1989)
Diplotaxis virgata ($n = 9$)	*B. rapa*	Takahata and Hinata (1983)
Enarthrocarpus lyratus ($n = 10$)	*B. rapa*	Gundimeda et al. (1992)
Erucastrum abyssinicum ($n = 16$)	*B. rapa*	Harberd and McArthur (1980)
Erucastrum canariense ($n = 9$)	*B. rapa*	Prakash et al. (2001); Bhasker et al. (2002)
Erucastrum cardaminoides ($n = 9$)	*B. rapa*	Chandra et al. (2004)
Eruca sativa ($n = 11$)	*B. rapa*	Matsuzawa et al. (1999)
Moricandia arvensis ($n = 14$)	*B. rapa*	Takahata and Takeda (1990)
Raphanus sativus ($n = 9$)	*B. rapa*	Matsuzawa et al. (2000)
Sinapis arvensis ($n = 9$)	*B. rapa*	Momotaz et al. (1998)
Diplotaxis catholica ($n = 9$)	*B. juncea*	Mohanty (1996)
Diplotaxis erucoides	*B. juncea*	Vyas et al. (1995)
Diplotaxis siifolia ($n = 10$)	*B. juncea*	Batra et al. (1990); Ahuja et al. (2003)
Diplotaxis tennuifolia ($n = 11$)	*B. juncea*	Salisbury (1989)
Diplotaxis virgata ($n = 9$)	*B. juncea*	Harberd and McArthur (1980)
Erucastrum abyssinicum ($n = 16$)	*B. juncea*	Rao et al. (1996)
Erucastrum gallicum ($n = 15$)	*B. juncea*	Batra et al. (1989); Rao et al. (1998)
Moricandia arvensis ($n = 14$)	*B. juncea*	Takahata et al. (1993)
Raphanus sativus ($n = 9$)	*B. juncea*	Rhee et al. (1997)
Sinapidendron frutescens ($n = 10$)	*B. juncea*	Harberd and McArthur (1980)
Brassica cossoneana ($n = 16$)	*B. napus*	Harberd and McArthur (1980)
Diplotaxis erucoides	*B. napus*	Delourme et al. (1989); Vyas et al. (1995)
Diplotaxis muralis ($n = 21$)	*B. napus*	Fan et al. (1985); Ringdhal et al. (1987)
Diplotaxis siifolia ($n = 10$)	*B. napus*	Batra et al. (1990)
Diplotaxis viminea ($n = 10$)	*B. napus*	Mohanty (1996)
Enarthrocarpus lyratus ($n = 10$)	*B. napus*	Gundimeda et al. (1992)
Erucastrum gallicum ($n = 15$)	*B. napus*	Batra et al. (1989)
Hirschfeldia incana ($n = 7$)	*B. napus*	Kerlan et al. (1993)
Moricandia arvensis ($n = 14$)	*B. napus*	Takahata et al. (1993); Meng et al. (1997)
Moricandia nitens ($n = 14$)	*B. napus*	Meng et al. (1999)
Raphanus sativus ($n = 9$)	*B. napus*	Kerlan et al. (1993); Warwick et al. (2003)
Sinapis arvensis ($n = 9$)	*B. napus*	Mathias (1991)
Sinapis alba ($n = 12$)	*B. napus*	Ripley and Arnison (1990); Brown et al. (1997)
Erucastrum abyssinicum ($n = 16$)	*B. carinata*	Rao et al. (1996)
Erucastrum lyratus ($n = 10$)	*B. carinata*	Gundimeda et al. (1992)

TABLE 13.2
Sexual Hybridization with *Brassica* Oilseed Species as the Maternal Parent

Brassica Species	Wild Species	Ref.
B. rapa	*B. atlantica* ($n = 9$)	Mithen and Herron (1991)
	B. barrelieri ($n = 10$)	Mattsson (1988)
	B. fruticulosa ($n = 8$)	Mizushima (1968)
	B. spinescens ($n = 8$)	Prakash et al. (1982)
	B. tournefortii ($n = 10$)	Prakash and Narain (1971)
	Capsella bursa-pastoris ($n = 16$)	Chen et al. (2007)
	Coincya pseuderucastrum ($n = 12$)	Inomata (2004)
	Eruca sativa ($n = 11$)	Mizushima (1950); Harberd and McArthur (1980)
	Erucastrum leucanthum ($n = 8$)	Harberd and McArthur (1980)
	Hirschfeldia incana ($n = 7$)	Mizushima (1968)
	Moricandia arvensis ($n = 14$)	Takahata and Takeda (1990)
	Orychophragmus violaceus ($n = 12$)	Li and Heneen (1999)
	Raphanus sativus ($n = 9$)	Harberd and McArthur (1980); Matsuzawa et al. (2000)
	Sinapis alba ($n = 12$)	Jandurova and Dolezel (1995)
	Sinapis arvensis ($n = 9$)	Mizushima (1950)
B. juncea	*B. gravinae* ($n = 10$)	Nanda Kumar et al. (1989)
	B. tournefortii ($n = 10$)	Yandav et al. (1991)
	Crambe abyssinica ($n = 45$)	Wang and Luo (1998)
	Diplotaxis muralis ($n = 21$)	Bijral and Sharma (1995)
	Diplotaxis virgata ($n = 9$)	Inomata (2003)
	Enarthrocarpus lyratus ($n = 10$)	Gundimeda et al. (1992)
	Orychophragmus violaceus ($n = 12$)	Li et al. (1998a, b)
	Raphanus sativus ($n = 9$)	Fukushima (1945)
	Sinapis alba ($n = 12$)	Mohapatra and Bajaj (1990)
	Sinapis arvensis ($n = 9$)	Mizushima (1950); Harberd and McArthur (1980)
B. napus ($n = 19$)	*B. cossoneana* ($n = 16$)	Harberd and McArthur (1980)
	B. cretica ($n = 9$)	Inomata (2002)
	B. gravinae ($n = 10$)	Nanda Kumar et al. (1989)
	B. montana ($n = 9$)	Inomata (2002)
	Capsella bursa-pastoris ($n = 16$)	Chen et al. (2007)
	Diplotaxis erucoides ($n = 7$)	Harberd and McArthur (1980)
	Diplotaxis harra ($n = 12$)	Inomata (2005)
	Erucastrum gallicum ($n = 15$)	Lefol et al. (1997)
	Hirschfeldia incana ($n = 7$)	Kerlan et al. (1993)
	India roroppa ($n = 16$)	Dai et al. (2005)
	Lesquerella fendleri ($n = 6$)	Du et al. (2008)
	Matthiola incana ($n = 7$)	Peng et al. 2003
	Orychophragmus violaceus ($n = 12$)	Li et al. (1995); Hua and Li (2006)
	Raphanus raphanistrum ($n = 9$)	Kerlan et al. (1993); Lefol et al. (1997)
	Raphanus sativus ($n = 9$)	Takeshita et al. (1980)
	Sinapis pubescens ($n = 9$)	Inomata (1991, 1994)
	Sinapis arvensis ($n = 9$)	Kerlan et al. (1993)

TABLE 13.2
Sexual Hybridization with *Brassica* Oilseed Species as the Maternal Parent

Brassica Species	Wild Species	Ref.
B. carinata	*B. fruticulosa* (*n* = 8)	Harberd and McArthur (1980)
	Diplotaxis assurgens (*n* = 9)	Harberd and McArthur (1980)
	Diplotaxis tenuisiliqua (*n* = 9)	Harberd and McArthur (1980)
	Diplotaxis virgata (*n* = 9)	Harberd and McArthur (1980)
	Erucastrum gallicum (*n* = 15)	Harberd and McArthur (1980)
	Orychophragmus violaceus (*n* = 12)	Hua et al. (2006)
	Raphanus sativus (*n* = 9)	Harberd and McArthur (1980)
	Sinapis alba (*n* = 12)	Momotaz et al. (1998)
	Sinapis arvensis (*n* = 9)	Bing et al. (1991); Momotaz et al. (1998)
	Sinapis pubescens (*n* = 9)	Harberd and McArthur (1980)

SOMATIC HYBRIDIZATION

Although the sequential culture technique is quite effective in overcoming cross barriers in many cross combinations of *Brassica* and alien species, the majority of the wild species in the secondary and tertiary gene pools are still inaccessible for *Brassica* crop improvement due to sexual incompatibility. In such situations, somatic hybridization is the first choice used for the introduction of desirable traits from alien species. As *Brassica* and related species are very amenable for tissue culture techniques, somatic hybridization has been extensively used in the *Brassicaceae*, with the additional merit of inducing cytoplasmic variability and recombination of cytoplasmic and nucleic genomes (reviewed by Glimelius, 1999; Christey, 2004; Liu et al., 2005). After the first successful report of protoplast fusion in Cruciferae, which involved *Brassica napus* and *Glycine max* (Kartha et al., 1974), the intertribal hybrid regenerated from the fusion of *B. rapa* and *Arabidopsis thaliana* marked the breakthrough of protoplast fusion technology in *Brassica* (Gleba and Hoffmann, 1980). Since then, a large number of somatic hybrids have been obtained between *Brassica* oilseed crops and alien species of interspecific, intergeneric, and intertribal combinations. The alien species used include 26 species of 18 genera, with 5 species successfully hybridized with *B. rapa*, 11 with *B. juncea*, 17 with *B. napus,* and 1 with *B. carinata* (Table 13.3). The desirable traits targeted in protoplast fusion experiments include disease resistance (blackleg, clubroot, alternaria leaf spot, beet cyst nematode); fatty acid composition (high nervonic acid content, high erucic acid content, high lesquerolic acid content, high linoleic and palmitic acid content); resistance to cold tolerance, photosynthetic capacity, zinc and cadmium hyperaccumulation; and cytoplasmic male sterility as well as the restoration of cytoplasmic male sterility.

INTROGRESSION OF CYTOPLASMIC TRAITS

Plastids and mitochondria contain genomes encoding genes for some important traits. These traits are maternally inherited, and sexual reproduction can only transmit cytoplasmic traits from the maternal parent. Genetic modification of these traits can be achieved by cell fusion, in which a parasexual biparental inheritance of organelle genomes is obtained systematically. Unique combinations of cytoplasmic traits may arise by the sorting out of organelles or by the recombination of organelle DNA molecules. Intergenomic recombination of mitochondrial DNA is common in somatic hybrids. Somatic hybridization thus provides possibilities for new associations between the nucleus

TABLE 13.3
Somatic Hybridization between *Brassica* Oilseed Species and Wild Aliens

Somatic Hybrids	Ref.
Arabidopsis thaliana ($n = 5$) + *B. rapa*	Gleba and Hoffmann (1980)
Barbarea vulgaris ($n = 8$) + *B. rapa*	Oikarinen and Ryöppy (1992)
Barbarea stricta ($n = 8$) + *B. rapa*	Oikarinen and Ryöppy (1992)
Isatis indigotica ($n = 7$) + *B. rapa*	Tu et al. (2008)
Moricandia nitens ($n = 14$) + *B. rapa*	Meng et al. (1999)
Arabidopsis thaliana ($n = 5$) + *B. juncea*	Ovcharenko et al. (2004)
Brassica spinescens ($n = 8$) + *B. juncea*	Kirti et al. (1991)
Diplotaxis catholica ($n= 9$) + *B. juncea*	Kirti et al. (1995a)
Diplotaxis harra ($n = 13$) + *B. juncea*	Begum et al. (1995)
Diplotaxis muralis ($n = 21$) + *B. juncea*	Chatterjee et al. (1988)
Eruca sativa ($n = 11$) + *B. juncea*	Sikdar et al. (1990)
Moricandia arvensis ($n = 14$) + *B. juncea*	Kirti et al. (1992a)
Raphanus sativus ($n = 9$) + *B. juncea*	Muller et al. (2001)
Sinapis alba ($n = 12$) + *B. juncea*	Gaikwad et al. (1996)
Thlaspi caerulescens ($n = 7$) + *B. juncea*	Dushenkov et al. (2002)
Trachystoma ballii ($n = 8$) + *B. juncea*	Kirti et al. (1992b, 1995b)
Arabidopsis thaliana ($n = 5$) + *B. napus*	Bauer-Weston et al. (1993); Forsberg et al. (1994, 1998); Yamagishi et al. (2002)
Barbarea vulgaris ($n = 8$) + *B. napus*	Fahleson et al. (1994a)
Brassica tournefortii ($n = 10$) + *B. napus*	Stiewe and Röbbelen (1994); Liu et al. (1995)
Crambe abyssinica ($n = 45$) + *B. napus*	Wang et al. (2003, 2004)
Descurainia sophia ($n = 14$) + *B. napus*	Guan et al. (2007)
Diplotaxis harra ($n = 13$) + *B. napus*	Klimaszewska and Keller (1988)
Eruca sativa ($n = 11$) + *B. napus*	Fahleson et al. (1988, 1997)
Lesquerella fendleri ($n = 6$) + *B. napus*	Skarzhinskaya et al. (1996, 1998); Schröder-Pontoppidan et al. (1999); Nitovskaya et al. (2006b)
Lunaria annua ($n = 14$) + *B. napus*	Craig and Millam (1995)
Moricandia arvensis ($n = 14$) + *B. napus*	O'Neill et al. (1996)
Moricandia nitens ($n = 14$) + *B. napus*	Meng et al. (1999)
Orychophragmus violaceus ($n = 12$) + *B. napus*	Hu et al. (2002a); Sakhno et al. (2007)
Raphanus sativus ($n = 9$) + *B. napus*	Sakai and Inamura (1990); Sundberg and Glimelius (1991); Lelivelt and Krens (1992)
Sinapis arvensis ($n = 8$) + *B. napus*	Hu et al. (2002b)
Sinapis alba ($n = 12$) + *B. napus*	Primard et al. (1988)
Thlaspi perfoliatum ($n = 21$) + *B. napus*	Fahleson et al. (1994b)
Thlaspi caerulescens ($n = 7$) + *B. napus*	Brewer et al. (1999)
Camelina sativa ($n = 20$) + *B. carinata*	Narasimhulu et al. (1994)

of one fusion partner and organelles from another, and for interactions at the DNA level between the brought-together genomes that may create novel genome-conferring traits of importance.

Cytoplasmic male sterility (CMS) is a consequence of the incompatibility between nuclear and mitochondrial genomes. It has been a major pollination control system for the heterosis application in *Brassica* oilseed crops. Alloplasm generally brings together the genomes from both organelles and nuclear genomes of different origin and thus causes incompatibility. In addition to somatic hybridization, repeated backcrossing with an alien species as cytoplasmic donor has also resulted in cytoplasmic male sterility due to induction of the cytoplasm from a wild species into *Brassica* crops. Several CMS systems in *Brassica* were derived from interspecific hybridization, viz., Ogura CMS (Bannerot et al., 1974; 1977); *tour* CMS (Stiewe and Röbbelen, 1994); *Diplo* CMS (Rao et al., 1994); *Trachy* CMS and *Mori* CMS (Prakash et al., 1998); *oxy* CMS (Banga et al., 1998); and NSa CMS (Hu et al., 2004). Both somatic and sexual hybridization have played an important role in the transfer of novel CMS systems from wild species to *Brassica* oilseed crops.

Tour CMS, originally found in *Brassica juncea* and then transferred to *B. napus* (Mathias, 1985), is a potential CMS source for hybrid production in *Brassica*. Its sterility is highly stable, and both maintainer lines and restorer genes are available within the *Brassica*. *B. juncea tour* CMS lines were identified to contain a mitochondrial genome of *B. tournefortii*, and it was proposed that CMS derived from the *B. tournefortii* cytoplasm (Pradhan et al., 1991; Stiewe and Röbbelen, 1994). Induction of CMS by protoplast fusion between fertile *B. tournefortii* and *B. napus* has also been achieved (Stiewe and Röbbelen, 1994; Liu et al., 1996). A lot of sterile or partially sterile plants were obtained (Stiewe and Röbbelen, 1994), where they recovered 1674 regenerants from fusing x-ray irradiated *B. tournefortii* protoplasts with those of nontreated *B. napus*. Six of the plants transmitted sterility to progenies, and five of the lines segregated for male sterility. One line (25-143) was completely male sterile from the beginning, and so were all its descendants from backcrosses. The occurrence of CMS hybrids derived from the fusion of fertile *B. tournefortii* and *B. napus* not only provides plant materials for the promising CMS system development, but also confirms the cytoplasmic origin of this CMS.

Introduction of cytoplasm from *Trachystoma ballii* to *Brassica juncea* nuclear background by protoplast fusion resulted in CMS (Kirti et al., 1995b). Male sterile plants with the nucleus completely replaced by *B. juncea* after several backcrosses closely resembled the normal fertile *B. juncea* in general morphology. Molecular analysis of organelle genomes indicated extensive mitochondrial DNA recombinations in the CMS line. Preliminary analysis of the chloroplast genome of the CMS line also indicated cpDNA recombination, which is a rare event in cybrids (Baldev et al., 1998). A dominant gene restoring fertility to this CMS line was also transferred to *B. juncea* in the same fusion event (Kirti et al., 1997).

Somatic hybrids derived from the fusion of *Brassica juncea* and *Moricandia arvensis* yielded stable sterile offspring that are allocytoplasmic CMS also (Kirti et al., 1992; Prakash et al., 1998). The original hybrid containing the sum of chromosomes of the parental species was sterile and backcross progeny was obtained by *in vitro* culture. After six backcrosses, the CMS line became stable and female fertility increased to above 90%. The CMS plants are similar to normal *B. juncea* although the leaves exhibit severe chlorosis, resulting in delayed flowering. Introgression of genetic information for fertility restoration was achieved following the development of a *M. arvensis* monosomic addition line on CMS *B. juncea*, which was isolated from the BC$_3$ fertile progeny of the somatic hybrid. The additional chromosome paired allosyndetically with one of the *B. juncea* bivalents and allowed introgression. The CMS (*Moricandia*) *B. juncea*, the restorer, and fertility restored F$_1$ plants possessing similar cytoplasmic organellar genomes as *M. arvensis* were revealed by Southern analysis. Rectification of chlorosis in the CMS lines has made this CMS suitable for commercial hybrid seed production (Kirti et al., 1998).

A combination somatic and sexual hybridization method established Ogura CMS for oilseed *Brassica* heterosis application. It was originally transferred from Japanese radish into *Brassica* by sexual cross with *Raphanus sativus* as the female parent and many generations of backcross to *B. napus* (Bannerot et al., 1977). The sterility of the CMS line is highly stable independent of environmental factors. However, the presence of radish chloroplasts of the cytoplasm in *Brassica* nuclear background had a side effect due to incompatibility. Plants of this cytoplasm were yellowish in color (caused by chlorophyll deficiency below 12°C), rendering the system not practically usable. Correction of the cold-temperature chlorosis proved possible by means of intraspecific somatic hybridization (Pelletier et al., 1983). Alloplasmic cybrids/hybrids were obtained from fusion between such CMS lines and normal lines in *B. napus* (Menczel et al., 1987; Jarl and Bornman, 1988). Male sterile plants with normal green leaves under cold conditions were selected, and transmission of both cytoplasmic phenotypes was observed. Stabilized chromosome number ($2n = 38$, typical of *B. napus*) was restored after two backcrosses of one hybrid with a doubled chromosome number ($2n = 76$). Other defects related to flower malformation of the CMS were improved in the progeny of some of the cybrids (Pelletier, 1993). These plants produced sufficient nectars and had normal seed set.

The introduction of cytoplasm from different *Diplotaxis* species resulted in male sterile lines. Hinata and Konno (1979) obtained CMS lines from a cross of *D. muralis* and *Brassica rapa*, with *D. muralis* as the maternal parent. The CMS was then transferred to *B. napus*, but the maintainers were rare in *B. napus* lines (Pellan-Delourme and Renard, 1987). The cytoplasm of *D. siettiana* also induced CMS in *B. rapa* background. In the hybrids of *D. siettiana* × *B. rapa*, both CMS plants and their restorer could be identified in the BC_2 generation (Nanda Kumar and Shivanna, 1993). *B. juncea* CMS lines with *D. siifolia*, or *D. erucoides*, or *D. berthautii*, or *D. catholica* were produced from crosses of the *Diplotaxis* species and *B. juncea*, with the former as cytoplasm donor (Malik et al., 1999; Pathania et al., 2003). The CMS lines with *D. erucoides* or *D. berthautii* cytoplasm were better than with other *Diplotaxis* species because they showed normal flowers with good nectars and stable sterility as well as good seed set.

Another novel CMS system established in oilseed *Brassica* by intergeneric hybridization is NSa CMS, which derived from somatic hybrids of *B. napus* and a *Sinapis arvensis* accession of Chinese origin. The first generation of somatic hybrids segregated on male fertility but the female fertility was adequate for maintaining the hybrids. After several generations of backcrossing, a stable CMS line was established and the restorer was also identified only from offspring of the fertile hybrids from the same fusion combination (Hu et al., 2004; Hu et al., 2007). Characterization of the mitochondrial genome showed that the cytoplasm of the CMS line comes from the wild species.

Transfer of the cytoplasm of *Eruca sativa* to *Brassica rapa* by intergeneric cross also resulted in cytoplasmic male sterility (Matsuzawa et al., 1999). After six generations of open pollination, the plants were crossed with *B. rapa* lines and the backcrossed hybrids showed different types of male sterility, such as no pollen grain release or pollen grain without viability. Repeated backcross of the sexual hybrids between *Enarthrocarpus lyratus* and *B. rapa* for the substitution of cytoplasm also resulted in *B. rapa* CMS with *E. lyratus* cytoplasm (Deol et al., 2003). CMS plants were also identified from the offspring of somatic hybrids between *Arabidopsis thaliana* and *B. napus*.

Cytoplasm from most of the wild species used in either sexual or somatic hybridization with *Brassica* species thus far can induce male sterility as discussed. The use of these CMS systems, however, is not frequently heard of in practical breeding programs. To our knowledge, only the Ogura CMS induced by *Raphanus* is widely used for heterosis application. Problems associated with these CMS systems include instability of sterility, lack of a stable restorer, and reduced female fertility. In any case, with more effort focused on the improvement of CMS systems, diversification of the cytoplasm of oilseed hybrids can be expected with the use of hybridization of wild species with *Brassica* oilseed crops.

INTROGRESSION OF CHROMOSOMAL TRAITS

Most agronomically important traits are controlled by genes located on chromosomes in the nucleus instead of cytoplasm organelles. Chromosome introgression is an important approach to the transfer of genes between species. In *Brassica* oilseed crops, genes conferring agronomically important traits were introduced not only from wild aliens, but also from closely related *Brassica* species. Here, major achievements of the introgression of chromosome located gene(s) for specific traits from wild species to *Brassica* oilseed crops are reviewed.

GENES FOR DISEASE RESISTANCE

In *Brassica* species, five main diseases have been targeted in interspecific hybridization for the enrichment of the *Brassica* genetic base. They are blackleg, clubroot, black rot, black spot, and beet cyst nematode disease.

Blackleg

Also called stem canker, blackleg caused by *Leptosphaeria maculans* (Desm.) Ces. Et de Not. (imperfect stage *Phoma lingam)* is a serious disease of crucifers throughout the world. In rapeseed, the disease is more severe in Australia, Canada, the United Kingdom, and Germany, causing yield losses ranging from 13% to 50%, depending on the virulence of the pathogenic strain. Few resources in *Brassica napus, B. campestris,* or *B. oleracea* were found resistant to the pathogen. In contrast, *B. nigra* and most lines of species containing the B genome (e.g., *B. juncea* and *B. carinata*) demonstrated complete resistance to this pathogen (Sjödin and Glemelius, 1988). Transfer of resistance was achieved by conventional intergeneric hybridization (Roy, 1984; Struss et al., 1991). Somatic hybridization has also been very successful (Sjödin and Glemelius, 1989a, b; Gerdemann-Knorck et al., 1995; Liu et al., 1995).

To transfer blackleg resistance from the B genome, crosses were made between *Brassica* species and 68 lines from *B. nigra* × *B. napus,* 87 lines from *B. oleracea* × *B. juncea,* and 57 lines from *B. campestris* × *B. carinata,* by repeated backcrossing (four to five generations) to a blackleg-susceptible, spring-type rapeseed variety and subsequent selfings (Plieske et al., 1998; Struss et al., 1991). A line derived from an interspecific hybrid containing the B genome of *B. juncea* was identified to segregate for resistance to *L. maculans* at the seedling stage after screening of *B. napus* lines carrying B genome chromosome additions and introgressions from *B. nigra, B. juncea,* and *B. carinata.* Both euploids and aneuploids were found in the resistant plants. The resistance gene, termed *rjlm*2, is effective in spring and winter type oilseed rape backgrounds against all tested virulent pathotypes, including two isolates that were shown to overcome two dominant (race-specific) B genome-derived resistance genes in *B. napus* (Saal et al., 2004).

Symmetric fusion of protoplasts from *Brassica napus* (susceptible to *Phoma lingam*) and *B. nigra* (resistant) resulted in an artificial allohexaploid combining all three genome of brassicas, A, B, and C (Sjödin and Glemilius, 1989a). Most of the synthesized plants had expected chromosomes as the sum of parentals and retained high levels of resistance to *P. lingam.* A few hybrids, including one with fewer chromosomes, were susceptible, which might be caused by mutational event such as loss of a small piece of a chromosome or changes in the nuclear constitution, which repress expression of resistance. The hybrids were able to set seeds upon selfing and backcross with *B. napus* pollen.

Combined with toxin selection, asymmetric hybrids of *Brassica napus* + *B. nigra, B. napus* + *B. carinata,* and *B. napus* + *B. juncea* have been produced for transfer of resistance to blackleg (Sjödin and Glemilius, 1989b). The toxin sirodesmin PL, a compound produced by the fungus, was added to the selection medium, which only permits the growth of fusion products with resistant gene(s). Resistant hybrids with partial genome of the donor species were recovered in all three combinations, and their fertility makes them useful for rapeseed improvement. Their asymmetric nature reduced the effort required for getting rid of undesirable genetic tags. The selection procedure proved highly efficient because all unselected asymmetric hybrids were susceptible, whereas 19 of 24 selected were resistant. Stable inheritance and introgression of the resistant genes have been reported in lines derived from progeny of the original hybrid (Dixelius, 1999). The hybrids *B. naponigra* and *B. napojuncea* were backcrossed to *B. napus* for seven generations and their chromosome number was stable as in *B. napus.* Inheritance studies showed that one single dominant allele controls the resistance in *B. napojuncea,* whereas two independent dominant loci regulate the resistance in *B. naponigra.* After two generations of self-pollination, seven of eight determined resistant lines did not segregate for resistance.

Another procedure for selection resistant asymmetric hybrids in the fusion of *Brassica napus* and *B. nigra* involves using a hygromycin-resistant marker gene previously transferred into the *B. nigra* genome by genetic transformation (Gerdemann-Knorck et al., 1995). The *B. nigra* fusion partner was a polyploid transgenic clone (34 to 36 chromosomes) carrying the *hph* gene. Using medium containing hygromycin B combined with x-ray irradiation to *B. nigra* protoplasts, somatic hybrids with chromosomes ranging from 46 to 85 were recovered. Most of them were asymmetric, containing part of the *B. nigra* genome. A total of 44 of 78 (56.4%) hybrids had the same level of resistance as the resistant parent (*B. nigra*). There were four hybrid clones that displayed resistance to both *P. lingam* and *Plasmodiophora brassicae*, a pathogen causing clubroot. Plants with very high asymmetry (with donor/recipient-ratio of 0.3 to 0.5) were also obtained, and elimination of the donor genome proved dose dependent.

Incorporation of blackleg resistance into *Brassica napus* background from *B. tournefortii* has also been achieved by somatic hybridization (Liu et al., 1995). Among the true hybrids that contained nuclear DNA content corresponding to the sum of the parents or higher, 12 of the 16 determined showed resistance to the *Phoma lingam* pathogen while the other four were susceptible. Male fertility of the hybrids was very low. However, backcross with *B. napus* pollen gave rise to seeds, which could be used for the generation of more offspring. Because RFLP studies revealed a close phylogenetic relationship between *B. nigra* and *B. tournefortii* (Pradhan et al., 1992), the two species may be derived from the same ancestor, as did the resistant genes.

More hybridizations aimed at the transfer of genes for disease resistance, including blackleg, have also been performed, including protoplast fusion between *Brassica juncea* and *Sinapis alba* (Gaikwad et al, 1996), *B. napus* and *S. arvensis* (Hu et al., 2002b), and sexual cross of *B. napus* and *S. arvensis* (Snowdon et al., 2000). Formation of multivalents in meiosis of the resultant somatic hybrids and BC$_1$ plants indicated the intergenomic homoeology between *Brassica* and *Sinapis* genome, which would permit introgressing genes from *S. alba* to *B. juncea*. However, the symmetric somatic hybrids between *B. napus* and *S. arvensis* that showed reduced lesion size after inoculation with *Leptosphaeria maculans* spores (Hu et al., 2002b) only gave rise to disomic addition lines with 40 chromosomes (Hu et al., 2007). Lines derived from sexual hybrids of *B. napus* × *S. arvensis* with high resistance to blackleg also proved to be addition lines. The additional chromosomes in the resistant lines are likely to confer the resistant genes, making the development of resistant euploid *B. napus* difficult.

Clubroot

Clubroot is one of the most serious diseases of vegetable *Brassica* crops. Considerable yield loss can also occur in oilseed *Brassica* crops in major production countries such as Germany, India, Poland, Sweden, and the United Kingdom. It is a root infection caused by the fungus *Plasmodiophora brassicae*. Resistant varieties are available in *Brassica* species, but because of the highly variable nature of the pathogen-resistant sources used for breeding, a resistant variety in one country might be completely susceptible to the stains of pathogens from another country. And the resistance is easily lost after a few years of cultivation in an infected field. High resistance has been found in black mustard (*B. nigra*) and radish (*Raphanus sativus*).

Transfer of clubroot resistance by somatic hybridization was studied in fusions between *Brassica napus* and *B. nigra* (Gerdemann-Knörck et al., 1995) and between *B. oleracea* and *R. sativus* (Hagimori et al., 1992). In all cases, somatic hybrids with resistance retained were recovered and progenies from some of the hybrids contained plants with high resistance.

In the fusion between *Brassica napus* and *B. nigra* (Gerdemann-Knörck et al., 1995), combined with hygromycin selection using a marker gene in the *B. nigra* donor genome, several asymmetric hybrids were produced; 28% of them had a resistant level as high as the resistant donor *B. nigra*, and 21.2% showed only slight symptoms. Fertility of the asymmetric hybrids was low but a few seeds were obtained after backcross to *B. napus*. The percentage of plants resistant to clubroot was lower than those resistant to blackleg, probably due to the different genetic distance between the marker gene and the loci controlling these resistance traits. However, as discussed, four hybrid plants showed resistance to both pathogens.

Black Spot

Black spot, also known as *Alternaria* blight or *Alternaria* black spot, is caused primarily by *Alternaria brassicae* (Berk.) Sacc., although depending on environmental conditions, *A. brassicicola* (Schw.) Wilts., *A. raphani* Groves and Skolko, and *A. alternata* (Fr.) Keissler may also be associated with the disease in some areas. *A. brassicae* is the most prevalent causal agent of this disease on oilseed *Brassica* species. Yield losses due to this disease range from 10% to 70% in oilseed *Brassica* (Downey and Rimmer, 1993).

Highly resistant genetic resources have not been reported in *Brassica* cultivated species, although some varieties displayed higher resistance and *B. carinata* is the highest species in terms of resistance level. *Sinapis alba* appears to be the most resistant of species closely related to *Brassica* (Brun et al., 1987). More distant relatives of *Brassica, Camelina sativa, Capsella bursa-pastoris,* and *Eruca sativa,* showed very high resistant levels to black spot (Conn et al., 1988).

The first somatic hybrids produced for the transfer of resistance genes were obtained in fusions between *Brassica napus* and *Sinapis alba* (Primard et al., 1988). Preliminary results revealed that none of the hybrids were as tolerant to *Alternaria brassicae* as the *S. alba* parent. The resistance seemed to be semi-dominant and thus difficult to introduce into *B. napus*. Chevre et al. (1991) reported interspecific transfer of the *S. alba* resistance into *B. napus* by somatic hybridization, as well as manual pollination using reciprocal crosses. Subsequently, embryo rescue and cytogenetic analysis resulted in *B. napus* plants with 38 chromosomes and with a resistance to *A. brassicae* similar to that of *S. alba*.

Somatic hybrids were also produced in fusion combinations of *Brassica oleracea* + *Sinapis alba* and *B. oleracea* var. *botrytis* + *B. carinata* for transfer resistance to blackleg as well as black spot (Ryschka et al., 1996). Among the plants vegetatively multiplied from two hybrids lines of the former fusion combination, a high resistance level to *Alternaria brassicae* was determined. Whereas all plants of the cauliflower accession were susceptible, plants of the *S. alba* accession ranged from moderately to absolutely resistant. The hybrids were female fertile, and thus production of backcross seeds could be realized and analysis of subsequent generations may provide interesting results.

Camelina sativa (false flax) is a source of resistance to *Alternaria* pathogens. It is determined to be virtually immune to *A. brassicicola*. Its high resistance has been linked to the production of a phytoalexin (camalexin), which is an antibiotic and may thus inhabit pathogen infection. The first intertribal somatic hybrids for transfer of resistance were produced between *Brassica carinata* and *C. sativa* (Narasimhulu et al., 1994). The three hybrid shoots failed to produce roots and it was impossible to establish plants. Hansen (1998) reported plant establishment of somatic hybrids regenerated from fusions of protoplasts from rapid cycling *B. oleracea* and *C. sativa* after various approaches to aid root induction and plant survival from transfer. No further investigation was made on the resistance performance or fertility of the hybrid due to the failure of plant survival after 1 month out of culture.

In addition, *Eruca sativa* and *Capsella bursa-pastoris,* both found to be resistant to black spot, have been fused with *Brassica* spp. for generating somatic hybrids (Fahleson et al., 1988; Sikdar et al., 1990; Sigareva and Earle, 1997). Somatic hybrids of *E. sativa* + *B. napus* (Fahleson et al., 1988) and of *E. sativa* + *B. juncea* (Sikdar et al., 1990) showed fertility both on male and female parts. Some of the somatic hybrids produced in fusions with *B. juncea* had pollen viability as high as 82.9% and seed set 50%, which indicated that somatic hybridization has great potential for transfer traits from *E. sativa* to *Brassica* species.

Beet Cyst Nematode

Beet cyst nematode (*Heterodera schachtii* Schm.; BCN) does not cause damage to *Brassica* crops but does cause considerable damage to sugar beet crops. In sugar beet production regions, *B. napus* cannot be included in the rotation system if highly resistant varieties are not available because it is a good host for the multiplication of this nematode. Because the level of resistance to BCN within *B. napus* is too low to allow selection of resistant cultivars, the introduction of BCN resistance from other resistant relatives is the only way to achieve resistant variety.

Raphanus sativus and *Sinapis alba* have been involved as resistant sources for the transfer of this trait to *Brassica napus* by somatic hybridization (Lelivelt and Krens, 1992; Lelivelt et al., 1993). Intergeneric somatic hybrids were produced from both fusion combinations. Most of the somatic hybrids showed a high level of resistance. A few of them had a level of resistance as high as the resistant donor parent. However, the somatic hybrids all demonstrated high levels of sterility; no progeny was generated although embryo rescue was applied. In contrast, sexual hybrids between *B. napus* and *S. alba* also showed high resistance levels and were able to produce progeny for further study and selection. As the expression of resistance to BCN has been proven in somatic hybrids, and cytological studies on chromosome association at meiotic metaphase I in the sexual hybrids suggest partial homology between chromosomes of the AC and Sal genome and thus their potential for gene exchange, transfer of this trait by somatic hybridization into *B. napus* remains possible.

In addition, Voss et al. (2000) obtained nematode-resistant progenies from crosses of *Raphanus sativus* and *B. napus* via embryo rescue. Using genomic *in situ* hybridization (GISH), highly resistant progeny with a minimal *Raphanus* genome component were identified among plants of backcross generations. A BC_3 plant with a monosomic, acrocentric addition chromosome was backcrossed once again and then selfed to produce a stable disomic addition line. To obtain rapeseed lines containing the resistance on a small introgression, further backcross and selection is still needed. Using backcross progeny plants of *Raphanobrassica* that segregated for all nine chromosomes of the *Raphanus* genome in a genetic background of synthetic rape, individuals with a resistance level comparable to that of the radish donor parent were identified. The existence of a single radish chromosome d is sufficient for the plants to be resistant.

INTROGRESSION OF GENES FOR FATTY ACID COMPOSITION

Practical introgression of genes responsible for quality traits by somatic hybridization has been mainly directed at modification of oil components. Oil production is the ultimate goal for oilseed *Brassica* planting. The quality of oil, which is determined by a composition of fatty acids, is one of the most important traits in oilseed *Brassica* crops. A basic concern of oil quality in oilseed rape is its nutrition value. Oil with significant levels of oleic and linoleic acids and low levels of erucic and linolenic acids is desirable for edible purposes. For industrial purposes, oil containing high levels of erucic acid and other components such as lauric and nervonic acids is of particular value.

In the case of intergeneric somatic hybridization between *Brassica napus* and *Thlaspi perfoliatum*, the presence of high concentrations of nervonic acid, a *Thlaspi*-specific fatty acid that is an available ingredient in technical oils, was detected in some progeny obtained from the initial hybrid backcross to *B. napus* (Fahleson et al., 1994b). The *B. napus* fusion partner only contained trace amounts of this fatty acid (1.1%). Five of the twelve plants had average values significantly greater than that of *B. napus*. The highest level in hybrids was determined to be 4.9%.

Somatic hybrids between *Brassica napus* and *Orychophragmus violaceus* showed increased levels of palmitic and linoleic acids, as well as reduced levels of erucic acid compared to *B. napus*, indicating the *B. napus* fatty acid profile was altered after the introgression of *O. violaceus* genes (Hu et al., 2002a). From the progeny populations (S1, BC1, or BC2), seeds with higher contents of palmitic acid and linoleic acid, and lower contents of erucic acid than the *B. napus* parent, were identified. In the progenies of sexual hybrids derived from the cross between *B. napus* and *O. violaceus*, plants with higher contents of oleic, linoleic, and erucic acids were obtained, although the *O. violaceus* parent is characterized as having a high linoleic acid content but not high oleic and erucic acid contents. The highest linoleic acid content reached was 38.3%, and the highest oleic acid content reached was 73.84%, whereas the *B. napus* parent only contains 63.8% oleic acid and 15.43% linoleic acid in seed oil.

Similarly, progeny derived from asymmetric somatic hybrids between *Brassica napus* and *Crambe abyssinica* contained significantly greater amounts of erucic acid than *B. napus*, which is a result of the introduction of genes responsible for high erucic acid content from *C. abyssinica* into the *B. napus* genome (Wang et al., 2003).

Lesquerella fendleri Gray (Wats.) is an annual native to the arid and semiarid regions of the southwestern United States and possesses favorable fatty acid compositions containing high amounts of hydroxy fatty acids, about 60% lesquerolic acid (20:1D11, 14-OH) and 1.5% densipolic acid (18:2D9, D15, 12-OH). The sexual intertribal hybridizations between *B. napus* and *L. fendleri* were successfully produced by immature embryo rescue with the aim of introducing unusual fatty acids into *B. napus*. The fatty acid compositions in the seeds of most of the progenies by selfing or by backcrossing to *B. napus* parents showed fatty acid profiles biased toward *B. napus*. However, plants with decreased oleic acid levels and increased levels of linoleic, linolenic, and eicosanoic acids were observed. The highest levels of linolenic and eicosanoic acids reached 13.17% and 13.66%, respectively, higher than those of their female parents. Moreover, 4.26% to 15.66% erucic acid was detected in the progenies, whereas the content of erucic acid was nearly zero in their *B. napus* parents and *L. fendleri*. Similarly, in the offspring of asymmetric somatic hybridizations between *B. napus* (zero-erucic acid type) and *L. fendleri*, one plant was found to produce a seed containing up to 16.5% erucic acid and 15% eicosaenoic acid, as well as a seed having 4.3% hydroxylated ricinoleic acid (18:1D9, 12-OH) (Schroder-Pontoppidan et al., 1999). These results indicate that the elongase and hydroxylase genes may have been integrated and expressed from *L. fendleri* to the plants, which resulted in the production of very long chain and hydroxylated fatty acids.

INTROGRESSION OF GENES FOR FERTILITY RESTORATION

As discussed previously, somatic hybridization as well as conventional interspecific hybridization have made transfer of creating cytoplasmic sterility possible. Therefore, many CMS systems are available and more can be expected in *Brassica* species. However, the source of fertility restoration for the newly transferred CMS is not usually in the crop species. For example, in Ogura CMS, genes for fertility restoration have only been found in radish. Transfer of the genes by sexual cross was achieved by many backcrosses with the use of the bridge plant *Raphanobrassica* (Heyn, 1976). However, low seed productivity of the progeny and the requirement of several restorer genes precluded their agronomic use. This problem might be caused by extra genetic information from radish and the complexity of the original Ogura cytoplasm (Pellan-Delourme and Renard, 1988). Asymmetric somatic hybridization provides a good approach to overcome the genetic drag. Protoplast fusions were attempted between *B. napus* and a Japanese radish cultivar containing a restorer gene for Kosena CMS for the transfer of restorer genes to *B. napus* (Sakai et al., 1996). The radish donor genome was x-ray fragmented at high dose (60 krad). A hybrid with 47 chromosomes restored the fertility of a CMS *B. napus* line and backcrossing to *B. napus* line without the restorer gene yielded plants with fertility segregated. In one of the lines, fertility segregated in a 1:1 ratio in several backcross generations, which indicated the dominant monogenic nature of the restorer gene. After two backcrosses, the chromosome number and petal color became identical to that of *B. napus*. No female sterility was observed in the BC_3 generations.

The transfer of the restorer gene(s) was usually accomplished along with the transfer of malesterile cytoplasm either by sexual or somatic hybridization. These include the restorers of *Brassica juncea* CMS from *Moricandia arvensis* (Prakash et al., 1998) and from *Trachystoma ballii* (Kirti et al., 1997); *B. rapa* CMS from *Enarthrocarpus lyratus* (Deol et al., 2003) and from *Diplotaxis siettiana* (Nanda Kumar and Shivanna, 1993); and *B. napus* CMS from *Sinapis arvensis* (Hu et al. 2007) and from *Arabidopsis thaliana* (Leino et al. 2004). As an example, in the fusion of *B. juncea* with *Trachystoma ballii,* which resulted in *B. juncea* CMS with *T. ballii* cytoplasm, plants with the chromosome number of *B. juncea* but intermediate morphology were backcrossed to *B. juncea*. In BC$_3$ and BC$_4$ generations, plants that restored CMS were recovered (Kirti et al., 1997). The introgression of this restorer gene from *T. ballii* to *B. juncea* resulted from forced pairing between chromosomes of the recurrent cultivar and chromosomes of the fusion hybrid. Segregation ratios of this fertility restorer gene followed a monogenic pattern. The introgression of the fertility restorer gene did not cause any abnormalities, such as reduced fertility; and pollen and seed fertilities of the restored plants were greater than 90%. Also, for the NSa CMS in *B. napus* with *Sinapis arvensis* cytoplasm developed in this author's institute, the initial screening for restorers in the *B. napus* gene pool did not give any positive results (Hu et al., 2004). Test crosses with progenies derived from partially fertile somatic hybrids either by self-propagation or backcross with *B. napus* recovered six lines with a restoration rate ranging from 93.7% to 100%. The restored plants released stainable pollen grains and showed normal flowers. Selfing of the F$_1$ plants yielded seeds as normal as ordinary fertile lines (Hu et al., 2007).

The *Brassica juncea* CMS containing *Diplotaxis catholica* cytoplasm was initially established by sexual cross with *D. catholica* as the maternal parent. Transfer of the restorer gene to *B. juncea*, however, was fulfilled by somatic hybridization (Pathania et al., 2003). Only progeny plants derived from somatic hybrids of *B. juncea* and *D. catholica* could restore the male fertility of the CMS plants. Genetic segregation data indicated that a single, dominant nuclear gene governs this fertility restoration.

The integration of restorer gene(s) from the cytoplasm donor of alloplasmic *Brassica* CMS systems into the *Brassica* genome is more efficient in *B. juncea* and *B. rapa* than in *B. napus*. The restorer lines developed from intergeneric hybridization for the alloplasmic *B. rapa* and *B. juncea* CMS, including *Mori* CMS and *Trachy* CMS of *B. juncea* and *Enarthro* CMS of *B. rapa*, are all euploids that can be directly used in breeding programs (Kirti et al., 1997; Prakash et al., 1998 ; Deol et al., 2003). The restorer lines of alloplasmic *B. napus* CMS, on the other hand, happened to be aneuploid, as were those for CMS induced by *Alternaria thaliana* and *Sinapis arvensis* cytoplasm (Leino et al., 2004; Hu et al., 2007). The C and A genomes of *B. napus* may not pair with the genomes from *A. thaliana* and *S. arvensis*, which are more related to the B genome of *Brassica* species. The restorer gene(s) residing on the additional chromosomes render further development of restorer lines more difficult (Leino et al., 2004).

INTROGRESSION OF GENES FOR C$_3$-C$_4$ CHARACTER

Moricandia arvensis is a C$_3$-C$_4$ intermediate species in the Cruciferae family. The plants of this species have high photosynthetic capacity due to their low CO$_2$ compensation point (CCP) and low photorespiratory activity (Holaday and Chollet, 1984). This character could, if incorporated into crop species, improve water use efficiency and result in a yield advantage. Intergeneric hybridization between *Moricandia* and *Brassica* may help the introduction of this C$_3$-C$_4$ intermediate character into *Brassica* crops. A sexual hybrid of *B. alboglabra* and *M. arvensis* showed a CCP value intermediate to its parents, as well as those of *Diplotaxis muralis* and *M. arvensis*, which demonstrated a CCP value close to that of the *Moricandia* parent (Apel et al., 1984; Razmjoo et al., 1996). However, the cross compatibility between *Moricandia* and *Brassica* species is extremely low (Apel et al., 1984), and only one hybrid plant was obtained from 2000 pollinated flowers in a cross between *B. alboglabra* and *M. arvensis*. Somatic hybridization as an alternative has been applied for the possible transfer of this trait.

Somatic hybrids between red cabbage (*Brassica oleracea* var. *capitata*) or cauliflower (*B. oleracea* var. *botrytis*) and *Moricandia arvensis* did not express the C$_3$-C$_4$ intermediate character (Razmjoo et al., 1996). These hybrids all had the same CCP value and apparent photosynthesis rate as the C3 parents. The authors concluded that the expression of the C$_3$-C$_4$ intermediate photosynthetic character from *M. arvensis* appears to depend on the combination of other characters in the parental species.

In contrast, somatic hybrids between *Brassica napus* and *Moricandia arvensis* were determined to express the character (O'Neill et al., 1996). Among the hybrids derived from fusions of

M. arvensis and three *B. napus* varieties, 3 out of 17 plants showed significantly lower levels of CCP (Γ = 33.7–35.6 μL/L) than the *B. napus* parents (Γ = 41.4 μL/L), although the expression was not as efficient as in the *M. arvensis* parent (Γ = 15.94 μL/L). These three hybrids were from one callus generated in a fusion with UV-treated *B. napus* protoplasts. In addition, anatomy analysis also proved this expression. The C_3-C_4 character in *M. arvensis* is associated with the presence of a Kranz-like ring of bundle sheath cells around the vascular tissues of the leaf. Within these cells are large numbers of mitochondria arranged centripetally adjacent to the vascular bundle and overlain by an outer layer of chloroplasts. Such an anatomy does not exist in *B. napus*. Thin sections through leaves from the three hybrids revealed a distribution of mitochondria and chloroplasts within the bundle sheath cells, which resembled, although not as distinct as, that in leaves of *M. arvensis*. Seeds were obtained from one of the plants, which would permit further investigation of the transmission of the character through generation.

Production of somatic hybrids and their backcross progeny between *B. juncea* and *Moricandia arvensis* has also been achieved (Kirti et al., 1992a) with a primary aim of combining the cytoplasm of the two species to increase cytoplasmic variability and introgress desirable nuclear genes conferring agronomic advantages to *B. juncea*. Exploitation of these plant materials may also result in the detection of expression of the C_3-C_4 character in materials that are readily crossible with *Brassica* crops.

CONCLUSION

Interspecific, intergeneric, and intertribal hybridizations for the integration of valuable traits from wild crucifers can be traced back in the literature to 1950 for sexual cross and 1979 for somatic hybridization (Prakash et al., 2008). Traditional crosses have been significantly improved by the embryo culture technique, a technique that overcomes the post-fertilization barriers between distant related species. For combinations where pre-fertilization barriers exist, the somatic approach becomes the method of choice to realize hybridization. Although homologous pairing of chromosomes from distantly related species is not a frequent phenomenon, multivalents do form in some of the hybrids and ensure the exchange of genetic materials of the genomes involved. This has resulted in the introgression of chromosome genes from wild species to *Brassica* crops for disease resistance, cytoplasmic male sterility restoration, and fatty acid composition.

The transfer of cytoplasmic traits, especially cytoplasmic male sterility from a wild species to *Brassica* crops, seems to be more efficient by somatic hybridization than by conventional cross combinations, partly due to the difficulties in producing hybrids with the wild species as maternal parent, which is the cytoplasmic donor (Hu et al., 2002b). Without the combination of chloroplasts from *Brassica* with mitochondria from radish, it would never be possible to use Ogura CMS for commercial seed production due to the yield penalty of chlorosis. Moreover, induction of the rearrangement of mitochondrial DNA or chloroplast DNA not only provides possibilities for recovering novel agronomic traits caused by alloplasmization, but also creates a substantial basis for investigating the mechanism of cytoplasmic phenomena. The procedure has been well exploited by Indian scientists in which parallel selection of the CMS line and restorer genes can be achieved, and it is of particular importance in the development of new pollination control systems for hybrid breeding.

Despite the advantages, somatic hybridization also has some drawbacks. First, the technique needs to be improved. In many fusion experiments, only a limited number of hybrids are produced, thus reducing the possibility of selecting usable plants among the hybrids; whereas in the fusion between *Brassica napus* and *Raphanus sativus*, only one hybrid plant was obtained (Lelivelt and Krens, 1992). Productivity of somatic hybrids largely depends on the regeneration capacity of parental lines. Thus, the genotype dependence of the regeneration capacity limits the combination of fusion partners. A highly efficient fusion procedure and regeneration system for different genotypes is desirable. At the same time, the development of good selection systems, both for hybrids and for desirable traits, is necessary. Fusion experiments with a good selection system not only save lots of effort in the identification of hybrids, but also ensure the recovery of hybrids with the desired traits, as with toxin selection (Sjödin and Glemilius, 1989b), and a hygromycin-resistant marker gene (Gerdemann-Knorck et al., 1995) blackleg resistance. In addition, fertility of somatic hybrids is generally low, especially that of symmetric hybrids. These allopolyploids containing complete sets of chromosomes from both fusion partners fail to produce functional gametes. Male sterility in some cases is expected in order to produce lines with cytoplasmic sterility. But the sterile female part of somatic hybrids could restrict their use for generating offspring for further investigation. Induction of asymmetry in somatic hybrids has been reported as beneficial for the increase in hybrid fertility (Hu et al., 2002a).

Hybrids of distantly related species with traits of interest usually contain additional chromosomes that harbor the genes for the desired traits, as do the restorers for NSa CMS and nematode-resistant spring rapeseed lines (Hu et al., 2007; Peterka et al., 2004). Direct use of these lines in breeding programs is not feasible. Further fragmentation of the additional chromosome and integration of the segments conferring the genes of interest must be performed in order to obtain euploids for applicable use. Thus, irradiation and a fast determination method for discriminating the euploids (such as flow cytometric analysis) should be explored.

REFERENCES

Ahuja, I., P.B. Bhaskar, S.K. Banga, and S.S. Banga. 2003. Synthesis and cytogenetic characterization of intergeneric hybrids of *Diplotaxis siifolia* with *Brassica rapa* and *B. juncea*. *Plant Breed.*, 22: 447–451.

Apel, P., H. Bauwe, and H. Ohle. 1984. Hybrids between *Brassica alboglabra* and *Moricandia arvensis* and their photosynthetic properties. *Biochem. Physio. Pflanzen.*, 179:793–797.

Baldev, A., K. Gaikwad, P.B. Kirti, T. Mohapatra, S. Prakash, and V.L. Chopra. 1998. Recombination between chloroplast genome of *Trachystoma ballii* and *Brassica juncea* following protoplast fusion. *Mol. Gen. Genet.*, 260:357–361.

Banga, S.K., S.S. Banga, G. Thomas, and A.A. Monteiro. 1998. Attempts to develop fertility restorers for oxy CMS in crop brassicas. *Acta Horticulturae*, 459: 305–311.

Bannerot, H., L. Boulidard, and Y. Cauderon, 1977. Unexpected difficulties met with the radish cytoplasm. *Cruciferae Newslett.*, 2:16.

Bannerot, T., L. Boulidard, Y. Cauderon, and J. Tempe. 1974. Transfer of cytoplasmic male sterility from *Raphanus sativus* to *Brassica oleracea*. *Eucarpia Meeting Cruciferae*, Dundee, Scotland, p. 52–54.

Batra, V., K.R. Shivanna, and S. Prakash. 1989. Hybrids of wild species *Erucastrum gallicum* and crop *Brassica*s. In: *Proc. 6th Int. Congr. Society for the Advancement of Breeding Researches in Asia and Oceania (SABRAO)*. Tsukuba, Japan, p. 443–446.

Batra, V., S. Prakash, and K.R. Shivanna. 1990. Intergeneric hybridization between *Diplotaxis siifolia* — a wild species and crop *Brassica*s. *Theor. Appl. Genet.*, 80:537–541.

Bauer-Weston, W.B., W. Keller, J. Webb, and S. Gleddie. 1993. Production and characterization of asymmetric somatic hybrids between *Arabidopsis thaliana* and *Brassica napus*. *Theor. Appl. Genet.*, 86:150–158.

Begum, F., S. Paul, N. Bag, S.R. Sikdar, and S.K. Sen. 1995. Somatic hybrids between *Brassica juncea* L. and *Diplotaxis harra* (Forsk.) Boiss. and the generation of backcross progenies. *Theor. Appl. Genet.*, 91:1167–1172.

Bhasker, P.B., I. Ahuja, H.S. Janeja, and S.S. Banga. 2002. Intergeneric hybridization between *Erucastrum canariense* and *Brassica rapa*-Genetic relatedness between E and A genomes. *Theor. Appl. Genet.*, 105:754–758.

Bijral, J.S.and T.R. Sharma. 1995. A sexual hybrid between *Brassica juncea* and *Diplotaxis muralis*. *Indian J. Genet. Plant Breed.*, 55(2):170–172.

Bing, D.J., R.K. Downey, and G. Rakow. 1991. Potential of gene transfer among oilseed *Brassica* and their weedy relatives. In: *Proc. 9th Int. Rapeseed Congr.*, Saskatoon, Canada, p. 1022–1027.

Brewer, E.P., J.A. Saunders, J.S. Angle. R.L. Chaney, and M.S. Mcintosh, 1999. Somatic hybridization between the zinc accumulator *Thlaspi caerulescens* and *Brassica napus*. *Theor. Appl. Genet.*, 99:761–771.

Brown, J., A.P. Brown, J.B. Davis, and D. Erickson. 1997. Intergeneric hybridization between *Sinapis alba* and *Brassica napus*. *Euphytica*, 93:163–168.

Brun, H., J. Plessis, and M. Renard, 1987. Resistance of some crucifers to *Alternaria brassica* (Berk.) Sacc. *Proc. 7th Int. Rapeseed Conf.*, p. 1222–1227.

Chandra, M., L. Gupta, S.S. Banga and S.K. Banga, 2004. Production of an interspecific hybrid between *Brassica fruticulosa* and *B. rapa*. *Plant Breed.*, 123: 497–498.

Chatterjee, G., S.R. Sikdar, S. Das, and S.K. Sen. 1988. Intergeneric somatic hybrid production through proto-plast fusion between *Brassica juncea* and *Diplotaxis muralis*. *Theor. Appl. Genet.*, 76:915–922.

Chen, H.F., W. Hua, and Z.Y. Li. 2007. Production and genetic analysis of partial hybrids in intertribal crosses between *Brassica* species (*B. rapa, B. napus*) and *Capsella bursa-pastoris*. *Plant Cell Rep.*, 26:1791–1800.

Chevre, A.M., P. This, F. Eber, M. Deschamps, M. Renard, M. Delseny, and C.F. Quiros. 1991. Characterization of disomic addition lines of *Brassica napus-Brassica nigra* by isozyme, fatty acids and RFLP markers. *Theor. Appl. Genet.*, 81:43–49.

Chevre, A.M., F. Eber, P. This, P. Barret, X. Tanguy, H. Brun, M. Delseny, and M. Renard. 1996. Characterization of *Brassica nigra* chromosomes and of black leg resistance in *B. napus-B. nigra* addition lines. *Plant Breed.*, 115:113–118.

Chevre, A.M., P. Barret, F. Eber, P. Dupuy, H. Brun, X. Tanguy, and M. Renard. 1997. Selection of stable *Brassica napus-B. juncea* recombinant lines resistant to black leg (*Leptosphaeria maculans*). 1. Identification of molecular markers, chromosomal and genomic origin of the introgression. *Theor. Appl. Genet.*, 95:1104–1111.

Choudhary, B.R. and P. Joshi. 2001. Crossability of *Brassica tournefortii* and *B. rapa*, and morphology and cytology of their F1 hybrids. *Theor. Appl. Genet.*, 102:1123–1128.

Christey, M.C. 2004. *Brassica* protoplast culture and somatic hybridization. In: E.C. Pua and C.J. Douglas (Eds.), *Biotechnology in Agriculture and Forestry*. Springer, New York, 54:119–148.

Craig, A. and S. Millam. 1995. Modification of oilseed rape to produce oils for industrial use by means of applied tissue culture methodology. *Euphytica*, 85:323–327.

Dai, X., C. Cheng, L. Song, et al. 2005. Germplasm enhancement through wide crosses of *Brassica napus* × *India rorippa*, *J. Plant Genetic Resources*, 6(2):242–244.

Delourme, R., F. Eber, and A.M. Chevre. 1989. Intergeneric hybridization of *Diplotaxis erucoides* with *Brassica napus*. I. Cytogenetic analysis of F1 and BC1 progeny. *Euphytica*, 41:123–128.

Deol, J.S., K.R. Shivanna, S. Prakash, and S.S. Banga. 2003. *Enarthrocarpus lyratus*-based cytoplasmic male sterility and fertility restorer system in *Brassica rapa*. *Plant Breed.*, 122:438–440.

Dixelius, C. 1999. Inheritance of the resistance to *Leptosphaeria maculans* of *Brassica nigra* and *B. juncea* in near-isogenic lines of *B. napus*. *Plant Breed.*, 118: 151–156.

Downey, R.K. and S.R. Rimmer, 1993. Agronomic improvement in oilseed *Brassicas*. *Adv. Agron.*, 50:1–50.

Du, X.Z., X.H. Ge, Z.G. Zhao, and Z.Y. Li. 2008. Chromosome elimination and fragment introgression and recombination producing intertribal partial hybrids from *Brassica napus* × *Lesquerella fendleri* crosses. *Plant Cell. Rep.*, 27:261–271.

Dushenkov, S., M. Skarzhinskaya, K. Glimelius, D. Gleba, and I. Raskin. 2002. Bioengineering of a phytore-mediation plant by means of somatic hybridization. *Int. J. Phytoremediation*, 4:117–126.

Fahleson J., I. Eriksson, M. Landgren, S. Stymne, and K. Glimelius. 1994a. Intertribal somatic hybrids between *Brassica napus* and *Thlaspi perfoliatum* with high content of the *T. perfoliatum*-specific nervonic acid. *Theor. Appl. Genet.*, 87: 795–804.

Fahleson, J., I. Eriksson, and K. Glimelius. 1994b. Intertribal somatic hybrids between *Brassica napus* and *Barbarea vulgaris* — production of *in vitro* plantlets. *Plant Cell Rep.*, 13: 411–416.

Fahleson, J., L. Råhlén, and K. Glimelius. 1988. Analysis of plants regenerated from protoplast fusions between *Brassica napus* and *Eruca sativa*. *Theor. Appl. Genet.*, 76:507–512.

Fahleson, J., U. Lagercrantz, A. Mouras, and K. Glimelius. 1997. Characterization of somatic hybrids between *Brassica napus* and *Eruca sativa* using species-specific repetitive sequences and genomic *in situ* hybrid-ization. *Plant Sci.*, 123:133–142.

Fan, Z.,., W. Tai, and B.R. Stefansson. 1985. Male sterility in *Brassica napus* L. associated with an extra chro-mosome. *Can. J. Genet. Cytol.*, 27:467–471.

Forsberg, J., C. Dixelius, U. Lagercrantz, and K. Glimelius. 1998. UV dose-dependent DNA elimination in asymmetric hybrids between *Brassica napus* and *Arabidopsis thaliana*. *Plant Sci.*, 131:65–76.

Forsberg, J., M. Landgren, and K. Glimelius. 1994. Fertile somatic hybrids between *Brassica napus* and *Arabidopsis thaliana*. *Plant Sci.*, 95:213–223.

Fukushima, E. 1945. Cytogenetic studies on *Brassica* and *Raphanus*. I. Studies on inter-generic F1 hybrids between *Brassica* and *Raphanus*. *J. Dep. Agr. Kyushu Imp. Univ.*, 7:281–400.

Gaikwad, K., P.B. Kirti, A. Sharma, S. Prakash, and V.L. Chopra. 1996. Cytogenetical and molecular investigations on somatic hybrids of *Sinapis alba* and *Brassica juncea* and their backcross progeny. *Plant Breed.*, 115:480–483.

Gerdemann-Knörck, M., S. Nielen, C. Tzscheetzsch, J. Iglisch, and O. Schieder. 1995. Transfer of disease resistance within the genus *Brassica* through asymmetric somatic hybridization. *Euphytica*, 85:247–253.

Gleba, Y.Y. and F. Hoffmann. 1980. *Arabidobrassica*: a novel plant obtained by protoplast fusion. *Planta*, 149:112–117.

Glimelius, K. 1999. Somatic hybridization. In: C. Gómez-Campo (Ed.), *Biology of Brassica Coenospecies*. Elsevier Science, Amsterdam, p. 107–148.

Guan, R.Z., S.H. Jiang, R.Y. Xin, and H.S. Zhang H. S. 2007. Studies on rapeseed germplasm enhancement by use of cruciferous weed *Descurainia sophia*. *Proc. 12th Int. Rapeseed Congr.*, Wuhan, I:261–265.

Gundimeda, H.R., S. Prakash, and K.R. Shivanna. 1992. Intergeneric hybrids between *Enarthrocarpus lyratus*, a wild species and crop *Brassicas*. *Theor. Appl. Genet.*, 83:655–662.

Hagimori, M., M. Nagaoka, N. Kato, and H. Yoshikawa. 1992. Production and characterization of somatic hybrids between the Japanese radish and cauliflower. *Theor. Appl. Genet.*, 84:819–824.

Hansen, L.N. 1998. Intertribal somatic hybridization between rapid cycling *Brassica oleracea* L. and *Camelina sativa* (L.) Crantz. *Euphytica*, 104:173–179.

Harberd, D.J. and E.D. McArthur. 1980. Meiotic analysis of some species and genus hybrids in the *Brassiceae*. In: S. Tsunoda, K. Hinata, and C. Gómez-Campo (Eds.), *Brassica Crops and Wild Allies: Biology and Breeding*, Japan Sci. Soc. Press, Tokyo, p. 65–87.

Heyn, F.W. 1976. Transfer of restorer genes from *Raphanus* to cytoplasmic male sterile *Brassica napus*. *Cruciferae Newslett.*, 1: 15–16.

Hinata, K. and N. Konno. 1979. Studies on a male-sterile strain having the *Brassica campestris* nucleus and the *Diplotaxis muralis* cytoplasm. *Japan J. Breed.*, 29:305–311.

Holaday, A.S. and R. Chollet. 1984. Photosynthetic/photorespiratory characteristics of C3-C4 intermediate species. *Photosynth. Res.*, 5:307–323.

Hosoda, T., H. Namai, and J. Goto. 1963. On the breeding of *Brassica napus* obtained from artificially induced amphidiploids. III. On the breeding of synthetic rutabaga (*Brassica napus* var. rapifera). *Japan J. Breed.*, 13:99–106.

Hu, Q., L.N. Hansen, J. Laursen, C. Dixelius, and S.B. Andersen. 2002a. Intergeneric hybrid between *Brassica napus* and *Orychophragmus violaceus* containing traits of agronomic importance for oilseed rape breeding. *Theor. Appl. Genet.*, 105:834–840.

Hu, Q., S.B. Andersen, C. Dixelius, and L.N. Hansen. 2002b. Production of fertile intergeneric somatic hybrids between *Brassica napus* and *Sinapis arvensis* for the enrichment of the rapeseed gene pool. *Plant Cell Rep.*, 21:147–152.

Hu, Q., Y. Li, D. Mei, X. Fang, L.N. Hansen, and S.B. Andersen. 2004. Establishment and identification of cytoplasmic male sterility in *Brassica napus* by intergeneric somatic hybridization. *Sci. Agric. Sin.*, 37:333–338.

Hu, Q., Y.C. Li, D.S. Me, Y.D. Li, and Y.S. Xu. 2007. Development of three-line system with a novel alloplasmic male sterility in *Brassica napus* L. *Proc. 12th Int. Rapeseed Congr.*, Wuhan, I:30–32

Hua, Y. W. and Z.Y. Li. 2006a. Genomic *in situ* hybridization analysis of *Brassica napus* × *Orychophragmus violaceus* hybrids and production of *B. napus* aneuploids. *Plant Breed.*, 125:144–149.

Hua, Y.W., M. Liu, and Z.Y. Li. 2006b. Parental genome separation and elimination of cells and chromosomes revealed by GISH and AFLP analysis in intergeneric hybrids between *Brassica carinata* and *Orychophragmus violaceus*. *Ann. Bot.*, 97:993–998.

Inomata, N. 1991. Intergeneric hybridization in *Brassica juncea* × *Sinapis pubescens* and *B. napus* × *S. pubescens* and their cytological studies. *Cruciferae Newslett.*, 14-15: 10–11.

Inomata, N.1994. Intergeneric hybrids between *Brassica napus* and *Sinapis pubescens* and the cytology and crossability of their progenies. *Theor. Appl. Genet.*, 89:540–544.

Inomata, N. 2003. Production of intergeneric hybrids between *Brassica juncea* and *Diplotaxis virgata* through ovary culture and the cytology and crossability of their progenies. *Euphytica*, 133:57–64.

Inomata, N. 2004. Intergeneric hybridization between *Brassica rapa* and *Coincya pseuderucastrum*, and the meiotic behaviour. *Cruciferae Newslett.*, 25:19–20.

Inomata, N. 2005. Intergenomic hybrids between *Brassica napus* and *Diplotaxis harra* through ovary culture and the cytogenetic analysis of their progenies. *Euphytica*, 145:87–93.

Jandurova, O.M. and J. Dolezel. 1995. Cytological study of interspecific hybrid *Brassica campestris* × *B. hirta* (*Sinapis alba*). *Sexual Pl. Reprod.*, 8:37–43.

Jarl, C.I. and C.H. Bornman. 1988. Correction of chlorophyll-defective, male sterile winter oilseed rape (*Brassica napus*) through organelle exchange: phenotypic evaluation of progeny. *Hereditas,* 108:97–102.

Kameya, T. and K. Hinata. 1970. Test tube fertilization of excised ovules in *Brassica. Japan J. Breed.,* 20:253–260.

Kartha, K. K., O.L. Gamborg, F. Constabel, and K.N. Kao. 1974. Fusion of rapeseed and soybean protoplasts and subsequent division of heterokaryocytes. *Can. J. Bot.,* 52:2435–2436.

Kerlan, M.C., A.M. Chevre, and F. Eber. 1993. Interspecific hybrids between a transgenic rapeseed (*Brassica napus*) and related species: cytogenetical characterization and detection of the transgene. *Genome,* 36:1099–1106.

Kirti, P.B., S. Prakash, and V.L. Chopra. 1991. Interspecific hybridization between *Brassica juncea* and *B. spinescens* through protoplast fusion. *Plant Cell Rep.,* 9:639–642.

Kirti, P.B., S.B. Narasimhulu, S. Prakash, and V.L. Chopra. 1992a. Somatic hybridization between *Brassica juncea* and *Moricandia arvensis* by protoplast fusion. *Plant Cell Rep.,* 11:318–321.

Kirti, P.B., S.B. Narasimhulu, S. Prakash, and V.L. Chopra. 1992b. Production and characterization of somatic hybrids of *Trachystoma ballii* and *Brassica juncea. Plant Cell Rep.,* 11:90–92.

Kirti, P.B., T. Mohapatra, H. Khanna, S. Prakash, and V.L. Chopra. 1995a. *Diplotaxis catholica* + *Brassica juncea* somatic hybrids: molecular and cytogenetic characterization. *Plant Cell Rep.,* 14:593–597.

Kirti, P.B., T. Mohapatra, S. Prakash, and V.L. Chopra. 1995b. Development of a stable cytoplasmic male sterile line of *Brassica juncea* from somatic hybrid *Trachystoma balli* + *Brassica juncea. Plant Breed.,* 114:434–438.

Kirti, P.B., A. Baldev, K. Gaikwad, S.R Bhat, V. Dineshkumar, S. Prakash, and V.L. Chopra. 1997. Introgression of a gene restoring fertility to CMS (*Trachystoma*) *Brassica juncea* and the genetics of restoration. *Plant Breed.,* 116:259–262.

Kirti, P.B., S. Prakash, K. Gaikwad, S.R. Bhat, V. Dineshkumar, and V.L. Chopra. 1998. Chloroplast substitution overcomes leaf chlorosis in *Moricandia arvensis* based cytoplasmic male sterile *Brassica juncea. Theor. Appl. Genet.,* 97:1179–1182.

Klimaszewska, K. and W.A. Keller. 1988. Regeneration and characterization of somatic hybrids between *Brassica napus* and *Diplotaxis harra. Plant Sci.,* 58:211–222.

Lefol, E., G Séguin-Swartz, and R.K. Downey. 1997. Sexual hybridization in crosses of cultivated *Brassica* species with the crucifers *Erucastrum gallicum* and *Raphanus raphanistrum*: potential for gene introgression. *Euphytica,* 95:127–139.

Leino, M., R. Teixeira, M. Landgren, and K. Glimelius. 2003. *Brassica napus* lines with rearranged *Arabidopsis* mitochondria display CMS and a range of developmental aberrations. *Theor. Appl. Genet.,* 106:1156–1163.

Leino, M., S. Thyselius, M. Landgren, and K. Glimelius. 2004. *Arabidopsis thaliana* chromosome III restores fertility in a cytoplasmic male-sterile *Brassica napus* line with *A. thaliana* mitochondrial DNA. *Theor. Appl. Genet.,* 109:272–279.

Lelivelt, C.L.C., and F.A. Krens. 1992. Transfer of resistance to the beet cyst nematode (*Heterodera schachtii* Schm) into the *Brassica napus* L. gene pool through intergeneric somatic hybridization with *Raphanus sativus* L. *Theor. Appl. Genet.,* 83:887–894.

Lelivelt, C.L.C., E.H.M. Leunissen, H.J. Frederiks, J.P.F.G. Helsper, and F.A. Krens. 1993. Transfer of resistance to the beet cyst nematode (*Heterodera schachtii* Schm.) from *Sinapis alba* L. (white mustard) to the *Brassica napus* L. gene pool by sexual and somatic hybridization. *Theor. Appl. Genet.,* 85:688–696.

Li, Z. and W.K. Heneen. 1999. Production and cytogenetics of intergeneric hybrids between the three cultivated *Brassica* diploids and *Orychophragmus violaceus. Theor. Appl. Genet.,* 99:694–704.

Li, Z., H.L. Liu, and P. Luo. 1995. Production and cytogenetics of intergeneric hybrids between *Brassica napus* and *Orychophragmus violaceus. Theor. Appl. Genet.,* 91:131–136.

Li, Z., J.G., Wu, Y. Liu, H.L. Liu, and W.K. Heneen. 1998a. Production and cytogenetics of intergeneric hybrids between *Brassica juncea* × *Orychophragmus violaceus* and *B. carinata* × *O. violaceus. Theor. Appl. Genet.,* 96:251–265.

Li, Z., X.M. Liang, X.M. Wu, and W.K. Heneen. 1998b. Morphology and cytogenetics of F3 progenies from intergeneric hybrids between *Brassica juncea* and *Orychophragmus violaceus. Hereditas,* 129:143–150.

Liu, J., X. Xu, and X. Deng. 2005. Intergeneric somatic hybridization and its application to crop genetic improvement. *Plant Cell Tissue Organ Cult.,* 82:19–44.

Liu, J.H., C. Dixelius, I. Eriksson, and K. Glimelius. 1995. *Brassica napus* (+) *B. tournefortii*, a somatic hybrid containing traits of agronomic importance for rapeseed breeding. *Plant Sci.,* 109:75–86.

Liu, J.H., M. Landgren, and K. Glimelius. 1996. Transfer of the *Brassica tournefortii* cytoplasm to *B. napus* for the production of cytoplasmic male sterile *B. napus*. *Physiol. Plant.,* 96:123–129.

Malik, M., P. Vyas, N.S. Rangaswamy, and K.R. Shivanna. 1999. Development of two new cytoplasmic male-sterile lines of *Brassica juncea* through wide hybridization. *Plant Breed.,* 118:75–78.

Mathias, R. 1985. Transfer of cytoplasmic male sterility from brown mustard (*Brassica juncea* L. Czern.) into rapeseed (*Brassica napus* L.). *Z. Pflanzenzüchtg,* 95:371–374.

Mathias, R. 1991. Improved embryo rescue technique for intergeneric hybridization between *Sinapis* species and *Brassica napus*. *Cruciferae Newslett.,* 14/15:90–91.

Matsuzawa, Y., S. Mekiyanon, Y. Kaneko, S.W. Bang, K. Wakui, and Y. Takahata. 1999. Male sterility in alloplasmic *Brassica rapa* L. carrying *Eruca sativa* cytoplasm. *Plant Breed.,* 118:82–-84.

Matsuzawa, Y., T. Funayama, M. Kamibayashi, M. Konnai, S.W. Bang, and Y. Kaneko. 2000. Synthetic *Brassica rapa-Raphanus sativus* amphidiploid lines developed by reciprocal hybridization. *Plant Breed.,* 119:357–359.

Mattsson, B. 1988. Interspecific crosses within the genus *Brassica* and some related genera. *Sveriges Utsadesforen Tidskr.,* 98:187–212.

Menczel, L., M. Morgan, M. Brown, and P. Maliga. 1987. Fusion-mediated combinations of Ogura-type cytoplasmic male sterility with *Brassica napus* plastids using x-irradiated CMS protoplasts. *Plant Cell Rep.,* 6:98–101.

Meng, J.L., G. Li, and Y. Zhun. 1997. Hybridization and hybrids analysis between *Moricandia arvensis* and *Brassica napus*. *Cruciferae Newslett.,* 19:25–26.

Meng, J., Z. Yan, Z. Tian, R. Huang, and B. Huang. 1999. Somatic hybrids between *Moricandia nitens* and three *Brassica* species. In: *Proc. 10th Int. Rapeseed Congr.* Canberra, Australia, p. 26–29.

Mithen, R.F. and C. Herron. 1991. Transfer of disease resistance to oilseed rape from wild *Brassica* species. In: *Proc. 8th Int. Rapeseed Congr.,* Saskatoon, Canada, p. 244–249.

Mizushima, U. 1950. Karyogenetic studies of species and genus hybrids in the tribe Brassiceae Cruciferae. *Tohoku J. Agr. Res.,* 1:1–14.

Mizushima, U. 1968. Phylogenetic studies on some wild *Brassica* species. *Tohoku J. Agr. Res.,* 19:83–99.

Mohanty, A. 1996. Hybridization between Crop *Brassica*s and Some of Their Wild Allies. Ph.D. thesis, University of Delhi, India.

Mohapatra, D. and Y.P.S. Bajaj. 1990. Intergeneric hybridization in *Brassica juncea* × *Brassica hirta* using embryo rescue. *Euphytica,* 36:321–26.

Momotaz, A., M. Kato, and F. Kakihara. 1998. Production of intergeneric hybrids between *Brassica* and *Sinapis* species by means of embryo rescue techniques. *Euphytica,* 103:123–130.

Morinaga, T.1934. On the chromosome number of *Brassica juncea* and *B. napus*, on the hybrid between the two, and on offspring line of the hybrid. *Japan J. Genet.,* 9:161–163.

Muller, J., K. Sonntag, and E. Rudloff. 2001. Somatic hybridization between *Brassica* spp. and *Raphanus sativus*. *Acta Hort.,* 560:219–220.

Namai, H. 1971. Studies on the breeding of oil rape (*Brassica napus* var. oleifera) by means of interspecific crosses between *B. campestris* ssp. *oleifera* and *B. oleracea*. I. Interspecific crosses with the application of grafting method or the treatment of sugar solution. *Japan. J. Breed.,* 2:40–48.

Nanda Kumar, P.B.A. and K.R. Shivanna. 1993. Intergeneric hybridization between *Diplotaxis siettiana* and crop brassicas for the production of alloplasmic lines. *Theor. Appl. Genet.,* 85:770–776.

Nanda Kumar, P.B.A., K.R. Shivanna, and S. Prakash. 1988. Wide hybridization in *Brassica*: crossability barriers and studies on hybrids and synthetic amphidiploids of *B. fruticulosa* × *B. campestris*. *Sex. Plant Reprod.,* 1:234–239.

Narasimhulu, S.B., P.B. Kirti, S.R Bhat, S. Prakash, and V.L. Chopra. 1994. Intergeneric protoplast fusion between *Brassica carinata* and *Camelina sativa*. *Plant Cell Rep.,* 13:657–660.

O'Neill, C.M., T. Murata, C.L. Morgan, and R.J. Mathias. 1996. Expression of the C3-C4 intermediate character in somatic hybrids between *Brassica napus* and the C3-C4 species *Moricandia arvensis*. *Theor. Appl. Genet.,* 93:1234–1241.

Oikarinen, S. and P.H. Ryöppy. 1992. Somatic hybridization of *Brassica campestris* and *Barbarea* species. In: *Proc. XIIIth Eucarpia Congress: Reproductive Biology and Plant Breeding*, Angers, France, p. 261–262.

Ovcharenko, O.O., I.K. Komarnytskyi, M.M. Cherep, I.I. Hleba, and M.V. Kuchuk. 2004. Obtaining of intertribal *Brassica juncea* + *Arabidopsis thaliana* somatic hybrids and study of transgenic trait behaviour. *Tsitol Genet.,* 38:3–8.

Pathania, A., S.R. Bhat, V. Dinesh Kumar, Asutosh, S. Prakash, and V.L. Chopra. 2003. Cytoplasmic male sterility in alloplasmic *Brassica juncea* carrying *Diplotaxis catholica* cytoplasm: molecular characterization and genetics of fertility restoration. *Theor. Appl. Genet.,* 107:455–461.

Pellan-Delourme, R. and M. Renard. 1987. Identification of maintainer genes in *Brassica napus* L. for male sterility inducing cytoplasm of *Diplotaxis muralis* L. *Plant Breed.,* 99:89–97.

Pelletier, G., 1993. Somatic hybridization. In: Hayward, M.D., N.O. Bosemark, and I. Romagosa (Eds.), *Plant Breeding: Principles and Prospects.* Chapman & Hall, London, p. 93–106.

Pelletier, G., C. Primard, F. Vedel, P. Chétrit, R. Rémy, P. Rousselle, and M. Renard. 1983. Intergeneric cytoplasmic hybridization in Cruciferae by protoplast fusion. *Mol. Gen. Genet.,* 191:244–250.

Peng, L., H.L. Fu, Z.Q. Lan, S.D. Zhou, H.F. Zhou, and Q. Luo. 2003. Phyogenetic studies on intergeneric hybridization between *Brassica napus* and *Matthiola incana. Acta Bot. Sin.,* 45:432–436.

Peterka, H., H. Budhan, O. Schrader, R. Ahne, and W. Schütze. 2004. Transfer of resistance against the beet cyst nematode from radish (*Raphanus sativus*) to rape (*Brassica napus*) by monosomic chromosome addition. *Theor. Appl. Genet.,* 109:30–41.

Plieske, J., D. Struss, and G. Röbbelen. 1998. Inheritance of resistance derived from the B-genome of *Brassica* against *Phoma lingam* in rapeseed and the development of molecular markers. *Theor. Appl. Genet.,* 97:929–936.

Pradhan, A.K., A. Mukhopadhyay, and D. Pental, 1991. Identification of the putative cytoplasmic donor of a CMS system in *Brassica juncea. Plant Breed.,* 106:204–208.

Pradhan, A.K., S. Prakash, A. Mukhopadhyay, and D. Pental. 1992. Phylogeny of *Brassica* and allied genera based on variation in chloroplast and mitochondrial DNA patterns: molecular and taxonomical classifications are incongruous. *Theor. Appl. Genet.,* 85:331–340.

Prakash, S. and A. Narain. 1971. Genomic status of *Brassica tournefortii* Gouan. *Theor. Appl. Genet.,* 41:203–204.

Prakash, S. and V.L. Chopra. 1990. Male sterility caused by cytoplasm of *Brassica oxyrrhina* in *B. campestris* and *B. juncea. Theor. Appl. Genet.,* 79:285–287.

Prakash, S., I. Ahuja, H.C. Uprety, V.D Kumar, S.R. Bhat, P.B. Kirti, and V.L. Chopra. 2001. Expression of male sterility in alloplasmic *Brassica juncea* with *Erucastrum canariense* cytoplasm and development of fertility restoration system. *Plant Breed.,* 120:178–182.

Prakash, S., P.B. Kirti, S.R. Bhat, K. Gaikwad, V. Dineshkumar, and V.L Chopra. 1998. A *Moricandia arvensis* based cytoplasmic male sterility and fertility restoration system in *Brassica juncea. Theor. Appl. Genet.,* 97:488–492.

Prakash, S., S. Tsunoda, R.N. Raut, and S. Gupta. 1982. Interspecific hybridization involving wild and cultivated genomes in the genus *Brassica. Cruciferae Newslett.,* 7:28–29.

Prakash, S., S.R. Bhat, C.F. Quiros, P.B. Kirti, and V.L. Chopra. 2009. *Brassica* and its close allies: cytogenetics and evolution. *Plant Breed. Rev.* (in press).

Primard, C., F. Vedel, C. Mathieu, G. Pelletier, and A.M. Chevre, 1988. Interspecific somatic hybridization between *Brassica napus* and *Brassica hirta* (*Sinapis alba* L.). *Theor. Appl. Genet.,* 75:546–552.

Rao, G.U., Y.K. Aksha, and A.K. Pradhan. 1998. Isolation of useful variants in alloplasmic crop brassicas in the cytoplasmic background of *Erucastrum gallicum. Euphytica,* 103:301–306.

Rao, G.U., M. Lakshmikumaran, and K.R. Shivanna. 1996. Production of hybrids, amphiploids and backcross progenies between a cold-tolerant wild species, *Erucastrum abyssinicum* and crop brassicas. *Theor. Appl. Genet.,* 92:786–790.

Rao, G.U., V.S Batra, S. Prakash, and K.R. Shivanna. 1994. Development of a new cytoplasmic male sterile system in *Brassica juncea* through wide hybridization. *Plant Breed.,* 112:171–174.

Razmjoo, K., K. Toriyama, R. Ishii, and K. Hinata. 1996. Photosynthetic properties of hybrids between *Diplotaxis muralis* DC., a C3 species, and *Moricandia arvensis* (L.) DC., a C3-C4 intermediate species in Brassicaceae. *Genes Genet. Syst.,* 71:189–192.

Rhee, W.Y., Y.H. Cho, and K.Y. Paek. 1997. Seed formation and phenotype expression of intra- and interspecific hybrids of *Brassica* and of intergeneric hybrids obtained by crossing with *Raphanus. J. Korean Soc. Hort. Sci.,* 38:353–360.

Ringdahl, E.A., P.B.E. McVetty, and J.L. Sernyk. 1987. Intergeneric hybridization of *Diplotaxis* ssp. with *Brassica napus*: a source of new CMS systems? *Can. J. Pl. Sci.,* 67:239–243.

Ripley, V.L. and P.G. Arnison. 1990. Hybridization of *Sinapis alba* and *Brassica napus* L. via embryo rescue. *Plant Breed.,* 104:26–33.

Roy, N.K. 1984. Interspecific transfer of *Brassica juncea* type high black leg resistance to *Brassica napus. Euphytica,* 33:95–303.

Ryschka, U., G. Schumann, E. Klocke. P. Scholze, and M. Neumann. 1996. Somatic hybridization in *Brassiceae*. *Acta Hortic.*, 407:201–208.

Saal, B., H. Brun, I. Glais, and D. Struss. 2004. Identification of a *Brassica juncea*-derived recessive gene conferring resistance to *Leptosphaeria maculans* in oilseed rape. *Plant Breed.*, 123:505–511.

Sageret, M. 1826. Considerations sur la production des variants et des variétiés en general, et sur celled de la famille de Cucurbitaceés en particulier. *Ann. Sci. Nat.*, 8:94–314.

Sakai, T. and J. Inamura. 1990. Intergeneric transfer of cytoplasmic male sterility between *Raphanus sativus* (CMS line) and *Brassica napus* through cytoplast-protoplast fusion. *Theor. Appl. Genet.*, 80:421–427.

Sakai, T., H.J. Liu, M. Iwabuchi, J. Kohno-Murase, and J. Inamura. 1996. Introduction of a gene from fertility restored radish (*Raphanus sativus*) into *Brassica napus* by fusion of x-irradiated protoplasts from a radish restorer line and iodacetoamide-treated protoplasts from a cytoplasmic male-sterile cybrid of *B. napus*. *Theor. Appl. Genet.*, 93:73–379.

Salisbury, P.A. 1989. Potential utilization of wild Crucifer germplasm in oilseed *Brassica* breeding. In: *Proc. ARAB 7th Workshop.* Toowoombu, Queensland, Australia, p. 51–53.

Schröder-Pontoppidan, M., M. Skarzhinskaya, C. Dixelius, S. Stymne, and K, Glimelius. 1999. Very long chain and hydroxylated fatty acids in offspring of somatic hybrids between *Brassica napus* and *Lesquerella fendleri*. *Theor. Appl. Genet.*, 99:108–114.

Shivanna, K.R. 1996. Incompatibility and wide hybridization. In: V.L. Chopra and S. Prakash (Eds.), *Oilseed and Vegetable Brassicas: Indian Perspective*. Oxford and IBH, New Delhi, p. 77–102.

Sigareva, M.A. and E.D. Earle. 1997. Direct transfer of a cold-tolerant Ogura male sterile cytoplasm into cabbage (*Brassica oleracea* ssp. Capitata) via protoplast fusion. *Theor. Appl. Genet.*, 94: 213–220.

Sikdar, S.R., G. Chatterjee, S. Das, and S.K. Sen. 1990. "Erussica" the intergeneric fertile somatic hybrid developed through protoplast fusion between *Eruca sativa* Lam. and *Brassica juncea* (L.) Czern. *Theor. Appl. Genet.*, 79:561–567.

Sjödin, C. and K. Glimelius. 1989a. *Brassica naponigra*, a somatic hybrid resistant to *Phoma lingam*. *Theor. Appl. Genet.*, 77: 651–656.

Sjödin, C. and K. Glimelius. 1989b. Transfer of resistance against *Phoma lingam* to *Brassica napus* by asymmetric somatic hybridization combined with toxin selection. *Theor. Appl. Genet.*, 78:513–520.

Skarzhinskaya, M., J. Fahleson, K. Glimelius, and A. Mouras. 1998. Genome organization of *Brassica napus* L. and *Lesquerella fendleri* and analysis of their somatic hybrids using genomic *in situ* hybridization. *Genome*, 41:691–701.

Skarzhinskaya, M., M. Landgren, and K. Glimelius. 1996. Production of intertribal somatic hybrids between *Brassica napus* L. and *Lesquerella fendleri* (Gray) Wats. *Theor. Appl. Genet.*, 93:1242–1250.

Snowdon, R.J., H. Winter, A. Diestal, and M.D. Sacristán. 2000. Development and characterisation of *Brassica napus-Sinapis arvensis* addition lines exhibiting resistance to *Leptosphaeria maculans*. *Theor. Appl. Genet.*, 101:1008–1014.

Stiewe, G. and G. Röbbelen. 1994. Establishing cytoplasmic male sterility in *Brassica napus* by mitochondrial recombination with *B. tournefortii*. *Plant Breed.*, 113:294–304.

Struss, D., U. Bellin, and G. Röbbelen. 1991. Development of B-genome chromosome addition lines of *Brassica napus* using different interspecific *Brassica* hybrids. *Plant Breed.*, 106:209–214.

Sundberg, E. and K. Glimelius. 1991. Effects of parental ploidy level and genetic divergence on chromosome elimination and chloroplast segregation in somatic hybrids within *Brassiceae*. *Theor. Appl. Genet.*, 83:81–88.

Takahata, Y. and K. Hinata. 1983. Studies on cytodemes in the subtribe *Brassicineae*. *Tohoku J. Agr. Res.*, 33:111–124.

Takahata, Y., T. Takeda, and N. Kaizuma. 1993. Wide hybridization between *Moricandia arvensis* and *Brassica* amphidiploid species (*B. napus* and *B. juncea*). *Euphytica*, 69:155–160.

Takeshita, M., M. Kato, and S. Tokumasu. 1980. Application of ovule culture to the production of intergeneric hybrids in *Brassica* and *Raphanus*. *Jap. J. Genet.*, 55:373–387.

Tu, Y., J. Sun, Y. Liu, X. Ge, Z. Zhao, X. Yao, and Z. Li. 2008. Production and characterization of intertribal somatic hybrids of *Raphanus sativus* and *Brassica rapa* with dye and medicinal plant *Isatis indigotica*. *Plant Cell Rep.*, 27:873–883.

Voss, A., R.J. Snowdon, W. Lühs, and W. Friedt. 2000. Intergeneric transfer of nematode resistance from *Raphanus sativus* into the *Brassica napus* genome. *Acta Hortic.*, 539:129–134.

Vyas, P., S. Prakash, and K.R. Shivanna. 1995. Production of wide hybrids and backcross progenies between *Diplotaxis erucoides* and crop *Brassica*s. *Theor. Appl. Genet.*, 9:549–553.

Wang, Y.P. and P. Luo. 1998. Intergeneric hybridization between *Brassica* species and *Crambe abyssinica*. *Euphytica,* 101:1–7.

Wang, Y.P., K. Sonntag, and E. Rudloff. 2003. Development of rapeseed with high erucic acid content by asymmetric somatic hybridization between *Brassica napus* and *Crambe abyssinica. Theor. Appl. Genet.,* 106:1147–1155.

Wang, Y.P., R.J. Snowdown, E. Rudloff, P. Wehling, W. Friedt, and K. Sonntag. 2004. Production and characterization of asymmetric somatic hybrids between *Brassica napus* and *Crambe abyssinica. Votr. Pflanzenzüchtung,* 64:85–86.

Warwick, S.I., M.J. Simard, and H.J. Beckie 2003. Hybridization between transgenic *Brassica napus* L and its wild relatives *Brassica rapa* L. *Raphanus raphanistrum* L., *Sinapis arvensis* L. and *Erucastrum gallicum. Theor. Appl. Genet.,* 107:528–539.

Yamagishi, H., M. Landgren, J. Forsberg, and K. Glimelius. 2002. Production of asymmetric hybrids between *Arabidopsis thaliana* and *Brassica napus* utilizing an efficient protoplast culture system. *Theor. Appl. Genet.,* 104:959–964.

Yandav, R.C., P.K. Sareen, and J.R. Chowdhury. 1991. Interspecific hybridization in *Brassica juncea -Brassica tournefortii* using ovary culture, *Cruciferae Newslett.,* 14/15:84

14 Biotechnology

Vinitha Cardoza and C. Neal Stewart, Jr.

CONTENTS

INTRODUCTION

The family *Brassicaceae*, formerly called Cruciferae, contains many economically important oilseed crops, vegetables and their wild relatives, as well as the genomic model, *Arabidopsis thaliana*. The members of the family are principally distributed in Europe, Asia, North America, and the Middle East, and comprise approximately 350 genera and 3000 species. The four petals of the flowers of this family have a distinct cross-like arrangement, hence the name Cruciferae. *Brassica* vegetables mostly belong to the species *B. oleracea* and consist of cauliflower, broccoli, cabbage, collards, Brussels sprouts, kale, and kohlrabi. The oilseed *Brassica* crops are found in *B. napus*, *B. rapa*, *B. juncea*, and *B. nigra*. Other important vegetables of this family are *Raphanus sativum* (radish), *Nasturtium officinalis* (watercress), *Wasabia japonica* (wasabi), and *Armoracia rusticana* (horseradish). Other than food crops, many *Brassicas* are also used as forage crops because of their high dry matter digestibility and high protein content. Kale, oilseed rape, and turnips are the ones also used as forage crops. With regard to scientific research, the species that has yielded the most data in the plant kingdom is *Arabidopsis thaliana*. Indeed, those data have translated to increasing our knowledge about economically important members of the family.

A very interesting group of biotypes in the Crucifer family is known as the rapid cycling *Brassicas* (also known as "fast plants") whose life cycle ranges between 20 and 60 days (Williams and Hill, 1986). With a small genome size, just three- to fourfold larger than *Arabidopsis thaliana*, they have proven useful as laboratory models in their own right. Also of interest in this group are their high female fertility, rapid seed maturation, and absence of seed dormancy.

Biotechnology has been important in improving *Brassicas* for agronomic and horticultural production, but they have also served as models for the development of biotechnological methods. Indeed, several *Brassicas* have served as biotechnology risk models (see, e.g., Halfhill et al., 2005) because of their interesting taxonomy and reproductive biology, but also because they are relatively easy to genetically transform.

TISSUE CULTURE AND TRANSFORMATION

Tissue culture and transformation studies are fairly well researched in the crucifers and have been reported in most of the economically important species such as *Brassica juncea* (Barfield and Pua,

1991); *B. napus* (Moloney et al., 1989); *B. rapa* (Radke et al., 1992); *B. oleracea* (De Block et al., 1989); *B. nigra* (Gupta et al., 1993).

Organogenesis is the most commonly used tissue culture technique for plant regeneration and transformation. Organogenesis has been accomplished using primarily hypocotyl explants (Yang et al., 1991; De Block et al., 1989; Cardoza and Stewart, 2003), although other tissues such as cotyledons (Hachey et al., 1991; Sharma et al., 1990; Ono et al., 1994), peduncle segments (Stringham, 1977; Eapen and George, 1997), and leaves (Radke et al., 1988) have also been used.

Members of the *Brassicaceae* have been amenable to *in planta* transformation, with the technique being most successfully developed in *Arabidopsis thaliana*, which is now routinely transformed by immersing the flowers in *Agrobacterium tumefaciens* cultures — the floral dip method (Pelletier and Bechtold, 2003). Subsequent to *in planta* transformation, seeds are harvested and then transformants selected. The other cruciferous crops transformed *in planta* include pakchoi (*Brassica rapa* ssp. *chinensis*) (Fan et al., 1996; Qing et al., 2000), radish (*Raphanus sativus* L. *longipinnatus* Bailey) (Curtis and Nam, 2001), and canola (Wang et al., 2003), but the efficiencies are low and the methods are far from routine.

Transformation has been reported in almost all of the economically important species, including *Brassica juncea* (Barfield and Pua, 1991); *B. napus* (Moloney et al., 1989); *B. rapa* (Radke et al., 1992); *B. oleracea* (De Block et al., 1989); *B. nigra* (Gupta et al., 1993), and *B. carinata* (Narasimhulu et al., 1992). Transformation has mostly targeted improvement in oil quality or developing herbicide, insect, and disease resistance. With increased awareness and interest in health, healthier vegetable oil choices are in higher demand. Canola has long been the most widely grown oilseed crop in the crucifer family. Its low saturated fatty acid content and intermediate levels of PUFA (polyunsaturated fatty acid) are the most desirable oil qualities that have been achieved via conventional breeding. However, biotechnology has proven useful in improving the oil quality in canola, and research in this direction has been carried out, in particular, with regard to increasing oleic acid levels (Stoutjesdijk et al., 2000), γ-linolenic acid (Liu et al., 2001).

Canola cultivation has been impacted greatly by engineered herbicide resistance. Canola tolerant to herbicides such as imidazoline, glufosinate, and glyophosate is available commercially in the United States and Canada. Herbicide-resistant canola is the fourth most planted transgenic crop in the world. However, herbicide resistance has been transformed, both in the vegetable and oilseed brassicas such as broccoli (Waterer et al., 2000) *B. rapa* (Cao et al., 1999) for glufosinate resistance, sulfonylurea resistance in *B. napus* (Blackshaw et al., 1994), and bromoxynil resistance in *B. napus* (Zhong et al., 1997).

It has been possible to develop insect-resistant *Brassica* crops using transformation. Among insects, the diamondback moth (*Plutella xylostella*) is a major pest of crucifer plants and amounts to millions of dollars of crop loss. It is necessary to develop resistance to this pest. Using transgenic approaches, insect resistance in *Brassicas* has been developed using the insecticidal crystal protein Bt genes. Bt *cry*1A(c) has been introduced in *Brassica napus* (Stewart et al., 1996); Chinese *B. napus* cultivars (Li et al., 1999); rutabaga (Li et al., 1995); cabbage (Metz et al., 1995a); broccoli (Metz et al., 1995b; Cao et al., 1999); and Chinese cabbage (Cho et al., 2001).

Although there has been a fair amount of research conducted in the area of insect resistance, not much transformation work has been performed to endow disease resistance in the crucifers, and this avenue requires further exploration. The major diseases that affect crucifers are bacterial leaf spot caused by *Pseudomonas syringae*, bacterial black rot and leaf spot (*Xanthomonas campestris*), yellowing (*Fusarium oxyspororum*), white rust (*Albugo candida*), downy mildew (*Peronospora parasitica*), ring spot (*Mycosphearella brassicicola*), *Plasmodiophora brassicae*, powdery mildew (*Erysiphe polygoni*), cauliflower mosaic virus (CaMV), and radish mosaic virus (RMV). It is necessary to develop methods to develop disease-resistant plants. Disease resistance is an important area that could benefit from the use of natural disease resistance properties of wild crucifer members such as *Camelina sativa*, which is naturally disease resistant and is a potential candidate for a pool of disease-resistant genes. Wild and weedy crucifers have the potential of disease resistance;

however, their potential has not been significantly exploited. Westman et al. (1999) identified potential Eurasian weedy crucifers that were resistant to North American crop pathogens that might be utilized in the development of disease resistance in North American crucifers.

Not only do wild crucifers have disease-resistance traits but also they possess many more valuable traits that are commercially important. It is possible to transfer these genes from the wild crucifers to cultivated species and thus tap into the valuable potential of these wild crucifer species; either in breeding, but perhaps more in cisgenic approaches using genetic transformation.

Wild non-*Arabidopsis* species of the *Brassicaceae* have generally not been manipulated. One exception has been wild *Brassica rapa*, which was genetically engineered as part of a larger biotechnology risk assessment project aimed at comparing a directly transformed transgene in *B. rapa* with one that was introgressed into *B. rapa* from *B. napus*. In this particular experiment, a *B. napus* transformation protocol (Stewart et al., 1996) was used to produce transgenic *B. rapa* for GFP and Bt *Cry1Ac* (Moon et al., 2007). The wild *B. rapa* used in this study was from original populations collected in California and Quebec, Canada.

CRUCIFERS IN PHYTOREMEDIATION

Many Cruciferae members are known phytoremediators. The most important members that have been extensively used in phytoremediation studies are *Brassica juncea* and *Thlaspi caerulescens,* two of the most important remediation plants identified in the plant kingdom. Genomics and biotechnology seem to hold promise in enhancing phytoremediation capabilities. *Arabidopsis thaliana* has been used to overexpress and evaluate various genes involved in phytoremediation because it serves as a model crop for all plant research. The phytoremediation capabilities of crucifers have been enhanced by engineering various genes that lead to increased metal tolerance or uptake. For example, by overexpressing Υ-glutamylcysteine synthetase and glutathione synthetase, cadmium tolerance and accumulation has been enhanced in *B. juncea* (Zhu et al., 1999a, b). *B. juncea* and *A. thaliana* engineered with selenocysteine methyltransferase had increased selenium tolerance and accumulation (LeDuc et al., 2004). In another example, A. *thaliana* plants engineered with yeast *YCF1* showed resistance to cadmium and lead (Song et al., 2003). Not only do engineered *Arabidopsis* plants show the potential of phytoremediation, but they have also been shown to serve as metal bioindicators. When *A. thaliana* plants were engineered with a nickel inducible promoter fused to the GUS gene, and plants were grown in media containing nickel, they conditionally expressed GUS. These experiments showed the feasibility of using these plants as environmental phytosensors (Krizek et al., 2003). Studies on phytodetoxification in *A. thaliana* have also shown promising results. By engineering *merA* and *merB,* which play a role in mercury detoxification, it has been possible for *A. thaliana* plants to uptake mercury and then volatilize it, thus cleansing the soil of mercury contamination. Finally, in *B. juncea,* the hairy roots obtained by engineering *Agrobacterium rhizogenes* were used in the phytoremediation of uranium (Eapen et al., 2003).

SOMATIC HYBRIDIZATION

Protoplast fusion has been used to transfer genes between various *Brassicaceae* members for the purposes of increased yield and disease resistance. Using protoplast fusion, it is possible to create hybrid and cybrid combinations of species that are sexually incompatible. Interspecific and intergeneric hybrids have been produced. *Brassica rapa* and *B. napus* have been fused to transfer yield genes (Qiang et al., 2003). *B. rapa* from China was found to be the best for transferring yield genes to *B. napus*. Rapid cycling *B. napus* lines with novel fatty acid compositions have been produced by protoplast fusion between rapid cycling *B. oleracea* and *B. rapa* (Hansen and Earle, 1994). Hu et al (2002) fused *B. napus* and *Sinapis arvensis* (wild mustard) to enrich the rapeseed gene pool. Protoplast fusion has facilitated the production of low linolenic acid *B. napus* lines (Heath and Earle, 1997)

The genus *Moricandia* possesses C3-C4 photosynthesis intermediates. Protoplast fusion between *Brassica oleracea* and *Moricandia nitens* resulted in the production of intergeneric hybrids that expressed a gas-exchange character that was intermediate between the two parents (Yan et al., 1999). By fusing *B. napus* and *Moricandia arvensis,* C3-C4 intermediates were produced (O'Neill et al., 1996). Ishikawa et al. (2003) fused *M. arvensis* and *B. oleracea* and observed that the somatic hybrid again showed intermediate traits between C3-C4 as *M. arvensis.* Thus, it is possible to modify the photosynthetic rate of cultivated C3 *Brassica* species by transferring genes from its wild relatives. Protoplast fusion of disparate species physiologies is a powerful tool to better understand C3 and C4 photosynthesis mechanisms in a comparative perspective.

Zinc tolerance was achieved in fused protoplasts between *Thlaspi caerulescens,* a natural zinc hyperaccumulator, and *Brassica napus* (Brewer et al, 1999). Thus, this technique could be powerful in examining phytoremediation and metal metabolism constraints at the cellular level.

Interspecific and intergeneric protoplast fusion has been useful in the study of plant-pathogen interactions. In one example, somatic hybrids resistant to bacterial soft rot have been produced by the fusion of *Brassica rapa* and *B. oleracea* protoplasts (Ren et al., 2000). In a second example, fusing protoplasts from a disease-resistant *B. nigra* with a susceptible *B. napus* yielded hybrids that were resistant to rapeseed pathogens (Gerdemann-Knörck et al., 1995). In addition, *Camelina sativa,* a wild crucifer, is known to be resistant to *Alternaria brassicae.* Somatic hybrids have been produced between *B. carinata* and *C. sativa* in an effort to transfer disease blight resistance genes into *B. carinata* (Narasimhulu et al., 1994).

Protoplast fusion has been used as a tool to produce very wide crosses. One particularly wide somatic hybrid was formed by the fusion of mesophyll protoplasts of *Trachystoma ballii* and *Brassica juncea* (Kirti et al., 1992). The hybrids were of intermediate morphology between the parents. The plants were pollen sterile; however, when these plants were backcrossed with *B. juncea* as the female parent, viable seeds were produced. Similarly, somatic hybrids between *B. napus* and *Lesquerella fendleri* were produced by fusing mesophyll protoplasts (Skarzhinskaya et al., 1996).

Cytoplasmic male sterility (CMS) plays an important role in hybrid seed production. Male sterile lines have been developed by protoplast fusion. Male-sterile, cold-tolerant *Brassica napus* somatic hybrids were produced by crossing an Ogura male-sterile, cold-sensitive cauliflower inbred (*B. oleracea* var. *botrytis* inbred NY7642A) and a cold-tolerant, fertile canola (*B. rapa* cv. Candle) (Heath and Earle, 1996). Male-sterile cybrids have also been produced by the fusion of protoplasts of *B. napus* and *B. tournefortii* (Liu Clarke et al., 1999). A cytoplasmic male-sterile cybrid of *Brassica oleracea* was produced by the transfer of the sterile 'Anand' cytoplasm (originally from wild species of *B. tournefortii*) from *B. rapa* to *B. oleracea* (Cardi and Earle, 1997). Cold-tolerant cytoplasmic male-sterile (CMS) cabbage (*B. oleracea* ssp. *capitata*) was produced by the fusion of leaf protoplasts from fertile cabbage and a cold-tolerant Ogura male-sterile broccoli (Sigareva and Earle, 1997). An interesting development using protoplast fusion is the combination of the development of male-sterile and fertility restoration, which is a perfect system for the production of heterotic hybrids. This was done in *B. juncea* by protoplast fusion with *Moricandia arvensis,* and fertility restoration of this male-sterile *B. juncea* was achieved by introgression (Prakash et al., 1998). However, these CMS lines were chlorotic. Protoplast fusion of chlorotic male-sterile *B. juncea* with green male-sterile *B. juncea* resulted in green male-sterile plants (Kirti et al., 1998). Protoplast fusion between *A. thaliana* and *B. napus* has resulted in the production of asymmetric hybrids that included three male sterile hybrids (Yamagishi et al., 2002). The male-sterile plants are excellent candidates for the study of genes involved in CMS.

MOLECULAR MARKERS

The vast knowledge of molecular markers available has greatly aided breeding in cruciferous crops. Molecular markers have been used in mapping important traits in crucifer plants. The most commonly used markers are RAPD (randomly amplified polymorphic DNA), RFLP (restriction fragment length polymorphism), AFLP (amplified fragment length polymorphism), and SSR (simple sequence repeats). Furthermore, fine genetic mapping is possible using SNP (single nucleotide polymorphism), making it possible to uncover allelic variation directly within expressed sequences of candidate genes (Snowdon and Friedt, 2004). Genetic diversity studies using marker technology have also been conducted in various *Brassicaceae* members. Breeding programs are successfully incorporating marker-assisted selection in economically important crops, especially in the oilseed *Brassicas*. Using various markers, genes affecting quantitative trait loci (QTL) for various traits have been mapped in several *Brassicas*. QTL have been used to identify genes controlling the levels of erucic acid and linoleic acids in *Brassica napus* (Thormann et al., 1996). Other than oil content, QTL controlling traits such as disease resistance (Pilet et al., 1998; Manzanares-Dauleux et al., 2000; Zhao and Meng, 2003, Soengas et al., 2007), flowering time (Ferreira et al., 1995), and fertility restoration (Jean et al., 1997; Hansen et al., 1997) have also been identified.

The sequencing of the *Arabidopsis* genome has proven extremely beneficial in marker-assisted breeding. *A. thaliana* recombinant inbred lines have been used in QTL mapping (Koornneef et al., 2004). Several insect resistance QTLs have been mapped and genes linked to these QTLs have been cloned (Jander et al., 2001; Lambrix et al., 2001; Kliebenstein et al., 2002; Kroymann et al., 2003; Zhang et al., 2006). It has also been possible to map QTL resistant to the crucifer specialist herbivore *Pieris brassicae* using *Arabidopsis* inbred lines (Pfalz et al., 2007)

CONCLUSIONS

Biotechnology plays a very important role in modern agriculture as continuous work is being carried out in the crucifer members. Although there has been a good research effort in this direction, biotechnology has yet to reach its full potential in this very important plant family. Research and technologies have aided the development of vegetable and oilseed *Brassicas* with novel edible traits. Tissue culture, transformation, and molecular markers together can be of use in the development of plants with novel desirable traits. Molecular markers and transformation should be synergistic for plant improvement. In addition to the extensive genomics work in *Arabidopsis thaliana*, the mapping and genomic characterization of the *Brassica rapa* genome is underway. It is possible to enhance crucifer breeding by taking advantage of the sequenced genomes along with utilizing the gene pool resource and transferring the economically important traits from wild crucifers.

REFERENCES

Barfield DG, Pua EC (1991) Gene transfer in plants of *Brassica juncea* using *Agrobacterium tumefaciens* mediated transformation. *Plant Cell Rep* 10:308–314.

Blackshaw RE, Kanashiro D, Moloney MM, Crosby WL (1994) Growth, yield and quality of canola expressing resistance to acetolactate synthase inhibiting herbicides. *Can J Plant* Sci 74:745–751.

Brewer EP, Saunders JA, Angle JS, Chaney RL, McIntosh MS (1999) Somatic hybridization between the zinc accumulator *Thlaspi caerulescens* and *Brassica napus. Theor Appl Genet* 99:761–771.

Cao J, Tang JD, Strizhov N, Shelton AM, Earle ED (1999) Transgenic broccoli with high levels of *Bacillus thuringiensis Cry1C* protein control diamondback moth larvae resistant to *Cry1A* or *Cry1C. Mol Breed* 5:131–141.

Cardi T, Earle ED (1997) Production of new CMS *Brassica oleracea* by transfer of Anand cytoplasm from *B. rapa* through protoplast fusion. *Theor Appl Genet* 94:204–212.

Cardoza V, Stewart CN Jr (2003) Increased *Agrobacterium* mediated transformation and rooting efficiencies in canola (*Brassica napus* L.) from hypocotyl explants. *Plant Cell Rep* 21:599–604.

Cho HS, Cao J, Ren JP, Earle ED (2001) Control of *Lepidopteran* insect pests in transgenic Chinese cabbage (*Brassica rapa* ssp *pekinensis*) transformed with a synthetic *Bacillus thuringiensis Cry1C* gene. *Plant Cell Rep* 20:1–7.

Curtis IS, Nam HG (2001) Transgenic radish (*Raphanus sativus* L. *longipinnatus* Bailey) by floral-dip method — plant development and surfactant are important in optimizing transformation efficiency. *Trans Res* 10:363–371.

De Block M, De Brower D, Tenning P (1989) Transformation of *Brassica napus* and *Brassica oleracea* using *Agrobacterium tumefaciens* and the expression of the *bar* and *neo* genes in the transgenic plants. *Plant Physiol* 91:694–701.

Eapen S, George, L (1997) Plant regeneration from peduncle segments of oil seed *Brassica* species: influence of silver nitrate and silver thiosulfate. *Plant Cell Tiss Org Cult* 51:229–232.

Eapen S, Suseelan KN, Tivarekar S, Kotwal SA and Mitra R (2003) Potential for rhizofiltration of uranium using hairy root cultures of *B. juncea* and *Chenopodium amaranticolor*. *Environ Res* 91:127–133.

Fan L, Qing CM, Lei Y, Robaglia C, Tourneur, C (1996) *In planta* transformation of pakchoi (*Brassica campestris* L. ssp. *chinensis*) by infiltration of adult plants with *Agrobacterium*. *Acta Hort* 467:187–192.

Ferreira ME, Satagopan J, Yandell BS, Williams PH, Osborn TC (1995) Mapping loci controlling vernalization requirement and flowering time in *Brassica napus*. *Theor Appl Genet* 90:727–732.

Gerdemann-Knörck M, Nielen S, Tzscheetzsch C, Iglisch J, Schieder O (1995) Transfer of disease resistance within the genus *Brassica* through asymmetric somatic hybridization. *Euphytica* 85:247–253.

Gupta, V, Sita GL, Shaila MS, Jagannathan, V (1993) Genetic transformation of *Brassica nigra* by *Agrobacterium* based vector and direct plasmid uptake. *Plant Cell Rep* 12:418–421.

Hachey JE, Sharma KK, Moloney M.M (1991) Efficient shoot regeneration of *Brassica campestris* using cotyledon explants cultured *in vitro*. *Plant Cell Rep* 9:549–554.

Halfhill MD, Moon HS, Sutherland JP, Poppy GM, Warwick SI, Rufty TW, Weissinger AK, Raymer PL, Stewart CN, Jr. (2005) Growth, productivity, and competitiveness of introgressed weedy *Brassica rapa* hybrids selected for the presence of Bt *cry1Ac* and *gfp* transgenes. *Mol Ecol* 14:3177–3189.

Hansen LN, Earle ED (1994) Novel flowering and fatty acid characters in rapid cycling *Brassica napus* L. synthesized by protoplast fusion. *Plant Cell Rep* 14:151–156.

Hansen M, Hallden C, Nisson NO, Sall T (1997) Marker-assisted selection of restored male-fertile *Brassica napus* plants using a set of dominant RAPD markers. *Mol Breed* 6:449–456.

Heath DW, Earle ED (1996) Synthesis of Ogura male sterile rapeseed (*Brassica napus* L.) with cold tolerance by protoplast fusion and effects of atrazine resistance on seed yield. *Plant Cell Rep* 15:939–944.

Heath DW, Earle E (1997) Synthesis of low linolenic acid rapeseed (*Brassica napus* L) through protoplast fusion. *Euphytica* 93:339–343.

Hu Q, Anderson SB, Dixelius C, Hansen LN (2002) Production of fertile intergeneric somatic hybrids between *Brassica napus* and *Sinapis arvensis* for the enrichment of the rapeseed gene pool. *Plant Cell Rep* 21:147–152.

Ishikawa S, Bang SW, Kaneko Y, Matsuzawa Y (2003) Production and characterization of intergeneric somatic hybrids between *Moricandia arvenis and B. oleracea*. *Plant Breed* 122:233–238.

Jander G, Cui J, Nhan B, Peirce NE, Ausubel Fm (2001) The TASTY locus on chromosome 1 of *Arabidopsis* affects feeding of the insect herbivore *trichoplusia ni*. *Plant Physiol* 126:890–898.

Jean M, Brown GG, Landry BS (1997) Genetic mapping of nuclear fertility restorer genes for the 'polima' cytoplasmic male sterility in canola (*Brassica napus* L.) using DNA markers. *Theor Appl Genet* 95:321–328.

Kirti PB, Narasimhulu SB, Prakash S, Chopra VL (1992) Production and characterization of intergeneric somatic hybrids of *Trachystoma ballii* and *Brassica juncea*. *Plant Cell Rep* 11:90–92.

Kirti PB, Prakash S, Gaikwad K, Dinesh Kumar V, Bhat SR, Chopra VL (1998) Chloroplast substitution overcomes leaf chlorosis in a *Moricandia arvensis*-based cytoplasmic male sterile *Brassica juncea*. *Theor Appl Genet* 97:1179–1182.

Kliebenstein DJ, Pedersen D, Barker B, Mitchell-Olds T (2002) Comparative analysis of quantitative trait loci controlling glucosinolates, myrosinase and insect resistance in *Arabidopsis thaliana*. *Genetics* 161:325–332.

Koornneef M, Alonso-Blanco C, Vreugdenhill D (2004) Naturally occurring genetic variation in *Arabidopis thaliana*. *Annu Rev Plant Biol* 55:141–172.

Krizek BA, Prost V, Joshi RM, Stoming, Glenn TC (2003) Developing transgenic *Arabidopsis* plants to be metal-specific bioindicators. *Environ Toxicol Chem* 22:175–181.

Lambrix V, Reichelt M, Mitchell-Olds T, Kliebenstein DJ, Gershenzon J (2001) The *Arabidopsis* epithiospecifier protein promotes the hydrolysis of glucosylonates to nitriles and influences *Trichoplusia ni* herbivory. *Plant Cell* 13:2793–2807.

LeDuc DL, Tarun AS, Montes-Bayon M, Meija J, Malit MF, Wu CP, AbdelSamie M, Chiang CY, Tagmount A, deSouza M, Neuhierl B, Böck A, Caruso J, Terry N (2004) Overexpression of selenocysteine methyltransferase in *Arabidopsis* and Indian mustard increases selenium tolerance and accumulation. *Plant Physiol* 135:377–383.

Li XB, Mao HZ, Bai YY (1995) Transgenic plants of rutabaga (*Brassica napobrassica*) tolerant to pest insects. *Plant Cell Rep* 15:97–101.

Li XB, Zheng SX, Dong WB, Chen GR, Mao HZ, Bai YY (1999) Insect-resistant transgenic plants of *Brassica napus* and analysis of resistance in the plants. *Acta Genet Sin* 26:262–268.

Liu JW, DeMichele S, Bergana M, Bobik E, Hastilow C, Chuang LT, Mukerji P, Huang YS (2001) Characterization of oil exhibiting high gamma-linolenic acid from a genetically transformed canola strain. *J Am Oil Chem Soc* 78:489–493.

Liu Clarke JH, Chevre AM, Landgren M, Glimelius K (1999) Characterization of sexual progenies of male-sterile somatic cybrids between *Brassica napus* and *Brassica tournefortii*. *Theor Appl Genet* 99:605–610.

Manzanares-Dauleux MJ, Delourme R, Baron F, Thomas G (2000) Mapping of one major gene and of QTLs involved in resistance to clubroot in *Brassica napus*. *Theor Appl Genet* 101:885–891.

Metz TD, Dixit R, Earle ED (1995a) *Agrobacterium tumefaciens*-mediated transformation of broccoli (*Brassica oleracea* var *italica*) and cabbage (*B. oleracea* var capitata). *Plant Cell Rep* 15:287–292.

Metz TD, Roush RT, Tang JD, Shelton AM, Earle ED (1995b). Transgenic broccoli expressing a *Bacillus thuringiensis* insecticidal crystal protein — Implications for pest resistance management strategies. *Mol Breed* 1:309–317.

Moloney MM, Walker JM, Sharma KK (1989) High efficiency transformation of *Brassica napus* using *Agrobacterium* vectors. *Plant Cell Rep* 8:238–242.

Moon HS, Halfhill MD, Good LL, Raymer PL, Stewart CN Jr. (2007) Characterization of directly transformed weedy *Brassica rapa* and introgressed *B. rapa* with Bt *cry1Ac* and *gfp* genes. *Plant Cell Rep* 26:1001–1010.

Narasimhulu SB, Kirti PB, Mohapatra T, Prakash S, Chopra VL (1992) Shoot regeneration in stem explants and its amenability to *Agrobacterium tumefaciens* mediated gene transfer in *Brassica carinata*. *Plant Cell Rep* 11:359–362.

Narasimhulu SB, Kirti PB, Bhatt SR, Prakash S, Chopra VL (1994) Intergeneric protoplast fusion between *B. carinata* and *Camelina sativa*. *Plant Cell Rep* 13:657–660.

Ono Y, Takahata Y, Kaizuma N (1994) Effect of genotype on shoot regeneration from cotyledonary explants of rapeseed (*Brassica napus* L.) *Plant Cell Rep* 14:13–17.

O'Neill Cm, Murata T, Morgan CL (1996) Expression of C3-C4 intermediate character in somatic hybrids between *Brassica napus* and C3-C4 species *Moricandia arvensis*. *Theor Appl Genet* 93:1234–1241.

Pelletier G, Bechtold N (2003) *In Planta* Transformation. In: Stewart, C.N. Jr. (Ed) *Transgenic Plants: Current Innovations and Future Trends*. Wymondham, UK: Horizon Scientific Press, p. 65–82.

Pfalz M, Vogel H, Mitchell-Olds T, Kroymann J (2007) Mapping of QTL for resistance against the crucifer specialist herbivore *Pieris brassicae* in a new *Arabidopsis* inbred line population, Da (1)-12 x Ei-2. PLoS ONE 2(6):e578.

Pilet ML, Delourme R, Foisset N, Renard M (1998) Identification of loci contributing to quantitative field resistance to blackleg disease, casual agent *Leptosphaeria maculans* (Desm.) Ces. et de Not., in winter rapeseed (*Brassica napus* L.). *Theor Appl Genet* 96:23–30.

Prakash S, Kirti PB, Bhat SR (1998) A *Moricanda arvensis*-based cytoplasmic male sterility and fertility restoration system in *Brassica juncea*. *Theor Appl Genet* 97:488–492.

Qiang W, Liu R, Meng J (2003) Genetic effects on biomass yield in interspecific hybrids between *B.napus* and *B. rapa*. *Euphytica* 134:9–15.

Qing CM, Fan L, Lei Y, Bouchez D, Tourneur C, Yan L, Robaglia, C (2000) Transformation of Pakchoi (*Brassica rapa* L. ssp *chinensis*) by *Agrobacterium* infiltration. *Mol Breed* 6:67–72.

Radke SE, Andrews BM, Moloney MM, Crouch ML, Krid JC, Knauf VC (1988) Transformation of *Brassica napus* L. using *Agrobacterium tumefaciens*: developmentally regulated expression of a reintroduced napin gene. *Theor Appl Genet* 75:685–694.

Radke SE, Turner JC, Facciotti, D (1992) Transformation and regeneration of *Brassica rapa* using *Agrobacterium tumefaciens*. *Plant Cell Rep* 11:499–505.

Ren JP, Dickson MH, Earle ED (2000) Improved resistance to bacterial soft rot by protoplast fusion between *Brassica rapa* and *B. oleracea*. *Theor Appl Genet* 100:810–819.

Sharma, KK, Bhojwani SS, Thorpe TA (1990) Factors affecting high frequency differentiation of shoots and roots from cotyledon explants of *Brassica juncea* (L.) Czern. *Plant Sci* 66:247–253.

Sigareva MA, Earle ED (1997) Direct transfer of a cold-tolerant Ogura male-sterile cytoplasm into cabbage (*Brassica oleracea* ssp. *capitata*) via protoplast fusion. *Theor Appl Genet* 94:213–220.

Skarzhinskaya M, Landgren M, Glimelius K (1996) Production of intertribal hybrids between *Brassica napus* L. and *Lesquerella fendleri* (Gray) Wats. *Theor Appl Genet* 93:1242–1250.

Snowdon RJ, Friedt W (2004) Molecular markers in *Brassica* oilseed: current status and future possibilities. *Plant Breed* 123:1–8.

Soengas P, Hand P, Vicente JG (2007) Identification of quantitative trait loci for resistance to *Xanthmonas campestris* pv. *Campestris* in *Brassica rapa*. *Theor App Genet* 114:637–645.

Song WY, Sohn EJ, Martinoia E, Lee YJ, Yang,YY, Jasinki M, Forestier C, Hwang I, Lee Y (2003) Engineering tolerance and accumulation of lead and cadmium in transgenic plants. *Nat Biotechnol* 21:914–919.

Stewart CN, Adang MJ, All JN, Raymer PL, Ramachandran S, Parrott WA (1996) Insect control and dosage effects in transgenic canola containing a synthetic *Bacillus thuringiensis cry1Ac* gene. *Plant Physiol* 112:115–120.

Stoutjesdijk PA, Hurlestone C, Singh SP, Green AG (2000). High-oleic acid Australian *Brassica napus* and *B. juncea* varieties produced by co-suppression of endogenous delta 12-desaturases. *Biochem Soc Trans* 28:938–940.

Stringham GR (1977) Regeneration in stem explants of haploid rapeseed (*Brassica napus* L.) *Plant Sci Lett* 9:115–119.

Thormann CE, Romero J, Mantet J. Osborn TC (1996) Mapping loci controlling the concentrations of erucic acid and linolenic acids in seed oil of *Brassica napus* L. *Theor Appl Genet* 93:282–286.

Wang WC, Menon G, Hansen G (2003) Development of a novel *Agrobacterium*-mediated transformation method to recover transgenic *Brassica napus* plants. *Plant Cell Rep* 22:274–281.

Waterer D, Lee S, Scoles G, Keller W (2000) Field evaluation of herbicide-resistant transgenic broccoli. *Hort Science* 35:930–932.

Westman AL, Kresovich S, Dickson MH (1999) Regional variation in *Brassica nigra* and other weedy crucifers for disease reaction to *Alternaria brassicicola* and *Xanthomonas campestris* pv. Campestris. *Euphytica* 106(3):253–259.

Williams PH, Hill CB (1986) Rapid cycling populations of *Brassica*. *Science* 232:1385–1389.

Yamagishi H, Landgren M, Forsberg J, Glimelius K (2002) Production of asymmetric hybrids between *Arabidopsis thaliana* and *Brassica napus* utilizing an efficient protoplast culture system. *Theor Appl Genet* 104:959–964.

Yan Z, Tian Z, Huang B, Huang R, Meng J (1999) Production of somatic hybrids between *Brassica oleracea* and the C3-C4 intermediate species *Morticandia nitens*. *Theor Appl Genet* 99:1281–1286.

Yang MZ, Jia SR, Pua EC (1991) High frequency of plant regeneration from hypocotyl explants of *Brassica carinata* A.Br. *Plant Cell Tiss Org Cult* 24:79–82.

Zhang Z, Ober JA, Kliebenstein DJ (2006) The gene controlling the quantitative trait locus *Epithiospecifier Modifieri* alters glucosinolate hydrolysis and insect resistance in *Arabidopsis*. *Plant Cell* 18:1524–1536.

Zhao J, Meng J (2003) Genetic analysis of loci associated with partial resistance to *Sclerotinia sclerotiorum* in rapeseed (*Brassica napus* L.). *Theor Appl Genet* 106:759–764.

Zhong R, Zhu F, Liu YL, Li SG, Kang LY, Luo P (1997) Oilseed rape transformation and the establishment of a bromoxynil-resistant transgenic oilseed rape. *Acta Bot. Sin.* 39:22–27.

Zhu YL, Pilon-Smits EAH, Tarun AS, Webber SU, Jouanin L, Terry N (1999a) Cadmium tolerance and accumulation in Indian mustard is enhanced by overexpressing ϒ-glutamylcysteine synthetase. *Plant Physiol* 121:1169–1177.

Zhu YL, Pilon-Smits EAH, Jouanin L, Terry N (1999b) Overexpression of glutathione synthetase in Indian mustard enhances cadmium accumulation and tolerance. *Plant Physiol* 119:73–79.

15 Microsporogenesis and Haploidy Breeding

Aditya Pratap, S.K. Gupta, and Y. Takahata

CONTENTS

INTRODUCTION

Haploids induced by *in vitro* culture of gametophytic cells, particularly male gametophytes, are of tremendous importance in crop breeding programs. A comprehensive utilization of doubled-haploid (DH) production has been involved in *Brassica* breeding and also in gene transfers, biochemical and physiological studies, and other manipulations (Palmer et al., 1996; Takahata, et al., 2005). DH breeding enables breeders to develop completely homozygous genotypes from heterozygous parents in a single generation and allows fixing the recombinant gametes directly as fertile homozygous lines. DH lines may be used for developing mapping populations for linkage maps using molecular markers, in addition to their use in mutation breeding and genetic engineering. Above all, *in vitro* screening for complex traits such as drought, cold, and salinity tolerance can be done during the culture process (Pratap et al., 2007; Figure 15.1). The advantages of DH technology have been long recognized by breeders globally and these have resulted in more than 280 varieties produced with the use of various haploid production methods in many crops with the further possibility of addition of several other varieties (Szarejko and Forster, 2006).

Oilseed *Brassica* are the world's third most important source of vegetable oils after palm and soybean, and contribute significantly to the economy of many countries (Gupta and Pratap, 2007a). In addition to improvements in the nutritional profile of the *Brassica* oil and its meal, conventional breeding in combination with modern biotechnological tools such as the DH technology has led to improvements in various agronomically important quantitative and qualitative characters in rapeseed. The first report on anther culture in *Brassica* was by Kameya and Hinata (1970), who described callus and haploid plants from cultured anthers of *B. oleracea*. This was later followed by microspore embryogenesis reports in *B. napus* (Thomas and Wenzel, 1975) and *B.*

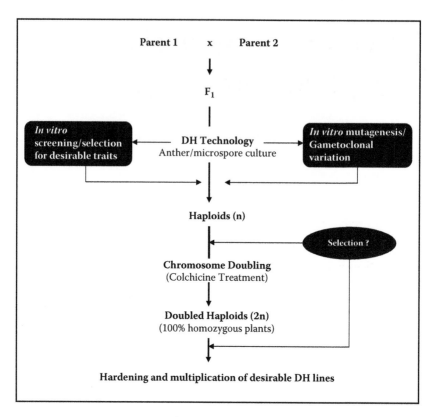

FIGURE 15.1 Pathway showing rapid development of doubled haploid (DH) plants in *Brassica* spp. through anther/microspore culture. Selection for desirable traits can be done at any stage of the process. (*Source:* From Pratap et al., 2007.)

campestris (Keller et al., 1975). After this, due to improvisation of the technique, haploids from several *Brassica* species were reported through anther culture (Jain et al., 1989; Keller and Armstrong, 1978, 1979; Sharma and Bhojwani, 1985). For DH production, the first success with the production of microspore-derived embryos from *Brassica* anthers was reported by Keller et al. (1975) and Thomas and Wenzel (1975), and later from microspores by Lichter (1982), which provided breeders with a new tool for breeding improved cultivars of rapeseed. However, in the earlier experiments, embryo production frequencies were low and sporadic. Since then, significant improvements have been made in the efficiency of both anther and microspore culture techniques, and have made these more reliable and more productive.

The microspore culture technique is relatively simpler and easier than that of anther culture and it has widespread applications in *Brassica* breeding due to its great efficiency in haploid and doubled haploid production, mutation and germplasm regeneration, and gene transformation (Xu et al., 2007). Microspores provide uniform, synchronous, and easily accessible populations of cells for understanding the basic biochemical and physiological aspects of embryogenesis (Kott, 1996; Palmer et al., 1996). Using the isolated microspore cultures, yields in excess of 150,000 embryoids per 100 anthers have been reported (Swanson et al., 1987). Siebel and Pauls (1989) found that microspore culture was ten times more effective than anther culture for embryo production in *B. napus*. Owing to its great potential, the microspore culture technique has been used to generate haploid and DH plants in many *Brassica* species (Takahata and Keller, 1991; Ferrie et al., 1995; Gu et al., 2003a, b, 2004 a, b; Zhang et al., 2003). Subsequently, extensive research has been carried out to investigate embryogenesis in anther- and microspore cultures and, as a result, DH technology has been developed to its present form (Charne and Beversdorf, 1988; Yu and Liu, 1995; Wang et al., 1999, 2002; Shi et al., 2002).

CULTURE TECHNIQUE

The technique of microspore culture is easier and simpler than that of anther culture (Takahata, 1997) in *Brassica*. For microspore isolation, buds of appropriate size, which contain responding microspores (late uninucleate to early binucleate stage), are used. These are surface sterilized in 0.5% sodium hypochlorite or 0.1% HgCl$_2$ solution prior to washing thoroughly in sterilized water. The appropriate developmental stage of microspores is determined by DAPI (4,6-diamidino-2-phenylindole) staining using fluorescence microscopy (Fan et al., 1988). The optimum bud size in different *Brassica* species depends on the growing condition, age, and genotype of the donor plant (Takahata et al., 1993). For release of microspores, the surface-sterilized buds are macerated in a beaker containing plant growth regulator free B5 medium, supplemented with 13% sucrose at pH 6.0 (B5-13), and are subsequently filtered through a 40- to −60-μm mesh filter into a centrifuge tube. For mass microspore isolation, a micro blender is reported to be useful (Swanson et al., 1987; Swanson, 1990). The filtered microspores are then washed thoroughly with B5-13 medium via centrifugation at 750 rpm for 5 minutes, followed by their suspension on modified Lichter's medium (Lichter, 1982) minus potato extract and plant growth regulators, and added with 13% sucrose at pH 6.0 (NLN-13) or ½ NLN medium at a density of 3–10 × 10^4/mL.

After the first step of microspore isolation, 2 to 3 mL of the suspension is plated on a Petri dish of appropriate size, followed by sealing with Parafilm and incubation in the dark at 32 to 33°C for 3 to 4 days in the case of *Brassica napus* (Chuong and Beversdorf, 1985; Pechan and Keller, 1988), and for 1 day in the case of *B. campestris* (Sato et al., 1989a, b) and *B. oleracea* (Takahata and Keller, 1991). Subsequently, the plated microspores are incubated at 25°C. For better results, medium refreshing can be done after 1 day of incubation following microspore isolation (Xu et al., 2007). For this, the suspension could be recentrifuged at 750 rpm for 5 minutes. The medium should then be changed with the same amount of fresh NLN medium, and cultured as described above (Gu et al., 2003 b, 2004a). Cell division starts within 2 days in the responding microspores; and after 2 to 3 weeks, different developmental stages of embryos are observed, such as globular to torpedo to cotyledonary stage (Figure 15.2). To regenerate plantlets, the large embryos (at late torpedo stage) are transferred to plant growth regulator free B5 or MS medium solidified with agar. Initially, these are cultured at low temperature (2°C) for about 10 days, followed by their incubation in a growth chamber at 24°C with a 14-hour day length regime. These could be further subcultured on solidified MS medium if required. Regeneration is observed within approximately 20 to 25 days. The developing shoots should be cut from any callus and hypocotyls tissues and subcultured on fresh solid MS medium.

Cotyledonary embryos are reported to convert into plantlets more easily than the earlier stage embryos (Takahata et al., 1991a). Transferring the embryos to filter paper placed on top of agar medium (Takahata and Keller, 1991) and/or to high concentrations of gelling agent (1.6% agar) (Peng et al., 1994) was reported to enhance the frequency of plant regeneration.

MECHANISM OF MICROSPORE EMBRYOGENESIS

Direct embryogenesis is usually observed in cultured microspores. For induction of embryogenesis, microspores at late uninucleate to early binucleate stage are most suitable. In pollen development *in vivo*, late uninucleate microspores divide asymmetrically to form a large vegetative cell and a small lens-shaped generative one (Figure 15.3A). This asymmetrical division of first pollen meiosis is considered important for the maintenance of gametophytic development. The generative cell then divides to form two sperm cells. However, several morphological changes were observed in microspores that were induced to enter in a sporophytic developmental pathway (Takahata, 1997). A

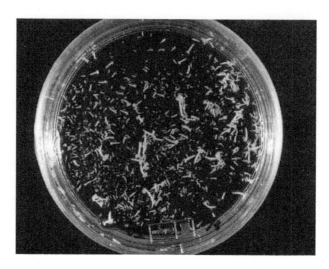

FIGURE 15.2   Embryos derived from microspores of *B. napus* cultivar Topas after 18 days of culture. (Source: From Takahata, 1997.)

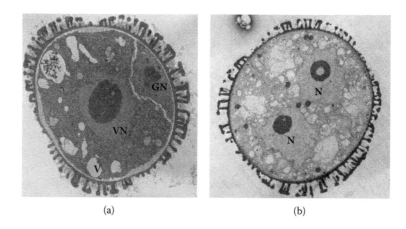

FIGURE 15.3   Transmission electron micrographs of *B. napus* microspores: (a) Early binucleate microspore in anther showing asymmetrical division, GN: Generative nucleus, VN: Vegetative nucleus, V: Vacuole. (b) Potentially embryogenic microspore *in vitro* culture showing symmetric division, N: Nucleus. (*Source:* From Takahata, 1997.)

remarkable change associated with embryo induction was symmetrical division of the uninucleate microspores (Figure 15.3B) (Fan et al., 1988; Zaki and Dickinson, 1990; Telmer et al., 1993). The disruption of asymmetrical division by colchicine, which suppresses the microtubular cytoskeleton involved in the asymmetrical positioning of the nucleus, increased the frequency of embryogenesis (Zaki and Dickinson, 1991; Iqbal et al., 1994). Hause et al. (1993) reported that the new arrangement of microtubules played a major role in inducing symmetrical division. The anti-microtubule activity of colchicine was also expected to change the localization of organelles. How a shift to sporophytic development pathway is caused by symmetric division is unknown, but the symmetry of division is believed to be a prerequisite in embryogenesis.

Prior to symmetrical division, several events associated with embryo induction were observed by Zaki and Dickinson (1990); these are: (1) nucleus migration from periphery to a central area, (2) the synthesis of starch in plastids, (3) development of a thick "somatic type cell wall," and (4) the appearance of a globular domain with the cytoplasm. Contrary to their findings of starch synthesis,

Telmer et al. (1993) reported that accumulation of starch was related to nonembryonic development. However, starch was synthesized in responsive microspores but the microspores accumulating excessive starch ceased to continue embryo development (Takahata, 1997).

A microspore wall of reticulosite structure is formed on the microspores undergoing embryogenesis, which is considered an obstacle to the developing embryo (Takahata, 1997). Scanning electron microscopy revealed two patterns of morphological changes in early embryo development (Nitta et al., 1997). In one pattern, the increase in cell volume was concentrated on the germination aperture, which is probably a weak structure and in the other, the enlargement of volume occurred over the entire surface of the microspore. Approximately 67% of microspores showing morphological changes were of the former pattern.

Increased temperature is a crucial factor to develop pollen into an embryo. Pechan et al. (1991) indicated that the alteration to embryogenic development occurred within 8 hours of microspore culture. Furthermore, from an analysis of proteins and *in vitro* translation products, a number of mRNAs and proteins were identified as unique in potentially embryogenic microspores. They suggested that some of them were heat shock proteins, which have an important role in cell proliferation. Similar results were obtained by Custers et al. (1994). It was considered that heat-shock proteins synthesized in microspores inhibited other cellular proteins associated with pollen development; while at the same time activating proteins involved in cell proliferation to induce embryogenesis (Pechan et al., 1991; Telmer et al., 1993).

Boutilier et al., (2002) reported that *BABY BOOM* gene, which was isolated from an early stage of microspore embryogenesis, plays a role in promoting cell proliferation and morphogenesis during embryogenesis. Recently, a large number of ESTs expressed in early microspore embryogenesis and/or microspore-derived embryos were identified in *B. napus* (Tsuwamoto et al., 2007; Malik et al., 2007; Tsuwamoto and Takahata, 2008). Genetic analysis of microspore embryogenesis in *B. napus* and *B. campestris* (syn. *B. rapa*) indicated that both additive and dominant effects contribute to embryogenesis with high heritability (Zhang and Takahata, 2001). DNA markers linked to the ability of microspore embryogenesis were reported in *B. napus* (Cloutier et al., 1995; Zhang et al., 2003) and *B. rapa* (Ajisaka et al., 1991).

FACTORS AFFECTING MICROSPORE EMBRYOGENESIS

DONOR PLANT GENOTYPE

Various factors could be responsible for microspore embryogenesis, including donor plant genotype and physiology, anther and microspore developmental stage, culture conditions and environment, culture media and pretreatments, etc. (Takahata et al., 1991b; Dunwell, 1996; Gu et al., 2003a, 2004c; Zhang et al., 2006; Pratap et al., 2007). Among these, genotype is considered the most important (Xu et al., 2007). The genotype effect is important not only in *Brassica*, but also in wheat, triticales, and other cereals (Pratap et al., 2006). The effect of genotypes has been described in all species of *Brassica* where development of haploids through microspore and anther culture has been attempted (Arnison and Keller, 1990; Baillie et al., 1992; Thurling and Chay, 1984; Wang et al., 2004). Although most genotypes respond, large genotypic variations in embryo yield have been reported in *B. napus* (Chuong and Beversdorf, 1985; Chuong et al., 1988a, b; Ohkawa et al., 1987; Gland-Zwerger et al., 1988; Hansen and Sivnnset, 1990); *B. campestris* (Kuginuki et al., 1990; Burnett et al., 1992; Baillie et al., 1992; Cao et al., 1994); *B. oleracea* (Cao et al., 1990; Takahata and Keller, 1991; Duijis et al., 1992); and *B. juncea* (Hiramatsu et al., 1994). Even plant-to-plant variation within a genotype has also been reported in many anther- and microspore culture experiments (Phippen and Ockendon, 1990; Seguin-Scwartz et al., 1983). Significant differences with respect to embryogenesis have also been reported among the winter and spring genotypes. Winter types were reported to have a higher response than the spring types in *B. napus* anther culture experiments (Keller et al., 1987). However, Ohkawa et al. (1987) reported that all high embryogenic genotypes in their microspore culture experiments on 96 genotypes of *B. napus* were spring type. Similar results were obtained in *B. campestris* (Kuginuki et al., 1990) and *B. juncea* (Hiramatsu et al., 1994).

Zhang and Takahata (2001) examined the inheritance of microspore embryogenic ability by diallel analysis using cultivars showing different response in *B. napus* and *B. campestris*. They indicated that genetic factors estimated for microspore embryogenic ability were similar between both species, that is (1) dominant genes had positive effects on microspore embryogenic ability, (2) both additive and dominant genetic effects were significant, (3) no significant maternal effects were observed, and (4) the heritability was high, especially broad-sense heritability was higher than 0.9. This suggests that microspore culture response can be improved genetically in brassicas.

Furthermore, plant growth conditions and environment also exert a significant effect on success. Field-grown plants are more responsive in comparison to glasshouse-grown plants, suggesting that some genotypes have specific environmental requirements to produce better results with respect to embryogenic microspores. The situation can be further complicated by genotype/environment interactions (Dunwell et al., 1985; Roulund et al., 1990; Thurling and Chay, 1984). It also is well demonstrated that plants grown in growth chambers or at low temperatures yield the most responsive microspores (Dunwell et al., 1985).

Donor Plant Condition

The genotype and growth and physiological conditions of donor plants play a vital role in *in vitro* embryogenesis and regeneration. In general, the plants grown under controlled conditions such as in glasshouse, polyhouse, or growth chamber appear to respond better in comparison to field-grown plants. However, somatic embryo yield was twice as high from plants grown in the growth chamber compared with those grown in the greenhouse (Guan, 1995). This is because flower control is better in a growth chamber and manual control is easy to exercise each day. Anther culture response is associated with high genotype dependency, which sometimes becomes a major constraint in the success of nonresponsive genotypes. A significant genotypic effect was observed for microspore embryogenesis by Prem et al. (2004). Genotype-dependent response was reported previously by Lichter (1989).

The prevailing temperature during the growth of donor plants is reported to play a crucial role in microspore embryogenesis. Embryo induction was not observed in microspores derived from *Brassica napus* plants grown at 25 to 30°C in summer. However, when these cultivars were grown at 15 to 20°C in the autumn, embryos developed from the microspores of the spring cultivars on B5 liquid medium (Aslani et al., 2005). A high frequency of embryogenesis was consistently obtained in donor plants grown at low temperatures (Keller et al., 1987; Dunwell et al., 1985). The effect of low temperature is thought to increase the number of microspores suitable for embryogenesis due to slow pollen development, and it also prolongs the duration for which culturable microspores are available in a crop.

Embryo yield was closely correlated with the age of the donor plant in both medium and low responsive material but was not correlated with the age of highly responsive donor plants. In addition, the most suitable stage for induction of embryogenesis occurred when the anther color was yellowish-green (Chunyun and Guan, 2005). In another study, Yang et al. (2005) observed that the optimal time for isolating microspores was 7 days before and after initial flowering. Although effective embryogenesis was obtained from buds of the main raceme taken from young plants in the anther culture of many plants (Maheshwari et al., 1980), the age of the plants and bud position (main vs. lateral raceme) were observed to have no influence on embryo production in *Brassica napus* microspore culture if selected at the appropriate stage of microspore development (Chuong et al., 1988; Takahata et al., 1991; Takahata and Keller, 1991). However, Burnett et al. (1992) observed that better embryo production was obtained from older plants in *B. campestris*. Similarly, Guo and Pulli (1996) also recommended older (>5 weeks) plants for more success in regeneration.

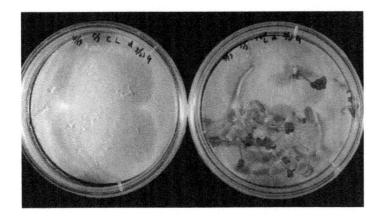

FIGURE 15.4 Plant regeneration from desiccated embryos of *Brassica napus* cv. Topas 4 weeks after rehydration. Embryos were treated with 0 (left) and 100 μM (right) ABA. (*Source:* From Takahata, 1997)

Microspore Stage

The high success of embryogenesis is related to the age of microspores and pollen development. The most responsive pollen stage may range between early uninucleate to late binucleate stage, although late uninucleate stage is considered the best in *Brassica napus* (Kott et al., 1988). Late uninucleate stage is also considered the best in cereal crops such as wheat and triticale (Pratap et al., 2006). However, in some studies, the mid-late to very late uninucleate stage is also considered the optimum timing for microspore culture (Guo and Pulli, 1996). Cytological studies in *B. napus* indicated that embryos were induced from the microspores of late uninucleate to early binucleate stage, and the late uninucleate exhibited higher response (Fan et al., 1988; Kott et al., 1988). However, the highest embryo yield was obtained from microspore populations containing both uninucleate and binucleate stage microspores in *B. napus* (Telmer et al., 1992) and *B. oleracea* (Duijis et al., 1992; Takahata et al., 1993). Takahata et al. (1991) reported higher frequencies of embryogenesis in *B. napus* with older inflorescences in comparison to younger ones. Guo and Pulli (1996) recommended older (>5 weeks) plants and a cold pretreatment for greater success. Embryo yield was closely correlated with the age of donor plant in both medium- and low-responsive material but was not correlated with the age of highly responsive donor plants. The most appropriate stage of anthers for induction of embryogenesis was when anther color was yellowish-green.

It has also been observed that partial desiccation of the embryos could also increase the rate of germinated embryos and plantlet development, and drying of embryos for 1 or 2 days could significantly improve the development of plants from microspore-derived embryos (Au et al., 2007). However, desiccation of embryos beyond 4 days led to the loss of viability and death (Zhang et al., 2006).

Culture Medium

Culture medium is one of the most crucial factors for plantlet regeneration from anthers in any crop. A wide range of culture media has been adopted in *Brassica* anther- and microspore cultures with varying degrees of success. However, Lichter's medium (1982) with slight modifications of lacking potato extract and growth regulators (NLN medium) has been widely accepted for *Brassica* microspore culture. Gland-Zwerger (1995) observed that ½ MS (Murashige and Skoog, 1962) and ½ B5 (Gamborg et al., 1968) media had a better effect on plant regeneration than full MS or B5 medium. Exogenous plant growth regulators are not required in the embryogenesis of *Brassica* (Keller et al., 1987). The inhibitory effect was observed in *B. napus* (Keller et al., 1987) and *B. campestris* (Sato et al., 1989b). On the other hand, low concentrations of BAP were optionally used to support normal embryo development (Swanson, 1990; Cao et al., 1994) and to increase embryo yield (Charne and

Beversdorf, 1988; Montfort et al., 1994).Tang et al. (2003) observed that benzylaminopurine (BAP) was helpful for the growth and development of shoots of *Brassicas*. Zhang et al. (2006) also found the ½ MS medium with 2.0 mg/L BAP added to it to be the best. Abscisic acid (ABA) was also reported to increase plant development (Hansen, 2000).

PRETREATMENT

Pretreatment of the plants, inflorescences, or buds before the culture of anthers or microspores could significantly affect microspore embryogenesis. Although the exact explanation for how pretreatment helps in improving embryogenesis is still unknown, it is believed to trigger embryogenesis and enhance the regeneration rate through osmotic stress and the formation of bicellular stage microspores. Pretreatment is found effective not only in anther and microspore culture, but also in embryo and ovule culture of wheat and triticale (Sethi et al., 2003; Pratap et al., 2006). A cold pretreatment at 4°C for 24 hours of cultured haploid embryos in wheat and triticale significantly enhances the regeneration rate (Pratap et al., 2006). In *Brassica* microspore culture, a short heat shock or cold shock treatment is reported to be effective to stimulate microspore embryogenesis. High-temperature (30 to 35°C) pretreatment for 1 to 3 days of the culture before transferring to 25°C is also reported to markedly stimulate embryogenesis in *Brassica* species (Constantine et al., 1996; Takahata, 1997). Although 32 or 33°C is used as the optimal temperature in many studies, the exposure period depends on the species (Takahata, 1997).

Treatment for 3 to 4 days brought about effective embryo production in *Brassica napus* (Chuong and Beversdorf, 1985; Pechan and Keller, 1988), *B. carinata* (Chuong and Beversdorf, 1985), and *B. juncea* (Ohkawa et al., 1988), while 1-day exposure was optimum in *B. campestris* (Sato et al., 1989a, b; Burnett et al., 1992; Cao et al., 1994) and *B. oleracea* (Takahata and Keller, 1991). Other pretreatments include decreased atmospheric pressure (Klimaszewska and Keller, 1983), γ-irradiation (MacDonald et al., 1988), colchicine treatment (Zaki and Dickinson, 1991), and ethanol stress (Pechan and Keller, 1989).

Cold treatment has been followed successfully in many studies, specifically in cereals (Constantine et al., 1996; Pratap et al., 2005, 2006), although compared to heat shock stress, cold stress pretreatment has been less frequently used in *Brassicas*. Sato et al. (2002) observed that in the microspore cultures of *Brassica rapa*, low-temperature treatment of buds or inflorescences distinctly improved the ability of microspore embryogenesis. In many crops, including *B. rapa*, cold pretreatment increases the percentage of bicellular stage microspores with two equal nuclei, which is supposed to be the necessary developmental stage for the induction of embryogenesis. It has been suggested that the induction effect of osmotic stress under cold pretreatment could be the reason behind increased microspore embryogenesis. Further, prolonged cold pretreatment of flower buds also showed a promoting effect, suggesting that the flower buds or inflorescences could be stored for fixed periods of time under low temperatures and could be used in the off-season without any loss in regeneration capacity. It has also been suggested that cold pretreatment could be combined with other treatments such as heat shock to increase the efficiency of microspore embryogenesis (Pechan and Smykal, 2001). A one to several-fold increase in microspore embryogenesis was observed in *B. napus* with appropriate duration of cold treatment at 2 to 4 days, while it was less effective in *B. rapa* and not effective in *B. oleracea* (Gu et al., 2003c). In general, it has been observed that cold pretreatment significantly enhances microspore embryogenesis in *B. napus,* which may be one to several-fold. However, it is comparatively less effective in *B. rapa* and *B. oleracea* (Xu et al., 2007).

High-temperature pretreatment disrupts the normal integrated development of somatic anther tissue and subsequently may synchronize the physiological status of the tissues, thereby stimulating the induction process (Dunwell et al., 1985). However, thermal shock pretreatment on the callogenesis process appears to be genotype dependent. Burbulis et al. (2004) observed that the frequency of embryo formation from anthers treated at high temperature was higher than the control. In some experiments, colchicine was also tried *in vitro* by adding it to the induction medium with freshly isolated microspores within the first 3 days of culture and it reportedly produced better embryogenesis (Iqbal et al., 1994).

Agarwal et al. (2006) observed that high-temperature treatment coupled with an antiauxin PCIB (*p*-chlorophenoxyisobutyric acid) treatment significantly enhanced microspore embryogenesis in *Brassica juncea* as compared to colchicine. An increase in temperature from 32 to 35°C increased embryogenesis by tenfold. However, they obtained highest embryogenesis when PCIB was added to the culture at 35°C 1 day after its initiation. PCIB (20 µM) could enhance microspore embryogenesis by fivefold. The effects of PCIB and other hormones on zygote embryo development in *B. juncea* have been observed earlier (Hadfi et al., 1998). In fact, PCIB acts as an auxin antagonist by inhibiting auxin action (Kim et al., 2000; Forster et al., 1995).

Colchicine Application

Colchicine has been reported to enhance microspore embryogenesis. Colchicine at high concentrations (>10 mg/L) for 24 hours proved convenient for direct recovery of haploid plants as observed by Agarwal et al. (2006). Colchicine can be applied at any stage during the culture process of *Brassicas*, that is, from isolated microspores to the regenerated plants. Short-term exposure of microspores to colchicine significantly increases symmetric cell division and ultimately leads to increased frequency of embryogenesis, the optimal time being the first 12 to 15 hours after microspore isolation. Colchicine treatment (50 and 500 mg/L) for the first 15 hours of isolated microspores in culture stimulated embryogenesis and produced a large number of healthy-looking embryos (Zhou et al., 2002a, b). These normal embryos germinated well at 24°C after being transferred to a solid regeneration medium and an exposed to initial period of low temperature (2°C) for 10 days, and could directly and rapidly regenerate into vigorous plants. Colchicine added *in vitro* to the induction medium with freshly isolated rapeseed microspores within the first 3 days of culture also produced improved frequencies of embryogenesis (Zaki and Dickinson, 1991).

The right combination of colchicine concentration and treatment duration is more important for increased embryogenesis and diploidization. Zhou et al. (2000) observed that in microspore cultures of *Brassica napus*, efficient embryogenesis and diploidization could be achieved by immediately treating the cultures with colchicine. Further, high doubling efficiency was obtained from 500-mg/L colchicine treatment for 5 hours with a low frequency of polyploids and chimeric plants. Also, increased duration and colchicine concentration exerted a negative impact on embryogenesis and chromosome doubling.

DIPLOIDIZATION

Several species, including some of the cereals and *Brassicas*, have an inherent tendency toward spontaneous diploidization in varying degrees. In addition to haploids, diploids and sometimes polyploids are also regenerated spontaneously from isolated microspores. Spontaneous diploidization is advantageous for haploid breeding because it omits the need for doubling treatments, thereby saving considerable time and resources. Gu et al. (2003) reported a high frequency of spontaneous doubled haploids without any subculture or pretreatment with colchicine. In *Brassica rapa* ssp. *oleifera*, about 30% spontaneous DHs were reported in comparison to 5 to 50% in *B. napus* (Möllers et al., 1994; Zhou et al., 2002a, b). Siebel and Pauls (1989) reported that 80% of the regenerants were haploid in the F_1 hybrid of Regent × Golden in *B. napus*, while the haploid-diploid ratio was nearly equal in the F_1 of Reston × G231. The frequencies of diploids in *B. campestris* seem to be higher than those of *B. napus* and *B. oleracea*. The exact mechanism of diploidization in microspore culture is still unknown. Keller et al. (1987) opined that diploidization and polyploidization could be due to nuclear fusion and endomitosis during the early stage of embryogenesis. However, spontaneous diploidization is highly erratic in *Brassica* and cannot be fully relied upon for successful crop improvement plans.

Methods have been devised that can induce up to 100% diploidization using chemicals such as colchicine. Chromosome doubling of haploid plants is generally achieved in *Brassica* haploids by immersing the roots of haploid plantlets in 0.2% colchicine solution for 2 to 6 hours (Huang and Keller, 1989; Swanson, 1990). Between 80 and 90% diploid embryos have been obtained by *in vitro* treatment of microspores with 50 mg/L colchicine for 24 hours by Möllers et al. (1994). In *B. napus*, 70% to 90% diploidization has been reported as a result of colchicine addition immediately after culture for a short duration (Zhou et al., 2002a, b).

In comparison to colchicine treatment of microspore-derived plants and embryos, immediate colchicine treatment of isolated microspores is reported to result in increased embryogenesis and diploidization (Zhou and Hagberg, 2000). This also leads to low chimeric percentages. In addition, the duration of colchicine treatment is also a crucial factor in diploidization and should always be correlated with the concentration used. In *Brassica juncea* anther culture, spontaneous diploidization has been reported to the extent of about 20%. However, this could be increased up to 43% through colchicine treatment. According to them, longer treatment (for 24 hours) with colchicine improved the diploidization rate but significantly reduced the embryogenic frequency. Zhou et al. (2002a) found that a longer treatment (30 hours) of *B. napus* with colchicine reduced embryo development, although a 15-hour treatment proved best.

It has been observed that the treatment of regenerated haploid plants with colchicine is associated with several problems, including the development of chimeras with small and uneven sectors of diploidization, delayed plant growth and development, reduced seed production, and generation of colchicine waste, which is an expensive chemical and also requires specialized handling. However, colchicine treatment of isolated microspores is free of these problems and is also more efficient, reliable, safer, and quicker. Application of some other chemicals, such as 1 to 10 μM trifluralin, during the first 18 hours of microspore culture (Zhao and Simmonds, 1995) was also reported to be effective for doubled haploid production.

CONCLUSION

Microspore culture is undoubtedly the most reliable and fastest approach for developing homozygous populations in the *Brassica*. However, the problems associated with microspore embryogenesis, such as low regeneration rate, genotype-specific response, and low recovery of DH plants, could limit the application of this technology as a routine breeding tool for *Brassica* improvement programs, particularly in those involving nonresponsive genotypes.

In the rapidly changing global vegetable oils scenario, the focus of *Brassica* researchers has shifted toward quality improvement and specialty products. Newer and more practical goals in rapeseed breeding — such as the development of herbicide-tolerant varieties, development of male-sterile and fertility restorer lines for use in hybrid seed production, modification of the fatty acid composition to suit industrial and domestic needs, development of insect-pest–resistant varieties, improvement in the feeding quality of the meal, and drug production (Gupta and Pratap, 2007b) — are now gaining increased importance. The microspore embryogenesis technique is comparatively simpler and more reliable among other methods of DH production in *Brassicas*, and it could be employed in tandem with conventional breeding approaches. Advancements in DH technology, particularly microspore embryogenesis, are required in such a way that a maximum number of DH plants per cycle are obtained with desirable yield and quality characteristics. Specialized efforts are required toward increasing the efficiency and utility of microspore culture techniques by minimizing genotype effects, *in vitro* selection for desirable traits, incorporation of molecular markers, partial desiccation of embryos, directed *in vitro* mutagenesis, modification or replacement of culture media, and modification in colchicine application methods. Because the ability of microspore embryogenesis is considered to be controlled by nuclear gene(s), embryogenic ability could be transferred from high-responsive to low-responsive or unresponsive genotypes to increase the efficiency of microspore culture protocols. A synergy of conventional and biotechnological approaches will definitely unleash the power of *Brassica* improvement programs for global food security.

REFERENCES

Agarwal, P.K., Agarwal, P., Custers, J.B.M., Liu, C.M., and Bhojwani, S.S. 2006. PCIB an antiauxin enhances microspore embryogenesis in microspore culture of *Brassica juncea*. *Plant Cell Tiss. Organ Cult.* 86:210–210.

Ajisaka, H., Kuginuki, Y., Shiratori, M., Ishiguro, K., Enomoto, S., and Hirai, M. 1999. Mapping of loci affecting the cultural efficiency of microspore culture of *Brassica rapa* L. syn. *campestris* L. using DNA polymorphism. *Breed. Sci.*, 49:187–192.

Arnison, P.G. and Keller, W.A. 1990. A survey of anther culture response of *Brassica oleracea* L. cultivars grown under field conditions. *Plant Breed.*, 104:125–133.

Aslani, F., Mozaffari, J., Ghannadha, M.R., and Attari, A.A. 2005. Microspore culture for producing haploid plants in different rapeseed (*Brassica napus*) cultivars. *Iranian J. Agric. Sci.*, 36:331–339.

Baillie, A.M.R., Epp, D.J., Hutcheson D., and Keller, W.A. 1992. *In vitro* culture of isolated microspores and regeneration of plants in *Brassica campestris*. *Plant Cell Rep.*, 11:234–237.

Boutilier, K., Offringa, R., Sharma, V.K., Kieft, H., Ouellet, T. Zhang, L., Hattori, J., Liu, C., and Lammeren, A.A.M. 2002. Extopic expression of *BABY BOOM* triggers a conversion from vegetative to embryonic growth. *Plant Cell*, 14:1737–1749.

Burbulis, N, Kupriene, R., and Zilenaite, L. 2004. Embryogenesis, callogenesis and plant regeneration from anther cultures of spring rape (*Brassica napus* L.) *Acta Universitasis, Biology,* 676:153–158.

Burnett, L., Yarrow, S., and Huang, B. 1992. Embryogenesis and plant regeneration from isolated microspores of *Brassica rapa* ssp. *oleifera*. *Plant Cell Rep.*, 11:215–218.

Cao, M.Q., Charlot, F., and Dore, C. 1990. Embryogenesis and plant regeneration of sauerkraut cabbage (*Brassica oleracea* L. ssp. *capitata*) via *in vitro* isolated microspore culture. *C.R. Acad. Sci., Paris,* 310, Series III:203–209.

Cao, M.Q., Li, Y., Lire, F., and Dore, C. 1994. Embryogenesis and plant regeneration of pakchoi (*Brassica rapa* ssp. *chinensis*) via *in vitro* isolated microspore culture. *Plant Cell Rep.*, 13:447–450.

Charne, D.G. and Beversdorf, W.D. 1988. Improving microspore culture as a rapeseed breeding tool: the use of auxins and cytokinins in an induction medium. *Can. J. Bot.*, 66:1671–1675.

Chunyun, L.X.G and Guan, M. 2005. Pollen fertility, anther culture, transfer embryo mode and isolated microspore culture of F_1 hybrid between *Brassica napus* and *Brassica juncea*. *Brassica,* 7:39–45.

Chuong, P.V. and Beversdorf, W.D. 1985. High frequency embryogenesis through isolated microspore culture in *Brassica napus* L. and *B. carinata* Braun. *Plant Sci.*, 39:219–226.

Chuong, P.V., Deslauriers, C. Kott, K.S., and Beversdorf, W.D. 1988b. Effects of donor genotype and bud sampling on microspore culture of *Brassica napus*. *Can. J. Bot.*, 66:1653–1657.

Chuong, P.V., Pauls, K.P., and Beversdorf, W.D. 1988a. High frequency embryogenesis in male sterile plants of *Brassica napus* through microspore culture. *Can. J. Bot.*, 66:1676–1680.

Cloutier, S., Cappadocia, M., and Landry, B.S. 1995. Study of microspore culture responsiveness in oilseed rape (*Brassica napus* L.) by comparative mapping of a F_2 population and two microspore-derived populations. *Theor. Appl. Genet.*, 91:841–847.

Constantine, E.P., Keller, W.A., and Arnison, P.G. 1996. Experimental haploidy in *Brassica* species, p. 143–172. In: Jain, S.M., Sopory, S.K., and Veilleux, R.E. (Eds.), *In vitro Haploid Production of Higher Plants*, Vol. 2, Kluwer Academic Publishers, Dordrecht, the Netherlands.

Custers, J.B.M., Cordewener, J.H.G., Hause, G., van Lammeren, A.A.M., Dons, J.J.M., and van Lookeren Campagne, M.M. 1994. Regulation of the inductive phase of microspore embryogenesis in *Brassica napus*. *Abstracts of ISHS Symposium on Brassicas, Ninth Crucifer Genetics Workshop,* Lisbon.

Duijis, J.G., Voorrips, R.E., Visser, D.L., and Custers, J.B.M. 1992. Microspore culture is successful in most crop types of *Brassica oleracea* L. *Euphytica*, 60:45–55.

Dunwell, J.M. 1996. Microspore culture. In: Jain S.M., Sopory S.K., and Veilleux R.E. (Eds.), *In Vitro Haploid Production in Higher Plants*, Vol 1, p. 205–216. Kluwer Academic Publishers, Dordrecht, the Netherlands.

Dunwell, J.M., Cornish, L.M., and Decourcel, A.G.L. 1985. Influence of genotype, plant growth, temperature and anther incubation temperature on microspore embryo production in *Brassica napus* ssp. *Oleifera*. *J. Exp. Bot.*, 36:679–689.

Fan, Z., Armstrong, K.C., and Keller, W.A. 1988. Development of microspores *in vivo* and *in vitro* in *Brassica napus* L. *Protoplasma*, 147:191–199.

Ferrie, A.M.R., Epp, D.J., and Keller, W.A. 1995. Evaluation of *Brassica rapa* L. genotypes for microspore culture response and identification of a highly embryogenic line. *Plant Cell Rep.*, 14:580–584.

Forster, R.J., McRae, D.H., and Bonner, J. 1995. Auxin-antiauxin interaction at high auxin concentrations. *Plant Physiol.*, 30:323–327.

Gamborg, O.L., Miller, R.A., and Ojima, K. 1968. Nutrient requirements of suspension cultures of soybean root cells. *Exp. Cell Res.*, 50:151–158.

Gland-Zwerger, A. 1995. Culture conditions affecting induction and regeneration in isolated microspore cultures of different *Brassica* species. In: *GCIRC Proc. 9th International Rapeseed Congr.*, p. 79–801. GCIRC, Cambridge, U.K.

Gland-Zwerger, A., Lichter, R., and Schweiger, H.G. 1988. Genetic and exogenous factors affecting embryogenesis in isolated microspore cultures of *Brassica napus* L. *J. Plant Physiol.*, 132:613–617.

Gu, H.H., Hagberg, P., and Zhou, W.J. 2004c. Cold pretreatment enhances microspore embryogenesis in oilseed rape (*Brassica napus* L.). *Plant Growth Regulation*, 42:137–143.

Gu, H.H., Tang, G.X., Zhang, G.Q., and Zhou, W.J. 2004a. Embryogenesis and plant regeneration from isolated microspores of winter cauliflower (*Brassica oleracea* var. *botrytis*). *J. Zhejiang University (Agricultural Life Science)*, 30:34–38.

Gu, H.H., Zh, J.Q., Zhang, G.Q., and Zhou, W.J. 2003c. Microspore culture and ploidy identification of regenerated plant in Chinese flowering cabbage (*Brassica rapa* ssp. *parachinensis*). *J. Agric. Biotechnol.*, 11:572–576.

Gu, H.H., Zhang, D.Q., and Zhou, W.J. 2004b. Effects of medium renovation and colchicine treatment on embryogenesis of isolated microspores in *Brassica rapa* ssp. *chinensis*. *Acta Agron. Sin.*, 30:78–81.

Gu, H.H., Zhang, D.Q., Zhang, G.Q., and Zhou, W.J. 2003a. Advances on *in vitro* microspore technology for mutation breeding in rapeseed. *Acta Agric. Zhejiangen*, 15:318–322.

Gu, H.H., Zhou, W.J., and Hagberg, P. 2003b. High frequency spontaneous production of doubled haploid plants from microspore culture in *Brassica rapa* ssp. *chinensis*. *Euphytica*, 134:239–245.

Guan, C.Y. 1995. Studies of microspore culture and doubled haploid breeding in rape. I. Effect of donor plant and microspore density on microspore culture in Swede rape. *Acta Agron. Sin.*, 21:665–670.

Guo, Y.D. and Pulli, S. 1996. High-frequency embryogenesis in *Brassica campestris* microspore culture. *Plant Cell Tiss. Organ Cult.*, 46:219–225.

Gupta, S.K. and Pratap, A. 2007a. History, Origin and Evolution, p. 1–20. In: Gupta, S.K. (Ed.), *Advances in Botanical Research-Rapeseed Breeding*, Vol. 45. Academic Press, London.

Gupta, S.K. and Pratap, A. 2007b. Recent trends in oilseed Brassicas, p. 284–299. In: Nayyar, H. (Ed.), *Crop Improvement: Challenges and Strategies*. I.K. International, New Delhi, India.

Hadfi, K., Speth, V., and Neuhaus, G. 1998. Auxin-induced developmental patterns in *Brassica juncea* embryos. *Development*, 125:879–887.

Hansen, M. 2000. ABA treatment and desiccation of microspore-derived embryos of cabbage (*Brassica oleracea* ssp. *capitata* L.) improves plant development. *J. Plant Physiol.*, 156:164–167.

Hansen, M. and Svinnset, K. 1990. Microspore culture of swede (*Brassica napus* ssp. *rapifera*) and the effects of fresh and conditioned media. *Plant Cell Rep.*, 12:496–500.

Hause, B., Hause, G., Pechan, P., and van Lammeren, A.A.M. 1993. Cytoskeletal changes and induction of embryogenesis in microspore and pollen cultures of *Brassica napus* L. *Cell Biol. Int.*, 16:153–168.

Hiramatsu, M., Odahara, K., and Matsu, Y. 1994. A survey of microspore embryogenesis in leaf mustard (*Brassica juncea*). *Abstr. of XXIVth Int. Hort. Congr.*, Kyoto, Japan.

Huang, B. and Keller, W.A. 1989. Microspore culture technology. *J. Tiss. Cult. Meth.*, 12:171–178.

Iqbal, M.C.M., Möllers, C., and Röbbelen, G. 1994. Increased embryogenesis after colchicine treatment of microspore cultures of *Brassica napus* L. *J. Plant Physiol.*, 143:222–226.

Jain, R.K., Brune, N., and Friedt, W. 1989. Plant regeneration from *in vitro* cultures of cotyledon explants and anthers of *Sinapis alba* and its implications on breeding of crucifers. *Euphytica*, 43:153–163.

Kameya, T. and Hinata, K. 1970. Induction of haploid plants from pollen grains of *Brassica*. *Jap. J. Breed.*, 20:82–87.

Keller, W.A. and Armstrong, K.C. 1978. High frequency production of microspore derived plants from *Brassica napus* anther cultures. *Z. Pflanzenzchtg.*, 80:100–108.

Keller, W.A. and Armstrong, K.C. 1979. Stimulation and embryogenesis and haploid production in *Brassica campestris* anther cultures by elevated temperature treatments. *Theor. Appl. Genet.*, 55:65–67.

Keller, W.A., Arnison, P.G., and Cardy, B.J. 1987. Haploids from gametophytic cells: recent developments and future prospects. In:*Plant Tissue and Cell Culture*, p. 233–241. Allan R. Liss, New York.

Keller, W.A., Rajhathy, R., and Lacapra, J. 1975. *In vitro* production of plants from pollen in *Brassica campestris*. *Can. J. Genet. Cytol.,* 17:655–666.

Kim, S.K., Chang, S.C., Lee, E.J., Chung, W.S., Kim, Y.S., Huang, S., and Lee, J.S. 2000. Involvement of brassinosteroids in the gravitropic response of primary root of maize. *Plant Physiol.,* 123:997–1004.

Klimaszewska, K. and Keller, W.A. 1983. The production of haploids production from *Brassica hirta* Moench (*Sinapis alba*) anther cultures. *Z. Pflanzenphysiol.,* 109:235–241.

Kott, L.S. 1996. Production of mutants using the rapeseed doubled haploid system. In: "Induced Mutation and Molecular Techniques for Crop Improvement. IAEA/FAO Proc. Int. Symp. Use of Induced Mutations and Molecular Techniques for Crop Improvement," p. 505–515, Vienna, Austria.

Kott, L.S., Polsoni, L., and Beversdorf, W.D. 1988. Cytological aspects of isolated microspore culture of *Brassica napus*. *Can. J. Bot.,* 66:1658–1664.

Kuginuki, Y., Nakamura, K., and Yoshikawa, H. 1990. Microspore culture in Chinese cabbage (*Brassica rapa* L. ssp. *pekinensis*). 1. Varietal differences in embryogenic ability. *J. Japan Soc. Hort. Sci.,* 60(Suppl. 2):212–213.

Lichter, R. 1982. Induction of haploid plants from isolated pollen of *Brassica napus*. *Z. Pflanzenphysiol.,* 103:229–237.

Lichter, R. 1989. Efficient yield of embryoids by culture of isolated microspores of different *Brassicaceae* species. *Plant. Breed.,* 103:119–123.

McDonald, M.V., Newsholme, D.M., and Ingram, D.S. 1988. The biological effects of gamma irradiation on secondary embryoids of *Brassica napus* ssp. *oleifera* Metzg. (Sinsk) winter oilseed rape. *New Phytologist,* 110:255–259.

Maheshwari, S.C., Tyagi, A.K., and Malhotra, K. 1980. Induction of haploidy from pollen grains in angiosperms-the current status. *Theor. Appl. Genet.,* 58:193–206.

Malik, M.R., Wang, F., Dirpaul, J.M., Zhou, N., Polowick, P.L., Ferrie, A.M.R., and Krochko, J.E. 2007. Transcript profiling and identification of molecular markers for early microspore embryogenesis in *Brassica napus*. *Plant Physiol.,* 144:134–154.

Möllers, C., Iqbal, M.C.M, and Röbbelen, G. 1994. Efficient production of doubled haploid *Brassica napus* plants by colchicine treatment of microspores. *Euphytica,* 75:95–104.

Montfoort, J., Brochec, J., Chatelet, P., and Herve, Y. 1994. Effects of cytokinins in microspore cultures of *Brassica oleracea* L. concar *botrytis* (L.) Alef. *Abstr. of ISHS Symposium on Brassicas, 9th Crucifer Genetics Workshop,* p. 93. Lisbon.

Murashige, T. and Skoog, F. 1962. A revised medium for rapid growth and bioassays with tobacco tissue cultures. *Physiol. Planata.,* 15:473–479.

Nitta, T., Takahata, Y., and Kaizuma, N. 1997. Scanning electron microscopy of microspore embryogenesis in *Brassica* species. *Plant Cell Rep.,* 16:406–410.

Ohkawa, Y., Bevis, E., and Keller, W.A. 1988. Validity study and microspore culture method in *Brassica napus*. *Cruciferae Newslett.,* 13:75.

Ohkawa, Y., Nakajima, K., and Keller, W.A. 1987. Ability to induce embryoids in *Brassica napus* cultivars. *Jap. J. Breed.,* (Suppl. 2):44–45.

Palmer, C.E., Keller, W.A., and Arnison, P.G. 1996. Utilization of *Brassica* haploids In: Jain, S.M., Sopory, S.K., and Veilleux, R.E. (Eds.), *In vitro Haploid Production in Higher Plants*, Vol. 3, p. 173–192. Kluwer Academic Publishers, Dordrecht, The Netherlands.

Pechan, P.M. and Keller, W.A. 1988. Identification of potentially embryogenesis microspores in *Brassica napus*. *Physiol. Plant.,* 74:377–384.

Pechan, P.M. and Keller, W.A. 1989. Induction of microspore embryogenesis in *Brassica napus* L. by gamma irradiation and ethanol stress. *In Vitro Cellular and Developmental Biol.,* 25:1073–1074.

Pechan, P.M. and Smykal, P. 2001. Androgenesis: affecting the fate of the male gametophyte. *Physiol. Plantarum,* 111:1–8.

Pechan, P.M., Bartels, D., Brown, D.C.W., and Schell, J. 1991. Messenger-RNA and protein changes associated with induction of *Brassica* microspore embryogenesis. *Planta,* 184:161–165.

Peng, S., Takahata, Y., Hara, M., Shono, H., and Ito, M. 1994. Effects of the matric potential of culture medium on plant regeneration of embryo derived from microspore of *Brassica napus*. *J. SHITA,* 5/6:8–14.

Phippen, C. and Ockendon, D.J. 1990. Genotype, plant, bud size and media factors affecting anther culture of cauliflower (*Brassica oleracea* var. *botrytis*). *Theor. Appl. Genet.,* 79:33–38.

Pratap, A., Gupta, S.K., and Vikas. 2007. Advances in doubled haploid technology of oilseed rape. *Ind. J. Crop Sci.,* 2:267–271.

Pratap, A., Sethi, G.S., and Chaudhary, H.K. 2005. Relative efficiency of different Gramineae genera for haploid induction in triticale and triticale × wheat hybrids through the chromosome elimination technique. *Plant Breed.*, 124:147–153.

Pratap, A., Sethi, G.S., and Chaudhary, H.K. 2006. Relative efficiency of anther culture and wheat × maize techniques for haploid induction in triticale × wheat and triticale × triticale hybrids. *Euphytica*, 150:339–345.

Prem, D., Gupta, K., and Agnihotri, A. 2004. Development of an efficient high frequency microspore embryo induction and doubled haploid generation system for Indian mustard (*Brassica juncea*). *Proc. 4th Int. Crop Sci. Congr.*, Brisbane, Australia, 26 Sept.–1 Oct, 2004.

Roulund, N., Hansted, L., Anderson, S.B., and Farestveit, B. 1990. Effect of genotype, environment and carbohydrate on anther culture response in head cabbage (*Brassica oleracea* L. convar. *capitata* Alef.). *Euphytica*, 49:237–242.

Sato, S., Katoh, N., Iwai, S., and Hagimori, M. 2002. Effect of low temperature pretreatment of buds or inflorescence on isolated microspore culture in *Brassica rapa* (syn. *B. campestris*). *Breed. Sci.*, 52:23–26.

Sato, T., Nishio, T., and Hirai, M. 1989a. Plant regeneration form isolated microspore culture of Chinese cabbage (*Brassica campestris* ssp. *pekinensis*). *Plant Cell Rep.*, 8:486–488.

Sato, T., Nishio, T., and Hirai, M. 1989b. Culture conditions for the initiation of embryogenesis from isolated microspores in Chinese cabbage (*Brassica campestris* L.). *Bull. Nat. Res. Inst. Veg. ornam. Plnts Tea Japan Ser.*, A 3:55–65.

Seguin-Swartz, G., Hutcheson, D.S., and Downey, R.K. 1983. Anther culture in *Brassica campestris*. In: *Proc. 6th Intl. Rapeseed Congr.*, pp. 246–251, Paris.

Sethi, G.S., Chaudhary H.K., Singh, S., Pratap, A., and Sharma, S. 2003. Genetic enhancement of productivity, quality and adaptability of bread wheat through doubled haploidy breeding following androgenesis and wheat x maize system. In: Behl, R.K. and Chhabra, A.K. (Eds.), *Enhancing Production and Food Value of Plants: Genetic Options. Proceed. Nat. Symp. Food and Nutritional Security: Technological Interventions and Genetic Options.* Vol. I, p. 33–41.

Sharma, K.K. and Bhojwani, S.S. 1985. Microspore embryogenesis in anther cultures of two Indian cultivars of *Brassica juncea* (L.) Czern. *Plant Cell Tiss. Organ Cult.*, 4:235–239.

Shi, S.W., Wu, J.S., Zhou, Y.M., and Liu, H.L. 2002. Diploidization techniques of haploids from *in vitro* culture microspores of rapeseed (*Brassica napus* L.). *Chinese J. Oil Crop Sci.*, 24:1–5.

Siebel, J. and Pauls, K.P. 1989. A comparison of anther and microspore culture as a breeding tool in *Brassica napus*. *Theor. Appl. Genet.*, 78:473–479.

Swanson, E.B. 1990. Microspore culture in *Brassica*. In: Pollard, J.W. and Walker, J.M. (Eds.), *Methods in Molecular Biology, Vol. 6, Plant Cell and Tissue Culture.* The Humana Press, p. 159–169.

Swanson, E.B., Coumans, M.P., Wu, S.C., and Barsby, T.L., and Beversdorf, W.D. 1987. Efficient isolation of microspore-derived embryos from *Brassica napus*. *Plant Cell Rep.*, 6:94–97.

Szarejko, I. and Forster, B.P. 2006. Doubled haploidy and induced mutation. *Euphytica*, 158:359–370.

Takahata, Y. 1997. Microspore culture, p. 162–181, In: Kalia, H.R. and Gupta, S.K. (Eds.), *Recent Advances in Oilseed Brassicas*, Kalyani Publishers, Ludhiana, India.

Takahata, Y. and Keller, W.A. 1991a. High frequency embryogenesis and plant regeneration in isolated microspore culture of *Brassica oleracea* L. *Plant Sci.*, 74:235–242.

Takahata, Y., Brown, D.C.W., and Keller, W.A. 1991b. Effect of donor plant age and inflorescence age on microspore culture of *Brassica napus* L. *Euphytica*, 58:51–55.

Takahata, Y., Takani, Y., and Kaizuma, N. 1993. Determination of microspore population to obtain high frequency embryogenesis in broccoli (*Brassica oleracea* L.). *Plant Tissue Cult. Lett.*, 10:49–53.

Takahata, Y., Fukuoka, H., and Wakui, K. 2005. Utilization of microspore-derived embryos. p. 153–169. In: Palmer, C.F., Keller, W.A., and Kasha, K.J. (Eds.), *Biotechnology in Agriculture and Forestry Vol. 56. Haploid in Crop Improvement II*, Springer-Verlag, Berlin Heidelberk.

Tsuwamoto, R., Fukuike, H., and Takahata, Y. 2007. Identification and characterization of genes expressed in early embryogenesis from microspores of *Brassica napus*. *Planta*, 225:641–652.

Tsuwamoto, R., and Takahata, Y. 2008. Idenification of genes specifically expressed in andorogenesis-derived embryo in rapeseed (*Brassica napus* L.) *Breed. Sci.*, 58:251–259.

Tang, G.X., Zhou, W.J., Li, H.Z., Mao, B.Z., He, Z.H., and Yoneyama, K. 2003. Medium, explant and genotype factors influencing shoot regeneration in oilseed *Brassica* spp. *J. Agron. Crop Sci.*, 189:351–358.

Telmer, C.A., Newcomb, W., and Simmonds, D.H. 1993. Microspore development in *Brassica napus* and the effect of high temperature on division *in vivo* and *in vitro*. *Protoplasma*, 172:154–165.

Telmer, C.A., Simmonds, D.H., and Newcomb, W. 1992. Determination of developmental stage to obtain high frequencies of embryogenic microspores in *Brassica napus*. *Physiol. Plant.*, 84:417–424.

Thomas, E. and Wenzel, G. 1975. Embryogenesis from microspores of *Brassica napus*. *Z. Pflanzenzucht.*, 74:77–81.

Thurling, N. and Chay, P.M. 1984. The influence of donor plant genotype and environment on production of multicellular microspores in cultured anthers of *Brassica napus* spp. *oleifera*. *Annals Bot.*, 54:681–695.

Wang, H.Z, Liu, G.H., Zheng, Y.B, Wang, X.F., and Yang, Q. 2002. Breeding of *Brassica napus* cultivar Zhongshuang No. 9 with resistance to *Sclerotinia sclerotiorum*. Chinese *J. Oil Crop Sci.*, 24:71–73.

Wang, H.Z., Wang, X.F., Liu, G.H., Zheng, Y.B. and Yang, Y. 2004. Studies on microspore culture of hybrid parents in *Brassica napus*. *Proc. 4th Int. Crop Sci. Congr.*, Brisbane, Australia, 26 Sept.–1 Oct., 2004.

Wang, M., Farnham, M.W., and Nannes, J.S.P. 1999. Ploidy of broccoli regenerated from microspore culture versus anther culture. *Plant Breed.*, 118:249–252.

Xu, L., Najeeb, U., Tang, G.X., Gu, H.H., Zhang, G.Q., He, Y., and Zhou, W.J. 2007. Haploid and doubled haploid technology, p. 182–216 In: Gupta, S.K. (Ed.), *Advances in Botanical Research-Rapeseed Breeding*, Vol. 45. Academic Press, London.

Yang, L.Y., Fan, Z.X., and Yang, G.S. 2005. Improvement of several microspore culture techniques for *Brassica napus* L. *Chinese J. Oil Crop Sci.*, 27:14–18.

Yu, F.Q. and Liu, H.L. 1995. Effects of donor materials and media on microspore embryoid yield of *Brassica napus*. *J. Huazhong Agric. Univ.*, 14:327–332.

Zaki, M. and Dickinson, H. 1991. Microspore-derived embryos in *Brassica:* the significance of division symmetry in pollen mitosis to embryogenic development. *Sexual Plant Reprod.*, 4:48–55.

Zaki, M.A.M. and Dickinson, H.G. 1990. Structural changes during the first divisions of embryos resulting from anther and free microspore culture in *Brassica napus*. *Protoplasma*, 156:149–162.

Zhang, F.l, Aoki, S., and Takahata, Y. 2003. RAPD markers linked to microspre embryogenic ability in *Brassica* crops. *Euphytica*, 131:207–213.

Zhang, F.L. and Takahata, Y. 2001. Inheritance of microspore embryogenic ability in *Brassica* crops. *Theor. Appl. Genet.*, 103:254–258.

Zhang, G.Q., Tang, G.X., Song, W.J., and Zhou, W.J. 2004. Resynthesizing *Brassica napus* from interspecific hybridization between *Brassica rapa* and *B. oleracea* through ovary culture. *Euphytica*, 140:181–187.

Zhang, G.Q., Zhang, D.Q., Tang, G.X., He, Y. and Zhou, W.J. 2006. Plant development from microspore-derived embryos in oilseed rape as affected by chilling, desiccation and cotyledon excision. *Biol. Planatarum*, 50:180–186.

Zhang, G.Q., Zhou, W.J., Gu, H.H., Song, W.J., and Momoh, E.J.J. 2003. Plant regeneration from the hybridization of *Brassica juncea* and *B. napus* through embryo culture. *J. Agron. Crop Sci.*, 189:347–350.

Zhao, J.P. and Simmonds, D.H. 1995. Application of trifluralin to embryogenic microspore cultures to generate doubled haploid plants in *Brassica napus*. *Physiol. Plant*, 95:304–309.

Zhao, J.P., Simmonds, D.H., and Newcomb, W. 1996. Induction of embryogenesis with colchicine instead of heat in microspores of *Brassica napus* L. cv. Topas. *Planta.*, 198:433–439.

Zhou, W.J. and Hagberg, P. 2000. High frequency production of doubled haploid rapeseed plants by direct colchicine treatment of isolated microspores. *J. Zhejiang University (Agric. and Life Sci.)*, 26:125–126.

Zhou, W.J., Hagberg, P. and Tang, G. X. 2002a. Increasing embryogenesis and doubling efficiency by immediate colchicine treatment of isolated microspores in spring *Brassica napus*. *Euphytica*, 128:27–34.

Zhou, W.J., Tang, G. X. and Hagberg, P. 2002b. Efficient production of doubled haploid plants by immediate colchicine treatment of isolated microspores in winter *Brassica napus*. *Plant Growth Regul.*, 37:185–192.

16 Genetic Improvement in Vegetable Crucifers

Pritam Kalia

CONTENTS

INTRODUCTION

Vegetable crucifers include a wide array of vegetable crops that span numerous genera and species in the family *Brassicaceae*. Of this family, two species — *Brassica oleracea* and *Brassica campestris* — are the source of edible crops. They originate in Europe by way of Asian countries (i.e., Iran, Afghanistan, and Pakistan). Vegetables in *B. oleracea* include cabbages, collards, cauliflower, broccoli, Brussels sprouts, kales, and kohlrabi; *B. campestris* includes Pak Choi, Chinese cabbage, Siberian kale, turnip, mustards, rape, rutabaga, and radish.

In the Western Hemisphere, including Europe, the predominance goes to *Brassica oleracea*. In Asia, *B. campestris* is the most cultivated species, owing to the great importance of Chinese cabbage, although both cabbage and cauliflower are extensively grown on the Indian subcontinent. Turnip and turnip greens, which also belong to *B. campestris*, are cultivated worldwide but have much less economic importance. *B. juncea*, which includes the vegetable mustard, has some economic relevance in the Far East but is a minor crop at the world level.

The center of origin of *Brassica oleracea* is the Mediterranean region. Vegetable *Brassica* have been cultivated in Europe since ancient times and subsequently spread to other parts of the world (Nieuwhof, 1969). *Brassica* include many different morphotypes, which are well adapted to temperate climates, require quite simple cultivation techniques, and produce abundant and nutritious food for mankind.

Breeding strategies and targets depend on market trends. Successful breeders anticipate changes in the market by developing new varieties that are ready for release to the growers when the demand increases. It is, therefore, interesting to see how breeding is reacting to eventual changes in *Brassica* consumption and to evaluate the potential influence that the *Brassica* market and growing systems may have on the definition of breeding targets and priorities.

BREEDING OBJECTIVES IN COLE VEGETABLES

The main criteria for crop improvement are higher production and productivity, biotic and abiotic stress resistance, and uniformity and continuity of cropping. Breeding for appearance, commercial quality, shelf life, taste, and nutritional value are parts of product improvement.

The most important objective of breeding is crop uniformity, because a uniform *Brassica* vegetable field makes grading much easier and reduces harvest time. High uniformity has been almost impossible to achieve with open-pollinated varieties owing to the cross-pollination habit of *Brassica*. The introduction of F_1 hybrids, which can produce a genetically uniform population, has progressed slowly. Until recently, *Brassica* hybrid breeding has used the sporophytic self-incompatibility mechanism because there is no cytoplasmic male sterility in *B. oleracea*. The instability and complex inheritance of the self-incompatibility mechanism makes its use difficult and conducive to low-quality F_1 hybrids.

However, the production of *Brassica* F_1 hybrids is now developing faster, albeit with some technical difficulties, using doubled haploid (DH) parent lines obtained through microspore culture and cytoplasmic male sterility introduced from *Raphanus sativus* into *B. oleracea*. Diverse sterile cytoplasm from wild *Brassica* are also being attempted into *B. oleracea* using embryo rescue at the Division of Vegetable Science, IARI, New Delhi (Kalia, 2008).

Disease resistance is another very important breeding objective. Sources of resistance to important diseases — for example, clubroot (*Plasmodiophora brassicae*), black rot (*Xanthomonas campestris* pv. *campestris*), fusarium yellows (*Fusarium oxysporum* f. *conglutinans*), and downy mildew (*Hyaloperonospora parasitica*) — were identified (Chiang et al., 1993) but have not been widely transferred into commercial varieties yet. Appearance, including color and shape, is an important extrinsic trait and the only major breeding objective that is addressed to consumers. The rapid transformation of the vegetable market with improvements in packing and display facilities, the large offering of commodities all year round, and the increased presence of colorful and appealing fruit and salad vegetables have forced *Brassica* producers to raise their presentation and quality standards. Growers are looking for high commercial quality, including adequate size and shape, good color, firmness, and appearance, to have the produce readily accepted by the trade. Qualitative traits have relatively high importance in vegetable breeding compared to that of crop yield, the traditional first priority for breeders. Nutritional quality is slowly gaining popularity with *Brassica* vegetable breeders, as this aspect will provide a marketing advantage in the future.

HISTORY OF *BRASSICA OLERACEA*

Brassica oleracea grows wild along the Atlantic coasts of Europe, where it might have been cultivated by Celts in its primitive form (kales). When it was eventually brought to the eastern Mediterranean region (estimated by the limit between the first and second millennia BC), it became fully domesticated and started an explosive diversification that gave rise to an enormous range of cultivated forms. The most widely known forms or groups of forms include:

1. Kales, which develop a strong main stem and are used for their edible foliage. These are old cultivated forms and include green curly kales, narrow stem kale, and giant jersey kale. Land races of these kales are widely scattered.
2. Cabbages are characterized by the formation of heads of tightly packed leaves and are represented by head cabbage, Savoy cabbage, and Brussels sprouts.
3. Kohlrabi is grown for its thickened stem.
4. Inflorescence kales are used for their edible inflorescences. Major forms are cauliflower, broccoli, and Calabrese.
5. Chinese alboglabra kale, which is used for its leaves.

Earlier, it was widely held that wild *oleracea* kales found along the Mediterranean coasts were progenitors of the cultivated forms, but the current belief is the opposite. Mediterranean kales are mere escapees from early cultivations. More recently, a polyphyletic origin by incorporation of genes into the *Brassica oleracea* genome from different wild Mediterranean species (Gustafsson, 1979; Snogerup, 1980) was suggested. The species preferentially considered in this respect were *B. cretica, B. rupestris, B. insularis,* and *B. montana*. However, Hosaka et al. (1990) did not find any molecular evidence to suggest specific wild ancestors for the different *B. oleracea* types.

The tendency now is to minimize the possible introgression by other species, thus returning again to the Atlantic *Brassica oleracea,* which has the main role in the development of cultivated forms but admitting that introgression of genes from wild species has probably been responsible for increasing the variability and adaptability of cultivated *B. oleracea*. The earliest cultivated *B. oleracea* was most likely a leafy kale that gave rise to a wide variety of kales along the coasts of the Mediterranean and Atlantic from Greece to Wales (Song et al., 1990). Diverse forms were developed in different areas primarily due to selection in different climates, natural hybridization, and gene introgression. Macro- and micro-mutational events and chromosomal changes also played a substantial role (Chiang and Grant, 1975; Kianian and Quiros, 1992). Song et al. (1988b, 1990) considered Chinese kale (var. alboglabra) to be very close to the primitive type that spread to the center of the eastern Mediterranean and eventually reached China.

Cauliflower and broccoli evolved in the eastern Mediterranean (Hyams, 1971; Snogerup, 1980). Although sprouting forms of cabbage were mentioned by early Greeks and Romans, a distinction was not apparent between cauliflower and broccoli, indicating that the differences did not exist or that the two were considered as variants of the same forms.

A Spanish-Arabic author, Ibn-al-Awam (circa 1140) made the first clear distinction between heading and sprouting forms in his book *kitab-al-falaha,* wherein he devoted a separate chapter to cauliflower (Hyams, 1971). He used the name *quarnabit,* the present-day Arabic word for cauliflower. Herbalist Dodoens (1578) referred to cauliflower as *B. cypria,* indicating its origin in Cyprus. Hyams (1971), based on the observations of Ibn-al-Awam who referred to it as Syrian or Mosul cabbage, considered Syria as the place of its origin. Cauliflower is generally regarded as derived from broccoli (Crisp, 1982; Gray, 1982). Crisp (1982), based on his work on hybridization between broccoli and cauliflower, concluded that a single major gene mutation in broccoli gave rise to cauliflower.

Broccoli is an Italian word derived from the Latin *brachium,* which means an arm or branch (Boswell, 1949). It includes heading forms with a single large terminal inflorescence. Another form is sprouting broccoli, a branched type in which the young edible inflorescences are referred to as sprouts: Broccoli probably originated between 400 and 600 years BC when the ancestral forms of modern varieties were selected (Schery, 1972). Dalechamp (1587) described it in the sixteenth-century herbal *Historia Generalis Plantarum,* which constitutes the first documented report. Broccolis were introduced into Italy from the eastern Mediterranean where diversification took place and many forms, including heading and sprouting ones, arose. The regular handsome inflorescence of recently commercialized "Romanesco" broccoli is often cited by the students of fractal geometry. Brussels sprouts were developed near Brussels in Belgium during the fourteenth century. A kale resembling Brussels sprouts, but with finely dissected leaves and numerous buds, is described by Gerarde (1597) as "Persil cabbage" (Henslow, 1908). These sprouts were reportedly served at a wedding feast in 1481 (Hyams, 1971).

The diverse *Brassica* genus comprises oilseed rape or canola (*B. napus* L. and hybrids with *B. campestris*), the turnip and Chinese cabbage (*B. campestris* L.), rutabaga or Swede (*B. napus* L.), black mustard (*B. nigra* Koch), brown mustard (*B. juncea* (L.) Czern), and Ethiopian mustard (*B. carinata*). The single cruciferae species, *Brassica oleracea,* has by itself yielded a remarkable array of vegetables, including the cabbage, kale, Brussels sprouts, cauliflower, broccoli, and kohlrabi.

The leafy kales were the second *Brassica* species to be cultivated, as the Greeks recorded growing them at least 2500 years ago (Helm, 1963). They derived from *Brassica oleracea* ssp. *Oleracea,* found naturally across the coast of the Mediterranean from Greece to England. In an early stage of their domestication there must have been a reduction in the bitter-tasting glucosinolates, which are found at high levels in the wild species (Josefsson, 1967). The other types of *B. oleracea* came into being much later, as humans began to actively select for the enlargement of different plant parts (Gray, 1982). Early types of cabbages were first grown in ancient Rome and Germany over 1000 years ago. Broccoli, cauliflower, and Brussels sprouts are a more recent development, appearing within the past 500 years — cauliflower from northern Europe and broccoli from the eastern Mediterranean. Brussels sprouts first appeared as a spontaneous mutation in France in 1750.

The different types of *Brassica oleracea* developed primarily by disruptive selection on the polymorphisms already available in wild *B. oleracea,* except for Brussels sprouts. This was demonstrated by Buckman in 1860 at the Royal Agricultural College in Southern England by selecting broccoli like cuttings from wild plants in only a few generations. Cauliflower is most closely related to broccoli, while cabbage is more closely related to the kales (Song et al., 1990). As they emerged, the new *Brassica* types spread rapidly throughout Europe and the Mediterranean countries, and hybridization with wild congeners undoubtedly played a role in the development of crop types. Helm (1963) reported that *B. cretica* contributed heavily to the development of cauliflower. Chloroplast DNA was reported in two populations of *B. napus,* probably the result of recent unprogressive hybridization with *B. oleracea* and *B. campestris* (Palmer et al., 1983).

Wild Relatives of *Brassica oleracea*

True wild types of *Brassica oleracea* are found along the European Atlantic coasts and have often been referred to as var. *sylvestris* L. They are close to cultivated forms (Gustafsson and Lanner-Herrera, 1997a) and almost morphologically indistinguishable from many kales.

The other two closely related taxa originally considered as species, namely *Brassica bourgeaui* and *B. alboglabra,* should probably merit intraspecific status within *B. oleracea* variation. *B. bourgeaui,* from the Canary Islands, was included in the genus *Sinapidendron* in the early literature and thought to be extinct for many years. It was re-found by Borgen et al. (1979), who ascribed it to the *n* = 9 *Brassica* group. They only found two plants on La Palma Island. Today, most opinions emphasize its close relation to *B. oleracea* (Lanner, 1998), although perhaps some possible differential characters in leaves and pod beaks should be studied more carefully. A second population was found (Marrero, 1989) on El Hierro Island. *B. alboglabra,* an old cultivated form from China, should be kept in *B. oleracea* (Bothmer et al., 1995). It had reached China from the Mediterranean region, which is evident from its local cultivation in southern Italy (Hammer et al., 1992). Its flowers are normally white although they may be distinctively yellow appearing in early wintertime.

Brassica oleracea is an Atlantic plant; it has a number of closely related species that grow around the Mediterranean basin. All have *n* = 9 chromosomes and are more or less interfertile among themselves and with *B. oleracea.* The taxonomy and geography of these taxa were described by Snogerup et al. (1990), who also proposed nomenclatural readjustments for *B. cretica* subspecies. The maximum morphological diversity of the group clearly occurs in Sicily where at least four species — *B. macrocarpa, B. villosa, B. rupestris,* and *B. incana* from west to east — are present. Raimondo and Mazzola (1997) confer the rank of subspecies to some other Sicilian taxa that are closely related to either *B. rupestris* or *B. villosa.*

The affinities among this group of species were studied from a different point of view as Stork et al. (1980) described a number of seed characters and micro-characters (such as testa layers, surface structures, and mucilage cells) throughout the species complex. Gomez-Campo et al. (1999) compared another micro-character, the morphology of epicuticular wax columns, and found a distinctive type for the other wild species.

Mithen et al. (1987) analyzed eighteen populations belonging to eleven specific or subspecific taxa for their contents of nine glucosinolates. Some taxonomically meaningful differences were found for wild vs. cultivated *Brassica oleracea,* as well as among wild Sicilian species and also for Sardinian vs. Tunisian populations of *B. insularis.*

Aguinagalde et al. (1992) used flavonoids, seed proteins and five isozyme systems and found a comparatively high phytochemical diversity in *Brassica cretica,* even higher than that for the Sicilian group of species. No differences were found between wild and cultivated *B. oleracea.* In turn, *B. bourgeaui* closely resembles the group formed by *B. oleracea* and *B. montana* (all lacking isorhamnetin), while *B. oleracea* and *B. alboglabra,* considered co-specific by most authors, can be distinguished by their flavonoid patterns.

Studies with RAPDs (random amplified polymorphic DNA) and isozymes on 22 populations belonging to 15 specific and sub-specific taxa (Lazaro and Aguinagalde, 1998a, b) clearly distinguish a group with the Sicilian species except *B. incana;* a second Western group including *B. oleracea, B. montana* and also *B. incana,* and *B. bourgeaui;* and finally, a third group including *B. alboglabra* (suggesting some gene influx) together with *B. cretica* and *B. hilarionis.* Other important studies comparing either species or populations are those by Hosaka et al. (1990), Gustafsson and Lanner-Herrera (1997a), and Lanner (1998). Evaluations of interfertility among all these taxa (Snogerup, 1996) show that reduced fertility measured by pollen stainability often occurs but is only loosely correlated to morphological differences.

Aguinagalde et al. (1991) studied eight populations belonging to all three existing subspecies of *Brassica cretica* in a single species focus. Seed storage proteins and five isoenzyme systems analyses indicated high interpopulation diversity. The ssp. *laconica,* however, was found to be much more homogeneous.

Maselli et al. (1996) studied five enzyme systems on seven populations of *Brassica insularis* from Tunisia, Sardinia, and Corsica. Widler and Bocquet (1979) had, however, recognized four differentiated varieties of this species in Corsica alone; a few years prior, Gustafsson and Lanner-Herrera (1997b) compared wild *B. oleracea* populations using different criteria, while Lanner-Herrera et al.

(1996) performed a similar comparison on 18 populations using isozyme and RAPD analyses. Intra- and interpopulation variation results were interesting but there was no correlation with geographic origin (Spain, France, and Great Britain). The truly wild status of *B. oleracea* populations in northern Spain has been supported by Fernandez-Prieto and Herrera-Gallastegui (1992) on phytosociological grounds. A detailed account of the French populations of *B. montana* with morphological analyses and a study of several enzymatic systems were produced.

Imperfectly known variability roughly related to *Brassica incana* existing in Adriatic coasts and islands was understood due to studies of Rac and Lovric (1991) and Eastwood (1996). Trinajsic and Dubravec (1986) recognized four different taxa with the range of subspecies, one part ascribed to *B. incana* itself and another part to *B. botteri,* a differentiated form that is recognized to merit specific rank.

Snogerup and Persson (1983) studied *Brassica balearica* cytologically and found that the $n = 9$ genome is present as part of its $2n = 32$ chromosome complement. They suggested that an ancient hybridization involving *B. insularis* gave rise to this species by amphidiploidy. Gomez-Campo (1993a) suggested that the other parent might have been a member of Subgen *Brassicaria*. Seed and seedling morphology, the general appearance of the plant, and some phytogeographical considerations do favor this hypothesis, although it is necessary to account for some chromosome losses. *B. balearica* belongs to Sect. *Brassica*, together with another hybrid species, *B. carinata,* for which one of the parents belongs to the $n = 9$ genome group. Although *B. napus* is a similar case, it is placed in Sect. *Rapa* due to its capacity to form tuberous roots.

WILD TAXA GENETIC DIVERSITY AND DOMESTICATION

There has been a "parallel evolution" of forms of *Brassica* in Europe and Asia. *B. oleracea* is a European crop species that has been selected to give a wide range of crop types. Vegetables have been developed where leaves, terminal buds, axillary buds, swollen stems, or floral tissues develop precociously and are harvested. *B. napus* L. occurs in Europe as a vegetable or animal food with swollen roots (the Swede) or as an oilseed crop. A similar pattern can be seen with *B. campestris* and *B. juncea* in Asia. Asiatic *B. campestris* contains vegetable forms that have been selected for large terminal buds, swollen stems, swollen roots, or large inflorescences. *B. juncea* includes forms with swollen stems or swollen roots. Both species have also been developed as leafy vegetables analogous to the kales of *B. oleracea,* and as oilseed crops analogous to that of *B. napus.*

Lazaro and Aguinagalde (1998b) evaluated genetic diversity in 29 populations of wild taxa of the *Brassica oleracea* L. group ($2n = 18$) and two cultivars, using RAPDs as molecular markers. The results of two molecular markers (isozyme and RAPDs) were compared. DNA from 10 individuals per population was analyzed using six different primers; the 151 detected bands were polymorphic, 11 were common to all species, 6 to all taxa, only 1 to every population, and no bands were shared by every individual. The dendrogram obtained using genetic distances clusters *B. oleracea* populations with *B. bourgeaui, B. alboglabra, B. montana* and *B. incana. B. insularis, B. macrocarpa, B. villosa,* and *B. rupestris* populations form another cluster. Populations of *B. cretica* and *B. hilarionis* form the third cluster. Genetic diversity in *B. oleracea* populations, the *B. rupestris* complex, and *B. cretica* subspecies was estimated using the AMOVA program, with the latter being the most diverse.

The distinct morphologies exhibited by *Brassica oleracea* subspecies represent one of the most spectacular illustrations of structural evolution in plants under domestication. *B. oleracea* is a perennial herb found largely in Europe and the Mediterranean (Tsunoda et al., 1980; Song et al., 1988a; Kalloo and Bergh, 1993) and is an extremely polymorphic species that includes at least six cultivated and one wild subspecies. Wild, perennial forms of *B. oleracea,* designated subspecies *oleracea* (wild cabbage), grow in coastal rocky cliffs of the Mediterranean, northern Spain, western France, and southern and southwestern Britain (Tsunoda et al., 1980). Selection for different characteristics during domestication, however, has resulted in extreme morphological divergence among cultivated subspecies. Of six domesticated taxa, two subspecies — *B. oleracea* ssp. *botrytis* (cauliflower) and *B. oleracea* ssp. *italica* (broccoli) — are characterized by the evolutionary modification of the inflorescence into large dense structures. The precociously large, undifferentiated inflorescence forming the curd is the defining characteristic of *B. oleracea* ssp. *botrytis*. The cauliflower curd consists of a dense mass of arrested inflorescence meristems, only ~10% of which will later develop into floral primordia and normal flowers (Sadik, 1962).

The cauliflower phenotype characteristic of *Brassica oleracea* ssp. *botrytis* has been observed in mutants of the related crucifer *Arabidopsis thaliana* (Bowman et al., 1993; Weigel, 1995; Yanofsky, 1995). In *Arabidopsis*, the early acting floral meristem identity genes are a class of flower developmental regulatory loci that specify the identity of the floral meristem in developing reproductive primordia. Members of this class include the genes APETALA 1 (AP1; Mandel et al., 1992; Gustafson-Brown et al., 1994) and CAULIFLOWER (CAL; Kempin et al., 1995). Both the APETALA 1 and CAULIFLOWER loci have also been shown to control the specification of floral meristem identity. *Arabidopsis* individuals that are mutant for both AP1 and CAL are arrested in development at the inflorescence meristem stage (Kempin, et al., 1995). In these plants, a dense mass of inflorescence meristems develops, similar to the *B. oleracea* ssp. *botrytis* curd. Genetic analyses in *B. oleracea* suggest the involvement of the *B. oleracea* CAL gene, referred to as BoCAL, in the formation of altered inflorescence in *B. oleracea* ssp. *botrytis* (Kempin et al., 1995). It has been demonstrated already that the BoCAL allele in domesticated cauliflower has a premature stop codon at position E151 in exon 5C (Kempin et al., 1995). This nonsense mutation appears to have arisen fairly recently within *B. oleracea*.

Studies by Purugganan et al. (2000) suggested that the evolution of plant morphologies during domestication events provides clues to the origin of crop species and the evolutionary genetics of structural diversification. The CAL gene, a floral regulatory locus, has been implicated in the cauliflower phenotype in both *Arabidopsis thaliana* and *Brassica oleracea*. Molecular population genetic analysis indicates that alleles carrying a nonsense mutation in exon 5 of the *B. oleracea* cauliflower (BoCAL) gene are segregating in both wild and domesticated *B. oleracea* subspecies. Alleles carrying this nonsense mutation are nearly fixed in *B. oleracea* ssp. *botrytis* (domestic cauliflower) and *B. oleracea* ssp. *italica* (broccoli), both of which show evolutionary modifications of inflorescence structures. Tests for selection indicate that the pattern of variation at this locus is consistent with positive selection at BoCAL in these two subspecies. This nonsense polymorphism, however, is also present in both *B. oleracea* ssp. *acephala* (kale) and *B. oleracea* spp. *oleracea* (wild cabbage). It is evident from these results that specific alleles of BoCAL were selected by early farmers during the domestication of modified inflorescence structures in *B. oleracea*.

KARYOTYPIC INVESTIGATION

Despite the fact that interspecific hybrids were reported much earlier (*Raphanus sativus* × *B. oleracea*, Sageret, 1826; *B. napus* × *B. rapa*, Herbert, 1847), the beginning of cytogenetical research in *Brassica* was marked by the determination of chromosome number for *Brassica rapa* (syn. *B. campestris*) by the Japanese researcher Takamine in 1916.

Hybridizations between different species and the study of their meiotic behavior led Morinaga (1928, 1934) to unravel the genetic architecture of crop *Brassica*. Around that time, Manton (1932) carried out an extensive genosystematical survey of cruciferae and determined the chromosome numbers for a number of taxa.

During the 1970s, wild germplasm of *Brassica* and related genera was extensively collected and cytogenetical studies were initiated. While the theme of research in the early phase centered on polyploidy breeding, priorities later shifted to the exploitation of wild allies for introgression of nuclear genes for desirable agronomic traits, cytoplasmic substitutions, and construction of chromosome maps, taking advantage of developments in the areas of somatic cell and molecular genetics in recent years.

The use of molecular techniques has considerably helped in constructing linkage maps by applying restriction fragment length polymorphism (RFLP), random amplified polymorphic DNA

(RAPD), and DNA fingerprinting. These investigations have yielded important information on genome organization, extent of gene duplications, chromosome structural changes, and intergenomic gene introgression.

There has been spectacular development in *in vitro* techniques such as ovary and embryo culture, and somatic hybridization in the recent past. Somatic hybridization not only overcomes reproductive barriers, but also generates cytoplasmic variability, which is not possible through conventional methods of sexual hybridization.

Previously, it was believed that *Brassicas* had evolved from a common prototype with x = 6 (Mizushima, 1950a; Röbbelen, 1960). However, the investigations based on nuclear, mitochondria, and chloroplast DNA restriction fragment length polymorphism (RFLP) established their evolution from two prototypes:(1) *Brassica nigra* from one prototype and (2) *B. oleracea* and *B. rapa* from the other. This evolutionary diversion is also reflected in their cytoplasm (Palmer, 1988; Warwick and Black, 1991; Pradhan et al., 1992). *B. oleracea* and *B. rapa* cytoplasms are closer to each other than either is to *B. nigra* (Palmer, 1988). Although variations were observed in the cytoplasms of *B. rapa* and *B. oleracea* in cp and mt DNA patterns (Song and Osborn, 1992), the mitochondrial genome of *B. rapa* was more variable than that of *B. oleracea*.

Pachytene chromosome analysis by Röbbelen (1960) and Venkoteswarlu and Kamala (1973) revealed that diploids have six basic types of chromosomes. The *Brassica rapa* genome is represented by AABCDDEFFF (tetrasomic for chromosomes A and D, and hexasomic for chromosome F); *B. oleracea* by ABBCCDEEF (tetrasomic for chromosome B, C, and E); and *B. nigra* by AABCDDEF (tetrasomic chromosomes A and D). Chromosomes of each type have lost homology due to structural and/or genetic alterations during their long evolutionary history.

GENETIC TOOLS FOR FACILITATING HYBRID BREEDING

SELF-INCOMPATIBILITY

Self-incompatibility (SI) is an elaborate breeding system for securing outcrossing and maximum recombination in angiosperms. In the *Brassicaceae*, it is classified into the homomorphic sporophytic type. According to Bateman's survey (1955), out of 182 species in the *Brassicaceae*, 80 species express self-incompatibility. In *Brassica* species and their closest allies, 50 species of the 57 species examined expressed self-incompatibility (Hinata et al., 1994). The self-incompatibility system has likely played an important role in the differentiation and adaptation of species in this group. A number of economically important vegetables are included in the *Brassicaceae* [*Brassica oleracea*, *B. rapa* (syn. *B. compestris*), *Raphanus sativus*, etc.], and hybrid variety seeds made using self-incompatibility are highly evaluated.

Morphology and Physiology of Self-Incompatibility

The stigma of *Brassicaceae* belongs to the so-called "dry stigma" group; it is covered with one layer of papilla cells on its surface. The papilla is a typical secretory cell in which the endoplasmic reticulum and Golgi apparatus are developed (Kishi-Nishizawa et al., 1990). The cell wall of the papilla cell consists of an inner cellulose layer and an outer cuticle layer on which waxes are deposited.

Yellow-colored pollen grains are generally transferred by bees. They have three germ slits, and the exine is covered with a lipoidal coating, which is called a pollen kit, tryphine, or pollen coat (Dickinson and Lewis, 1973; Roggen, 1974). The coating is believed to include some other substances that may provide the signal of pollen identity.

In the sporophytic system, the behavior of pollen tubes is determined by the phenotype of the sporophyte that produced the pollen. Therefore, the phenotype of the pollen and the stigma of heterozygous plants depend on the outcome of complex dominant, recessive allelic interactions (Thompson and Taylor, 1966). Dominance relationships between *S* alleles have been investigated for several species in the family *Brassicaceae*: *Iberis amara* (Bateman, 1954, 1955); *Brassica olera-cea* and *Raphanus sativus* (Haruta, 1962); *B. oleracea* (Thompson and Taylor, 1966; Ockendon, 1975; Visser et al., 1982; Wallace, 1979); *R. raphanistrum* (Sampson, 1964, 1967); *Sinapis arvensis* (Stevens and Kay, 1989); and more recently in *B. rapa* (Hatakeyama et al., 1998a). Among the characteristic features of dominance relationships in these species are (1) codominant relationships occur more frequently than dominant/recessive ones, (2) dominant/recessive relationships occur more frequently in the stigma, (3) dominant/recessive relationships are not identical for S alleles between stigma and pollen, and (4) nonlinear dominance relationships are observed more frequently in the stigma than in the pollen. Numbers of S alleles were estimated by several studies (Sampson, 1967; Ockendon, 1974, 1980; Ford and Kay, 1985; Stevens and Kay, 1988, 1989; Karron et al., 1990; Nou et al., 1991, 1993a, b). More than a hundred S alleles have been estimated in *B. rapa* throughout the world (Nou et al., 1993a).

Self-incompatibility (SI) promotes outcrossing in *Brassica* species and has been exploited for producing F_1 hybrids in cole crops. The SI locus (S-locus) is very polymorphic and cross compatibil-ity between plants depends on the allelic forms (S-haplotypes) of genes at the S-locus. Among these genes, the SLG (S-locus glycoprotein) gene is specifically expressed in the stigma. Two classes (class I and class II) of SLG genes have been described, and antibodies allowing a specific identification of class I or class II SLG are available. Ruffio-Chable and Gaude (2000) reported polymorphism in *B. oleracea* cultivars (cauliflower, broccoli, romanesco, cabbage, kale, and wild forms of the Normandy coast) as assessed using an isoelectric focusing and immunostaining combined analysis. Plants from an S-allele collection were also introduced in the analysis to allow the identification of S-haplotypes. From 950 plants analyzed, approx. 80 different immunostaining patterns were iden-tified using two class-specific SLG antibodies. However, specific immunostaining patterns could not be found for about 10% of the plants analyzed. Among these plants, nine S-haplotypes were identified by crossing experiments with lines of known S-haplotype; and three S-haplotypes (S_2, S_{29}, S_{63}) gave variable immunostaining patterns, where SLG bands were either clearly detectable, faintly detectable, or not present. This indicates that, in these particular S-haplotypes, the amount of SLG is highly dependent on the genetic background of the plants. All but one of the S-haplotypes described in the wild forms were also found in cultivated forms. The three class II haplotypes (i.e., S_2, S_5, S_{15}) were detected in wild *Brassica* but only S_{15} was common to all cultivated as well as wild *Brassica* plants. Moreover, S_{15} was the most frequent S-haplotype detected in the species.

Breeding for uniformity has resulted, notably, in the production of F_1 hybrids. The first reported commercial production of F_1 crucifers was a cabbage cultivar in Japan in 1938 (Sakata, 1973). This was based on self-incompatible parental material maintained vegetatively by cuttings rather than as inbred lines. Since that time, F_1 vegetables have come to dominate the Japanese crucifer seed trade. In *Brassica oleracea,* the Japanese produced F_1 hybrids of Calabrese, Brussels sprouts, cabbage, cauliflower, horticultural kale, and kohlrabi. In *B. campestris,* also Chinese cabbages and turnips are produced as F_1s. Radish F_1 cultivars are also developed.

There are two problems associated with the production of F_1 seed. First, parental material may have low yields due to inbreeding depression. Second, the procedure of breeding an inbred parent may result in considerable inadvertent selection in favor of self-fertile material within that inbred line. This results in self rather than cross fertilization between lines when used as parents for an F_1. Formation of higher quantities of selfed rather than hybrid seed in the F_1 can considerably reduce its value.

An alternative method of producing F_1 and other hybrids lies in maintaining male-sterile or self-incompatible parental material, not as recurrently seeding inbred lines but as vegetative clones. The original F_1 hybrid cabbage produced in Japan is believed to have had one vegetatively maintained, highly self-incompatible parent that was pollinated by a less incompatible parent maintained by seedling. This technique was also advocated for F_1 cabbage in United States. The overall trend among vegetable crucifers is that hybrids can be produced in a wide range of vegetables.

MALE STERILITY

Male-sterile plants have mainly arisen as spontaneous mutants. Single dominant gene male sterility in *Brassica oleracea* (Dunemann and Grunewaldt, 1991) was obtained through mutagenic treatment. A genic male sterility system was developed by genetic engineering by Mariani et al. (1990), who fused the tapetum-specific promoter TA29 from tobacco (Goldberg, 1988) to the coding regions of two genes encoding RNases: barnase from *Bacillus amyloliquefaciens* and RNase T_1 from *Aspergillus oryzae*. Male-sterile *Brassica napus* plants were produced by transferring these *Chimaeric* genes into the plant genome via *Agrobacterium tumefaciens*. The male-sterile plants with the barnase construction can be restored to fertility by crossing to restorer lines carrying the barstar gene, encoding a specific inhibitor for barnase (Mariani et al., 1992). The TA29-barnase construct was also introduced into *B. oleracea* var. *botrytis* and the male-sterile transformants were obtained (Reynaerts et al., 1993). In most histologically described male sterilities, microsporogenesis breakdown occurs at the tetrad stage and sometimes at the uninucleate vacuolate microspore stage (Table 16.1). These two stages of degeneration are observed in the dominant male sterility of Mathias (1985a) and Ruffio-Chable et al. (1993), depending on the temperature conditions (Theis and Röbbelen, 1990). The digenic male sterility of Li et al. (1988) shows a specific degeneration process. The two meiotic divisions and the subsequent cytodieresis are disturbed (micro-nuclei formation) or stopped (Zhou, 1990). When male sterility is complete, the meiosis is completely stopped and the disturbed nuclear envelope does not perfectly separate the nuclear chromatin from the surrounding cytoplasm. In the engineered male sterility of Mariani et al. (1990), the disruption of microsporogenesis is correlated with the activity of the chimeric TA 29-RNase or barnase genes (De Block and Debrouwer, 1993; Denis et al., 1993). Degeneration of the tapetal RNA takes place

TABLE 16.1
Genic Male Sterilities in the *Brassica oleracea*

Species	Ref.	Inheritance	Origin
Brassica oleracea var. *italica*	Anstey and Moore (1954); Cole (1959)	1 recessive gene ms-1	Spontaneous
	Dickson (1970)	1 recessive gene ms-6	Spontaneous
	Dunemann and Grunewaldt (1991)	1 dominant gene	Mutagenic treatment
B. oleracea var. *botrytis*	Nieuwhof (1961)	1 recessive gene ms-5	Spontaneous
	Borchers (1966)	1 recessive gene ms-4	Spontaneous with microsporogenesis breakdown stage after microspore release
	Chatterje and Swarup (1972)	1 recessive gene	Spontaneous
	Ruffio-Chable et al. (1993)	1 dominant gene	Spontaneous with microsporogenesis breakdown stage at tetrad microspore (Ruffio-Chable (1994)
	Reynaerts et al. (1993)	1 dominant gene	Genetic engineering
B. oleracea var. *capitata*	Nishi and Hiraoka (1958)	1 recessive gene	Spontaneous at tetrad
	Rundfeldt (1960)	1 recessive gene	Spontaneous
B. oleracea var. *gemmifera*	Johnson (1958)	1 recessive gene ms-2	Spontaneous at tetrad
	Nieuwhof (1961, 1968)	1 recessive gene	Spontaneous

after microspore release from the tetrads and is immediately followed by the disappearance of the RNA from microspores, after which they collapse. This confirms the intimate relationships of the tapetum and microspores during their development.

Cytoplasmic Male Sterility

In the *Brassicaceae* family, the first example of cytoplasmic male sterility (CMS) was described by Ogura (1968) in a Japanese population of radish. In France, using this source, several hybrid varieties manifesting considerable hybrid vigor were developed (Bonnet, 1975, 1977). The commercial production of triple-cross hybrids began in 1989 with the first variety, Serrida, registered on the official catalog in France. At the same time, extensive work was performed to transfer this "Ogura cytoplasm" to various *Brassica* crop species by interspecific crosses (Bannerot et al., 1974, 1977).

Many CMS systems have been elaborated in the *Brassica oleracea* (Table 16.2). Because CMS is the result of specific nuclear/mitochondrial interactions, the association of cytoplasm and nucleus from different species often results in total or partial male sterility, sometimes associated with other striking effects on floral morphology.

Ogura (1968) first described the male sterility system in radish. This male sterility inducing cytoplasm was successfully transferred to *Brassica oleracea* (Bannerot et al., 1974). The CMS system was obtained in *B. oleracea* with the cytoplasm of *B. nigra* (Pearson, 1972) through sexual hybridization. Protoplast fusion between more or less related species was used by Kameya et al. (1989) in *B. oleracea* to acquire new CMS systems. Protoplast fusion was used by Yarrow et al. (1990) to transfer the 'Polima' cytoplasm from *B. napus* to *B. oleracea* and by Cardi and Earle (1997) to transfer the *B. tournefortii* cytoplasm from *B. rapa* to *B. oleracea*. Protoplast fusion was also extensively used to induce new combinations of cytoplasmic traits. Male sterility has often been obtained by introducing the genome of one species into an alien cytoplasm and results from mitochondria/nucleus interactions, but at the same time defects may appear from chloroplast/nucleus or mitochondria/nucleus incompatibilities. When the male sterility inducing cytoplasm of radish (Ogura, 1968) was introduced into *B. oleracea,* male-sterile lines were obtained but they showed a severe chlorophyll deficiency and low nectar secretion. This CMS

TABLE 16.2
Cytoplasmic Male Sterilities in the *Brassica oleracea*

Species	Ref.	Cytoplasm Origin	Transfer Method
Brassica oleracea	Nishi and Hiraoka (1958)	*Brassica rapa*	Interspecific cross and fertility restoration with 2 dominant genes
	Pearson (1972	*B. nigra*	Interspecific cross and fertility restoration with 2 dominant genes
	Christey et al. (1991)	*B. nigra*	Protoplast fusion
	Chiang and Crete (1987)	*B. napus*	Interspecific cross
	Yarrow et al. (1990)	*B. napus* (Polima)	Protoplast fusion
	McCollum (1981)	*R. sativus* (Early scarlet globe)	Intergeneric cross
	Bannerot et al. (1974)	*R. sativus* (Ogura)	Intergeneric cross
	Pelletier et al. (1989)	*R. sativus* (Ogura)	Protoplast fusion
	Kao et al. (1992)	*R. sativus* (Ogura)	Protoplast fusion
	Walters et al. (1992)	*R. sativus* (Ogura)	Protoplast fusion
	Kameya et al. (1989)	*R. sativus* (Shougoin)	Protoplast fusion

system was then improved (Kao et al., 1992; Pelletier et al., 1989; Walters et al., 1992; Walters and Earle, 1993).

Improved 'Ogura' cytoplasms were then transferred to vegetable *Brassica rapa* (Heath et al., 1994) and cabbage (Sigareva and Earle, 1997). Protoplast fusion was also an efficient way of combining atrazine-resistant chloroplasts in *B. rapa* with the CMS trait of *B. nigra* cytoplasm in *B. oleracea* (Christey et al., 1991).

A New Source of Cytoplasmic Male Sterility through Cybridization in *Brassica oleracea*

In *Brassica oleracea*, two sources of male sterility were known to originate from interspecific crosses: (1) the *Brassica nigra* cytoplasm (Pearson, 1972) and (2) the radish Ogura cytoplasm. In the first case, flowers are abnormal, without nectaries. Pollination being dependent on honeybees that refuse to visit flowers without nectar and restorer genes for this system occurring very frequently thus making the stock redundant. In the second case, several defects were practically unusable; as in the previous case, male-sterile plants produced no nectar and, moreover, a chlorophyll deficiency was expressed at low (12°C) temperature and flower malformations led to reduced female fertility.

Protoplast fusion experiments were performed between Ogura-bearing cytoplasm and normal fertile *Brassica oleracea* material to correct these defects. Protoplasts were isolated from leaves of axenic shoot cultures initiated from seedlings and propagated on a synthetic medium. Fusions were performed using the polyethylene glycol (PEG) method and protoplasts were cultured to produce colonies and regenerate plantlets. After regeneration, plantlets were grown in growth chambers and submitted to a vernalization treatment at 5 to 7°C for 8 weeks. This treatment helped in identifying cold-sensitive (expressed by yellowing) or cold-resistant plants, presuming that they correspond to plants having, respectively, Ogura or normal plastids. This vernalization period was, in general, sufficient to induce further flowering, and each plant was observed for pollen production. From about 1700 plants originating from independent colonies and observed under these conditions, 18 cybrids showing both male sterility and normal greening were selected. Each cybrid was crossed by several *B. oleracea* varieties — sauerkraut cabbage, broccoli, cauliflower, savory cabbage, garden cabbage, kales — to confirm the maternal inheritance and the stability of the male-sterile trait, the absence of restorer genes in the species, and to have an evaluation, on a large genetic basis, of the potential of each cybrid from a commercial seed production perspective.

In contrast to uniform correction of chlorophyll deficiency in these cybrids, which can be attributed to the simple replacement of Ogura plastids by normal plastids, in *Brassica oleracea* plastids, as shown by restriction fragment analysis of cybrid chloroplast DNA, the other defects were corrected to different degrees. On the basis of a completely normal seed set under conditions of open pollination by bees, some were retained for further use in commercial hybrid production. These selected cytoplasms led to normal flower morphology, good nectar production, and perfect male sterility under all conditions tested and whatever the genotype. This morphological variability of cybrids can be correlated with the composition of their mitochondrial genomes. It was observed that these well-formed cybrids contained mitochondrial genomes resembling the *B. oleracea* type more than the Ogura one on the basis of restriction fragment profiles, and that some parts of the Ogura genome were systematically deleted.

Molecular Analysis of Ogura Male Sterility

Although the mitochondrial genome of the Ogura radish has been extensively analyzed and compared to the normal fertile radish (Makaroff and Palmer, 1988; Makaroff et al., 1989;,1990, 1991), the identification of the mitochondrial determinant for this CMS has been possible only after *Brassica* cybrids were obtained via protoplast fusion (Pelletier et al., 1983). The fusion experiments not only gave the *B. napus* and *B. oleracea* cytotypes, now most extensively used for the commercial production of hybrid seeds, but also offered the best material for the molecular characterization of the sterility determinant.

The occurrence of interparental mitochondrial recombination in cybrids suggested that a comparison of these cybrids could be used to define the region of the genome carrying the CMS determinant

(Bonhomme et al., 1991). *Brassica oleracea* cybrids as well as *B. napus* cybrids obtained in similar experiments were used in this study. A 2.5-kb *NcoI* restriction fragment of Ogura radish mitochondrial DNA was detected in all (except one) male-sterile cybrids and was never present in fertile cybrids or fertile revertants. This specific fragment hybridizes to a 1.4-kb transcript found only in male-sterile plants bearing an Ogura-derived cytoplasm and to a 1.3-kb transcript in the male-sterile cybrid devoid of the 2.5-kb *Nco I* restriction fragment. The sequence analysis of this fragment revealed a tRNAfMet sequence, a 138 amino acid open reading frame (*orf138*), and a 158 amino acid ORF (*orf158*) previously observed in mitochondrial genomes from several other plant species (Bonhomme et al., 1992). Transcription mapping showed that both ORFs are present on the 1.4-kb CMS-specific transcript. The *orf158* sequence is also transcribed in fertile plants on a different mRNA, and thus cannot be related to CMS. The *orf138* is transcribed in all male-sterile cybrids and only in them. The translation product of this gene has been detected, after isolation of mitochondria from male-sterile cybrids and *in organelle* protein synthesis, as a 19-kDa polypeptide that is absent in the mitochondria of fertile plants (Grelon et al., 1994). Antibodies against a glutathione S-transferase-ORF138 fusion protein were raised to establish that this 19-kDa polypeptide is the product of *orf138*. The anti-ORF138 serum was used to demonstrate that the *orf138* translation product occurs only in sterile cybrids and co-purifies with the mitochondrial membrane fraction. Expression of this mitochondrial gene has since been associated with the sterile phenotype of Ogura radish (Krishnaswamy and Makaroff, 1993).

BREEDING *BRASSICA OLERACEA* VEGETABLES

The highest variability in plant morphology in the genus *Brassica* is exhibited where the plants are consumed as vegetables with root, stems, leaves, and terminal or axillary buds sometimes drastically modified to attain high nutritive qualities. This is particularly true for *Brassica oleracea*. As early as fourth century BC, Theophrastus described the use of two different type of cabbage in Greece. Plinius described the cultivation of even six distinct *Brassica* cultivars in the Roman empire, including types that resembled varieties of cabbage, kohlrabi, curly kale, or broccoli of today (Korber-Grohne, 1987).

During the distribution of these novel cultivated forms, new genetic characters were generated, also presumably by genetic introgression through hybridization with other related wild *Brassica* species (Gustafsson, 1981). Most interestingly, a parallel evolution occurred in China with *B. rapa* and *B. juncea*. In these species also, ideotypes were developed by means of selective cultivation, which formed modified plant organisms with particular qualities for vegetable uses.

Alternatives to the pedigree method are the development of doubled haploid (DH) lines and the single seed descend (SSD) method. From a genetical point of view, these two methods are very similar — both are rapid approaches to reach homozygosity without any selection. In experimental comparisons, DH and SSD populations from the same crosses were similar in mean and variance of the lines (Charne and Beversdorf, 1991; Stringham and Thiagarajah, 1991).

COMMERCIAL HYBRID DEVELOPMENT

Three principal requirements must be fulfilled for breeding hybrid cultivars: (1) there must be a sufficient amount of heterosis; (2) a system for large production of hybrid seed must be available; and (3) there must be an efficient method to identify hybrids with high combining ability.

In principle, three different genetic mechanisms exist to produce hybrid seed: (1) self-incompatibility (SI), (2) nuclear male sterility (NMS), and (3) cytoplasmic male sterility (CMS).

Inbreds may be obtained from *in vitro* anther or microspore culture as doubled haploids (Keller and Armstrong, 1983), or by selfing of about five generations. In the latter case, stigmas are hand-pollinated either several (9 to 10) days prior to anthesis, when the self-incompatibility barrier in the stigma has not yet been developed (Tatebe, 1951), or after previous spraying of the open flowers with a solution of 3% NaCl (Monteiro et al., 1988). After establishing the SI lines, their mass propagation is usually performed by *in vitro* microcloning (Clare and Collin, 1973) or by cultivation of the flowering self-pollinated plants at an increased concentration (approx. 5%) of CO_2 gas (Nakanishi et al., 1969; Palloix et al., 1985), which reduces the SI reaction to such an extent that self-pollen can attain fertilization and seed set. Another method for mass propagation in inbred lines is based on the use of near-isogenic sublines that only differ with regard to their S allele constitution, that is, S_1S_1 versus S_2S_2 and S_3S_3 versus S_4S_4 parental lines, which are directly used for hybrid seed production (Kuckuck, 1979).

Occasionally, high temperatures, high air humidity, or similar environmental anomalies may disrupt SI functions. Therefore, seed lots harvested for F_1 plant cultivation are routinely tested for the purity of their hybrid character. All seeds which may have formed within the female by selfing or intracrossing, will generate weak inbred plants, which of course are undesired. The proportion of these seeds is determined using morphological seedling markers or by biochemical markers such as isozymes or PCR markers (Nijenhuis, 1971). The off types are mainly contained in the small seed fraction. Meanwhile, the combination of breeding procedures and seed production techniques in cabbage regularly produces almost 100% hybridity in the commercial seed of most cultivars.

GENIC MALE STERILITY

The use of genic male sterility is limited because in many cases the male sterility phenotype is unstable and because male-fertile plants must be discarded just before flowering in hybrid seed production fields. However, in cauliflower, the selected male-sterile plants can be vegetatively propagated (*in vivo* or *in vitro*). Thus, it has been possible to use recessive male sterility for the commercial production of F_1 hybrids, and dominant male sterility is now used in the same way (Ruffio-Chable, 1994).

CYTOPLASMIC MALE STERILITY

The commercial F_1 hybrid production based on CMS has been achieved in *Brassica oleracea* using the improved Ogura cytoplasm obtained by Pelletier et al. (1989). F_1 hybrids of various *B. oleracea* types (cauliflower, sauerkraut cabbage, garden cabbage, and savoy cabbage) have also recently been registered (Leviel, 1998). Other improved Ogura cytoplasms are available in *B. oleracea* (broccoli, cauliflower, and cabbage) as well as in vegetable *B. rapa* (Chinese cabbage and pak choi) and are being tested for use in commercial hybrid production in seed companies worldwide (Earle and Dickson, 1995).

In cabbage breeding, where the F_1 hybrid is harvested as vegetative produce and its production does not depend on fertility restoration, the ogu-CMS is well accepted and used as a valuable hybridizing system (Clause-Semences, 1994; De Melo and Giordano, 1994).

INTERSPECIFIC HYBRIDIZATION FOR DISEASE RESISTANCE

Tonguc and Griffiths (2004c) produced interspecific hybrid plants and backcross 1 (BC_1) progeny through sexual crosses and embryo rescue between *Brassica carinata* accession BI 360883 and *B. oleracea* cvs. 'Titleist' and 'Cecile' to transfer resistance to powdery mildew to *B. oleracea*. Four interspecific hybrids were obtained through the application of embryo rescue from crosses with *B. carinata* as the maternal parent, and their interspecific nature was confirmed through plant morphology and random amplified polymorphic DNA (RAPD) analysis. Twenty-one BC_1 plants were obtained through sexual crosses and embryo rescue, although it was not necessary to produce first backcross general plants between interspecific hybrids and *B. oleracea*. All interspecific hybrids and eight of the BC_1 plants were resistant to powdery mildew.

Tonguc and Griffiths (2004b) identified resistance to major black rot races 1 or 4 in related *Brassica* species, including *B. carinata* and *B. napus*. In this study, two *B. juncea* accessions (A 19182 and A 19183) that are resistant to races 1 and 4 of Xcc were used as maternal and paternal parents to generate interspecific hybrids with *B. oleracea* cultivars. Interspecific hybrids were recovered using the embryo rescue technique and confirmed through inheritance of paternal molecular markers. Twenty-six interspecific hybrid plants were obtained between A 19182 and *B. oleracea* cultivars, but no interspecific hybrids were obtained using A 19183. Although interspecific hybrid plants were male-sterile, they were used successfully as maternal parents to generate backcross plants using embryo rescue. All hybrid and BC_1 plants were resistant to black rot races 1 and 4.

Tonguc and Griffiths (2004a) evaluated *Brassica carinata* for resistance to black rot disease and generated interspecific crosses with *B. oleracea*. Fifty-four accessions and susceptible control plants were wound inoculated with four isolates of Xcc race 4 at the juvenile stage of the 54 accessions tested; A 19182 and A 19183 exhibited no symptoms when inoculated with Xcc for all plants tested; and the accessions including PI 199947, PI 199949, and PC 194256 segregated for resistance to Xcc.

Tonguc et al. (2003) developed three segregating F_2 populations by self-pollinating three black rot resistant F_1 plants, derived from a cross between black rot resistant parent line 11B-1-12 and the susceptible cauliflower cultivar 'Snow Ball.' Plants were wound inoculated using four isolates of *Xanthomonas campestris* pv. *campestris* (Xcc) race 4, and disease security ratings of F_2 plants from the three populations were scored. A total of 860 arbitrary oligonucleotide primers were used to amplify DNA from black rot resistant and susceptible F_2 plants and bulks. Eight RAPD markers amplified fragments associated with completely disease-free plants following black rot inoculation, which segregated in frequencies far lower than expected. Segregation of markers with black rot resistance indicates that a single dominant major gene controls black rot resistance in these plants. Stability of this black rot resistance gene in populations derived from 11 B-1-12 may complicate introgression into *Brassica oleracea* genotypes for hybrid production.

MOLECULAR MAPPING AND ITS APPLICATION IN BREEDING

The advent of modern molecular techniques is playing an important role in understanding the organization and relationships of the *Brassica* genomes. Results from these studies not only confirmed the origin of the amphidiploid species, but also suggested that the A and C genomes originated from a single lineage, whereas the B genome is genetically distant to both the A and C genomes, forming a separate lineage (Song et al., 1990; Warwick and Black, 1991). A common assumption is that the $n = 8, 9, 10$ cultivated species have evolved in an ascending diploid series from a common primitive genome, 'Urgenome' (Haga, 1938). Although there are no known *Brassica* species in nature with genomes of less than $n = 7$ chromosomes, Catcheside (1934), Sikka (1940), and Röbbelen (1960) postulated that the ancestral genome for these species consisted of five or six basic chromosomes, which originated through polysomy. The present-day cultivated genomes range from $n = 8$ to $n = 10$ chromosomes. Thus, as a corollary of these hypotheses, the cultivated diploids can be considered secondary polyploids (Prakash and Hinata, 1980).

Genetic maps in *Brassica* serve two purposes: (1) understanding the relationship among the genomes of the *Brassica* cultivated diploid species, and (2) utilization in applied genetics and breeding of the numerous *Brassica* crops.

LINKAGE MAPS

Several maps have been developed independently for this species involving crosses between different crops. Slocum et al. (1990) reported an extensive RFLP map of 258 markers covering 820 recombination units in nine linkage groups, with average intervals of 3.5 units. This is a proprietary map developed from a broccoli × cabbage cross. Landry et al. (1992) constructed a map consisting of 201 RFLP markers distributed on nine major linkage groups covering 1112 cM. The F$_2$ progeny used to construct this map was developed by crossing a cabbage line resistant to clubroot to a rapid cycling stock (Figure 16.1). The cabbage line was derived from an interspecific cross involving *Brassica oleracea* and *B. napus,* followed by a series of backcrosses to *B. oleracea.* The resultant cabbage line is likely to have two chromosome segments of the A genome carrying the disease-resistant genes. These had been introgressed or substituted in the C genome of the cabbage line. Kianian and Quiros (1992) developed a map comprising 108 markers, spread in 11 linkage groups and covering 747 cM.

The map in Figure 16.1 was based on three intraspecific populations and one interspecific F$_2$ population, namely; collard × cauliflower, collard × broccoli, kale × cauliflower, and kohlrabi × *B. insularis.* The majority of the markers in the map are RFLP loci, with morphological and isozyme markers. Later, a few RAPD markers were added and the map was redrawn based only on the three intraspecific crosses using JOINMAP (Stam, 1993). It consists of 82 markers distributed on eight major linkage groups covering 431 cM (Quiros et al., 1994). Recombinant inbreds by single seed descent are under development for the collard by cauliflower and collard by broccoli progenies.

Camargo et al. (1997) reported a map developed in an F$_2$ population of cabbage × broccoli. It includes 112 RFLP and 47 RAPD loci and a self-incompatibility locus on nine main linkage groups. Kearsey et al. (1996) and Ramsay et al. (1996) constructed a linkage map based on backcross progenies obtained from double haploid lines of broccoli and *Brassica alboglabra*. Recombination of

Pathways to haploid embryos from microspores or pollen in *Brassica*

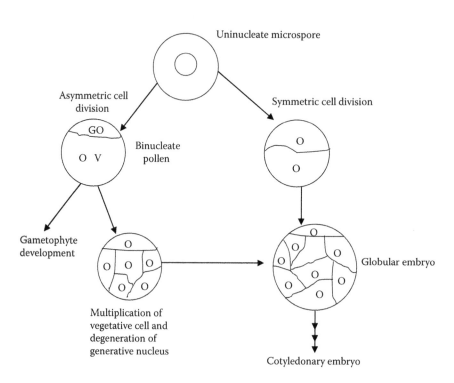

FIGURE 16.1   Linkage map of *Brassica oleracea* developed by an F$_2$ from a cabbage × broccoli cross. There are 112 RRFLP and 47 RAPD loci and a self-incompatibility locus on nine main linkage groups. Letters in parenthesis after some of the markers indicate recessive alleles originating from cabbage (A) or broccoli parent (B). Asterisks indicate loci with significant segregation distortion (*Source:* From Camargo et al., 1997.)

markers was higher in the BC_2 than in the BC_1 generation. These differences were attributed to differential chiasma frequency in female and male meiosis.

Bouhon et al. (1996) used this map for aligning the C genome linkage groups of *Brassica oleracea* and *B. napus*. Lagercrantz and Lydiate (1995) also observed sex-dependent recombination rates in *B. nigra*. In this species, however, recombination rates in male and female gametes proved to be a widespread phenomenon; this poses yet another complexity to consider when applying linkage information to breeding problems.

Most of the linkage groups in four of these maps (Hu et al., 1998) have been aligned. For this purpose, one linkage map was constructed from an F_2 population of 69 individuals with sequences previously mapped independently in three linkage maps of this species. These maps were published by Kianian and Quiros (1992), Landry et al. (1992), and Camargo et al. (1997). The base developed in this study consisted of 167 RFLP loci in nine linkage groups, plus eight markers in four linkage pairs, covering 1738 cM. Linkage group alignment was also possible with a fourth map published by Ramsay et al. (1996), containing common loci with the map of Camargo et al. (1997). In general, consistent linear order among markers was maintained, although often the distances between markers varied from map to map. A linkage group in Landry's map carrying a clubroot resistance QTL was found to be rearranged, consisting of markers from two other linkage groups. This was not surprising considering that the resistance gene was introgressed from *Brassica napus*. The extensively duplicated nature of the C genome was revealed by 19 sequences detecting duplicated loci within chromosomes and 17 sequences detecting duplicated loci between chromosomes. The variation in mapping distances between linked loci pairs on different chromosomes demonstrated that sequence rearrangement is a district feature of this genome. Although the consolidation of all linkage groups in the four *B. oleracea* maps compared was not possible, a high number of markers to corresponding linkage groups were added. Some chromosome segments were enriched with many markers that may be useful for future research in gene tagging or cloning.

The partial accomplishment of *Brassica oleracea* linkage groups to their respective chromosome task has been based on the development of alien addition lines allowing the construction of synteny maps. Two synteny maps for the nine C genome chromosomes, including isozymes RFLP and RAPD markers, were developed.

One map had approximately 194 markers and was constructed using a set of alien addition lines *Brassica rapa-oleracea* extracted from artificial *B. napus* 'Hakuran' (McGrath and Quiros, 1990; McGrath et al., 1990). The second map had approximately 103 markers and was assembled from a set of addition lines *B. rapa-oleracea* extracted from natural *B. napus* (Quiros et al., 1987). Chen et al. (1992) and Cheng et al. (1994a, b, 1995) also developed *B. rapa-oleracea* (alboglabra) alien addition lines, allowing the identification of the chromosome carrying genes for seed and flower color. Later on, various chromosomes of these lines were characterized by molecular markers (Jorgenson et al., 1996). Although it was possible in most cases to physically assign linkage groups to chromosomes using these sets of lines, two major complications arose from this activity. The first one is the frequent lack of polymorphism of interspecific markers of the alien chromosomes in the intraspecific crosses used to develop the linkage maps. The low coincidence of polymorphism between these two sets of materials makes chromosome assignment tedious and time consuming. The second complication is the instability of the alien chromosomes. This problem has been addressed by following groups of syntenic markers in the progeny of monosomic addition lines of *B. rapa-oleracea* (Hu and Quiros, 1991). Data have been obtained from two progenies of approximately 100 plants each, derived from two monosomic addition lines of *B. rapa-oleracea* for chromosomes C_4 and C_5 ($2n = 21$). After following several markers located on each chromosome, the alien chromosomes were found not to be always stable. All the expected markers were recovered in approximately 50% of the plants carrying the alien chromosome.

MARKER-ASSISTED BREEDING

The *Brassica* linkage maps are now extensively applied to tag genes of interest, including the quantitative trait log (QTL) of economic importance. Camargo and Osbourn (1996) performed QTL analysis of flowering time in F_3 populations of *Brassica oleracea* obtained by crossing cabbage and broccoli. A total of five QTLs were detected; one of these was linked to the S locus slgb. Another QTL was associated to petiole length. No conservation for flowering time QTLs were detected between *Brassica oleracea* and *B. napus* and *B. rapa*.

Clubroot (*Plasmodiophora brassicae*) resistance in *Brassica oleracea:* RAPD markers associated to this trait were detected by Grandclement et al. (1996). Previous work in *B. napus* by Figdore et al. (1993) had identified three QTLs for this trait. Landry et al. (1992) found two QTLs associated to resistance to this disease in *B. oleracea*.

Black rot (*Xanthomonas compestris*) in *Brassica oleracea*: Two QTLs were found associated to this trait by Camargo et al. (1995). One of these was also associated to petiole length.

Morphological traits: Genes or markers for 28 traits, some of which were associated to as many as five QTLs were determined in a *B. rapa* progeny of Chinese cabbage 'Michihili' × 'Spring' broccoli (Song et al., 1995).

HAPLOIDY BREEDING

The development of haploids in a number of plant species is now recognized as the most rapid route to the achievement of homozygosity and the production of pure lines. They occur in a number of families and genera of both angiosperms and gymnosperms. There are several ways in which haploidy can be achieved:

1. Spontaneous occurrence in plant populations where they are recognized by small narrow leaves, male sterility, and abnormal physical features. Such haploids occur at very low frequency and may be of maternal or paternal origin.
2. Anther and microspore cultures where the male gamete develops into an embryo by a process called androgenesis.
3. The culture of unfertilized ovules and ovaries leading to embryo development from one or more of the haploid cells within the unfertilized embryo sac. This process is called gynogenesis.
4. Chromosome elimination as a consequence of wide crosses, as is the case with the maize-mediated system of haploidy in wheat (Laurie and Bennett, 1988).

Haploids should be produced efficiently to be useful in crop improvement. With anther and microspore cultures, the potential exists for the recovery of large numbers of haploid plants because of the large number of microspores/pollen grains in each anther. As a consequence, this is the method of choice for haploid production in monocots and dicots. In *Brassica* and related species, this is the most widely used method of haploid plant production and although there are variations, such plants can be produced in a wide variety of genotypes, some of which are of commercial importance. With the available method for chromosome doubling, doubled haploid plants can be efficiently recovered in most cases.

HISTORICAL OVERVIEW OF HAPLOIDY

The process of microsporogenesis follows a precise pathway leading to the differentiation of the male gametophyte and haploid sperm cells. Environmental disturbances, especially high temperature, alter the differentiation pathway, resulting in a sporophytic pattern of development and the emergence of pollen embryos (Stow, 1930; LaCour, 1949). This process is referred to as androgenesis and occurs naturally, but at low frequencies, in some species where the male gamete develops into an embryo in the embryo sac (Kimber and Riley, 1963; Pandey, 1973). The value of haploids was realized in plant breeding when it became known that homozygous plants could be recovered from these through chromosome doubling. The research was then directed toward the development of experimental protocols for the production of such plants by anther culture.

The confines of the anther walls might qualitatively and quantitatively limit haploid embryo-genesis (Hoffmann et al., 1982) as only a small number of embryos were produced compared to a large number of microspores within each anther. As a consequence, methods were developed for the culture of isolated microspores that increased the frequency of embryogenesis severalfold. In *Brassicas*, an isolated microspore culture protocol was first reported by Lichter (1982) and with modification it is possible to recover haploid embryos from isolated microspores of a number of *Brassica* species. The embryos produced through isolated microspore culture are more efficient.

For the efficient recovery of haploid embryos from *Brassica* species, one of the most important controlling factors is the genetic make-up of the donor plant (Dunwell et al., 1985; Baillie et al., 1992; Ferrie and Keller, 1995). Despite genetics being a major factor, environmental interactions do modify this effect (Dunwell et al., 1985; Roulund et al., 1990; Thurling and Chay, 1984). The field grain plants of *B. oleracea* convar. capitata (L.) Alif. were generally found to be more responsive than greenhouse grown plants (Roulund et al., 1990).

The plant growth regulator response may be species specific as cytokinins suppressed embryo-genesis in anther cultures of cauliflower (Yang et al., 1992), while the response was genotype dependent with broccoli anther cultures (Arnison et al., 1990). Low levels of auxin enhanced microspore embryogenesis in *Brassica oleracea*, while high levels were inhibitory (Ockendon and McClenaghan, 1993; Yang et al., 1992). Callusing of anther tissues appears to increase with high levels of auxins, and this was correlated with reduced microspore embryogenesis (Yang et al., 1992). Exogenous ethylene and conditions that enhance endogenous ethylene production are inhibitory to pollen embryogenesis (Biddington and Robinson, 1990, 1991). The *B. oleracea* genotypes that are unresponsive to culture showed enhanced ethylene production (Biddington and Robinson, 1990; Biddington et al., 1992, 1993). Endogenous ethylene production is a limiting factor in *B. oleracea* microspore embryogenesis, and anther filament and wall tissues are rich sources of endogenous ethylene (Biddington, 1992). Exogenous abscisic acid inhibited embryogenesis in anther cultures of *B. oleracea* by stimulating ethylene production (Biddington et al., 1992). The pathways to embryo-genesis from the uninucleate microspore are indicated in Figure 16.2.

Figure 16.2 shows both symmetric and asymmetric cell division from the uninucleate microspores, and binucleate pollen may divide to produce embryos without gametophyte development.

ADVANTAGES OF MICROSPORE-DERIVED EMBRYOS

From a genetics and breeding point of view, an advantage of microspore culture technology in *Brassica* is the rapid production of homozygous lines from haploid embryos through chromosome doubling compared to conventional backcrossing. With doubled haploids, all loci are homozygous and all func-tional genes are expressed. The value of this technology is evident in outcrossing or self-incompatible species such as *Brassica oleracea* and *B. rapa* where it is a reliable method of rapid homozygous plant production (Duijs et al., 1992). With male-sterile genotypes, microspore culture is an advantageous method of achieving homozygosity where there is microspore development to the responsive stage (Sodhi et al., 1995). Microspore culture technology has therefore become an integral part of *Brassica* breeding programs (Keller et al., 1982; Hoffman et al., 1982; Keller et al., 1987; Chen and Beversdorf, 1990; Scarth et al., 1991; Stringham and Thiagarajah, 1991; Dewan et al., 1995).

From a mutation and selection angle in *Brassica oleracea*, doubled haploid plants were screened and resistance to clubroot disease identified (Voorrips and Visser, 1990). This was identified with-out *in vitro* selection pressure and represented gametoclonal variation that existed in the culture (Morrison and Evans, 1988).

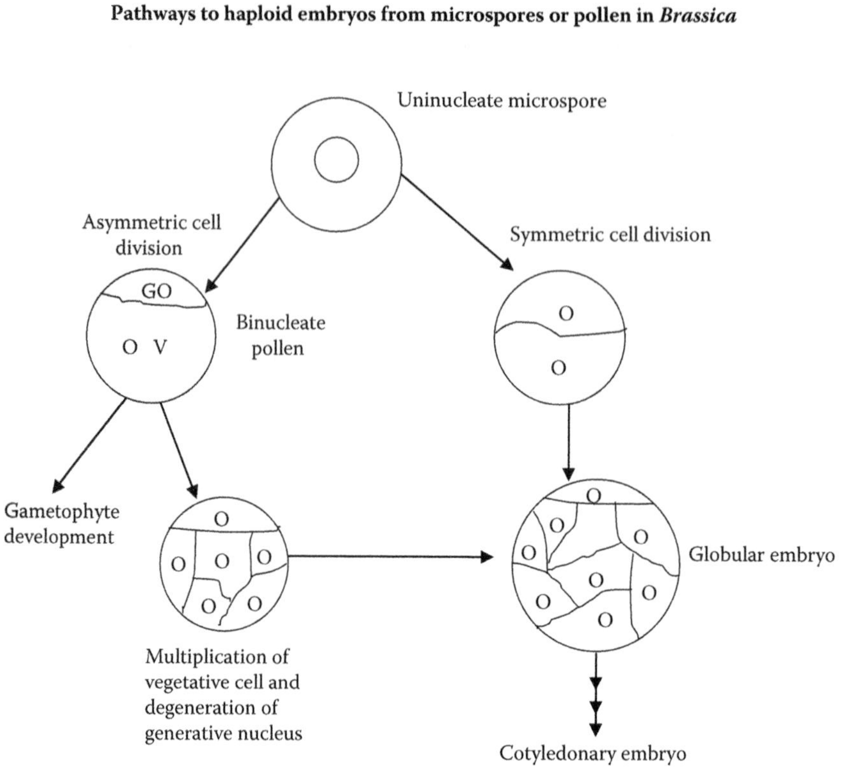

Pathways to haploid embryos from microspores or pollen in *Brassica*

FIGURE 16.2 Symmetric and asymmetric cell division from the uninucleate microspores, and binucleate pollen may divide to produce embryos without gametophyte development.

GENETIC ENGINEERING

Transformation technology has already greatly expanded the genetic diversity of *Brassica* crops. It is providing improved forms of vegetable crops and may also permit their new agricultural uses. *Brassica oleracea* transformants include all the major vegetables (broccoli, cauliflower, cabbage, kale, Brussels sprouts) as well as rapid cycling lines.

Methods of Transformation

Most research efforts have been directed at developing *Agrobacterium tumefaciens* mediated transformation, with little emphasis on direct transfer methods. A common procedure for all varieties of *Brassica oleracea* being unavailable, the protocols generally remain genotype specific (Puddephat et al., 1996). *A. rhizogenes* mediated transformation causes formation of hairy roots that can be induced to form shoots. However, the bacteria carry the *rol* gene that may have phenotypic effects, for instance on flowering. The method has recently proved efficient in transferring the NptII and other genes in 12 vegetable *Brassica* cultivars representing six varieties: broccoli, Brussels sprouts, cabbage, cauliflower, rapid cycling cabbage, and Chinese cabbage. However, fertility was often reduced and morphogenic changes were noted in a number of plants (Christey et al., 1997). For cauliflower, a number of protocols were developed using marker genes (Puddephat et al., 1996). The protocol of DeBlock et al. (1989) has been modified by Bhalla and Smith (1998) using a special combination of growth hormones, starting with cotyledons rather than hypocotyls, but still using silver nitrate, an inhibitor of ethylene action. Transgenic plants of three commercial genotypes were produced harboring an antisense BcpI gene encoding a protein essential for pollen functionality driven by a pollen-specific promoter. Another protocol, also using silver nitrate but starting with hypocotyls previously treated with 2,4-dichlorophenoxyacetic acid, gave very good regeneration rates and led to the production of 100 primary transformants putatively harboring a trypsin inhibitor gene conferring resistance to insects (Ding and Hu, 1998).

An *Agrobacterium tumefaciens*-mediated transformation method has been devised for broccoli by Metz et al. (1995) using flowering stocks. This method, derived from the protocol of Toriyama et al. (1991), has been used for transferring *Bacillus thuringiensis* genes into broccoli with a transformation efficiency of about 6.4% (Metz et al., 1995; Cao and Tang, 1999). An improved *A. rhizogenes* mediated transformation of broccoli has been reported by Henzi et al. (1999). This protocol has been used to generate plants producing low levels of ethylene (Henzi et al., 1999).

The Chinese cabbage (*Brassica campestris* ssp. *pekinensis*) is considered a recalcitrant species in plant regeneration. Procedures for the transformation and regeneration of transgenic plants have been reported by Jun et al. (1995) for the 'Spring flavor' genotype, and by Lim et al. (1998) for a number of other genotypes. However, the efficiency of the transformation was low and depended on the genotype.

TRANSFORMATION FOR HERBICIDE RESISTANCE

Putative transformants of Chinese cabbage have been shown to express *bar* (bialaphos resistance) gene (Lim et al., 1998). However, studies on the resistance of the transgenic plants to the herbicide and the inheritability of the transgene have not been performed.

RESISTANCE TO VIRUSES

Cauliflower

Transgenic cauliflower carrying the capsid gene and antisense gene VI of the cauliflower mosaic virus were generated through *Agrobacterium tumefaciens* mediated transformation. However, while the transcription of the transgenes was detected in all plants, the capsid protein was not present.

Chinese Cabbage

The tobacco mosaic virus 355S coat protein gene has been expressed in five regenerants of the 'Spring flavor' cultivar of Chinese cabbage (*Brassica campestris* ssp. *pekinensis*). Stable inheritance of the gene was shown in the progeny but virus resistance was not assessed (Jun and Kwon, 1995).

RESISTANCE TO INSECTS AND FUNGI

Cauliflower

Insect pests represent a serious problem for cauliflower cultivation. A trypsin inhibitor from the sweet potato has been transferred to Taiwan cauliflower cultivars that gave transgenic primary transformants substantial resistance to local insects upon in-planta feeding bioassays (Ding and Hu, 1998). Progeny behavior studies and field tests remain to be performed.

Broccoli

Metz et al. (1995) have generated a large number of transgenic broccoli lines carrying the Bt Cry1A© gene, most of them causing 100% mortality of first instar larvae of the diamondback moth, a major insect pest of crucifers. However, Cry1A-resistant larvae were able to survive on the transgenic plants. More recently, a synthetic Bt Cry1C gene was introduced also using the method developed by Metz et al. (1995). Lines producing high levels of Cry1C protein were protected not only from susceptible or Cry1A resistant diamondback moth larvae, but also from larvae selected for moderate levels of resistance to CrylC (Cao and Tang, 1999). In addition, the CrylC-transgenic broccoli was also resistant to other lepidopteran pests of crucifers such as cabbage looper and imported cabbage worm.

Cabbage

The CrylA© gene was likewise introduced into cabbage (Metz et al., 1995); however, the disadvantages of this gene in failing to control resistant insects are the same as already discussed for broccoli. The introduction of other synthetic Bt genes is awaited in this variety.

QUALITY TRAITS

Broccoli

Transgenic lines of broccoli containing a tomato antisense 1-cyclopropane-1-carboxylic acid oxidase gene showed a significant reduction in ethylene production in the florets (Henzi et al., 1999). However, the decrease in ethylene production was probably not enough (although sometimes it reached more than 90%) to slow down senescence and preserve the quality of the florets. Higher levels of reduction are required to obtained interesting phenotypes, in particular through the use of homologous genes.

TYPES OF GENES TRANSFERRED

In *Brassica oleracea,* studies have been limited to establishing and optimizing the transformation system, generally using the gas reporter gene or selectable marker genes, although a few genes of interest have been tested (Table 16.3).

FUTURE PROSPECTS OF TRANSGENICS

The impact of transgenic *Brassica* vegetables is very difficult to predict, especially because of opposition to GMOs worldwide and also due to less support for work on these diverse groups of vegetables. Obtaining the data required for regulatory and legal approval is onerous and expensive. On the other hand, transgenic resistance to herbicides, diseases, or insect pests could have important environmental benefits for vegetable production by allowing the use of modern herbicides and for a reduction in the application of insecticides. Such changes could be especially beneficial in developing countries where application of chemicals for insect pests such as the diamondback moth (DBM) is often high and poorly regulated. Opportunities for using other genes already identified in other systems include control of pollination, alteration of growth habits, or extension of shelf life if the cost comparisons to other approaches are favorable. Recent attention given to *Brassica* vegetables as anticarcinogens makes manipulation of nutrients and flavor components another attractive target for genetic manipulation.

REFERENCES

Aguinagalde, I., Gomez-Campo, C., and Sanchez-Yelamo, M.D. (1992). A chemosystematic survey on wild relatives of *Brassica oleracea* L. *Bot. J. Linean Soc.,* 109:57–67.

Aguinagalde, I., Sanchez-Yelamo, M.D., and Sagarodsky, T. (1991). Phytochemical diversity in *Brassica cretica* Lam. *Botanika Chronika,* 10:667–672.

Anstey, T.H. and Moore, J.F. (1954). Inheritance of glossy foliage and cream petals in green sprouting broccoli. *J. Hered.,* 45:39–41.

Arnison, P.G., Donaldson, P., Jackson, A., Semple, C., and Keller, W.A. (1990). Genotype-specific response of cultured broccoli (*Brassica oleracea* var. *italica*) anthers to cytokinins. *Plant Cell Tissue Organ. Cult.,* 20:217–222.

Bai, Y.Y., Mao, H.Z., Cao, X.L., Tang, T., Wu, D., Chen, D.D., Li, W.G., and Fu, W.J. (1993). Transgenic cabbage plants with insect tolerance. In You, C.B., Chen, Z., and Ding, Y. (Eds.), *Current Plnt Science and Biotechnology in Agriculture,* Kluwer, 15:156–159.

TABLE 16.3
Transformation of *Brassica oleracea* Subspecies

Species	Variety or Subspecie	Method	Gene (other than markers or *Agrobacterium* genes)	Ref.
B. oleracea	*italica*	*Agrobacterium transformation*	S-locus genes	Toriyama et al. (1991); Conner et al. (1997)
		Agrobacterium transformation	Bt[*cryIA(c)*]	Metz et al. (1995a, b)
		Agrobacterium rhizogenes		Hosoki et al. (1991)
		Agrobacterium rhizogenes	Antisense ACC	Henzi et al. (1998)
		Agrobacterium rhizogenes		Christey et al. (1997)
	botrytis	*Agrobacterium rhizogenes (onc.)*		David and Tempe (1988)
		Agrobacterium transformation (onc.)		Srivastava et al. (1988)
		Agrobacterium transformation	*bar*	DeBlock et al. (1989)
		Direct, to PP	*bar*	Mukhpadhyay et al. (1991)
		Agrobacterium rhizogenes		Christey et al. (1997)
		Agrobacterium transformation	Anti sense EFE	Bai et al. (1993)
			Bt[*cryIA(c)*]	Metz et al. (1995b)
	capitata	*Agrobacterium rhizogenes*		Christey et al. (1997)
				Berthomieu et al. (1994)
	Rapid cycling	*Agrobacterium transformation (onc. + disarmed)*		Berthomieu and Jouanin (1992)
	Rapid cycling	*Agrobacterium rhizogenes*		Sato et al. (1991)
	acephala	*Agrobacterium transformation*		Christey and Sinclair (1992)
		Agrobacterium rhizogenes	ALS	Hosoki et al. (1989, 1994)
		A.t and A.r.		Hamada et al. (1989);
	gemmifera	*A.r.*		Christey et al. (1997)
	Rapid cycling	*A.t.*		Millam et al. (1994)
B. rapa	*pekinensis*	*A.t.*	TMV coat protein	Jun et al. (1995)

–continued

TABLE 16.3 (Continued)
Transformation of *Brassica oleracea* Subspecies

Species	Variety or Subspecie	Method	Gene (other than markers or *Agrobacterium* genes)	Ref.
		A.r.		He et al. (1995); Christey et al. (1997)
	rapifera	A.r.	ALS	Christey and Sinclair (1992)
B. napobrassica		A.t.	Bt(cryIA)	Li et al. (1995)

Note: A.t.: Agrobacterium tumifaciens, disarmed except when noted as oncogenic (onc.); *A.r.: Agrobacterium rhizogenes; PP:* protoplast; *S*-locus gene: *Brassica* genes related to self- incompatibility; Bt: δ-endotoxin gene from *Bacillus thuringiensis;* ACC: ACC oxidase; bar: confers resistance to glufosinate (phosphinotricin) herbicides; EFE: ethylene forming enzyme; ALS: confers resistance to sulfonylurea herbicides.

Baille, A.M.R., Epp, D.J., Hutcheson, D., and Keller, W.A. (1992). *In vitro* culture of isolated microspores and regeneration of plants in *Brassica campestris*. *Plant Cell Reports,* 11:234–237.

Bannerot, H., Boulidard, L., and Chupeau, Y. (1977). Unexpected difficulties met with the radish cytoplasm. *Cruciferae Newslett.,* 2:16.

Bannerot, T., Boulidard, L., Cauderon, Y., and Tempe, J. (1974). Transfer of cytoplasmic male sterility from *Raphanus sativus* to *Brassica oleracea. Eucarpia Meeting on Cruciferae,* Dundee, Scotland, p. 52–54.

Bateman, A.J. (1954). Self incompatibility systems in angiosperms. II. *Iberis amara. Heredity,* 8:305–332.

Bateman, A.J. (1955). Self incompatibility systems in angiosperms. III. *Cruciferae. Heredity,* 9:52–68.

Berthomieu, P. and Jouanin, L. (1992). Transformation of rapid cycling cabbage (*Brassica oleracea* var. *vapitata*) with *Agrobacterium rhizogenes. Plant Cell Rep.,* 11:334–338.

Berthomieu, P., Beclin, C., Charlot, F., Dore, C., and Jouanin, L. (1994). Routine transformation of rapid cycling cabbage (*Brassica oleracea*) — Molecular evidence for regeneration of chimeras. *Plant Sci.,* 96:223–235.

Bhalla, L. and Smith, N.A. (1998). *Agrobacterium tumefaciens*-mediated transformation of cauliflower, *B. oleracea* var. *botrytis. Mol. Breed.,* 4:531–541.

Biddington, N.L. (1992). The influence of ethylene in plant tissue culture. *Plant Growth Regul.,* 11:73–87.

Biddington, N.L. and Robinson, H.T. (1990). Variations in response to high temperature treatments in anther culture of Brussels sprouts. *Plant Cell Tissue Organ. Cult.,* 22:48–54.

Biddington, N.L. and Robinson, H.T. (1991). Ethylene production during antherculture of brussels sprouts (B.O. var gemmiferra) and its relationship wih factors that affect embryo production *Plant Cell Tissue Organ. Cult.,* 25:169–177.

Biddington, N.L., Robinson, H.T., and Lynn, J.R. (1993). ABA promotion of ethylene production in anther culture of Brussels sprout (*Brassica oleracea* var. *gemmifera*) and its relevance to embryogenesis. *Physiol. Plant,* 88:577–582.

Biddington, N.L., Sutherland, R.A., and Robinson, H.T. (1992). The effects of gibberellic and fluridone, abscisic acid and pactobutrazol on anther culture of Brussels sprouts. *Plant Growth Regul.,* 11:81-84.

Bonhomme, S., Budar, F., Ferault, M., and Pelletier, G. (1991). A 2.5 kb Ncol fragment of Ogura radish mitochondrial DNA is correlated with cytoplasmic male sterility in *Brassica* cybrids. *Curr. Genet.,* 19:121–127.

Bonhomme, S., Budar, F., Lancelin, D., Small, I., Defrance, M.C., and Pelletier, G. (1992). Sequence and transcript analysis of the *NcoI* Ogura specific fragment correlated with cytoplasmic male sterility in *Brassica* cybrids. *Mol. Gen. Genet.,* 235:340–348.

Bonnet, A. (1975). Introduction et utilization d'une sterilite male cytoplasmique dans des varietes precoces europeennes de radis, *Raphanus sativus* L. *Ann. Amel. Plant,* 25:381–397.

Bonnet, A. (1977). Breeding in France of a radish F_1 hybrid obtained by use of cytoplasmic male sterility. *Eucarpia Cruciferae Newslett.* 2:5.

Borchers, E.A. (1966). Characteristics of a male sterile mutant in purple cauliflower (*Brassica oleracea* L.). *Proc. Am. Soc. Hort. Sci.,* 88:406–410.

Borgen, L., Rustan, O.H., and Elven, R. (1979). *Brassica bourgeaui* (Cruciferae) in the Canary Islands. *Norw. J. Bot.,* 26:255–264.

Boswell, V.R. (1949). Our vegetable travelers. *Nat. Geogr. Magaz.* 96:134–217.

Bothmer, R. von, Gustafsson, M., and Snogerup, S. (1995). *Brassica* sect. *Brassica (Brassicaceae).* II. Inter- and intraspecific crosses with cultivars of *B. oleracea. Genet. Resources and Crop Evolut.,* 42:165–178.

Bouhon, E.J.R., Keith, D.J., Parkin, I.A.P., Sharpe, A.G., and Lydiate, D. (1996). Alignment of conserved C genomes of *Brassica oleracea* and *B. napus. Theor. Appl. Genet.,* 93:833–839.

Bowman, J.L., Alvarez, J., Weigel, D., Meyerowitz, E.M., and Smyth, D. (1993). Control of flower development in *A. thaliana* by AP_1 and interacting genes. *Development,* 119:721–743.

Camargo, L.E.A. and Osborn, T.C. (1996). Mapping loci controlling flowering time in *Brassica oleracea. Theor. Appl. Genet.,* 92:610–616.

Camargo, L.E.A., Savides, L., Jung, G., Nienhuis, J., and Osborn, T.C. (1997). Location of the self incompatibility locus in an RFLP and RAPD map of *Brassica oleracea. J. Hered.,* 88:57–59.

Camargo, L.E.A., Williams, P.H., and Osborn, T.C. (1995). Mapping of quantitative trait loci controlling resistance of *Brassica oleracea* to *Xanthomonas campestris* pv *campestris* in the field and greenhouse. *Phytopathology,* 85:1296–1300.

Cao, J. and Tang, J.D. (1999). Transgenic broccoli with high levels of *Bacillus thuringiensis Cry 1C* protein control diamondback moth larvae resistant to Cry 1A or Cry 1C. *Mol. Breed.,* 5:131–141.

Cardi, T. and Earle, E.D. (1997). Production of new CMS *Brassica oleracea* by transfer of 'Anand' cytoplasm from *B. rapa* through protoplast fusion. *Theor. Appl. Genet.,* 94:202–212.

Catcheside, D.G. (1934). The chromosomal relationship in the swede and turnip groups *Brassica. Ann. Bot.,* 601:33–56.

Charne, D.G. and Beversdorf, W.D. (1991). Comparison of agronomic and compositional traits in microspore-derived and conventional populations of spring *Brassica napus.* In McGregor, D.I. (Ed.), *Proc. 8th Int. Rapeseed Congr,* Saskatoon, Canada, 1:64–69.

Chatterjee, S.S. and Swarup, V. (1972). Indian cauliflower has a still greater future. *Ind. Hortic.* 17:18–20.

Chen, B.Y., Simonsen, V., Lanner-Herrera, C., and Heneen, W.K. (1992). A *Brassica campestris-alboglabra* addition line and its use for gene mapping, intergenomic gene transfer and generation of trisomics. *Theor. Appl. Genet.,* 84:592–599.

Chen, J.L. and Beversdorf, W.D. (1990). A comparison of traditional and haploid derived breeding populations of oilseed rape (*Brassica napus* L.) for fatty acid composition of the seed oil. *Euphytica,* 51:59–65.

Cheng, B.F., Chen, B.Y., and Heneen, W.K. (1994a). Addition of *Brassica alboglabra* Bailey chromosomes to *B. campestris* L. with special emphasis to seed colour. *Heredity,* 73:185–189.

Cheng, B.F., Heneen, W.K., and Chen, B.Y. (1994b). Meiotic studies of a *Brassica campestris alboglabra* monosomic addition line and derived *B. campestris* primary trisomics. *Genome,* 37:584–589.

Cheng, B.F., Heneen, W.K., and Chen, B.Y. (1995). Mitotic karyotypes of *Brassica campestris* and *B. alboglabra* and identification of the *B. alboglabra* chromosome in an addition line. *Genome,* 38:313–319.

Chiang, B.Y. and Grant, W.F. (1975). A putative heterozygous interchange in the cabbage (*Brassica oleracea* var. *capitata*) cultivar 'Badger Shipper'. *Euphytica,* 24:581–584.

Chiang, M.S. and Crête, R. (1987). Cytoplasmic male sterility in *Brassica oleracea* induced by *B. napus* cytoplasm, female fertility and restoration of male fertility. *Can J. Plant Sci.,* 67:891–897.

Chaing, M.S., C. Chong, B.S. Landey, and R. Crête. (1993). Cabbage, pp. 113–115. In G. Kalloo and B.O, Bergh (eds.) *Genetic Improvement of Vegetable Crops.* Pergamon Press, Oxford.

Christey, M.C. and Sinclair, B.K. (1992). Regeneration of transgenic kale (*Brassica oleracea* var. *acephala*), rape (*B. napus*) and turnip (*B. campestris* var. *rapifera*) plants via *Agrobacterium rhizogenes* mediated transformation. *Plant Sci.,* 87:161–169.

Christey, M.C., Makaroff, C.A., and Earle, E.D. (1991). Atrazine resistant cytoplasmic male sterile-nigra-broccoli obtained by protoplast fusion between cytoplasmic male sterile *Brassica oleracea* and attrazine resistant *Brassica campestris. Theor. Appl. Genet.,* 83:201–208.

Christey, M.C., Sinclair, B.K., Braun, R.H., and Wyke, L. (1997). Regeneration of transgenic vegetable *Brassicas* (*B. oleracea* and *B. campestris*) via Ri mediated transformation. *Plant Cell Rep.,* 16:587–593.

Clare, M.V. and Collin, H.A. (1973). Meristem culture of Brussels sprouts. *Hort. Res.,* 13:111–118.

Clause-Semences, (1994). Kopfkohl-Die Hybritop-Sorten. *Industr. Obst-Gemuseverwertung,* 79:448–449.

Cole, K. (1959). Inheritance of male sterility in green sprouting broccoli. *Can. J. Genet. Cytol.,* 1:203-207.

Conner, J.A., Tantikanjana, T., Stein, J.C., Kandasamy, M.K., Nasrallah, J.B., and Nasrallah, M.E. (1997). Transgene-induced silencing of S locus genes and related genes in *Brassica. Plant J.,* 11:809–823.

Crisp, P. (1982). The use of an evolutionary scheme for cauliflowers in the screening of genetic resources. *Euphytica,* 31:725–734.

Cunha, C., Tonguc, M., and Griffiths, P.D. (2004). Discrimination of diploid crucifer species using PCR-RFLP of chloroplast DNA. *HortScience,* 39(7):1575–1577.

Dalechamp, J. (1587). *Historia Generalis Plantarum,* Lugduni, Batavia.

David, C. and Tempe, J. (1988). Genetic transformation of cauliflower (*Brassica oleracea* L. var. *botrytis*) by *Agrobacterium* rhizogenes. *Plant Cell Rep.,*7:88–91.

De Melo, P.E. and Giordano, L. de B. (1994). Effect of Ogura male sterile cytoplasm on the performance of cabbage hybrid variety. II. Commercial characteristics. *Euphytica,* 78:149–154.

DeBlock, M. and Debrouwer, D. (1993). Engineered fertility control in transgenic *Brassica napus* L.: histochemical analysis of anther development. *Planta,* 189:218–225.

DeBlock, M., deBrouwer, D., and Tenning, P. (1989). Transformation of *Brassica napus* and *B. oleracea* using *A. tumefaciens* and the expression of the bar and neo genes in the transgenic plants. *Plant Physiol.,* 91:694–701.

Denis, M., Delourme, R., Gourret, J.P., Mariani, C., and Renard, M. (1993). Expression of engineered nuclear male sterility in *Brassica napus. Plant Physiol.,* 101:1295–1304.

Dewan, D.B., Downey, R.K., and Rakow, G.F.W. (1995). Field evaluation of *Brassica rapa* doubled haploid. *Proc. 9th Int. Rapeseed Congr.,* Cambridge, U.K., 3:795–797.

Dickinson, H.G. and Lewis, D. (1973). The formation of the tryphine coating the pollen grains of *Raphanus,* and its properties relating to the self incompatibility system. *Proc. R. Soc. Lond. B.,* 184:149–165.

Dickson, M.H. (1970). A temperature sensitive male sterile gene in broccoli, *Brassica oleracea* L. var *italica. J. Am. Soc. Hort. Sci.,* 95:13–14.

Ding, L.C. and Hu, C.Y. (1998). Development of insect-resistant transgenic cauliflower plants expressing the trypsin inhibitor gene isolated from local sweet potato. *Plant Cell Rep.,* 17:854–860.

Duijs, J.G., Voorrips, R.E., Visser, D.L., and Custers, J.B.M. (1992). Microspore culture is successful in most crop types of *Brassica oleracea* L. *Euphytica,* 60:45–55.

Dunemann, F. and Grunewaldt, J. (1991). Identification of a monogenic dominant male sterility mutant in broccoli (*Brassica oleracea* L. var. *italica* Plenck). *Plant Breeding,* 106:161–163.

Dunwell, J.M., Cornish, L.M., and DeCourcel, A.G.F. (1985). Influence of genotype, plant growth temperature and anther incubation temperature on microspore embryo production in *Brassica napus* ssp. *oleifera, J. Expt. Bot.,* 36:679–689.

Earle, E.D. and Dickson, M.H. (1995). *Brassica oleracea* cybrids for hybrid vegetable production. In Terzi, M., Cella, R., and Falavigna, A. (Eds.), *Current Issues in Plant Molecular and Cellular Biology,* p. 171–176. Kluwer Academic Publishers, Dordrecht, The Netherlands.

Eastwood, A. (1996). The Conservation and Utilization of *Brassica incana* ten. in Croatia, with Particular Reference to Leaf Trichomes as a Potential Source of Insect Resistance. M.Sc. thesis. Faculty of Science. University of Birmingham.

Fernandez-Prieto, J.A. and Herrera-Gallastegui, M. (1992). *Brassica oleracea* L., distribution y ecologia en las costas atlanticas ibericas. *Lazaroa,* Sevilla, 13:121–128.

Figdore, S.S., Ferriera, M.E., Slocum, M.K., and Williams, P.H. (1993). Association of RFLP markers with trait loci affecting clubroot resistance and morphological characters in *Brassica oleracea. Euphytica,* 69:33–44.

Ford, M.A. and Kay, Q.O.N. (1985). The genetics of incompatibility in *Sinapis arvensis. Heredity,* 54:99–102.

Goldberg, R.B. (1988). Plants: novel development processes. *Science,* 25:1460–1467.

Gomez-Campo, C. (1993). *Brassica.* In Castroviejo, S. et al. (Eds.)., *Flora Iberica,* CSIC, Madrid, 4:362–384.

Gomez-Campo, C., Tortosa, M.E., Tewari, I., and Tewari, J.P. (1999). Epicuticular wax columns in cultivated *Brassica* species and in their close wild relatives. *Ann. Bot.,* 83:515–519.

Grandclement, C., Laurent, F., and Thomas, G. (1996). Detection and analysis of QTLs based on RAPD markers for polygenic resistance to *Plasmodiophora brassicae* Woron in *Brassica oleracea* L. *Theor. Appl. Genet.,* 93:86–90.

Gray, A.R. (1982). Taxonomy and evolution of broccoli (*Brassica oleracea* var. *italica*). *Econ. Bot.,* 36:397–410.

Grelon, M., Budar, F., Bonhomme, S., and Pelletier, G. (1994). Ogura cytoplasmic male sterility (CMS)-associated *orf138* is translated into a mitochondrial membrane polypeptide in male sterile *Brassica* cybrids. *Mol. Gen. Genet.,* 243:540–547.

Gustafson-Brown, C., Savidge, B., and Yanofsky, M.F. (1994). Regulation of the *Arabidopsis floral* homeotic gene APETALA1. *Cell,* 76:131–143.

Gustafsson, M. (1979). Biosystematics of the *Brassica oleracea* group. In *Eucarpia Cruciferae Conference,* Wageningen, p. 11–21.

Gustafsson, M. (1981). Biosystematic studies in the *Brassica oleracea* group. *Proc. Brassica Conference, EUCARPIA, Aas, Norway,* p. 112–116.

Gustafsson, M. and Lanner-Herrera, C. (1997a). Overview of the *Brassica oleracea* complex: their distribution and ecological specifications. In Valdes, B., Heywood, V., Raimondo, F.M., and Zohary, D. (Eds.), *Proc. of Three Workshops on "Conservation of the Wild Relatives of European Cultivated Plants." Bocconea,* 7:27–37.

Gustafsson, M. and Lanner-Herrera, C. (1997b). Diversity in natural populations of wild cabbage (*Brassica oleracea* L.). In Valdes, B., Heywood, V., Raimondo, F.M., and Zohary, D. (Eds.), *Proc. of Three Workshops on "Conservation of the Wild Relatives of European Cultivated Plants". Bocconea,* 7:95–103.

Haga, T. (1938). Relationship of genome to secondary pairing in *Brassica* (a preliminary note). *Jpn. J. Genet.,* 13:227–284.

Hamada, M., Hosoki, T.I., Kusabiraki, Y., and Kigo, T. (1989). Hairy root formation and plantlet regeneration from Brussels sprouts (*Brassica oleracea* var. *gemmifera*) mediated by *Agrobacterium rhizogenes. Plant Tiss. Cult. Lett.,* 6:130–133. (in Japanese with English summary).

Hammer, K., Knupfer, H., Lagheti, G., and Perrino, P. (1992). *Seeds from the Past. A Catalogue of Crop Germplasm in South Italy and Sicily.* Germplasm Institute. Consiglio Nazionale delle Ricerche, p. 29.

Haruta, T. (1962). Studies on the *genetics* of self and cross incompatibility in cruciferous vegetables. *Res. Bull. Takii Pl. Breeding Ex. Sta.,* Kyoto, Japan.

Hatakeyama, K., Watanabe, M., Takasaki, T., Ojima, K., and Hinata, K. (1998). Dominance relationships between S-alleles in self incompatible *Brassica campestris* L. *Heredity,* 79:241–247.

He, Y.K., Want, J.Y., Wei, Z.M., Xu, Z.H., and Gong, Z.H. (1995). Effects of whole Ri T-DNA and auxin genes alone on root induction and plant phenotype of Chinese cabbage. *Acta Hort.,* 402:418–422.

Heath, D.W., Earle, E.D., and Dickson, M.H. (1994). Introgressing cold tolerant Ogura cytoplasms from rapeseed into pak choi and Chinese cabbage. *Hortsci.,* 29: 202–203.

Helm, I. (1963). Morphologisch-taxonomische Gliederung der kultursippen von *Brassica oleracea. Kulturpflanze,* 11:92–210.

Henslow, G. (1908). History of the cabbage tribe. *J. Roy Hort. Soc.,* 34:15–23.

Henzi, M.X., Christey, M., McNeil, D.L., and Downs, C. (1998). Transgenic broccoli (*Brassica oleracea* L. var. *Italica*) plants containing an antisense ACC oxidase gene. *Acta Hort.,* 464:147–151.

Henzi, M.X., Christey, M.C., McNeil, D.L., and Davies, K.M. (1999). Agrobacterium rhizogenes-mediated transformation of broccoli (*B. oleracea* L. var. *italica*) with an antisense 1-aminocyclopropane-1-carboxylic acid oxidase gene. *Plant Sci.,* 143:55–62.

Henzi, M.X., McNeil, D.L., Christey, M.C., and Lill, R.E. (1999). A tomato antisense 1-aminocyclopropane-1-carboxylic oxidase gene causes reduced ethylene production in transgenic broccoli. *Australia J. Plant Physiol.,* 26:179–183.

Herbert, W. (1847). On hybridization amongst vegetables. *Journal Hort. Soc.,* 2:1–28 and 81–107.

Hinata, K., Isogai, A., and Isuzugawa, K. (1994). Manipulation of sporophytic self incompatibility in plant breeding. In Williams, E.G., Knox, R.B., and Clarke, A.E. (Eds.). *Genetic Control of Self Incompatibility and Reproductive Development in Flowering Plants.* p. 102–115, Kluwer, Dordrecht, The Netherlands.

Hoffmann, F., Thomas, E., and Wenzel, G. (1982). Anther culture as a breeding tool in rape. II. Progeny analysis of androgenic lines and induced mutant from haploid cultures. *Theor. Appl. Genet.,* 61:225–232.

Hosaka, K., Kianian, S.F., McGrath, J.M., and Quiros, C.F. (1990). Development and chromosomal localization of genome specific DNA markers of *Brassica* and the evolution of amphidiploids and n=9 diploid species. *Genome,* 33:131–142.

Hosoki, T., Kanbe, H., and Kigo, T. (1994). Transformation of ornamental tobacco and kale mediated by *Agrobacterium tumefaciens* and *A. rhizogenes* harboring a reporter, B-glucuronidase (GUS) gene. *J. Japan, Soc. Hort. Sci.,* 63:167–172.

Hosoki, T., Kigo, T., and Shiraishi, K. (1991). Transformation and regeneration of ornamental kale (*Brassica oleracea* var. *Italica*) mediated by *Agrobacterium rhizogenes. J. Japan Soc. Hort. Sci.,* 60:71–75.

Hosoki, T., Shiraishi, K., Kigo, T., and Ando, M. (1989). Transformation and regeneration of ornamental kale (*Brassica oleracea* var. *acephala* DC) mediated by *Agrobacterium rhizogenes. Scientia Hort.,* 40:259–266.

Hu, J. and Quiros, C.F. (1991). Molecular and cytological evidence of deletions in alien chromosomes for two monosomic addition lines of *Brassica campestris-oleracea. Theor. Appl. Genet.,* 81:221–226.

Hu, J., Sadowsky, J., Osborn, T.C., Landry, B.S., and Quiros, C.F. (1998). Linkage group alignment from four independent *Brassica oleracea* RFLP maps. *Genome,* 41:226–235.

Hyams, E. (1971). Cabbages and kings. *Plants in the Service of Man,* J.M. Dent & Sons, London, p. 33–61.

Johnson, A.G. (1958). Male sterility in *Brassica. Nature,* 24:97–105.

Jorgensen, R.B., Chen, B.Y., Cheng, B.F., Heneen, W.K., and Simonsen, V. (1996). Random amplified polymorphic DNA markers of the *Brassica alboglabra* chromosome of a *B. campestris-alboglabra* addition line. *Chromosome Res.,* 4:111–114.

Josefsson, E. (1967). Distribution of thioglucosides in different parts of *Brassica* plants. *Phytochemistry,* 32:151–159.

Jun, S.I. and Kwon, S.Y. (1995). Agrobacterium mediated transformation and regeneration of fertile transgenic plants of Chinese cabbage (*B. campestric* ssp. *pekinensis* cv. Spring flavor. *Plant Cell Report,* 14:620–625.

Kalia, Pritam (2008). Exploring Cytoplasmic Male Sterility for F_1 hybrid development in Indian cauliflower. *Cruciferae Newslett.,* 27:75–76.

Kalloo, G. and Bergh, B.O. (1993). *Genetic Improvement of Vegetable Crops.* Pergamon Press, Oxford.

Kameya, Y., Kanzaki, H., Toki, S., and Abe, T. (1989). Transfer of radish (*Raphanus sativus* L.) chloroplasts into cabbage (*Brassica oleracea* L.) by protoplast fusion. *Japan J. Genet.,* 64:27–34.

Kao, H.M., Keller, W.A., Gleddie, S., and Brown, G.G. (1992). Synthesis of *Brassica oleracea/Brassica napus* somatic hybrid plants with novel organelle DNA compositions. *Theor. Appl. Genet.,* 83:313–320.

Karron, J.D., Marshall, D.L., and Oliveras, D.M. (1990). Numbers of sporophytic self incompatibility alleles in populations of wild radish. *Theor Appl. Genet.,* 79:457–460.

Kearsey, M.J., Ramsay, L.D., Jennings, D.E., Lydiate, D.J., Bohuon, E.J.R., and Marshall, D.F. (1996). Higher recombination frequencies in female compared to male meiosis in *Brassica oleracea. Theor. Appl. Genet.,* 92:363–367.

Keller, W.A. and Armstrong, K.C. (1983). Production of haploids via anthesis culture in *Brassica oleracea* var. *italica. Euphytica,* 32:151–159.

Keller, W.A., Armstrong, K.C., and De la Roche, A.I. (1982). The production and utilization of microspore derived haploids in *Brassica* crops. In Sen, S.K. and Giles, K.L. (Eds.), *Plant Cell Culture in Crop Improvement.* p. 169–183, Plenum Press, New York.

Keller, W.A., Arnison, P.G., and Cardy, B.J. (1987). Haploids from gametophytic cell — recent development and future prospects. In *Plant Tissue and Cell Culture.* p. 233–241, Alan R. Liss, New York.

Kempin, S., Savidge, S., and Yanofsky, M.F. (1995). Molecular basis of the cauliflower phenotype in *Arabidopsis. Science,* 267:522–525.

Kianian, S.F. and Quiros, C.F. (1992). Generation of a *Brassica oleracea* composite RFLP map: linkage arrangements among various populations and evolutionary implications. *Theor. Appl. Genet.,* 84:544–554.

Kimber, G. and Riley, R. (1963). Haploid angiosperms. *Bot. Rev.,* 29:480–531.

Kishi-Nishizawa, N., Isogai, A., Watanabe, M., Hinata, K., Yamakawa, S., Shojima, S., and Suzuki, A. (1990). Ultrastructure of papillar cells in *Brassica campestris* revealed by liquid helium rapid freezing and substitution fixation method. *Plant Cell Physiol.,* 31:1207–1219.

Korber-Grohne, U. (1987). *Nutzpflanzen in Deutschland — Kulturgeschichte und Biologie.* p. 149–192, Thesis Verlag, Stuttgart.

Krishnasamy, S. and Makaroff, C.A. (1993). Characterization of the radish mitochondrial orfB locus: possible relationship with male sterility in Ogura radish. *Curr. Genet.,* 24:156–163.

Kuckuck, H. (1979). *Gartenbauliche Pflanzenzuchtung, 2nd ed. Parey,* Berlin, Hamburg, p. 64–72.

LaCour, L.F. (1949). Nuclear differentiation in the pollen grain. *Heredity,* 3:319–337.

Lagercrantz, U. and Lydiate, D.J. (1995). RFLP mapping in *Brassica nigra* indicates differing recombination rates in male and female meiosis. *Genome,* 38:255–264.

Landry, B.S., Hubert, N., Crete, R., Chang, M.S., Lincoln, S.E., and Etho, T. (1992). A genetic map of *Brassica oleracea* based on RFLP markers detected with expressed DNA sequences and mapping of resistance genes to race 2 of *Plasmodiophora brassicae* (Woronin). *Genome,* 35:409–420.

Lanner, C. (1998). Relationships of wild *Brassica* species with chromosome number 2n=18, based on comparison of the DNA sequence of the chloroplast intergenic region between trnL (UAA) and trnF (GAA). *Can. J. Bot.,* 76:228–237.

Lanner-Herrera, C., Gustafsson, M., Falt, A.S., and Bryngelsson, T. (1996). Diversity in natural populations of wild *Brassica oleracea* as estimated by isozyme and RAPD analysis. *Genet. Resour. Crop Evolut.,* 43:13–23.

Laurier, D.A. and Bennett, M.D. (1988). The production of haploid wheat plants from wheat and maize crosses. *Theor. Appl. Genet.,* 76:393–397.

Lazaro, A. and Aguinagalde, I. (1998a). Genetic diversity in *Brassica oleracea* L. and wild relatives (2n=18) using isozymes. *Ann. Bot.,* 82:821–828.

Lazaro, A. and Aguinagalde, I. (1998b). Genetic diversity in *Brassica oleracea* L. and wild relatives (2n=18) using RAPD markers. *Ann. Botany,* 82:829–833.

Leviel, R. (1998). La sterilite male chez le chou-fleur. *PHM, Revue Horticole,* 388:31–33.

Li, S.L., Quian, Y.X., Wu, Z.H., and Stefansson, B.R. (1988). Genetic male sterility in rape (*B. napus*) conditioned by interaction of genes at two loci. *Can. J. Plant Sci.,* 68:1115–1118.

Li, X., Mao, H.Z., and Bai, Y.Y. (1995). Transgenic plants of rutabaga (*Brassica napobrassiaca*) tolerant to pest insects. *Plant Cell Rep.,* 15:97–101.

Lichter, R. (1982). Induction of haploid plants from isolated pollen of *Brassica napus. Z. Pflanzen-Physiol.,* 105:427–434.

Lim, H.T., You, Y.S., Park, E.J., and Song, Y.N. (1998). High plant regeneration genetic stability of regenerants and genetic transformation of herbicide resistance gene (Bar) in Chinese cabbage (*B. campestris* ssp. *pekinensis*). *Aca Hort.,* 459:199–208.

Makaroff, C.A. and Palmer, J.D. (1988). Mitochondrial DNA rearrangements and transcriptional alterations in the male sterile cytoplasm of Ogura radish. *Mol. Cell. Biol.,* 8:1474–1480.

Makaroff, C.A., Apel, I.J., and Palmer, J.D. (1989). The atp6 coding region has been disrupted and a novel reading frame generated in the mitochondrial genome of cytoplasmic male sterile radish. *J. Biol. Chem.,* 264:11706–11713.

Makaroff, C.A., Apel, I.J., and Palmer, J.D. (1990). Characterization of radish mitochondrial atpA: influence of nuclear background on transcription of atpA-associated sequences and relationship with male sterility. *Plant Mol. Biol.*, 15:735–746.

Makaroff, C.A., Apel, I.J., and Palmer, J.D. (1991). The role of coxI-associated repeated sequences in plant mitochondrial DNA rearrangements and radish cytoplasmic male sterility. *Curr. Genet.*, 19:183–190.

Mandel, M.A., Gustafson-Brown, C., Savidge, B., and Yanofsky, M.F. (1992). Molecular characterization of the *Arabidopsis* floral homeotic gene APETALA1. *Nature*, 360:273–277.

Manton, I. (1932). Introduction to the general cytology of the *Cruciferae. Ann. Bot.*, 46:509–556.

Mariani, C., De Beucheleer, M., Truttner, S., Leemans, J., and Goldberg, R.B. (1990). Induction of male sterility in plants by a chimearic ribonuclease gene. *Nature*, 347:737–741.

Mariani, C., Gossele, V., De Beucheleer, M., De Block, M., Goldberg, R.B., De Greef, W., and Leemans, J. (1992). A chimaeric ribonuclease inhibitor gene restores fertility to male sterile plants. *Nature*, 357:384–387.

Marrero, A. (1989). Dos citas de interes en la Flora Canaria. *Botanica Macaronesica*, 18:89–90.

Maselli, S., Diaz-Lifante, Z., and Aguinagalde, I. (1996). The Mediterranean populations of *Brassica insularis*, a biodiversity study. *Biochem. Syst. Ecol.*, 24(2):165–170.

Mathias, R. (1985). A new dominant gene for male sterility in rapeseed, *Brassica napus* L. *Z. Pflanzenzuchtg.*, 94:170–173.

McCollum, G. (1981). Induction of an alloplasmic male sterile *Brassica oleracea* by substituting cytoplasm for "Early Scarlet Globe" (*Raphanus sativus*). *Euphytica*, 30:855–859.

McGrath, J.M. and Quiros, C.F. (1990). Generation of alien addition lines from synthetic *B. napus*: morphology, cytology, fertility and chromosome transmission. *Genome*, 33:374–383.

McGrath, J.M., Quiros, C.F., Harada, J.J., and Landry, B.S. (1990). Identification of *Brassica oleracea* monosomic alien chromosome addition lines with molecular markers reveals extensive gene duplication. *Mol. Gen. Genet.*, 223:198–204.

Metz, T.D., Dixit, R., and Earle, E.D. (1995). *Agrobacterium tumefaciens* mediated transformation of broccoli (*Brassica oleracea* var. *Italica*) and cabbage (*B. oleracea* var. *capitata*). *Plant Cell Rep.*, 15:287–292.

Metz, T.D., Tang, J.D., Shelton, A.M., Roush, R.T., and Earle, E.D. (1995b). Transgenic broccoli expressing a *Bacillus thuringiensis* insecticidal crystal protein: implications for pest resistance management strategies. *Mol. Breed.*, 1:309–317.

Millam, S., Lanying, W., Whitty, P., Fryer, S. Burns, A.T.H., and Hocking, T.J. (1994). An efficient transformation system for rapid cycling *Brassica oleracea. Cruciferae Newslett.*, 16:65–66.

Mithen, R.F., Lewis, B.G., Heaney, R.K., and Fenwick, G.R. (1987). Glucosinolates of wild and cultivated *Brassica* species. *Phytochemistry*, 26(7):1969–1973.

Mizushima, U. (1950). Karyogenetic studies of species and genus hybrids in the tribe *Brassiceae* of *Cruciferae. Tohoku J. Agr. Res.*, 1:1–14.

Monteiro, A.A., Gabelman, W.H., and Williams, P.H. (1988). The use of sodium chloride solution to overcome self incompatibility in *Brassica campestris. Cruciferae Newslett.*, 13:122–123.

Morinaga, T. (1928). Preliminary note on interspecific hybridization in *Brassica. Proc. Imp. Acad.*, 4:620–622.

Morinaga, T. (1934). Interspecific hybridization in *Brassica* VI. The cytology of F_1 hybrids of *B. juncea* and *B. nigra. Cytologia*, 6:62–67.

Morrison, R.A. and Evans, D.A. (1988). Haploid plants from tissue culture: new varieties in a shortened time frame. *Biotechnology*, 6:684–690.

Mukhopadhyay, A., Topfer, R., Pradhan, A.K., Sodhi, Y.S., Steinbiss, H.H., Schell, J., and Pental, D. (1991). Efficient regeneration of *Brassica oleracea* hypocotyls protoplasts and high frequency genetic transformation by direct DNA uptake. *Plant Cell Rep.*, 10:375–379.

Nakanishi, T., Esashi, Y., and Hinata, K. (1969). Control of self incompatibility by CO_2 gas in *Brassica. Plant Cell Physiol.*, 10:925–927.

Nieuwhof, M. (1961). Male sterility in some cole crops. *Euphytica*, 10:351–356.

Nieuwhof, M. (1968). Effects of temperature on the expression of male sterility in Brussels sprouts (*Brassica oleracea* L. var. *gemmifera* DC). *Euphytica*, 17:265–273.

Nieuwhof, M. (1969).*Cole Crops*. Leonard Hill, London.

Nijenhuis, B.T. (1971). Estimation of the proportion of inbred seed in Brussels sprouts hybrid seed by acid phosphate isoenzyme analysis. *Euphytica*, 20:498–507.

Nishi, S. and Hiraoka, T. (1958). Studies on F_1 hybrid vegetable crops. (1) Studies on the utilization of male sterility on F_1 seed production I. Histological studies on the degenerative process of male sterility in some vegetable crops. *Bull. Natl. Inst. Agric. Jpn.*, Ser. E. 6:1–41.

Nou, I.S., Watanabe, M., Isogai, A., and Hinata, K. (1993b). Comparison of S-alleles and S-glycoproteins between two wild populations of *Brassica campestris* in Turkey and Japan. *Sex. Plant Reprod.*, 6:79–86.

Nou, I.S., Watanabe, M., Isogai, A., Shiozawa, H., Suzuki, A., and Hinata, K. (1991). Variation of S-alleles and S-glycoproteins in a naturalized population of self incompatible *Brassica campestris* L. *Japan J. Genet.*, 66:227–239.

Nou, I.S., Watanabe, M., Isuzugawa, K., Isogai, A., and Hinata, K. (1993a). Isolation of S-alleles from a wild population of *Brassica campestris* L. at Balcesme, Turkey and their characterization by S-glycoproteins. *Sex. Plant Reprod.*, 6:71–78.

Ockendon, D.J. (1974). Distribution of self incompatibility alleles and breeding structure of open pollinated cultivars of Brussels sprouts. *Heredity*, 33:159–171.

Ockendon, D.J. (1975). Dominance relationships between S-alleles in the stigma of Brussels sprouts (*Brassica oleracea* var. *gemmifera*). *Euphytica*, 24:165–172.

Ockendon, D.J. (1980). Distribution of self incompatibility alleles and breeding structure of cape broccoli (*Brassica oleracea* var. *italica*). *Theor. Appl. Genet.*, 58:11–15.

Ockendon, D.J. and McClenaghan, R. (1993). Effect of silver nitrate and 2,4-D on anther culture of Brussels sprout (*Brassica oleracea* var. *gemmifera*). *Plant Cell Tissue Organ. Cult.*, 32:41–46.

Ogura, H. (1968). Studies of the new male sterility in Japanese radish with special reference to the utilization of this sterility towards the practical raising of hybrid seeds. *Mem. Fac. Agric. Kagoshima Univ.*, 6:39–78.

Palloix, A., Herve, Y., Knox, R.B., and Dumas, C. (1985). Effect of carbon dioxide and relative humidity on self incompatibility in cauliflower, *Brassica oleracea*. *Theor. Appl. Genet.*, 70:628–633.

Palmer, J.D. (1988). Intraspecific variation and multicircularity in *Brassica* mitochondrial DNAs. *Genetics*, 118:341–351.

Palmer, J.D., Shields, C.R., Cohen, D.B., and Orton, T.J. (1983). Chloroplast DNA evolution and the origin of amphidiploid *Brassica* species. *Theor. Appl. Genet.*, 65:181–189.

Pandey, K.K. (1973). Theory and practice of induced androgenesis. *New Phytol.*, 72:1129–1140.

Pearson, O.H. (1972). Cytoplasmically inherited male sterility characters and flavor components from the species cross *Brassica nigra* (L.) Koch × *B. oleracea* L. *J. Am. Soc. Hort. Sci.*, 97:397–402.

Pelletier, G., Primard, C., Vedel, F., Chetrit, P., Remy, R., Rousselle, and Renard, M. (1983). Intergeneric cytoplasmic hybridization in Cruciferae by protoplast fusion. *Mol. Gen. Genet.*, 191:244–250.

Pradhan, A.K., Prakash, S., Mukhopadhyay, A., and Pental, D. (1992). Phylogeny of *Brassica* and allied genera based on variation in chloroplasts and mitochondrial DNA patterns: molecular and taxonomic classifications are incongruous. *Theor. Appl. Genet.*, 85:331–340.

Prakash, S. and Hinata, K. (1980). Taxonomy, cytogenetics and origin of crop *Brassicas*, a review. *Opera Bot.*, 55:1–57.

Puddephat, I.J., Riggs, T.J., and Fenning, T.M. (1996). Transformation of *Brassica oleracea* L. A critical review. *Mol. Breed.*, 2:185–210.

Purugganan, M.D., Boyles, A.L., and Suddith, J.I. (2000). Variation and selection at the CAULIFLOWER floral Homeotic Gene Accompanying the Evolution of Domesticated *Brassica oleracea*. *Genetics*, 155:855–862.

Quiros, C.F., Hu, J., and Truco, M.J. (1994). DNA-based marker *Brassica* maps. In Phillips, R.L. and Vasil, I.K. (Eds.), *Advances in Cellular and Molecular Biology of Plants. Vol. I: DNA based Markers in Plants.* Kluwer Academic Publ. p. 199–222, Dordrecht, The Netherlands.

Quiros, C.F., Ochoa, O., Kianian, S.F., and Douches, D. (1987). Analysis of the *Brassica oleracea* genome by the generation of *B. campestris-oleracea* chromosome addition lines, characterization by isozymes and rDNA genes. *Theor. Appl. Genet.*, 74:758–766.

Rac, M. and Lovric, A.Z. (1991). Insular woody endemics of *Brassica* and related ancient cultivars in the Adriatic archipelago. *Botanika Chronika*, 10:673–678.

Raimondo, F.M. and Mazzola, P. (1997). A new taxonomic arrangement in the Sicilian members of *Brassica* L. Sect. *Brassica. Lagascalia*, 19(1-2):831–838.

Ramsay, L.D., Jennings, D.E., Bohuon, E.J.R., Arthur, A.E., Lydiate, D.J., Kearsey, M.J., and Marshall, D.F. (1996). The construction of a substitution library of recombinant backcross lines in *Brassica oleracea* for the precision mapping of quantitative loci. *Genome*, 39:558–567.

Reynaerts, A., Vandewiele, H., Desutter, G., and Janssens, J. (1993). Engineered genes for fertility and their application in hybrid seed production. *Scientia Hort.*, 55:125–139.

Röbbelen, G. (1960). Beitrage zur Analyse des *Brassica* Genomes, *Chromosoma,* 11:205–228.

Roggen, H. (1974). Pollen washing influences (in) compatibility in *Brassica oleracea* varieties. In Linskens, H.F. (Ed.), *Fertilization in Higher Plants.* p. 273–278, North-Holland.

Roulund, N., Hansted, L., Anderson, S.B., and Forestveit, B. (1990). Effect of genotype, environment and carbohydrate on anther culture response in head cabbage (*Brassica oleracea* L. convar *capitata* Alef.). *Euphytica,* 49:237–242.

Ruffio-Chable, V. (1994). Les systemes d' Hybridations chez le Chou Fleur (*Brassica oleracea* L. var. *botrytis* L.). Application a Lamelioration Genetique, thesis, ENSA de Rennes.

Ruffio-Chable, V. and Gande, T. (2001). S-Haplotype Polymorphism in *Brassica oleracea. Acta Hort.,* 546:257–261.

Ruffio-Chable, V., Bellis, H., and Herve, Y. (1993). A dominant gene for male sterility in cauliflower (*Brassica oleracea* var. *botrytis*). Phenotype expression, inheritance and use in F_1 hybrid production. *Euphytica,* 67:9-17.

Rundfeldt, H. (1960). Untersuchungen zur Zuchtung des Kopfkohls (*B. oleracea* L. var. *capitata). Z. Pflanzenzucht,* 44:30–62.

Sadik, S. (1962). Morphology of the curd of cauliflower. *Am. J. Bot.,* 49:290–297.

Sageret, M. (1826). Considerations sur la production des variants et des varieties en general, et sur celles de la famille de cucurbitacees en particulier. *Ann. Sci. Nat.,* 8:294–314.

Sakata (1973). Seed catalogue for 1973/4. Sakata Seed Co., Japan.

Sampson, D.R. (1964). A one locus self incompatibility system in *Raphanus raphanistrum. Can. J. Genet. Cytol.,* 6:435–445.

Sampson, D.R. (1967). Frequency and distribution of self incompatibility alleles in *Raphanus raphanistrum. Genetics,* 56:241–251.

Sato, T., Thorsness, M.K., Kandasamy, M.K., Nishio, T., Hirai, M., Nasrallah, J.B., and Nasrallah, M.E. (1991). Activity of an S locus gene promoter in pistils and anthers of transgenic *Brassica. The Plant Cell,* 3:867–876.

Scarth, R., Seguin-Swartz, G., and Rakow, G.F.W. (1991). Application of doubled haploidy to *Brassica napus* breeding. In McGregor, E.I. (Ed.), *Proc. 8th Int. Rapeseed Congr.,* Saskatoon, Canada, 5:1449–1453.

Schery, R.W. (1972). *Plants for Man,* p. 508–509, Prentice Hall, Englewood Cliffs, NJ.

Sigareva, M.A. and Earle, E.D. (1997). Direct transfer of a cold tolerant Ogura male sterile cytoplasm into cabbage (*Brassica oleracea* ssp. *capitata*) via protoplast fusion. *Theor. Appl. Genet.,* 94:213–220.

Sikka, S.M. (1940). Cytogenetics of *Brassica* hybrids and species. *J. Genet.,* 40:441–509.

Slocum, M.K., Figdore, S.S., Kennard, W.C., Suzuki, J.Y., and Osborn, T.C. (1990). Linkage arrangement of restriction fragment length polymorphism loci in *Brassica oleracea. Theor. Appl. Genet.,* 80:57–64.

Snogerup, S. (1980). The wild forms of the *Brassica oleracea* group (2n=18) and their possible relations to the cultivated ones. In Tsunoda, S., Hinata, K. and Gomez-Campo, C. (Eds.), *Brassica Crops and Wild Allies, Biology and Breeding,* Scientific Society Press, Tokyo, Japan, p. 121–132.

Snogerup, S. (1996). Cytogenetic differentiation and reproduction in *Brassica* L. Sect. *Brassica.* In: Demiriz, H. and Ozhatay, N. (Eds.), p. 377–393, *Proc. v. OPTIMA Meeting,* Istanbul.

Snogerup, S. and Persson, D. (1983). Hybridization between *Brassica insularis* Moris and *B. balearica* Pers. *Hereditas,* 99:187–190.

Snogerup, S., Gustafsson, M., and Von Bothmer, R. (1990). *Brassica* Cectio *Brassica (Brassicaceae).* I. Taxonomy and variation. *Willdenowia,* 19:271–365.

Sodhi, Y.S., Stiewe, G., and Mollers, C. (1995). Production of haploids in CMS plants and their use for the molecular analysis of male sterility. In *19th Int. Rapeseed Congr.,* Cambridge, U.K., 3:822–824.

Song, K.M. and Osborn, T.C. (1992). Polyphyletic origins of *Brassica napus:* new evidence based on organelle and nuclear RFLP analyses. *Genome,* 35:992–1001.

Song, K., Osborn, T.C., and Williams, P.H. (1988a). *Brassica taxonomy* based on nuclear restriction fragment length polymorphisms (RFLPs). *Thoer. Appl. Genet.,* 79:497–506.

Song, K.M., Lu, P., Tang, K., and Osborn, T.C. (1995). Rapid genome change in synthetic polyploids of *Brassica* and its implications for polyploidy evolution. *Proc. Nat. Acad. Sci.,* 92:7719–7723.

Song, K.M., Osborn, T.C., and Williams, P.H. (1988b). *Brassica* taxonomy based on nuclear restriction fragment length polymorphisms (RFLPs). 3. Preliminary analysis of subspecies within *B. rapa* (syn. *campestris*) and *B. oleracea. Theor. Appl. Genet.,* 76:593–600.

Song, K.M., Osborn, T.C., and Williams, P.H. (1990). *Brassica* taxonomy based on nuclear restriction fragment length polymorphisms (RFLPs). 2. Genome relationships in *Brassica* and related genera and the origin of *B. oleracea* and *B. rapa* (syn. *campestris*). *Theor. Appl. Genet.,* 79:497–506.

Srivastava, V., Reddy, A.S., and Guha-Mukherjee, S. (1988). Transformation and regeneration of *Brassica napus* mediated by a monogenic *Agrobacterium tumefaciens. Plant Cell Rep.,* 7:504–507.

Stam, P. (1993). Join Map Version 1.1. A computer program to generate genetic linkage maps. Center for plant breeding and reproduction research, CPRO-DLO. Wageningen, the Netherlands.

Stevens, J.P. and Kay, Q.O.N. (1988). The number of loci controlling the sporophytic self incompatibility system in *Sinapis arvensis*. *Heredity,* 61:411–418.

Stevens, J.P. and Kay, Q.O.N. (1989). The number, dominance relationships and frequencies of self incompatibility alleles in a natural population of *Sinapis arvensis* L. in South Wales. *Heredity,* 62:199–205.

Stork, A.L., Snogerup, S., and Wuest, J. (1980). Seed characters in *Brassica* section *Brassica* and some related groups. *Candollea,* 35:421–450.

Stow, I. (1930). Experimental studies on the formation of the embryosac-like giant pollen grains in the anther of *Hyacinthus orientalis. Cytologia,* 1:417–432.

Stringham, G.R.and Thiagarajah, M.R. (1991). Effectiveness of selection for early flowering in F$_2$ populations of *Brassica napus* L. — A comparison of doubled haploid and single seed descent methods. In McGregor, D.I. (Ed.), *Proc. 8th Int. Rapeseed Congr.,* Saskatoon, Canada, 1:70–75.

Takamine, N. (1916). Uber die rubenden und die prasynaptischen Phasen der Reduktionsteilung. *Bot. Mag.* Tokyo, 30:293–303.

Tatebe, T. (1951). Studies on the behaviour of incompatible pollen in *Brassica.* IV. *Brassica oleracea* L. var. *capitata* and var. *botrytis.* L. *J. Hort. As. Japan,* 20:–19–26.

Theis, R. and Robbelen, G. (1990). Anther and microspore development in different male sterile lines of oilseed rape (*Brassica napus* L.). *Angewandte Botanik,* 64:419–434.

Thompson, K.F. and Taylor, J.P. (1966). Non-linear dominance relationships between S alleles. *Heredity,* 21:345–362.

Thurling, N. and Chay, P.M. (1984). The influence of donor plant genotype and environment on production of multicellular microspores in cultured anthers of *Brassica napus* ssp. *oleifera. Ann. Bot.,* 54:681–695.

Tonguc, M. and Griffiths, P.D. (2004b). Development of black rot resistant interspecific hybrids between *Brassica oleracea* L. cultivars and *Brassica* accession A 19182, using embryo rescue. *Euphytica,* 136:313–318.

Tonguc, M. and Griffiths, P.D. (2004a). Evaluation of *Brassica carinata* accessions for resistance to black rot (*Xanthomonas campestris* pv. *campestris*). *HortScience,* 39(5):952–954.

Tonguc, M. and Griffiths, P.D. (2004c). Transfer of powdery mildew resistance from *Brassica carinata* to *Brassica oleracea* through embryo rescue. *Plant Breeding,* 123:587–589.

Tonguc, M., Earle, E.D., and Griffiths, P.D. (2003). Segregation distortion of *Brassica carinata* derived black rot resistance in *Brassica oleracea. Euphytica,* 134:269–276.

Toriyama, K., Stein, J.C., Nasrallah, M.E., and Nasrallah, J.B. (1991). Transformation of *B. oleracea* with an S-locus gene from *B. campestris* changes the self-incompatibility phenotype. *Theor Appl. Genet.,* 81:769–776.

Trinajstic, I. and Dubravec, K. (1986). *Brassica* L. In Trinajstic, I. (Ed.), *Analiticka Flora Jugoslavije,* 2:415–425.

Tsunoda, S., Hirata, K., and Gomez-Campo, C. (1980). *Brassica Crops and Wild Allies.* Japan Scientific Societies Press, Tokyo.

Venkateswarlu, J. and Kamala, T. (1973). Meiosis in double trisomic *Brassica campestris. Genetica,* 44:283–287.

Visser, D.L., Van Hal, J.G., and Verhoeven, W.H. (1982). Classification of S-alleles by their activity in S-heterozygotes of Brussels sprouts (*Brassica oleracea* var. *gemmifera* (DC.) Schultz). *Euphytica,* 31:603–611.

Voorrips, R.E. and Visser, D.L. (1990). Doubled haploid lines with clubroot resistance in *Brassica oleracea.* In McFerson, J.R., Kresovich, S., and Dwyer, S.E. (Eds.), *Proc. 6th Crucifer Genet. Workshop.* Geneva, NY, p. 40.

Wallace, D.H. (1979). Interactions of S-alleles in sporophytically controlled self incompatibility of *Brassica. Theor. Appl. Genet.,* 54:193–201.

Walters, T.W. and Earle, E.D. (1993). Organellar segregation, rearrangement and recombination in protoplast fusion-derived *Brassica oleracea* calli. *Theor. Appl. Genet.,* 85:761–769.

Walters, T.W., Mutschler, M.A., and Earle, E.D. (1992). Protoplast fusion derived Ogura male sterile cauliflower with cold tolerance. *Plant Cell Rep.,* 10:624–628.

Warwick, S.I.and Black, L.D. (1991). Molecular systematics of *Brassica* and allied genera (subtribe *Brassicinae, Brassiceae*) — Chloroplast genome and cytodeme congruence. *Theor. Appl. Genet.,* 82:81–92.

Weigel, D. (1995). The genetics of flower development. *Annu. Rev. Genet.,* 29:19–39.

Widler, B.E. and Bocquet, G. (1979). *Brassica insularis* Moris: Beispiel eines messinischen Verbreitunsmusters. *Candollea,* 34:133–151.

Yang, Q., Chauvin, J.E., and Herve, Y. (1992). Obtention d'ambryons androgenetiques par culture *in vitro* de boutons floraus chez le broccoli (*B. o.* var *italica*). *CR Acad. Sci. Paris*, t 314 Serie III, 145–152.

Yanofsky, M.F. (1995). Floral meristems to floral organs: genes controlling early events in *Arabidopsis* flower development. *Annu. Rev. Plant Physiol. Plant Mol. Biol.*, 46:167–188.

Yarrow, S.A., Burnett, L.A., Wildeman, R.D., and Kemble, R.J. (1990). The transfer of Polima cytoplasmic male sterility from oilseed rape (*Brassica napus*) to broccoli (*Brassica oleracea*) by protoplast fusion. *Plant Cell Rep.*, 9:185–188.

Zhou, X. (1990). La sterilite male digenique dominante du colza (*Brassica napus*). Recherche du gene restaurateur Rf et etude cytologique en microscopie electronique. Memoire D.E.A., Labo. De Biologie Cellulaire. Universite Rennes I.

17 Industrial Products

Carla D. Zelmer and Peter B.E. McVetty

CONTENTS

INTRODUCTION

Modern industry relies heavily on a wide range of products currently produced from petrochemicals. These products include greases and other lubricants, fuel oils, heating oils, printing inks, polymers and resins, plastics, paints and additives, and slip agents. As the cost of petroleum rises, many industries are looking for alternatives to these products but economics is not the only important motivator. Petroleum-based products are not renewable and can be harmful to the environment, owing to their toxicity to organisms, their poor biodegradability, and the damage caused by their extraction, processing, and refining. When burned, they actively contribute greenhouse gases to the atmosphere, and are carbon-positive. Many petroleum-based lubricants are released directly into the environment, a particular concern for aquatic systems, forestry, and farming, but also an important source of contamination on construction sites (Perez, 2005).

Vegetable oils, such as the seed oils produced by plants in the family *Brassicaceae*, offer an excellent alternative to petrochemicals. They are biodegradable and are produced from renewable sources. Products such as biodegradable plastics will reduce the landfill space required to accommodate waste in the future. The domestic production of vegetable oils as a replacement for petroleum-derived compounds, including fuel, could mean enhanced local and rural economies, and provide greater energy security. Vegetable-oil-based products are generally less toxic than their petroleum-based counterparts. For example, rapeseed oil has a lower mutagenicity than petrodiesel fuel (Knothe and Dunn, 2005). Marine engine lubricants and agricultural greases are released directly into the environment. The lower toxicity and higher biodegradability of plant-based oils are especially important in these cases (National Non-Food Crops Centre, 2005). While still a source of some greenhouse gases when burned, biodiesel has lower sulfur monoxide emissions than petrodiesel (Knothe and Dunn, 2005). When the crops that yield them are raised sustainably, plant-based oils can have a much reduced impact on carbon balance. Vegetable oils tend to be more viscous and less volatile than mineral oils. Along with these properties, their higher flashpoint is also an important safety consideration — reducing the risk of fires. They have lower friction coefficients than mineral oils, so they have excellent lubricity (National Non-Food Crops Centre, 2005). In addition, some unique fatty acids and other substances that are not easily or economically synthesized are also found in the seeds of the *Brassicaceae*, such as hydroxy fatty acids.

Numerous opportunities have been identified to replace many of the petrochemical-based industrial compounds with those based on more ecologically friendly vegetable oils. These include paints and coatings (Van De Mark and Sandefur, 2005), home heating oils (Tao, 2005), biofuels (Knothe and Dunn, 2005), printing inks (Erhan, 2005), hydraulic fluids, surfactants, protective coatings, drying agents, plastics, and engine oils (Dierig et al., 2006d). Lubricants derived from plant oils are currently used in drilling, metal working, and cutting fluids, as well as marine and two-stroke engine oils. They are also commonly found in hydraulic oils and chainsaw lubricants (National Non-Food Crops Centre, 2005). The *Brassicaceae* represent a rich source of vegetable oils for these uses. This chapter examines the *Brassicaceae* contribution to products that are currently used or being developed for industrial use, as well as some of the challenges to their development and acceptance, and also highlights some of the traditional and emerging *Brassicaceae* industrial oilseed crops.

BRASSICACEAE OILS OF INTEREST TO INDUSTRY

Plants in the family *Brassicaceae* accumulate fatty acids as storage products within their seeds, primarily as triacylglycerols. The major fatty acid components range from 12 to 24 carbon atoms in chain length, and may be saturated, mono- or polyunsaturated. The oil composition of a typical *Brassica napus* rapeseed cultivar is shown in Table 17.1.

Some members of the *Brassicaceae* family, such as species of the genus *Lesquerella*, also produce hydroxy fatty acids such as lesquerolic acid (C20:1OH), densipolic acid (C18:2OH), and auricolic acid (C20:2OH). These hydroxy fatty acids are similar to ricinoleic acid (C18:1OH), produced by the economically important castor plant *(Ricinus communis).* At present, hydroxy fatty acids from castor oil are used in lubricants, plastics, paints, cosmetics, coatings, and shampoos, and for the production of polyurethane for moldings and foams. Uses also include greases, hydraulic fluids

TABLE 17.1
Fatty Acid Composition for the High Erucic Acid
Rapeseed (*B. napus*) Cultivar MillenniUM 03

Fatty Acid	Name	% Total Fatty Acids
C12:0	Lauric acid	ND
C14:0	Myristic acid	ND
C16:0	Palmitic acid	2.6
C16:1	Palmitoleic acid	0.2
C18:0	Stearic acid	0.8
C18:1	Oleic acid	11.5
C18:2	Linoleic acid	11.6
C18:3	Linolenic acid	8.9
C20:0	Arachidic acid	0.7
C20:1	Gadoleic acid	0.4
C22:0	Behenic acid	0.7
C22:1	Erucic acid	54.0
C24:0	Lignoceric acid	0.2
C24:1	Nervonic acid	1.0

Note: ND, not detected in this cultivar.

Source: Adapted from Ratnayake and Daun (2004).

and motor oils, nylon-11, drying agents, protective coatings, surfactants, cosmetics, and pharmaceuticals (Roetheli et al., 1991). Castor oil is imported into North America for industrial use. This is due to the difficulties associated with growing this crop, especially the highly toxic protein, ricin, produced by castor plants, and to the presence of a storage protein that causes illness and allergic reactions in castor farm workers (Dierig, 2006).

Although *Lesquerella* is a relatively new crop, the oil is already in use in biodegradable grease products (Adhavaryu et al. 2005). It may soon also form components of paints. *Lesquerella* oil and dehydrated *Lesquerella* oil compared favorably in tests to castor oil and dehydrated castor oil in drying time, flexibility, and corrosion resistance in alkyd-type coatings (Van De Mark and Sandefur 2005).

C18 fatty acids, such as oleic acid, are major components of many *Brassicaceae* seed oils, and these fatty acids are of great importance to industry in the replacement of mineral oils with feedstocks and for the production of biodiesel. Biodiesel is an alternative diesel fuel produced from vegetable oils or animal fats instead of petrochemicals (Knothe and Dunn, 2005). Processes to convert vegetable oils to biodiesel continue to advance. So-called "first-generation" processes use transesterification of the oils with methanol and either a sodium or potassium hydroxide catalyst (leaving glycerol as a byproduct). "Second-generation" biodiesel production involves partially combusting the products to form "syngas," a mixture of carbon monoxide, carbon dioxide, and hydrogen. After cleaning, the gas is reacted with a catalyst to form straight-chained paraffins that are hydrocracked to produce diesel via the Fischer-Tropsch reaction or used directly for some applications (National Non-Food Crops Centre, 2005). Low erucic acid oils such as canola (*Brassica* spp.) oil and *Camelina* have advantages in both the biodiesel/industrial and edible oil markets, while the high oil content of high erucic acid rapeseeds may improve the economics of the industrial oil applications for this oil type.

Many members of the *Brassicaceae* family produce the important industrial commodity erucic acid (C22:1). Erucic acid is an important lubricant for the extrusion of metal and for marine uses. It serves as a feedstock for the production of many products, including slip agents for polythene films (erucamide), nylon 13-13 (National Non-Food Crops Centre, 2005), and rubber additives (Nieschlag and Wolff, 2007). A discussion of the many industrial uses of erucic acid and its derivatives can be found in Nieschlag and Wolff (2007). In 2000, industry in the United States utilized approximately 18 million kg of high erucic acid oil, and most of this was imported. A large market exists for high erucic acid oil in North America (Bhardwaj and Hamama, 2000). At present, erucic acid from high erucic acid rapeseed (*Brassica napus*) and *Crambe abyssinica* oils supply this demand. Edible oil rapeseed varieties (canola-quality cultivars) contain negligible amounts of this fatty acid. Other high erucic acid crops include *Brassica* spp. (*B. carinata* and *B. juncea*,) *Sinapis alba, Eruca sativa* spp. *oleifera*, and *Crambe hispanica* (Lazzeri et al., 2004). *B. cretica, B. incana, B. rupestris*, and *B. villosa* are also *Brassica* species with greater than 45% to 50% erucic acid (Warwick and Francis, 1994) but they are not currently cultivated for this purpose.

The future of *Brassicaceae*-derived industrial oils appears bright. Some products have already found markets, while others are still in development. There is a large pool of potential *Brassicaceae* species for domestication into crops for the future or for use in breeding programs to enhance those already grown. There are, however, limitations to overcome before these and other vegetable-oil-based products can gain broad acceptance and marketability.

CHALLENGES IN THE DEVELOPMENT OF VEGETABLE-OIL-BASED INDUSTRIAL PRODUCTS

There are numerous challenges with regard to the development of vegetable-oil-based industrial products. There are many reasons, such as lower toxicity of the products, that may encourage the use of vegetable-oil-based products even if the price is somewhat higher than the price of a comparable petroleum-based product. For many markets, however, it will be the price of the vegetable-oil-based product relative to the more usual petroleum-based product that will determine the success of a biobased oil or its derivatives.

Historically, petroleum has been a less expensive feedstock than vegetable oils for the production of industrial lubricants and chemicals, but the situation is changing rapidly with the rise in petroleum prices. In 2006, the comparative price for processed soybean oil was USD$78 per barrel, on par with crude oil at USD$70 to $80 per barrel. In addition, the price of soybean oil was less volatile than that of crude oil (Epobio Workshop, 2006). As petroleum prices continue to rise, vegetable oils will become increasingly affordable — or will they? Because the cost of production for oilseed crops is tied to petroleum prices through expenses related to fueling machinery, fertilizer costs, and transportation of goods, their prices will also rise. Presently, biofuels are more expensive than fossil fuels due to the costs of new infrastructure, research and development, and the current small scale of production (National Non-Food Crops Centre, 2005).

Low-input crops may have the ability to make serious inroads into the industrial products market in the future by minimizing the costs of production. For imported products, such as castor oil, a domestically produced substitute such as *Lesquerella* oil also may relieve some of the long-distance transportation costs. Offsetting the initial higher cost, biolubricants may last longer than mineral lubricants, extending the life span of equipment (National Non-Food Crops Centre, 2005).

Development of crops to supply industrial oils may create some land-use conflicts with food resources. For example, *Lesquerella* will require at least 142,000 ha based on current usage of castor oil if it is grown as the replacement source for castor hydroxy fatty acids (Dierig et al., 2006d). Some of the "newer" oilseed crops from the *Brassicaceae* (such as *Lesquerella* and *Camelina*) may be suitable for growing on arid, marginal, or poorly fertile land, but pressing these areas into production may not be beneficial to wildlife or watersheds.

Although the meals and other products remaining after oil extraction are valuable as livestock feeds, there will be a need to develop additional value-added markets for these products. Falling prices of rapeseed meal and glycerol due to large-scale biodeisel production in Europe will result in an overall increase in biodiesel production costs (Epobio Workshop, 2006). New markets for these byproducts could offset the cost of biodiesel production. Breeding efforts may have to concentrate on improving the quality of the byproducts — not just the oils, as was done in the past with the reduction in glucosinolates from the meal of canola.

The industrial products developed from oilseed crops will have to consistently meet industry standards for performance and stability. There are particular concerns with the quality of high-performance fuel oils, and of lubricants. There are also concerns about the higher iodine values, and low-temperature performance of biodiesel fuels, because many vegetable oils become solids at temperatures that are higher than those experienced by vehicles in temperate climates. If a machine's warranty specifies a mineral or synthetic lubricant, the use of a biolubricant may render the warranty void. Oxidation of oils at high engine temperatures remains a problem. Breeding for improved stability of the oils may help solve this problem because antioxidants are found in most *Brassicaceae* oils. In addition, vehicles may need modifications to use the biobased products. New vegetable-oil-resistant seals may be required when using biolubricants. (National Non-Food Crops Centre, 2005).

Development of some new *Brassicaceae* oilseed specialty crops may be hampered by the lack of large markets for the products. When a product is in demand, there are often research funds available to speed breeding efforts. In contrast, the market may not be created until an improved product is both readily available *and* inexpensive. Investment is critical for the success of new oilseed crops, both to enhance breeding and agronomy of the crops and to help identify markets for the oils and byproducts.

For the purpose of supplying specific market requirements, the breeding of targeted oilseed cultivars may be necessary. Authors of the Epobio Workshop (2006) foundation paper for the plant oil flagship advocate the use of "platform" crops for industrial use. These would be ideally non-food crops that do not compete with or intercross easily with food supply crops, that are easily grown with existing technologies, and that are amenable to improvement and transformation for the customization of the oil products. This can involve the development of existing industrial crops, or the domestication of wild species, which often gives access to a diverse gene pool, including related species. Perennialism would be a favorable life history trait in some species because it may reduce production costs, erosion, and the need for irrigation. Plants that are self-fertile and grow readily in tissue culture are very useful. Those with known male sterility systems and reasonable yields hold great promise for improving the original stock.

SELECTED *BRASSICACEAE* OILSEED CROPS

Presented here are profiles of a selection of *Brassicaceae* crops now used by industry or in various stages of pre-industrial development.

CRAMBE

Interest in the *Crambe* genus as an oilseed species is a consequence of its high erucic acid content in the seed oil and the seed meal component left after oil extraction (Whitely and Rinn, 1963; Downey, 1971). *Crambe* has a high erucic acid (*cis*-13-docosenoic acid) content of 50% to 60% in the seed oil (Lessman and Berry, 1967). *Crambe* (Abyssinian mustard, Abysinian kale, colewart, or datran) is a member of the Cruciferae (mustard) family.

The *Crambe* genus consists of about 20 different species with different chromosome numbers (*C. hispanica* var. *glabarata*, $n = 15$; *C. hispanica* var. *hispanica*, $n = 30$; and *C. abyssinica*, $n = 45$) (White and Solt, 1978). The differences in ploidy level are thought to mask the genetic variability within the genus (Downey, 1971). *Crambe* originated in the eastern Mediterranean region (Turkey and Iran). It spread to Asia and Europe from its center of origin. *Crambe* is adapted to moderate rainfall, warm temperature regions of the world. It is drought tolerant, especially later in the growing season, but is not frost tolerant. *Crambe* species are erect, annual, indeterminate flowering habit crucifers with wide-ranging adaptation. Plant height varies between 0.5 and 2 m (Whitely and Rinn, 1963). *Crambe* is similar to mustard in growth form, but produces small white flowers borne on long racemes. The flowers produce seed capsules approximately 5 mm in diameter containing a single seed 1 to 2.5 mm in diameter. The flowers are primarily self-pollinated with some natural outcrossing (Beck et al., 1975). The main stem of *Crambe* branches profusely, starting near the ground to produce 30 or more secondary stems. Each branch ends in a terminal raceme.

Several *Crambe* species have been evaluated as potential oilseed crops in Canada, Denmark, Germany, Poland, Russia, Sweden, and Venezuela since circa 1932 (White and Higgins, 1966). *Crambe abyssinica* is the *Crambe* species with the most potential for commercial production (Hirsinger, 1989). In both Europe and North America, *C. abyssinica* flowers in approximately 7 weeks from seedling, flowers for approximately 3 weeks, and then reaches physiological maturity approximately 2 weeks after the end of flowering. *Crambe* reaches physiological maturity in approximately 12 to 15 weeks from planting (National Non-Food Crops Centre, 2005).

Crambe seeds weigh 7.0 to 7.5 g/1000 seeds with a capsule content of 15% to 40% (McGregor et al., 1961; Earl et al., 1966). The capsule stays on the seed at harvest (Papathanasiou and Lessman, 1966), resulting in a very light test weight of 34 kg/hectoliter (National Non-Food Crops Centre, 2005). At harvest, the *C. abyssinica* seed plus capsule contains 26% to 38% oil (32% oil on average), with the capsule making up 30% of the harvested product (Earle et al., 1966). Dehulled *Crambe* seed has an oil content of 33% to 54% and a protein content of 30% to 50% (Earle et al., 1966).

Crambe has been grown in small-scale commercial production in eastern and western Europe, Russia, North America, and South America (Lessman, 1990). Average seed yields of 1400 kg/ha in the United States in the early 1990s and 2500 kg/ha in the United Kingdom in the early 2000s for *Crambe abyssinica* were obtained in commercial production (Carlson et al., 1996; National Non-Food Crops Centre, 2005). Average seed yields in other European countries ranged from 450 to

2000 kg/ha (National Non-Food Crops Centre, 2005). There was small-scale commercial production of *C. abyssinica* in western Canada in the 1960s. A severe outbreak of *Alternaria brassicola* ended *Crambe* production there in the late 1960s. *C. abyssinica* commercial production began in North Dakota (United States) in 1990 and continues on a small scale (Knights, 2002). *C. abyssinica* commercial production began in the United Kingdom in 2001 and also continues on a small scale (National Non-Food Crops Centre, 2005).

The primary breeding objectives for *Crambe* are to increase seed yield, increase oil production, and improve protein meal quality (Lessman, 1990). There is limited genetic variability within *Crambe abyssinica,* which makes it difficult to produce improved cultivars. Lessman and Meier (1972) therefore suggested mutagenesis or interspecific crosses as the breeding approach most likely to produce new *Crambe* cultivars with agronomic and/or seed quality improvements.

Lessman, working at the USDA station at Peoria, Illinois,, used mass selection within *Crambe abyssinica* to produce the cultivars Prophet and Indy, while interspecific crosses between C. *abyssinica* and *C. hispanica* were used to produce the cultivar Meyer (Lessman, 1990). Campbell et al. (1986), working at the Plant Genetics and Germplasm Institute in Beltsville, Maryland, used introgression of wild *Crambe* germplasm into the cultivar Indy to produce the *Crambe* cultivars BelAnn and BelEnzian as well as new *Crambe* germplasm (C-22, C-29, and C-37) that is moderately tolerant to *Alternaria brassicola. Crambe* production in North Dakota in the 1990s utilized the *Crambe* cultivars Meyer, BelAnn, and BelEnzian. A *Crambe* breeding program was initiated at North Dakota State University in 1991 using previously released *Crambe* cultivars, germplasm from Lessman's *Crambe* breeding program, and *Crambe* germplasm from the North Central Plant Introduction Station at Ames, Iowa (Knights, 2002). The *Crambe* breeding program at North Dakota State University continues to develop new *Crambe* germplasm and cultivars.

HIGH ERUCIC ACID RAPESEED AND OTHER *BRASSICA* SPP.

There has been much interest in the *Brassica* genus as an oilseed for industrial oils due to its high erucic acid content in the oil and the high protein meal left over after oil extraction (Röbbelen, 1991). Several *Brassica* oilseed species, including *Brassica carinata* (Ethiopian mustard), *B. juncea* (mustard), *B. napus* (oilseed rape/rapeseed), and *B. rapa* (turnip rape/rapeseed), naturally produce seed oil that is moderate to high in erucic acid and contains moderate to high protein the seed meal after oil extraction (Downey and Röbbelen, 1989). Ranges of erucic acid content in these species have been reported by Velasco et al. (1998) as *B. carinata:* 29.6% to 51.0%; *B. juncea*: 15.5% to 52.3%; *B. napus*: 5.6% to 58.1%; and *B. rapa*: 6.5% to 61.5%. Although all these oilseed species could potentially be developed as sources of high erucic acid seed oil, only *B. napus* rapeseed has been successfully developed as an industrial oil species (Stefansson and Downey, 1995).

Brassica species are able to germinate and grow at low temperatures and are one of the few oilseeds adapted to the cooler temperate agricultural zones and to winter production. *Brassica carinata*, *B. napus,* and *B. juncea* are amphidiploids, combining chromosome sets of the diploid species *B. rapa*, *B. nigra* (black mustard), and *B. oleracea* (kale) (U, 1935). *B. rapa* rapeseed is thought to have a primary center of origin in the Indian subcontinent with secondary centers of origin in Europe, the Mediterranean area, and Asia (Hedge, 1976); *B. oleracea* and *B. napus* rapeseed both originated in the Mediterranean area (Hedge, 1976); *B. nigra* and *B. juncea* both originated in the Middle East (Prakash and Hinata, 1980); and *B. carinata* originated in northeastern Africa (Downey and Röbbelen. 1989). *B. napus* rapeseed and *B. rapa* rapeseed have both spring and winter annual forms, while *B. juncea* and *B. carinata* are exclusively spring types. The winter forms of *B. napus* rapeseed and *B. rapa* rapeseed are more productive than the spring forms but less winter hardy than winter cereals (Downey and Röbbelen, 1989). Rapeseed plant heights range from 0.5 to 2.5 m. The flowers are yellow, typical crucifer flowers. Flowering is indeterminate, lasting 2 to 4 weeks, beginning at the lowest bud on the main raceme. *B. napus* is a self-compatible species displaying a high degree of self-pollination, while most *B. rapa* is self-incompatible (except for the Indian subspecies yellow sarson, which is self-compatible) (Downey and Röbbelen, 1989). Each fertilized flower produces 25 or more seeds per pod, and each plant produces 40 or more pods resulting in a multiplication rate of 1000 to 1 or more in the *Brassica* rapeseed species (Downey and Röbbelen, 1989). Spring habit *B. napus* rapeseed is grown in Canada and northern China and matures in 90 to 140 days, while winter habit *B. napus* rapeseed is grown in most of Europe, as well as central and southern China, and matures in 320 to 360 days. The seed of rapeseed contains more than 40% oil and a meal containing 36% to 44% protein. Rapeseed meal is used as a fertilizer in Asia and as animal feed in the rest of the world (Downey and Röbbellen, 1989).

High erucic acid, low glucosinolate rapeseed (HEAR) cultivars are grown in both Europe and Canada. Spring habit HEAR cultivars are grown in the Canadian Prairie provinces and northern Europe, while winter habit HEAR cultivars are grown primarily in Germany (Downey and Röbbelen, 1989). Seed yields for the spring types are 1.0 to 1.5 tonnes/ha, while seed yields for the winter types are 2.5 to 3.0 tonnes/ha (Röbbelen, 1991).

The primary breeding objectives for HEAR cultivars are to increase seed yield, increase oil production, improve meal quality, increase erucic acid concentration, improve disease resistance, and, for the winter habit types, improve winter hardiness and frost tolerance (Downey and Röbbelen, 1989).

HEAR breeding programs have been conducted at Danisco Seed (Denmark), NPZ Lembke (Germany), and the University of Manitoba (Canada). Danisco Seed developed several spring habit HEAR cultivars, including Industry and Sheila, adapted to northern European growing areas in the 1990s and early 2000s prior to the sale of this breeding program to NPZ Lembke circa 2005. NPZ Lembke developed several winter habit HEAR cultivars adapted for production in Germany in the 1990s and 2000s. The first NPZ Lembke HEAR cultivars were open-pollinated populations, while the more recent NPZ Lembke HEAR cultivars are male-sterile Lembke (MSL) hybrids. Information on these proprietary private-sector breeding programs is limited. In contrast, there is considerable information available for the University of Manitoba HEAR breeding program. The University of Manitoba developed and released the world's first HEAR cultivar, Reston, in 1982 and has developed numerous spring habit, open-pollinated population HEAR cultivars since then.

High erucic acid, low glucosinolate *Brassica napus* cultivar development began at the University of Manitoba in the early 1970s as a parallel and complementary breeding program to the double low rapeseed (canola) breeding program (Scarth et al., 1992). The double low rapeseed (canola) cultivars Regent released in 1977 and Reston released in 1982 both originated from the same series of crosses (Stefansson and Downey, 1995). Reston had 40% to 45% erucic acid content and low glucosinolate content in the meal. Reston was a commercial success and was grown as the exclusive Canadian source of high erucic acid rapeseed oil for several years (Stefansson and Downey, 1995).

The second HEAR cultivar developed by the University of Manitoba, Hero, released in 1989 (Scarth et al., 1991) had an erucic acid content of over 50% and a low glucosinolate content in the meal. A series of high erucic acid, low glucosinolate rapeseed cultivars with incremental improvements in agronomic performance and/or seed quality was developed by the University of Manitoba following Hero. The HEAR cultivars Mercury (Scarth et al., 1994), Venus (McVetty et al., 1996a), Neptune (McVetty et al., 1996b), Castor (McVetty et al., 1998), MillenniUM 01 (McVetty et al., 1999), MillenniUM 02 (McVetty et al., 2000a), and MillenniUM 03 (McVetty et al., 2000b) were all developed and released by the University of Manitoba. Simultaneous improvements in seed yield, oil content, protein content, and erucic acid content occurred for the Reston to MillenniUM 03 series of HEAR cultivars. All these HEAR cultivars were grown on significant production areas and were commercially successful, although in many cases their commercial lifetime was short because they were replaced quickly by incrementally improved new HEAR cultivars.

Marker-assisted selection using sequence related amplified polymorphisms (SRAPs) (Li and Quiros, 2001) molecular markers has been incorporated into the HEAR breeding program in recent years. An ultra-dense genetic recombination map for *Brassica napus* has been developed (Sun et al., 2007). The development of genome-specific erucic acid gene molecular markers for the two erucic acid controlling genes in *B. napus* (Rahman, 2007) has permitted the selection of homozygous high erucic acid genotypes in segregating generations and backcross generation progeny of HEAR × canola crosses, greatly improving breeding efficiency.

The University of Manitoba HEAR breeding program is developing novel herbicide-tolerant HEAR cultivars. The world's first two glyphosate-tolerant HEAR cultivars, Red River 1826 (McVetty et al., 2006a) and Red River 1852 (McVetty et al., 2006b), were developed and released in 2006. They are the world's first commercially successful glyphosate-tolerant (Roundup Ready™) HEAR cultivars. The University of Manitoba is also developing glufosinate ammonium tolerant HEAR cultivars. The world's first two glufosinate-ammonium-tolerant HEAR lines were supported for registration in Canada in March 2008. These are the first glufosinate-ammonium-tolerant (Liberty Link™) HEAR lines in the world to reach this development stage.

The development of hybrid HEAR cultivars is also underway at the University of Manitoba. Cuthbert (2006) reported high parent heterosis for seed yield in HEAR hybrids exceeding 100%. The University of Manitoba is using the *ogu* INRA CMS system for hybrid HEAR cultivar development. In addition, Lembke Research Limited and the University of Manitoba are collaboratively developing hybrid HEAR cultivars using Lembke's proprietary Male Sterile Lembke (MSL) pollination control system (Frauen and Paulmann, 1999).

The University of Manitoba HEAR program is also developing super-high erucic acid, low glucosinolate rapeseed (SHEAR) germplasm with the goal of reaching an erucic acid content greater than 66% in the seed oil. Microspore mutagenesis in the progeny of resynthesized *Brassica napus* line crosses is being used to create doubled haploid lines with homozygous mutations for fatty acid profile variations, including possible SHEAR genotypes. A transgenic approach to SHEAR development is also being pursued in collaboration with the Plant Biotechnology Institute in Saskatoon, Saskatchewan (Canada).

The emerging emphasis on renewable energy, chemical feedstocks, industrial oils, and novel uses of vegetable oils, along with the steadily growing bio-economy, will provide significant growth opportunities for high erucic acid oil and for super-high erucic acid oil. Double-digit annual growth in demand for high erucic acid rapeseed oil and super-high erucic acid rapeseed oil is anticipated.

CAMELINA SATIVA

Camelina (*Camelina sativa* (L.) Crantz) is known by many names, including Gold of pleasure, German sesame, leindotter, Siberian oilseed, and false flax. It is native to eastern Europe and southwestern Asia, but is now widely distributed. It is a spring or winter annual crop adapted to cooler areas (Straton et al., 2007). An oilseed of some antiquity, it has been in use since Neolithic times (Putnam et al., 1993). It was widely grown until the 1940s when it was largely replaced by rapeseed oil, which was easier to hydrogenate (Crowley, 1999). Interest in *Camelina* oil has risen with the rise in petroleum prices because *Camelina* may have lower input (fertilizer, pest control costs) associated with its production (Crowley, 1995) than some of the more conventional oilseeds.

At present, *Camelina* has a limited market worldwide and is a small-scale commercial crop. It is a component of salad oils and margarines in some European countries. *Camelina* oil is registered as a food oil in Canada and many European countries, and is a major seed oil in Siberia (Francis and Campbell, 2003). Although it is a valuable byproduct of oil extraction, restrictions on the sale of *Camelina* meal has limited the use of *Camelina* in some countries (Matthaus, 2004). Although *Camelina* has a long association with humans, until recently there have been few intensive breeding efforts directed at this crop (Vollmann et al., 2005).

An essential element in the development of a crop plant is access to genetic variability. Several authors have examined the potential of *Camelina* breeding programs by estimating genetic diversity within germplasm collections. In Austrian trials, considerable genetic variation in linolenic and erucic acids was demonstrated within *Camelina* accessions. The authors believed that there was enough variation to allow good progress toward breeding goals of increasing oil content and seed yield (Vollmann et al., 2007).

Camelina cultivars evaluated in the Maritime provinces of Canada showed considerable variation in many different agronomic traits and oil quality traits. This variation was considered significant enough to determine the success of crop under the agronomic conditions of the areas (Urbaniuk et al., 2008). Vollmann et al. (2005) used both seed-quality-based data and DNA-based markers to determine genetic variation. Estimates for variation within the *Camelina* accessions differed between the two methods. Although both were useful in determining variability, they believed that for breeding purposes, phenotypic diversity may be the more valuable indicator.

Camelina is a spring annual suitable for cultivation in temperate climates (Fröhlich and Rice, 2005). In mild climates, it acts as a winter annual, forming a rosette of four to six leaves and then remaining dormant until spring (Crowley, 1999). As a winter sown crop, tillage is not required. It can be broadcast or drill sown on frozen ground in late November to early December. Spring sowing is usually done from mid-April to early May (Straton et al., 2007). Overwintering plants may flower again in the spring but are usually killed by spring frosts (BioMatNet, 2007). Mature height of the crop can vary from 30 to 90 cm. It produces small pale yellow or greenish flowers on branched stems. *Camelina* is a self-pollinating crop (Vollmann et al., 2005) with a flowering period of about 2 weeks (Akk and Ilumäe, 2005). Maturity is rapid (85 to 100 days), so it is well adapted to areas with short growing seasons.

The small, rounded siliques of *Camelina* contain a few seeds each. The seeds have little dormancy (Francis and Campbell, 2003). They are readily harvested without significant amounts of shattering (Francis and Campbell, 2003). *Camelina* seeds are small but variable in size, ranging from 800 to 1400 seeds per gram. In Australia, the seeds of the largest seeded line (1000 seed wt = 1.61 g) were nearly three times larger than the smallest seeded accessions (1000 seed wt = 0.66 g) (Francis and Campbell, 2003). Although Gugel and Falk (2006) recommend increased seed size and oil content as breeding goals for *Camelina*, Austrian trials indicated that large seed size was negatively correlated with oil content and yield (Vollmann et al., 2007).

Interest in *Camelina* as a crop is due in part to its low level of required inputs (Vollmann et al., 2005). Many of the cultivars tested in one Canadian study responded to lower levels (60 to 80 kg) N/ha⁻ of added nitrogen by increasing plant height, seed yield, oil content, seed protein, and total nitrogen. Significant increases in these parameters did not occur at higher levels of nitrogen supplementation (Urbaniuk et al., 2008) at these sites. As nitrogen supplementation increased, a reduction in oleic and eicosenoic acid content of the seed oil was seen. Day and Chalmers (2007) reported that a significant yield response to nitrogen was observed at only one out of three sites in their trials in western Canada. This site had lower initial soil nitrogen than the other two sites. *Camelina* does, however, respond well to nitrogen inputs on some sites. In Ireland, *Camelina* yields responded positively even to quite high levels (140 kg N/ha) at one site over the course of 2 site-years. On a more fertile site during the same time period, the higher levels of nitrogen produced greater lodging and disease, reducing yield (Crowley, 1999). Over-application of N is unlikely to be a problem, however, because it would remove one of the economic advantages of *Camelina* production. The crop well tolerates low-fertility soils (Francis and Campbell, 2003). In addition, *Camelina* can be used with no-tillage systems, reducing fuel costs, and is seeded at a relatively low rate.

Relative to the *Brassica* oilseeds used as controls, Gugel and Falk (2006) found that *Camelina* was more drought resistant, earlier maturing, and more resistant to flea beetle attack.

Camelina produces phytoalexins, which control some diseases and insect pests (Urbaniuk, 2006), such as *Alternaria brassicae* and flea beetles. *Camelina* is also reputed to be resistant to *Leptosphaeria maculans*, an important disease of oilseed rape crops. *Camelina* therefore may be very useful as a high-value crop where the incidence of blackleg is high, as in Australia. *Camelina* can be quite competitive with weeds due to its early vigorous rosette formation. In Manitoba, Canada, *Camelina* can be grown without broadleaf herbicides if a preseason burnoff is used.

Saucke and Ackerman (2006) explored the use of *Camelina* in an intercrop with peas and found that there was a significant suppressive effect on weeds in the intercrop relative to either monocrop. The early establishment of *Camelina* made it effective as an early-season competitor with annual weeds, but it did not prove effective in the control of perennial weeds. Akk and Ilumäe (2005) suggested the use of clover in rotations to suppress the perennial weeds in *Camelina* fields. The pea-*Camelina* intercrop yielded higher monocropped peas or *Camelina* due to increased weed pressures in the monocrops. Paulsen (2007) grew wheat, lupins, or peas intercropped with *Camelina*. He too found that the yield of the legume intercrops with *Camelina* could exceed that of the monocrops, but only when yields of the monocrops were low. Wheat yield in the intercrop did not surpass that of the monocropped wheat. However, other advantages were seen for the intercrops, including reduced lodging and increased weed control. If a substantial market was available for *Camelina* meal and oil, the economics of the intercrops would be favorable.

Controlling broadleaf weeds within *Camelina* crops presents some challenges, but progress has been made in determining *Camelina*'s tolerance to herbicides (Francis and Campbell, 2003; Crowley, 1999). Disease problems in *Camelina* may be a concern in some areas. An evaluation of the crop in Ireland noted that the wet winters encouraged weed growth, which formed a dense canopy within which winter-seeded *Camelina* suffered diseases and lodging. As a winter crop, *Camelina* was susceptible to both *Botrytis* and *Sclerotinia* in these wet conditions (Crowley, 1999). Although seeded as a spring crop in western Canada, diseases tentatively identified as *Fusarium* spp. caused the destruction of two *Camelina* trials in Manitoba (S. Chalmers, 2008, personal communication). In the drier areas of the Upper Midwest (United States), no disease problems were of sufficient concern to require control (Straton et al., 2007).

Camelina is high yielding for a relatively unimproved crop that has not undergone intensive breeding. An Austrian study reports seed yields up to 2.8 tonnes/ha (Vollmann et al., 2007). In Australia, yields of 1.7 tonnes/ha were achieved, outyielding canola at some sites (Francis and Campbell, 2003). In Denmark, trials by Zubr (1997) achieved yields of 2.6 and 3.3 tonnes dm/ha for summer and winter varieties, respectively. In a recent nitrogen response trial in western Canada (Day and Chalmers, 2007), *Camelina* yields ranged from 620.34 to 1179.10 kg/ha. Oil contents of the seeds varied with cultivar and environmental conditions. Oil contents of 19 lines ranged from 38% to 43% in trials (Gugel and Falk, 2006) carried out in western Canada. Oil contents were generally higher at their most northern site (i.e., Beaverlodge, Alberta). The oil content of *Camelina* seeds in Austrian trials (Vollmann et al., 2007) was reported to be up to 48%.

Camelina has a pale straw-color oil that does not require bleaching (Francis and Campbell, 2003). It is especially enriched with C18 fatty acids. Linolenic acid (C18:3), primarily α-linolenic acid (Francis and Campbell, 2003), comprises the largest proportion of the seed oil (30% to 42%), with substantial amounts of linoleic acid (C18:2) 16% to 25%, and oleic acid (C18:1) 13%, also present (Zubr, 2003, in Urbaniuk, 2006). Erucic acid content is subject to environmental influence and accounts for 2% to 6 % of the total oil content (Vollmann et al., 2007).

Polyunsaturated oils tend to be unstable but *Camelina* oil is relatively stable due to its high levels of antioxidants, such as tocopherols, that protect the oil from degradation (Zubr, 2003). *Camelina* oil also contains 13.5% to 16.7 % eicosenoic acid (C20:1) (Francis and Campbell 2003), which, owing to the position of the double bond, has enhanced oxidative stability. *Camelina* oil is subject to photooxidation, so it is best stored in darkness (Abramovic and Abram, 2005). The high linolenic and eicosenoic acid content flavors the oil (Akk and Ilumäe, 2005). For a vegetable oil, *Camelina* oil is unusually high in cholesterol (45 mg per 100 g) (Matthaus, 2004).

Although the high linolenic acid content positions *Camelina* oil well for the edible oil and skin-care markets, it also has excellent potential for biodiesel mixtures (BioMatNet, 2007). The antioxidants present in *Camelina* oil may also be useful for improving the stability of other vegetable oils for industrial use.

After *Camelina* is crushed and the oil is removed, a valuable meal remains, although markets for this product have not yet been fully developed. Meal protein for the 19 *Camelina* lines grown in a western Canadian study was reported to range from 21% to 33% and was inversely correlated with oil content (Gugel and Falk, 2006). Glucosinolate levels of the meal are low, and the fiber content of the meal is 10% to 11%. It has good potential as a high-quality animal feed. The levels of eicosenoic and erucic acids, as well as glucosinolates, restrict the use of its oil and meal for animal and human consumption (Akk and Ilumäe, 2005) in some countries.

Although an ancient crop, little intensive breeding work has been carried out with *Camelina* (Vollman et al., 2005) to improve the agronomic characteristics, yield, and oil content, or to select for particular fatty acid profiles. Recent research programs aimed at evaluating and selecting adapted lines include Vollmann et al.'s (2007) trials in Austria of advanced breeding lines, Gugel and Falk's (2006) agronomic trials of *Camelina* in western Canada, and Urbaniuk's (2006) trials in eastern Canada. Selection objectives included high oil content, increased 1000 seed weight, and specific fatty acid profiles.

While the variability inherent in *Camelina* has yet to be fully exploited using conventional breeding, transgenesis may be another option for developing suitable oil profiles and other desired traits. *Camelina* lends itself to crop improvement via transgenesis. Lu and Kang (2008) have reported the successful transformation of *C. sativa* using *Agrobacterium*-mediated transfer. Using this method, a castor-derived fatty acid hydroxylase gene was transferred to *Camelina*, resulting in the production of novel hydroxy fatty acids. Somatic embryogenesis and microspore-derived embryogenesis coupled with mutagenesis could speed the breeding efforts for particular fatty acid profiles.

The prospects for *Camelina* as a biodiesel feedstock were examined by Fröhlich and Rice (2005). They found that the yields of methyl esters from 6×350-kg laboratory test batches were similar to that of rapeseed oils. The cost of production of *Camelina* may be lower in some areas than that of rapeseed, suggesting that *Camelina* could be a viable alternative to rapeseed oil for biodiesel production. The iodine number of the esterified product, a measure of the degree of unsaturation of an oil sample, was in excess of EU standards for biodiesel; however, deterioration of the fuel was not observed during the study. The properties of the *Camelina* fuel were mostly acceptable but the behavior of the biodiesel in cold weather situations could be problematic. Pour point depressants could improve the performance of the fuel for cold weather areas, or *Camelina* biodiesel could be blended with petrodiesel. A reduction in fuel economy in vehicles using *Camelina* biodiesel was similar to that of other biodiesel products. Fröhlich and Rice (2005) recommended further vehicle testing to confirm the suitability of unblended *Camelina* biodiesel. A new joint venture in the United States has the objective of providing 100 million gallons of *Camelina*-based biodiesel by the year 2010 (Targeted Growth Inc., 2007).

Cold-pressed and filtered *Camelina* oil has also been used directly to test its use as a fuel for diesel engines (Bernardo et al., 2003). The performance of the *Camelina* oil was compared to that of mineral diesel fuel. *Camelina* oil was less efficient (12.57 km/L) than the mineral fuel (14.03 km/L), but the smoke opacity and carbon monoxide production was 50% less than with the mineral fuel. Nitric oxide was actually higher (6%) for the *Camelina* oil at higher engine rpm, but similar at less than 3500 rpm. Little nitrogen dioxide was produced by either fuel, and they were similar in their production of carbon dioxide and oxygen. The iodine value of the oil and its viscosity was found to be reduced by heating to 170°C.

Straton et al. (2007) suggested that because of its high linolenic acid content, *Camelina* oil may also be suitable for linoleum production. Another potential use of *Camelina* is as a replacement for petroleum oil in pesticides sprays.

Lesquerella spp.

Lesquerella is a genus of 90 to 100 species native to drier lands of the United States, Canada, and Mexico (Dierig et al., 2006d). It is being heralded as a domestic replacement for the similar hydroxy fatty acid in castor oil, ricinoleic acid, a USD$50 million import for the USA (Dierig, 2006). Hydroxy fatty acids are used commercially in lubricants, plastics, cosmetics, and coatings

(Reed et al., 1997). As additives to biofuels and ultra-low sulfur diesel, estiolides and 2-ethylhexyl esters derived from *Lesquerella* oil show promise in improving lubricity and may reduce pour points in low blend rates (Moser et al., 2008).

Three major groups of hydroxy fatty acids have been identified in the genus *Lesquerella* — namely, lesquerolic acid (C20:1OH), densipolic acid (C18:2OH), and auricolic acid (C20:2OH) (Dierig, 2006). For example, lesquerolic acid is produced by *L. fendleri* and *L. lindheimeri*, densipolic acid is produced by *L. lyrata,* and auricolic acid is found in the oil of *L. auriculata* (Tomasi et al., 2002). *Lesquerella* oil is used to replace petroleum-based products such as hydraulic fluids, surfactants, two-cycle engine oils, automotive oils, protective coatings, drying agents, and plastics (Dierig et al., 2006). Biodiesel can also be made from *Lesquerella* oil and may have some unusual uses. *Lesquerella* biodiesel has been tested as a broadleaf contact herbicide as a safer alternative for homeowners (Vaughn and Holser, 2007). It was more toxic to the two broadleaf weeds than to perennial ryegrass in this study.

Lesquerella fendleri (Gray) Wats. has been under development as a new oilseed crop since the mid-1980s (Senft, 1992). It is native to arid regions of the southwestern United States (Isbell and Cermak, 2005) where it grows as a winter annual or a perennial (Dierig et al., 1996). Small orange-brown seeds (6 to 25 in number) are formed in each round silique. For *L. fendleri,* the 1000 seed weights range from 0.5 to 1.2 g (Dierig, 1996). The small seeds of *L. fendleri* contain 30% oil; between 55 and 64% of this oil is lesquerolic acid (Isbell and Cermak, 2005). The plants are quite productive, yielding 1880 kg/ha for a synthetic population, and between 1000 and 1350 kg/ha for unselected populations (Thompson and Dierig, 1993).

Oil is just one of several marketable products from *Lesquerella* seeds. *Lesquerella* meal, a valuable byproduct of oil extraction, has high protein and lysine contents, as well as antioxidants derived from glucosinolates (Salywon et al., 2005). Seed meal of *Lesquerella* contains antioxidants and the oil contains estiolides that can be used as inexpensive additives to improve the oxidative stability and pour point of other industrial vegetable oils (Hayes et al., 1995; Isbell et al., 2006). Biofumigants (thiofunctionalized glucosinolates) from the meal can be degraded via myrosinase to produce compounds effective against some pathogenic fungi (Lazzeri et al., 2004). Gums are also components of *Lesquerella* seeds, and these may be as valuable a product as the seed oil for thickening or gelling edible and non-edible products (Salywon et al., 2005). *Lesquerella* gum has potential in industrial products as a thickener for paints and drilling products (Agricultural Research Service, USDA, 1999). Other markets for the gums may include the pharmaceutical industry and as binders in pelleted animal feeds.

Breeding efforts for *Lesquerella* are still at a relatively early stage, although it is already in commercial production. There are several registered cultivars in the United States, including a high oil cultivar of *Lesquerella fendleri* (WCL-LO3) registered in 2006 by Dierig et al. (2006a) and a mutant with a cream flower color that was also registered that year by Dierig et al. (2006b). There appears to be ample variability within the species (Dierig et al., 1996) to encourage development *of L. fendleri* and related species as crops.

For example, variation in seed oil composition was evaluated in a diverse germplasm collection using a half seed method by Dierig et al (2006c). Variability was detected for oil composition in the S-1 populations. At least two segregating genes were expected to be responsible for lesquerolic acid content.

Hydroxy fatty acids are also produced by other *Lesquerella* species (Reed et al., 1997), and these may be useful for introducing traits (such as temperature and elevation tolerance) into *L. fendleri* (Dierig et al., 2006c). There may be opportunities to develop perennial oil crops using relatives of *L. fendleri.* Ploschuk et al. (2005) compared reproductive allocation of *L. fendleri* and the perennial *L. mendocina,* and concluded that there was no difference in the seed biomass allocation of the two species. There are some incompatibility issues to be resolved in the formation of interspecific hybrids of *Lesquerella* species, but ovule culture helps to overcome some of these barriers (Tomasi et al., 2002). Intergeneric hybrids have also been attempted. Partial hybrids between *Brassica napus* and *L. fendleri* were produced by Du et al. (2008) and these may be of use to *B. napus* breeders.

Increased seed yields, oil content, and lesquerolic acid content are obvious breeding objectives for *Lesquerella fendleri*, but breeding efforts are also underway to develop yellow-seeded cultivars. The pigmentation of the oil presently limits the use of the oil for some applications. A lighter-colored oil would find a greater market in the cosmetics industry (Dierig et al., 1996). Autofertility may also be important. Pollination can be a limiting factor for seed set (Senft, 1992). *L. fendleri* plants are 86% to 95% outcrossing, with self-incompatibility that appears to be a sporophytic multiple allelic system (Dierig et al., 1996). However, autofertile accessions are known. Inbreeding depression appears to reduce seed yield in half-sib populations (Salywon et al., 2005). Hybrid cultivar development will be aided by the existence of male sterility systems for *L. fendleri*. Male sterility frequently occurs in this species (Dierig et al., 1996).

Lesquerella fendleri is well suited to arid conditions, so the major production centers are likely to be in Arizona, Texas, and New Mexico in the United States (Dierig, 2006). Thus far there are no major disease or insect problems for crops of *L. fendleri*, although these can easily arise once the crop is widely grown (Senft, 1992).

Harvesting can be done with existing machinery (Senft, 1992), which may aid in the adoption of the crop by growers. In the United States, *Lesquerella fendleri* is grown as a winter annual on well-drained soils. It is broadcast sown in October and emerges 2 weeks afterward. It remains dormant at a very small size until it resumes growth in February. At a seeding rate of 6 to 8 kg/ha, it reaches full ground cover in 2 months after resuming growth. Flowering begins in February, and seeds are mature by the end of May. Presently, there are no herbicides registered for use on *Lesquerella* crops, so weed control can be a problem. *L. fendleri* responds well to 60 to 120 kg/ha N fertilizer, and to irrigation. Plants are harvested in mid to late June (Dierig et al., 1996). Biomass and seed yields showed large changes in response to elevation (Dierig et al., 2006d). Although the oil content of *L. fendleri* is known to be affected by environmental conditions (Salywon et al., 2005), it does not appear to be sensitive to elevation (Dierig et al., 2006d).

For those areas not appropriate for the cultivation of *Lesquerella fendleri* or related species, there may be opportunities to use genetic engineering to transfer genes for the production of hydroxy fatty acids to other species. An FAD2 homolog from *L. lindheimeri* was shown to be responsible for the production of fatty acid hyroxylase. Yeast cells transformed with this gene produced ricinoleic acid along with small amounts of other di-unsaturated fatty acids. Transformation of *Arabidopsis* fad2/fae1 mutant lines with this gene resulted in the production of hydroxy fatty acids contents of up to 18% of total fatty acids. Oil content in the highest hydroxy fatty acid lines was reduced (Dauk et al., 2007). In the future, the expression of this gene in a well-adapted and high-yielding oilseed *Brassicaceae* species may enhance domestic supplies of hydroxy fatty acids without the inherent problems of castor oil production.

CONCLUSION

There is great demand in many industries for lubricants, slip agents, plasticizers, greases, and other products that are presently produced using petroleum oil. Due to the rising cost of petroleum, and to concerns regarding the toxicity and low biodegradability of petroleum-based products, there is increasing interest in replacing petrochemicals with vegetable-oil-based bioproducts. Bioproducts can be produced more sustainably, have lower toxicity, and are more easily biodegraded than petroleum-based products. In addition, certain characteristics, such as the high lubricity and low volatility of vegetable oils, make these oils preferred to mineral-based oils for some applications. Oils from the seeds of some members of the family *Brassicaceae* are excellent alternatives to petroleum oils. However, there are many hurdles to the large-scale commercialization of these oil-producing crops, including the marketing of byproducts, meeting industry standards, and the cost of the vegetable-oil-based products relative to their petroleum-based counterparts.

There are several areas of use developing for the seed oils of *Brassicaceae*. Biodiesel can be produced from the seeds of many species, including *Brassica* spp. and *Camelina sativa*. *Crambe abyssinica* and *Brassica* spp. (especially high erucic acid rapeseed) are producers of erucic acid, an industrially important commodity. Hydroxy fatty acids from *Lesquerella* may be poised to supplement or replace the demand for similar fatty acids from castor oil, an imported product in North America. All these crops are presently under development for industrial oil production, although some have a much longer history of cultivation and/or improvement. The oils of many of these plants have potential industrial markets (realized or future) as paint and printing ink additives, lubricants, greases, heating oils, polymers and resins, plastics, and in the production of cosmetics and pharmaceuticals. The future is promising for oil crops in the *Brassicaceae*.

REFERENCES

Abramovic, H. and V. Abram. 2005. Physico-chemical properties, composition and stability of *Camelina sativa* oil. *Food Technol. Biotechnol.*, 43:63–70.

Adhavaryu, A., B.K. Sharma, and S. Erhan. 2005. Current developments of biodegradable grease. In S.Z. Erhan (Ed.). *Industrial Uses of Vegetable Oils, United States Department of Agriculture,* AOCS Press, Illinois, p. 14–30.

Agricultural Research Service, USDA. 1999. A storybook future for *Lesquerella*? www.ars.usda.gov/is/AR/archive/nov99/story1199.htm?pf=1 (Accessed November 1, 2007).

Akk, E. and E. Ilumäe. 2005. Possibilities of growing *Camelina sativa* in ecological cultivation. http://72.14.205.104/search?q=cache:rdgHSPRI92sJ:www.eria.ee/public/files/Camelina%2520ENVIRFOOD.pdf+Akk+and+Ilumae&hl=en&ct=clnk&cd=1&gl=ca (Accessed March 1, 2008).

Beck, L.C., K.J. Lessman, and R.J. Buker. 1975. Inheritance of pubescence and its use in out crossing measurements between a *Crambe hispanica* type and *C. abyssinica* Hochst. ex. R.E. Fries. *Crop Sci.,* 15:221–224.

Bernardo, A., R. Howard-Hildige, A. O'Connell, R. Nichol, J. Ryan, B. Rice, E. Roche, and J.J. Leary. 2003. *Camelina* oil as a fuel for diesel transport engines. *Industr. Crops Prod.,* 17:191–197.

Bhardwaj, H.L and A.A. Hamama. 2000. Oil, erucic acid and glucosinolate contents in winter hardy rapeseed germplasms. *Industr. Crops Prod.,* 12:33–38.

BioMatNet. 2007. www.biomatnet.org/secure/Crops/S592.htm (Accessed March 6, 2008).

Campbell, T.A., J. Crock, J.H. Williams, et al. 1986. Registration of C-22, C-29, C-37 *Crambe* germplasm. *Crop Sci.,* 26:1088–1089.

Carlson, K.D., J.C. Gardner, V.L. Anderson, and J.J. Hanzel. 1996. *Crambe*: New crop success. In J. Janick (Ed.), *Progress in New Crops.* ASHS Press, Alexandria, VA, p. 306–322.

Crowley, J.K. 1995. Agronomy of *Camelina* sativa. EU Research Contract No. AIR3-CT 94 2178.

Crowley, J.G. 1999. Evaluation of *Camelina* sativa as an alternative oilseed crop. TEAGASC Irish Agriculture and Food Development Authority Project Report 4320 www.teagasc.ie/research/reports/crops/4320/eopr-4320.htm (Accessed March 6, 2008).

Cuthbert, R.D. 2006. Assessment of Selected Traits in Hybrid HEAR. M.Sc. thesis, University of Manitoba, Winnipeg, Manitoba, Canada.

Dauk, M., P. Lam, L. Kunst, and M.A. Smith. 2007. A FAD2 homologue from *Lesquerella lindheimeri* has predominantly fatty acid hydroxylase activity. *Plant Sci. ,* 173:43–49.

Day, S. and S. Chalmers. 2007. Westman Agricultural Diversification Organization Annual Report 2007. Manitoba Agriculture, Food and Rural Initiatives, Manitoba, Canada.

Dierig, D. 2006. *Lesquerella*: a domestic source of hydroxy fatty acids. Agronomy Abstracts (CD-ROM P24080). http://199.133.10.189/research/publications/publications.htm?seq_no_115=203671&pf=1 (Accessed March 31, 2008).

Dierig, D.A., A.M. Salywon, P.M. Tomasi, G.H. Dahlquist, and T.A. Isbell. 2006c. Variation of seed oil composition in parent and S-1 generation of *Lesquerella fendleri* (*Brassicaceae*). *Industr. Crops Prod.,* 24:274–279.

Dierig, D.A., N.R. Adam, B.E. Mackey, G.H. Dahlquist, and T.A. Coffelt, 2006d. Temperature and elevation effects on plant growth, development, and seed production of two *Lesquerella* species. *Industr. Crops Prod.*, 24:17–25.

Dierig, D.A., T.A. Coffelt, F.S. Nakayama, and A.E. Thompson. 1996. *Lesquerella* and *Vernonia*: oilseeds for arid lands. In J. Janick (Ed.) *Progress in New Crops*. ASHS Press, Alexandria, VA, p. 347–354.

Dierig, D.A., G.H. Dahlquist, and P. M. Tomasi. 2006a. Registration of WCL-LO3 high oil *Lesquerella fendleri* germplasm. *Crop Sci.*, 46:1832–1833.

Dierig, D.A., A.M. Salywon, and D.J. De Rodriquez. 2006b. Registration of a mutant *Lesquerella* genetic stock with cream flower color. *Crop Sci.*, 46:1836–1837.

Downey, R.K. 1971. Agricultural and genetic potentials of cruciferous oilseed crops. *J. Am. Oil Chem. Soc.*, 48:718–722.

Downey, R.K. and G. Röbbelen. 1989. *Brassica* species. In G. Röbbelen, R.K. Downey, and A. Ashri (Eds.) *Oil Crops of the World*. McGraw-Hill, New York, chap. 16, p. 339–362.

Du, X.Z., X.H. Ge, Z.G. Zhao, and Z.Y. Li 2008. Chromosome elimination and fragment introgression and recombination producing intertribal partial hybrids from *Brassica* napus and *Lesquerella fendleri* crosses. *Plant Cell Rep.*, 27:261–271.

Earle, F.E., J.E. Peters, I.A. Wolff, and G.A. White. 1966. Compositional differences among *Crambe* samples and between seed components. *J. Am. Oil Chem. Soc.*, 43:330–333.

Epobio Workshop 2006. Foundation Paper for the Plant Oil Flagship. Wageningen International Conference Center, The Netherlands, 22–24 May 2006.

Erhan, S.Z. 2005. Printing inks. In S.Z. Erhan (Ed.), *Industrial Uses of Vegetable Oils*, United States Department of Agriculture, AOCS Press, Illinois, p. 163–169.

Francis, C.M. and M.C. Campbell, 2003. New High Quality Oil Seed Crops for Temperate and Tropical Australia. Rural Industries Research and Development Corporation Publication No 03/045, RIRDC Project No UWA-47A.

Frauen, M. and W. Paulmann. 1999. Breeding of hybrids varieties of winter oilseed rape based on the MSL-system. *10th Int. Rapeseed Congr.*, Canberra Australia, p. 258–263.

Fröhlich, A. and B. Rice. 2005. Evaluation of *Camelina sativa* oil as a feedstock for biodiesel production. *Industr. Crops Prod.*, 21:25–31.

Gugel, R.K. and K.C. Falk. 2006. Agronomic and seed quality evaluation of *Camelina sativa* in western Canada. *Can. J. Plant Sci.*, 86:1047–1058.

Hayes, D.G., R. Kleiman, and B.S. Phillips. 1995. The triglyceride composition, structure, and presence of estiolides in the oils of *Lesquerella* and related species. *J. Am. Oil Chem. Soc.*, 72:559–569.

Hedge, I.C. 1976. A systematic and geographical survey of the old world cruciferae. In J.G. Vaughan, A.J. MacLeod, and B.M.G. Jones (Eds.), *The Biology and Chemistry of the Cruciferae*, Academic Press, New York, p. 10–45.

Hirsinger, F. 1989. New annual oil crops. In: G. Röbbelen, R.K. Downey, and A. Ashri (Eds.), *Oil Crops of the World*, McGraw-Hill, New York, p. 518–532.

Isbell, T.A., M.R. Edgcomb, and B.A. Lowry. 2001. Physical properties of estiolides and their ester derivatives. *Industr. Crops Prod.*, 13:11–20.

Isbell, T., B. Lowery, S. Dekeyser, M. Winchell, and S. Cermak. 2006. Physical properties of triglyceride estiolides from *Lesquerella* and castor oils. *Industr. Crops Prod.*, 23:256–263.

Knights, S. 2003. *Crambe*: A North Dakotan case study. The Australian Society of Agronomy. www.regional.org.au/au/asa/2003/c/11/knights.htm (Accessed March 6, 2008).

Knothe, G. and Dunn, R.O. 2005 Biodiesel: an alternative diesel fuel from vegetable oils or animal fats. In S.Z. Erhan (Ed.), *Industrial Uses of Vegetable Oils,* United States Department of Agriculture, AOCS Press, Illinois, p. 42–89.

Lazzeri, L., Errani, M., Leoni, O., and Venturi, G. 2004. *Eruca sativa* spp. *oleifera*: a new non-food crop. *Industr. Crops Prod.*, 20:67–73.

Lessman, K.J. 1990. *Crambe*: a new industrial crop in limbo. In J. Janick and J.E. Simon (Eds.), Advances in New Crops, Timber Press, Portland, OR, p. 217–222.

Lessman, K.J. and C. Berry. 1967. *Crambe* and *Vernonia* research results at the forage farm in 1966. Purdue Univ. Agr. Exp. Sta. Res. Prog. Rpt. 284. West Lafayette, IN.

Lessman, K.J. and V.D. Meier. 1972. Agronomic evaluation of *Crambe* as a source of oil. *Crop Sci.*, 12:224–227.

Li, G. and C.F. Quiros. 2001. Sequence-Related Amplified Polymorphism (SRAP), a new marker system based on a simple PCR reaction: its application to mapping and gene tagging in *Brassica. Theor. Appl. Genet.,* 103:455–461.

Lu, C.F. and J.L. Kang. 2008. Generation of transgenic plants of a potential oilseed crop *Camelina sativa* by Agrobacterium-mediated transformation. *Plant Cell Rep.,* 27:273–278.

Matthaus, B. 2004. *Camelina sativa* — revival of an old vegetable oil? *Ernahrungs-Umschau,* 51:12–17.

McGregor, W.G., A.G. Plessers, and B.M. Craig. 1961. Species trials with oil plants. I. *Crambe. Can. J. Plant Sci.,* 41:716–719.

McVetty, P.B.E., R. Scarth, and S.R. Rimmer. 2000b. MillenniUM 03 high erucic low glucosinolate summer rape. *Can. J. Plant Sci.,* 80:611–612.

McVetty, P.B E., R. Scarth, S.R. Rimmer, and C.G.J. Van Den Berg. 1996a. Venus high erucic low glucosinolate summer rape. *Can. J. Plant Sci.,* 76:341–342.

McVetty, P.B.E., R. Scarth, and S.R. Rimmer. 1999. MillenniUM 01 high erucic low glucosinolate summer rape. *Can. J. Plant Sci.,* 79:251–252.

McVetty, P.B.E., R. Scarth, and S.R. Rimmer. 2000a. MillenniUM 02 high erucic low glucosinolate summer rape. *Can. J. Plant Sci.,* 80:609–610.

McVetty, P.B.E., R. Scarth, S.R. Rimmer, and C.G.J. Van Den Berg. 1996b. Neptune high erucic low glucosinolate summer rape. *Can. J. Plant Sci.,* 76:343–344.

McVetty, P.B.E., S.R. Rimmer, and R. Scarth. 1998. Castor high erucic low glucosinolate summer rape. *Can. J. Plant Sci.,* 78:305–306.

McVetty, P.B.E., W.G.D. Fernando, R. Scarth, and G. Li. 2006a. Red River 1826 Roundup Ready high erucic acid low glucosinolate summer rape. *Can. J. Plant Sci.,* 86:1179–1180.

McVetty, P.B.E., W.G.D. Fernando, R. Scarth, and G. Li. 2006b. Red River 1852 Roundup Ready high erucic acid low glucosinolate summer rape. *Can. J. Plant Sci.,* 86:1181–1182.

Moser, B.R., Cermak, S.C., and T.A. Isbell. Evaluation of castor and *Lesquerella* oil derivatives as additives in biodiesel and ultralow sulfur diesel fuels. *Energy and Fuels,* 22:1349–1352.

National Non-Food Crops Centre. 2005. www.nnfcc.co.uk/metadot/index.pl?iid=2193&isa=Category (Accessed November 1, 2007).

Nieschlag, H.J. and I.A. Wolff. 2007. Industrial uses of high erucic oils. J. *Am. Oil Chem. Soc.,* 48:1558–9331.

Papathanasiou, G.A. and K.J. Lessman. 1966. *Crambe. Purdue Univ. Agr. Expt. Res. Sta. Bull. 819,* West Lafayette, IN.

Paulsen, H.M. 2007. Organic mixed cropping system with oilseeds 1. Yields of mixed cropping systems of legumes or spring wheat with false flax (*Camelina sativa* L. Crantz). *Landbauforschung Volkenrode,* 57:107–117.

Perez, J.M. 2005. Vegetable oil-based engine oils: are they practical? In S.Z. Erhan (Ed.), *Industrial Uses of Vegetable Oils,* United States Department of Agriculture, AOCS Press, Illinois, p. 31–41.

Ploschuk, E.L., G.A. Slafer, and D.A. Ravetta. 2005. Reproductive allocation of biomass and nitrogen in annual and perennial *Lesquerella* crops. *Ann. Bot.,* 96:127–135.

Prakash, S. and K. Hinata. 1980. Taxonomy, cytogenetics and origin of crop *Brassica* — a review. *Opera Bot.,* 55:1–57.

Putnam, D.H., J.T. Budin, L.A. Fields, and W.M. Breene. 1993. A promising low input oilseed. In *New Crops,* J. Janick and J.E. Simon (Eds.), Wiley, New York, p. 314–322.

Rahman, M. 2007. Development of Molecular Markers for Marker Assisted Selection for Seed Quality Traits in Oilseed Rape. Ph.D. thesis, University of Manitoba, Winnipeg, Manitoba Canada.

Ratnayake, W.M.N. and J.K. Daun. 2004. Chemical composition of canola and rapeseed oils. In F.D. Gunstone (Ed.), *Rapeseed And Canola Oil: Production, Processing, Properties and Uses,* CRC Press, Boca Raton, FL, p. 37–78.

Reed, D.W., D.C. Taylor, and P.S. Covello. 1997. Metabolism of hydroxy fatty acids in developing seeds in the genera *Lesquerella* (*Brassicaceae*) and *Linum* (Linaceae). *Plant Physiol.,* 114:63–68.

Röbbelen, G. 1991 rapeseed in a changing world: plant production potential. In GCIRC (Ed.), Proc. 8th Int. Rapeseed Congr., Saskatoon, Canada, 9–11 July 1991, p. 29–38.

Roetheli, J.C., K.D. Carlson, R. Kleiman, A.E. Thompson, D.A. Dierig, L.K. Glaser, M.G. Blase, and J. Goodell. 1991. *Lesquerella* as a source of hydroxy fatty acids for industrial products. USDA-CSRS Office of Agricultural Materials. Growing Industrial Materials Series (unnumbered), Washington, D.C.

Salywon, A.M., D.A. Dierig, J.P. Rebman, and D.J. de Rodriguez. 2005. Evaluation of new *Lesquerella* and *Physaria* (*Brassicaceae*) oilseed germplasm. *Am. J. Bot.,* 92:53–62.

Saucke, H. and K. Ackermann. 2006. Weed suppression in mixed cropped grain peas and false flax. *Weed Res.,* 46:453–461.

Scarth, R., P.B.E. McVetty, S.R. Rimmer, and B.R. Stefansson. 1991. Hero summer rape. *Can. J. Plant Sci.,* 71:865–866.

Scarth, R., P.B E. McVetty, S.R. Rimmer, and J.K. Daun. 1992. Breeding for specialty oil quality in canola rapeseed: the University of Manitoba program. In S.L. MacKenzie and D.C. Taylor (Eds.), *Seed Oils for the Future,* AOCS Press, Champaign, IL, chap. 17, p. 171–176.

Scarth, R., P.B.E. McVetty, and S.R. Rimmer. 1994. Mercury high erucic low glucosinolate summer rape. *Can. J. Plant Sci.,* 74:205–207.

Senft, D. 1992. Let's hear it for *Lesquerella*! Agricultural Research. http://findarticles.com/p/articles/mi_m3741/is_n9_v40/ai_12799103/print (Accessed March 31, 2008).

Stefansson, B.R., and R.K. Downey. 1995. Rapeseed. In Slinkard A.E. and D.R. Knott, (Eds.), *Harvest of Gold,* University Extension Press, University of Saskatchewan, Saskatoon, SK, Canada, chap. 12, p. 140–152.

Straton, A., J. Kleinschmidt, and D. Keeley. 2007. *Camelina*. Institute for Agriculture and Trade Policy. Published January 2007. www.iatp.org/iatp/publications.cfm?accountID=258&refID=97279.

Sun, Z., Z. Wang., J. Tu, J. Zhang, F. Yu, P.B.E. McVetty, and G. Li. 2007. An ultra dense genetic recombination map for *Brassica napus*, consisting of 13551 SRAP markers. *Theor. Appl. Genet.,* 114:1305–1317.

Tao, B.Y. 2005. Biofuels for home heating oils. In S.Z. Erhan (Ed.), Industrial Uses of Vegetable Oils, United States Department of Agriculture, AOCS Press, Illinois, p. 90–109.

Targeted Growth, Inc. 2007. www.targetedgrowth.com/sites/targetedgrowth/upload/filemanager/Press-Releases/Sustainable_Oils_Launches-11-20-07.pdf Press Release (Accessed March 11 2008).

Seedtec Terramax. 2000. Research and Crop Development. http://www.terramax.sk.ca/.

Tomasi, P., D. Dierig, and G. Dahlquist. 2002. An ovule culture technique for producing interspecific *Lesquerella* hybrids. In J. Janick and A. Whipkey (Eds.), *Trends in New Crops and New Uses*, ASHS Press, Alexandria, VA, p. 208–212.

Urbaniuk, S.D. 2006. An Agronomic Evaluation of *Camelina* and Solin for the Maritime Provinces of Canada. M.Sc. thesis. Nova Scotia Agricultural College, Truro, Nova Scotia and Dalhousie University, Halifax, Nova Scotia, Canada.

Urbaniuk, S.D., C.D. Caldwell, V.D. Zheljazkov, R. Lada, and L. Luan. 2008. The effects of cultivar and applied nitrogen on the performance of *Camelina sativa* L. in the Maritime Provinces of Canada. *Can. J. Plant Sci.,* 88:111–119.

Van De Mark, M.R. and K. Sandefur. 2005. Vegetable oils in paint and coatings. In S.Z. Erhan (Ed.), *Industrial Uses of Vegetable Oils*, United States Department of Agriculture, AOCS Press, Illinois, p. 143–162.

Vaughn, S.F. and R.A. Holser. 2007. Evaluation of biodiesels from several oilseed sources as environmentally friendly contact herbicides. *Industr. Crops Prod.,* 26:63–68.

Velasco, L., F.D. Goffman, and H.C. Becker. 1998. Variability for the fatty acid composition of the seed oil in a germplasm collection of the genus *Brassica*. *Genetic Resources Crop Evolut.,* 45:371–382.

Vollmann, J., H. Grausgruber, G. Stift, V. Dryzhruk, and T. Lelley. 2005. Genetic diversity in *Camelina* germplasm as revealed by seed quality characteristics and RAPD polymorphism. *Plant Breed.,* 124:446–453.

Vollmann, J., T. Moritz, C. Kargl, S. Baumgartner, and H. Wagentristl. 2007. Agronomic evaluation of *Camelina* genotypes selected for seed quality characteristics. Industr. Crops Prod., 26:270–277.

Warwick, S.I. and A. Francis. 1994. Guide to the Wild Germplasm of *Brassica* and Allied Crops, Parts I to V. Technical Bulletin 1994, Centre for Land and Biological Resources Research, Agriculture and Agri-Food Canada.

White, G.A. and J.J. Higgins. 1966. Culture of *Crambe*: a new industrial oilseed crop. Agr. Res. Serv., USDA, ARS Production Research Report 95.

White, G.A. and M. Solt. 1978. Chromosome numbers in *Crambe, Crambella* and *Hermicrame. Crop Sci.,* 18:160–161.

Whitely, E.L. and C.A. Rinn. 1963. *Crambe abyssinica*, a potential new crop for the Blackland. *Texas Agr. Progress,* 9(5):23–24.

Zubr., J. 1997. Oilseed crop: *Camelina* sativa. *Industr. Crops Prod.,* 6:113–119.

18 Collection, Preservation, and Distribution of Wild Genetic Resources

César Gómez-Campo

CONTENTS

EARLY COLLECTIONS AND DISTRIBUTION BY BOTANICAL GARDENS

For a few centuries, botanical gardens played a major role as suppliers of wild seed material either to institutions or to individuals. They covered many different purposes, including a wide range of practical uses: exhibition, pharmaceutical applications, and also botanical research. The annual publication of an "*index seminum*" has been a common policy in most gardens. Not only has this practice favored seed exchange among different gardens, but it has also promoted free availability of their seed material to the general public and therefore contributed to the knowledge and popularity of the gardens themselves.

However, the demands for wider diversity and improved reliability in taxonomic determinations soared during the twentieth century. By the middle of that century, it became evident that most botanical gardens were not performing well enough to meet such demands. Much of the research carried out during that time — such as chromosome number determinations, botanical observations, etc. — is in error because of too frequent mistakes in taxonomic determinations. Choosing the *Brassicaceae* (Cruciferae) as an example, Gómez-Campo (1969a) surveyed the contents of the indexes of 111 European botanical gardens and roughly reached the following conclusions:

1. Some common or easy-to-grow species such as *Arabis caucasica, Alyssum saxatile, Hesperis matronalis,* and *Peltaria alliacea* were present in most seed catalogs while rare, endangered, or endemic species were practically absent. It was obvious that the seeds came most often from cultivation and from an active exchange among gardens, while collections from the natural habitats seldom took place. As a result, the available total diversity was rather low. It appeared that the gardens acted as seed collectors in the wild only in a very limited way — they rather worked based on their own cultivated material.
2. Taxonomic errors were too frequent. Within a set of 200 accessions that could be evaluated, 20 were in error at genus level and 40 additional ones at species level (total 30%). Two accessions did not even belong to the family *Brassicaceae*. Within the erroneous accessions, weeds such as *Sisymbrium irio, Sinapis arvensis,* and *Rapistrum rugosum* were commonly found, thus pointing at a careless collection and/or sowing and subsequent mislabeling as the main cause of these errors. The lack of supervision by taxonomists was noticeable, and blind seed exchanges were often the reason that the same errors were repeated in many gardens.
3. At the time, there were almost no methods to extend seed longevity. The idea that low temperature and low moisture were positive factors for seed preservation already existed but had little practical application. In fact, seeds were most often kept within paper envelopes in drawers or closets at room temperature and with no humidity control. The usual policy was to sow and then to harvest from the garden — every year for annuals and every few years for perennials. Dead accessions were probably frequent but the lack of any distinction between real death and dormancy left many cases undefined in this respect.

After the 1970s, all these trends started to change. Botanical gardens have gradually increased their awareness of their potential role in plant preservation in accordance with the growing general concern for the conservation of nature. The introduction of the concept of biodiversity has been a strong incentive for many improvements in *ex situ* and *in situ* preservation policies. These days, many botanical gardens are safe suppliers of a wider diversity of seed material often collected in their natural habitats. They also have often adopted middle-term or long-term seed preservation methods, and it is interesting to note that their performance regarding this is of a much higher quality than that of most of the larger seed banks for crop species. In some countries, associations of gardens have also been a positive factor to improve their coordination and their effectiveness.

COLLECTIONS BY THE UNIVERSIDAD POLITÉCNICA DE MADRID, SPAIN, AND BY THE UNIVERSITY OF TOHOKU, JAPAN

Between 1960 and 1963, a set of approximately 60 accessions of wild members of the tribe Brassiceae was gathered by the Universidad Politécnica de Madrid, Spain (UPM) to investigate their comparative resistance to ionizing radiation (Gómez-Campo and Delgado, 1964, 1969b); 47 of them were used in this research. They were rather common accessions of the type present in the catalogs of botanical gardens, although they also included many accessions collected in their natural habitats — mostly in central Spain. When received from botanical gardens, they were immediately rejuvenated and their taxonomic identity was checked. Later on — sometimes recollected and sometimes rejuvenated — the seed material remaining from these experiments became the original nucleus of the UPM Crucifer Seed Bank that was officially started in 1966. In the following years, some of the accessions coming from gardens were replaced by others collected in their natural habitats.

In 1965, a Japanese team from the University of Tohoku, Sendai (Japan), visited the Mediterranean area and collected approximately 110 interesting seed accessions, most of which had never been introduced before to Japan. The major purpose was to explore sources of cytoplasmic male sterility

FIGURE 18.1 Clockwise from the left: fruiting stems of *Brassica deflexa* (it is not upside down!), *Guiraoa arvensis, Sisymbrium cavanillesianum,* and *Coincya rupestris.* The first species lives from Anatolia to Iran. All others are endemic to Spain.

to facilitate ulterior hybridization in several cultivated *Brassica* species, especially *B. rapa.* Although the collections were mainly undertaken on the basis of visits to the surroundings of some important cities, the number of countries the team visited was very high (Egypt, Turkey, Greece, Italy, Spain, Morocco, Algeria, and Tunisia). As a result, there were a considerable number of useful seed accessions (Mizushima and Tsunoda, 1967). On the way to the Mediterranean area, India and Ethiopia were also visited.

The Mediterranean area is rich in members of the family *Brassicaceae* (Figure 18.1). It is also, beyond any doubt, the center of diversity for the tribe Brassiceae — that is, the tribe of the family *Brassicaceae* where some of the genera of maximum economical importance are located (e.g., *Brassica, Sinapis, Raphanus, Eruca, Crambe,* etc.). This is the reason why most Cruciferae (= *Brassicaceae*) collectors have frequently put their emphasis on the tribe Brassiceae. If we restrict ourselves to the genera of this tribe that are endemic to a single country, Morocco holds six of them (*Ceratocnemon, Crambella, Fezia, Foleyola, Trachystoma, Rytidocarpus*), while Spain has three (*Boleum, Euzomodendron, Guiraoa*) and Algeria has only one (*Otocarpus*). When considering the species level, a similar pattern is obtained. In this author's opinion, these simple figures give a good idea about where the maximum diversity of this important tribe is geographically concentrated. Within the Mediterranean region, the southwestern side (i.e., Spain and Morocco) is comparatively richer. However, the genus *Brassica* itself is spread throughout the Mediterranean region.

The location of Madrid on this southwestern side of the Mediterranean region was a major stimulus to undertake a series of collecting missions that began in 1965 and continued during the succeeding years. These missions were sponsored first by the Instituto Nacional de Investigaciones Agronómicas (INIA) and later by the Universidad Politécnica de Madrid (UPM). Some seed requests from other researchers also made it clear to the team that establishing a seed bank based on this material might be of interest for many other people. The main geographic regions visited by UPM teams in the next few years were the following:

1965: Southern Spain (Andalucía, Sierra Nevada, Serranía de Ronda), eastern Spain (S. Aragon), Canary Islands
1966: Southeastern Spain (Murcia, Alicante), Canary Islands
1967: Central Morocco, Atlas, Antiatlas. Spain: Canary Islands, northeastern Spain (Pyrenees)
1968: Northern and northeastern Morocco, Rif, Beni Snassen
1970: Algeria: Oran, Alger, Spain: Mallorca Island
1971: Algeria: Sahara, down to El Golea, Tunisia
1972: Morocco

The main goal of the visits to the Canary Islands was not only to obtain endangered species in general, but also to collect species of the genus *Crambe* Sect. Dendrocrambe as well as other interesting species of *Descurainia*, *Lobularia*, etc. Morocco, in turn, yielded the singular endemic genera mentioned above, as well as many sub-endemic ones shared with Spain and/or Algeria as well as many species and subspecies belonging to the complex *Diplotaxis-Erucastrum-Brassica*, which are of direct interest for breeding programs of cultivated *Brassica* species. Algeria is interesting for species of several genera, including *Moricandia*, *Eruca* and *Enarthrocarpus*.

During this period, the intrinsic interest and the diversity of the collected material experienced an important leap forward. After several mimeographed catalogs, a printed catalog of the UPM Crucifer seed bank was jointly published by INIA and UPM in 1974 and already included 365 accessions. Seed requests from institutions of several countries were common by that year, and they provided an important complementary incentive for collecting and storing activities.

OTHER INTERNATIONAL AND SPANISH EFFORTS IN THE 1970s

In 1974, a three-person team from the USDA led by George A. White visited the Mediterranean region in an extensive search for *Crambe* species of Sect. Leptocrambe (*Crambe abyssinica*, *C. hispanica*, *C. kralikii*, and *C. filiformis*). *C. abyssinica* is an interesting oilseed species cultivated in the eastern part of the area, while the other members are closely related wild species. A search for high levels of erucic acid content followed. Oil with low erucic acid and glucosynolate levels are preferable for human consumption, while high erucic acid and glucosynolate levels are better for industrial uses.

The activities of the UPM included accompanying Dr. White in his travels through Morocco. On the way back to Madrid, seeds of *Diplotaxis siettiana* were collected on the Island of Alboran in southern Spain (Figure 18.2). No more than 150 plants of this narrow endemism could be seen. This collection served to save this species from extinction (Gómez-Campo, 1990; Moreno et al., 2005), because shortly thereafter the area was irrigated with seawater to avoid dust in a nearby helicopter alley and all the plants disappeared. The reintroduction of the plant in its natural habitat from the stored material is a very good example of the role that a seed bank can play.

1975 was a very active year for Crucifer collectors. This author visited Iran alone for 20 days. Iran is considered to belong to the Irano-Turanian region, with temperatures higher in summer and lower in winter compared to the Mediterranean region. The proportion of taxa belonging to the tribe Brassiceae is here noticeably lower, apart from *Brassica deflexa* and some interesting genera such as *Fortuynia*, *Pseudofortuyinia*, and *Physorrhynchus*. However, many other curious genera belonging to other tribes (e.g., *Diptychocarpus*, *Goldbachia*, *Elburzia*, *Parlatoria*, *Chorispora*, *Physoptichys*, *Pseudocamelina*, *Sameraria*, *Sterigmostemum*, etc.) are present in the area and were collected. Most of them looked rather exotic for a visitor from the West. The area is also rich in species of the genera *Arabis*, *Alyssum*, *Erysimum*, *Isatis*, and *Crambe* (Sect. Crambe, tribe Brassiceae).

Other members of the UPM team (Hernández-Bermejo and Clemente) traveled around the Mediterranean (Sicily, mainland Italy, and Greece) in search of close relatives of *Brassica oleracea*. They worked at species level and brought back accessions of *B. macrocarpa*, *B. villosa* ssp.

FIGURE 18.2 Some seeds of *Diplotaxis siettiana* collected in 1974 saved this species from extinction after it disappeared from its tiny natural habitat by 1986.

drepanensis, B. incana, B. rupestris, and *B. cretica* ssp. *aegaea.* Another team, this time from Sweden and including S. Snogerup and M. Gustafsson, also collected some similar material from the eastern part of the area and carried out their sampling from the point of view of a plant breeder.

Also in 1975, three joint collecting missions by the University of Tohoku (Sendai) and the Universidad Politécnica (Madrid) took place in the Mediterranean region with U. Mizushima, S. Tsunoda, and K. Hinata as members of the Japanese team, and C. Gómez-Campo on the Spanish side. The first mission developed in southern Spain and central Morocco, while the second was in northwestern Algeria (Oran area). The third mission involved a visit to the area of Alger and also a flight to Tamanrasset Hoggar, southern Algeria. Hoggar massif is located in the center of the Sahara Desert but the higher altitude creates a climate resembling the Mediterranean region. The Japanese-Spanish collaboration involved the exchange of seed material between both genebanks and also provided material for a book (Tsunoda et al., 1980) that soon was considered a classic in the subject. The final chapter in that book (Tsunoda et al., 1980b) summarizes the collected material during these missions, while in Chapter 2, Tsunoda describes the habitats of many of the species.

On his way back from a coordination visit to Japan in 1977, Gómez-Campo had a chance to collect some interesting accessions in Egypt viz., *Erucaria microcarpa*, and naturalized *Brassica juncea*, as well as some other taxa that already existed in the UPM bank but were of interest because of the distance of their localities from the western Mediterranean region (e.g., *Farsetia aegyptiaca, Brassica tournefortii, Enarthrocarpus lyratus, Zilla spinosa, Carrichtera annua,* etc.).

In the period between 1976 and 1980, UPM collecting missions were usually shorter because their itineraries were designed to obtain specific material, thus filling the gaps left within the more intensive activities of previous years. These missions were mostly conducted within Spain, although some short visits were made to Tunisia, Algeria, and Morocco. In Morocco, an interesting visit to the highlands of the Great Atlas region was held in the autumn.

Missions within Spain were often designed to support certain taxonomic studies or were intermixed with other projects to collect endemic and endangered species in general. Collection by some correspondents also played key roles in procuring taxa that were otherwise difficult to obtain.

BIOVERSITY (IPGRI/IBPGR) SPONSORED COLLECTING
MISSIONS OF WILD *N* = 9 *BRASSICA* SPECIES

The 1980s were marked by the a series of collecting missions sponsored by IBPGR/IPGRI (now Bioversity) to obtain relatives of *Brassica oleracea* on the Mediterranean coasts of Spain, Italy, Tunisia, and Greece, and of wild *B. oleracea* itself along the Atlantic coasts of northern Spain, the United Kingdom, and France (Figures 18.3, 18.4, and 18.5, respectively). Although Turkey was momentarily set aside, a Japanese-Turkish-Spanish mission sponsored by the University of Tohoku in 1983 amply filled this gap. All these collections were made acting at population level.

The places visited every year and the type of material collected are summarized below for every collecting season:

1982: Atica and the Peloponessos (Greece); *Brassica cretica* ssp. *aegaea* and *Brassica cretica* ssp. *laconica*.

1983: Eubea Island and Crete (Greece); *Brassica cretica* ssp. *aegaea* and *Brassica cretica* ssp. *cretica*

1984: Southern Italy and Sicily; *Brassica incana*, *B. rupestris* (three subspecies), *B. villosa* (three subspecies), and *B. macrocarpa*.

1985: Mediterranean coasts of northeastern Italy (Alpi Apuani, Riviera), southern France (Côte d'Azur), and northeastern Spain (Costa Brava); *Brassica montana*.

1986: Cyprus; *Brassica hilarionis*. Tunisia, Corse, and Sardinia; *Brassica insularis*.

1988: Coasts of northern Spain, northwestern France, and southern and southwestern United Kingdom Channel area and Wales; wild *Brassica oleracea*.

This series of missions was very rewarding not only because of the number of seed accessions collected, but also for the chorological and ecological data obtained, as well as for fruitful discussions on the general biology and the evolution of the entire group. Many previously undetected populations were detected, studied, and collected. Special attention was given to the demography

FIGURE 18.3 *Brassica montana* (left) is often seen within the shrubby vegetation while *Brassica cretica* (right) prefers to grow on calcareous cliffs. Both strategies are interpreted as a protection against grazing animals.

FIGURE 18.4 Roca Ficuzza in Sicily is a huge rocky system whose north face is inhabited by the chasmophyte *Brassica rupestris*.

FIGURE 18.5 Wild *Brassica oleracea* (inbox) grows on the cliffs of the Channel area. (Photo Yvort, France.)

and to the conservation status of each population. Seeds of some species such as *Brassica hilarionis* were collected for the first time.

Sicily was particularly interesting because of its high diversity. Four *Brassica* species are present (*B. incana*, *B. villosa*, *B. rupestris*, and *B. macrocarpa*). Among these, *B. villosa* and *B. rupestris* have three subspecies each. B. *macrocarpa* inhabits the small Egadi archipelago offshore of western Sicily. All this strongly suggests that Sicily might be the geographical area where the $n = 9$ *Brassica* group originated.

Several reports and articles describe in detail the accessions obtained, the observations made, and the new localities found (Gustafsson et al., 1983a, b, c, 1984, 1985; Gómez-Campo and Gustafsson, 1991).

According to Bioversity policies, the collected material was distributed into three parts. One part was stored in the UPM bank, Madrid, Spain, which had previously been designed by Bioversity as the base bank for wild Crucifers. A second duplicate was stored in the bank of Tohoku University, Sendai, Japan. The third duplicate was stored in a bank within the country where the material was collected, namely, the Conservatoire Botanique de Porquerolles in France, The Banco Nationale de Germoplasma in Bari (Italy), the North Greece Genebank of Thessaloniki (Greece), and the Kew Gardens in the United Kingdom.

When the entire project was coming to an end, a multiplication project was carried out in Madrid for those accessions where the amount of collected material had been low. The selected method consisted of using large cages of plastic web with pollinating bees.

A complementary expedition to Turkey, sponsored by the Tohoku University, was held in 1983, just before the Eubea collections. The participants included M. Ozturk from Ege University, S. Tsunoda, and K. Hinata from Tohoku University, and C. Gómez-Campo from the Universidad Politécnica de Madrid. The area covered was very large as it included the eastern part of Anatolia and the coasts of the Black Sea and the Mediterranean Sea, where the material collected was abundant and diverse within the family *Brassicaceae* (Ozturk et al., 1983). Among other material, several *Crambe* species of Sectio Crambe (syn. Sarcocrambe) were obtained. It was in the final days of this mission when some *Brassica cretica* populations were collected in the western coasts and islands. Duplicates were stored in Madrid (Spain), Sendai (Japan), and Izmir (Turkey).

OTHER COLLECTIONS

The period from 1991 to 1995 did not see many crucifer collections by the UPM because the time was primarily devoted to complete collections of endangered species in general. However, within the same decade, the years 1995 through 1997 were very active in collecting accessions of the species *Eruca vesicaria*. This species has a maximum diversity in the Iberian Peninsula with two subspecies — ssp. *vesicaria* and ssp. *sativa* — that grow on different types of soils. At least a hundred populations were first collected and then morphologically characterized (Gómez-Campo, 2003). Collections of wild *Eruca* have been also carried out by Pignone of the Institute of Plant Genetics (IGV) in Bari, Italy.

In the new millennium, at least two UPM extensive Crucifer collecting trips deserve mention. The first one, in 2002, targeted the northern coast of Spain, which was thoroughly covered by a team of nine people searching for new localities of *Brassica oleracea*. The already-known 19 localities became 42 after the summer of that year. Seed collections were also thoroughly carried out during this trip (Gómez-Campo et al., 2007a). In the following season, a similar exploration focused this time on the northeastern coast of Spain (Girona) in search of new localities and seeds of *B. montana* (Gómez-Campo et al., 2007). Both missions were financed with the help of a Spanish grant obtained by I. Aguinagalde.

SEED DRYING AND SEED STORAGE PROCEDURES IN THE UPM SEED BANK

The procedure used to preserve crucifer seeds in the UPM seed bank was designed with complete independence from the procedures used by the other eight or nine banks in existence in 1966. Instead of placing the emphasis on low temperatures, the emphasis was placed on low moisture content.

The low moisture content was obtained by placing some dehydrated silica gel together with the seeds in sealed glass tubes. The silica gel absorbs moisture from the seeds to levels down to approximately 2% moisture content, a value that falls within what is practically considered ultra-desiccation. The use of silica gel has several additional advantages. It is manufactured with an indicator that warns — by changing its color — if moisture accidentally enters the container. It also absorbs some toxic gases produced during seed aging, which eventually constitute the ultimate cause of seed death. Thus, silica gel protects the seeds by means of two independent mechanisms,

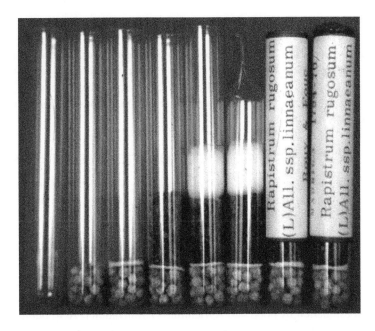

FIGURE 18.6 Flame-sealed glass tubes with silica gel inside have proved highly efficient for the long-term "*ex situ*" preservation of the Brassicaceae, as well as for other botanical groups with small seeds.

reducing water vapor and reducing toxic gases. Of particular practical interest is the possibility of a visual control from the outside of possible entrances of water vapor.

To keep the seeds ultra-dry, it is necessary to use reliable waterproof containers. Flame-sealed glass provides the most reliable solution (Figure 18.6). Standard laboratory glass vials are stretched and closed using the flame of a Bunsen burner (helped or not by a blower, depending on the size of the tubes). Some 100 of such ampoules are usually made for each accession.

To prevent breakage of the sealed glass tip, an external label is rolled around and a melted mixture of wax (2/3) and resin (1/3) is poured inside (Gómez-Campo, 2007b).

The ampoules are then stored in a cold room at moderately low temperature, usually between 5°C and −5°C.

DUPLICATION

Duplication of valuable stored material is strongly recommended to minimize all possible risks. UPM material is shared by at least five other banks, although this is in the form of "black box" samples, small in size but safely preserved for a maximum longevity.

SEED VIABILITY: HIGHLY ENCOURAGING RESULTS AFTER 40 YEARS OF STORAGE

In 2006, the UPM seed bank became 40 years old and germination tests combined with anti-dormancy treatments were performed to evaluate the viability of the oldest accessions. The final average viability is 98.4%, not significantly different from 100% (Pérez-García et al., 2007). Therefore, the silica gel preservation method designed more than 40 years ago (1966; see Gómez-Campo, 1969a, b, 1972) and described above has worked very effectively.

Other evaluations made in the past (Ellis, 1993; Ramiro, 1995; Maselli, 1999) also produced satisfactory results but the figures were lower because dormancy was not taken into account — as should always be done when working with wild species. When this factor is not taken into account, germination tests are largely meaningless because viability — and not germinability — will be measured.

Still more striking were the results obtained with a set of ultra-dried seeds that had also been stored for almost 40 years but within a closet at room temperature. Their average viability (excluding three accessions where the tests could not be satisfactorily completed because of the low number of seeds) was 93.8%. This means that low temperatures might not deserve the relative importance they have been usually assigned.

GENERAL IMPLICATIONS FOR EFFICIENT LONG-TERM SEED PRESERVATION

After the above results were obtained, a search for parallel viability records from other seed gene-banks of similar age was done for comparison. Nothing comparable could be found either in the scientific literature or on the Internet. By 1960, only three large seed banks existed in the world; that is, one in St. Petersburg later ascribed to the Institute N.I. Vavilov (former Soviet Union), a second in the Leibnitz Institute in Gatersleben (East Germany), and the third in Fort Collins, Colorado (United States, USDA). However, deep silence, results limited to short time periods, or poor results could only be found. The higher value (54.8 %) corresponded to the Crucifers stored in Fort Collins (Walters et al., 2005). Since that time, the number of world seed banks has skyrocketed to 1300 by the year 2000 (an estimation by Food and Agriculture Organization [FAO]) but the poor preservation methods widely used have meant that large parts of the original material have been lost. For many, seed banks are large "cemeteries of germplasm," a statement that might be exaggerated but is not too far from reality. Most seed banks, even relatively young seed banks, are now rejuvenating their material in the field or deeply worried at the very idea of rejuvenating because the inconveniences are multiple (e.g., rejuvenation reduces genetic variability, thus generating genetic erosion; it consumes large amounts of time, labor, space, etc.; it is also a source of unwanted selection, unwanted crosses, mislabeling, and other possible mistakes).

What has failed? In the opinion of the author of this chapter (Gómez-Campo) and in line with recent UPM results, poor preservation in crop seed banks has been caused by the following factors:

1. An excessive emphasis has been placed on low temperatures (which are not as critical as initially thought) while the importance of low moisture content in the seeds has most often been neglected (Pérez-García et al., 2007).
2. The possibility of using low seed moisture content — ultra-drying between 1 and 3% — has been ignored. For a time, this has even been considered harmful. However, for most orthodox seeds this is not true. Ultra-drying can extend longevity by a factor of 8 to 16 (Harrington, 1973).
3. In the vast majority of cases, the effect of any measures to maintain lower seed moisture content has been ruined by the widespread use of inadequate nonhermetic containers, which allowed water vapor intake.
4. In summary, seeds have most often been preserved in very cold environments but with higher than optimum moisture content. A lower, well-controlled moisture content might have produced much better results, even at higher temperatures.

Rejuvenation (regeneration) cycles for the unpreserved seed collections of old-time plant breeders were between 6 and 8 years for accessions of cereals, oil crops, etc., and somewhat longer for legumes. Modern seed banks extended these cycles to approximately 25 to 35 years, meaning that although low temperatures are not the main factor, they still play a certain role.

If the strict control of very low seed moisture contents (1% to 3%) produces an average viability that is not significantly different from 100% after 40 years, the use of this method for crop species with orthodox seeds might well extend rejuvenation cycles to one to two centuries or, perhaps longer. It is noteworthy that a new breed of small seed banks established in botanical gardens has used the silica gel method and can expect to have their seeds fully alive when they reach 40 years old.

One-day free workshops on "efficient long-term seed preservation" to explain the high efficiency of the silica gel method are being held around the world. By April 2008, some 39 workshops had been held in 13 countries. They are financed by Grupo Santander and Fundación Marcelino Botín (Spain).

SEED CATALOGS AND SEED DISTRIBUTION: THE IMPACT OF THE UPM SEED BANK ON BASIC AND APPLIED RESEARCH

Collection, storage, and distribution are the three main activities expected from a seed bank. However, the goal of storage is not only the distribution of the samples, but also the prevention of possible extinction of genetic material in the future. In this respect, the concept of "black box" samples (i.e., only aimed at preventing such extinction) is important and should be added to the concepts of base and active collections, although it is of very limited use in the practical management of seed banks.

Regarding the distribution of the seeds, the UPM Crucifer material has been fully and freely available from many published catalogs for many years, first mimeographed, later printed (see Gómez-Campo, 1969a, b, 1990) and then on the Internet. Many hundreds of fully viable seed samples have been freely distributed during the past 40 years.

Although it is normally difficult to obtain some feedback on how the distributed accessions of a seed bank have been used, some of the UPM seed bank characteristics have converged to make this possible. On the one hand, it is a specialized collection where the users are all interested in the same botanical material. They usually send their reprints, and their work is easy to trace from the literature. On the other hand, the continued work of the UPM staff on the Crucifer family made it easier by creating a compilation of published literature. The results of a survey of this kind have been recently published (Gómez-Campo, 2007b). Some 312 papers, having used the material stored and distributed by the UPM Crucifer Seed Bank, are referenced. The fields of anatomy, taxonomy, cytogenetics, plant breeding, conservation, molecular biology, cancer research, physiology, phytopathology, and weed research, among many others, have benefited from the distributed seeds.

However, this chapter finishes with a double-sided thought. On the one hand, the benefits of the UPM seed bank for basic and applied research would have been much less significant — in fact, almost impossible to offer today — if the collections had to be made with the limitations that started at the Convention on Biological Diversity in 1992 (Río de Janeiro summit). Fortunately, most UPM collections were performed before that time. On the other hand, the seeds themselves would not be fully alive after 40 years but rather halfway to their death if preservation methods had been simply borrowed from other existing seed banks at that moment in time.

Collecting and storing Crucifer accessions by many other researchers in relation to their particular research areas are probably very abundant but impossible to reflect here.

ACKNOWLEDGMENTS

The following persons participated in one or more of the collecting missions described above: S. Abderrazhak, I. Aguinagalde, P. Arús, L. Ayerbe, P. Ballesteros, M. Casas-Builla, J.L. Ceresuela, S. Ciafarelli, M. Clemente, J. Fernández, J.A. Fernández-Prieto, M. Gustafsson, E. Hernández-Bermejo, K. Hinata, J. Lesouef, S. Linington, J.P. Martín-Clemente, J. Martínez-Laborde, R. Mithen, U. Mizushima, L. Olivier, M. Oztürk, M. Parra-Quijano, P. Perrino, D. Pignone, P. Ponce, A. Prina, M.D. Sánchez-Yélamo, E. Simonetti, S. Snogerup, E. Sobrino, G. Thomas, E. Torres, M.E. Tortosa, S. Tsunoda, A. Zamanis, and J.L. On behalf of the UPM, many other collectors could have been cited but they have been omitted because their activities were not specifically directed to the family *Brassicaceae* but rather to endemic or endangered species in general.

REFERENCES

Ellis, R.H., Hong, T.D., Martin, M.C., Pérez-García, F., and Gómez-Campo, C. 1993. The long-term storage of seeds of seventeen crucifers at very low moisture contents. *Plant Varieties and Seeds*, 6, 75–81.

Gómez-Campo, C. 1969a. The availability of Crucifer seed from European Botanic Gardens. *Plant Introduction Newsletter*, 22, 25–32.

Gómez-Campo, C. 1969b. A germ plasm collection of crucifers. *Catálogos del Instituto Nacional Investigaciones Agrarias* (mimeographed).

Gómez-Campo, C. 1972. Preservation of West Mediterranean members of the cruciferous tribe Brassiceae. *Biological Conservation*, 4, 355–360.

Gómez-Campo C. 1990. A germ plasm collection of crucifers. 9th ed. *Catálogos Instituto Nacional de Investigaciones Agrarias*, Madrid, 22, 1–55.

Gómez-Campo, C. 2003. Morphological characterisation of *Eruca vesicaria* (DC.) Cav. (Cruciferae) germplasm. *Bocconea*, 16(2), 615–624.

Gómez-Campo, C. 2007a. A guide to efficient long term seed preservation. *Monographs ETSIA, Univ. Politécnica de Madrid*, 170, 1–17 (www.seedcontainers.net).

Gómez-Campo, C. 2007b. Assessing the contribution of genebanks. The case of the UPM seed bank in Madrid. *Plant Genetic Resources Newsletter*, Bioversity, Rome. 151, 40–49.

Gómez-Campo, C. and Delgado, L. 1964. Radioresistance in crucifers. *Radiation Botany*, 4, 479–483.

Gómez-Campo, C. and Gustafsson, M 1991. Germplasm of wild n=9 Mediterranean species of *Brassica*, *Botanika Chronika*, 10, 429–434.

Gómez-Campo, C., Aguinagalde, I. Ceresuela, J.L., Lázaro, A., Martínez-Laborde, J.B., Parra-Quijano, M., Simonetti, E., Torres, E., and Tortosa, M.E. 2005. An exploration of wild *Brassica oleracea* L. germplasm in Northern Spain. *Genetic Resources and Plant Evolution*, 52, 7–13.

Gómez-Campo, C., Aguinagalde, I., Arús, P., Jiménez-Aguilar, C., Lázaro, A., Martín-Clemente, J.P., Parra-Quijano, M., Sánchez-Yélamo, M.D., Simonetti, E., Torres, E., Torcal, L., and Tortosa, M.E. 2007. Geographical distribution and conservation status of *Brassica montana* in NE Spain. *Cruciferae Newsletter*, 27(2), 32–34..

Gustafsson, M., Gómez Campo, C., and Zamanis, A. 1983a. Report from the first *Brassica* germ-plasm exploration in Greece 1982. *Sveriges Utsädesför. Tidskrift*, 93, 151–160.

Gustafsson, M., Gómez-Campo, C., and Zamanis, A. 1983b. Germplasm conservation of wild Mediterranean *Brassica* species. Collecting missions in 1983. *Sveriges Utsädesför. Tidskrift*, 95, 137–143.

Gustafsson, M., Gómez-Campo, C., and Zamanis, A. 1983c. *Brassica cretica* Lam. Germplasm collection in Greece. *EUCARPIA Cruciferae Newsletter*, 8, 2–4.

Gustafsson, M., Gómez-Campo, C., and Perrino, P. 1984. Germplasm Conservation of Wild Mediterranean *Brassica* Species. Report from explorations in Italy 1984 IBPGR-FAO.

Gustafsson, M., Gómez-Campo, C., and Perrino, P. 1985. Germplasm Conservation of Wild Mediterranean *Brassica* Species. Report from explorations in Italy, France and Spain 1985 IBPGR-FAO.

Harrington, J.F. 1972. Seed storage and longevity. p. 145–245 in Kozlowsli, T.T. (Ed.) *Seed Biology. Vol. 3. Insects, And Seed Collection, Storage, Testing and Certification*. New York, Academic Press.

Harrington, J.F. 1973. Biochemical basis of seed longevity. *Seed Sci. Tech.* 1:453–461.

Maselli, S., Pérez-García, F., and Aguinagalde, I.. 1999. Evaluation of seed storage conditions and genetic diversity of four crucifers endemic to Spain. *Annals of Botany*, 84, 207–212.

Mizushima, U. and Tsunoda, S. 1967. A plant exploration in *Brassica* and allied genera. *Tohoku Journal Agricultural Research*, (17) 4, 249–277.

Moreno J.C., Mota F., and Gómez-Campo, C. 2005. *Diplotaxis siettiana*. In: B. Montmollin and W. Strahm (Eds.). *The Top 50. Mediterranean Island Plants*. International Union for the Conservation of Nature. p. 8–9.

Ozturk, M., Hinata, K., Tsunoda, S., and Gómez-Campo, C. 1983. A general account of the distribution of cruciferous plants in Turkey. *Ege University Faculty of Science Journal* (Izmir), 6, 87–98.

Përez-Garcia, F., Gonzalez-Benito, M.E., and Gómez-Campo, C. 2007. High viability recorded in ultra-dry seeds of 37 species of *Brassicaceae* after almost 40 years of storage. *Seed Science and Technology*, 35, 143–153.

Ramiro, M.C., Pérez-García, F., and Aguinagalde, I. 1995. Effect of different seed storage conditions on germination and isozyme activity in some *Brassica* species. *Annals of Botany*, 75, 579–585.

Tsunoda, S., Hinata, K., and Gómez-Campo, C. (Eds.) 1980a. *Brassica Crops and Wild Allies*. Japan Scientific Societies Press, Tokyo.

Tsunoda, S., Hinata, K., and Gómez-Campo, C. 1980b. Preservation of Genetic Resources. In Tsunoda, S., Hinata, K., and Gómez-Campo, C., Eds. *Brassica Crops and Wild Allies.* Japan Scientific Societies Press, Tokyo, p. 339–354.

Walters, C., Wheeler, L.M., and Grotenhuis, J.M. 2005. Longevity of seeds stored in a genebank: species characteristics. *Seed Science Research,* 15, 1–20.

Index

Printed and bound by CPI Group (UK) Ltd, Croydon, CR0 4YY

18/10/2024

01776249-0009